Java®

All-IN-ONE

for
dummies®
A Wiley Brand

Java®

All-IN-ONE

6th Edition

by Doug Lowe

A Wiley Brand

Java® All-in-One For Dummies®, 6th Edition

Published by: **John Wiley & Sons, Inc.,** 111 River Street, Hoboken, NJ 07030-5774, www.wiley.com

Copyright © 2020 by John Wiley & Sons, Inc., Hoboken, New Jersey

Published simultaneously in Canada

For general information on our other products and services, please contact our Customer Care Department within the U.S. at 877-762-2974, outside the U.S. at 317-572-3993, or fax 317-572-4002. For technical support, please visit https://hub.wiley.com/community/support/dummies.

Wiley publishes in a variety of print and electronic formats and by print-on-demand. Some material included with standard print versions of this book may not be included in e-books or in print-on-demand. If this book refers to media such as a CD or DVD that is not included in the version you purchased, you may download this material at http://booksupport.wiley.com. For more information about Wiley products, visit www.wiley.com.

Library of Congress Control Number: 2020941824

ISBN 978-1-119-68045-1 (pbk); ISBN 978-1-119-68048-2 (ebk); ISBN 978-1-119-68051-2 (ebk)

Manufactured in the United States of America

SKY10020298_080520

Contents at a Glance

Table of Contents

Introduction

Welcome to *Java All-in-One For Dummies*, 6th Edition — the one Java book that's designed to replace an entire shelf full of the dull, tedious titles you'd otherwise have to buy. This book contains all the basic information you need to know to get going with Java programming, starting with writing statements and using variables and ending with techniques for writing programs that use animation and play games. Along the way, you find plenty of not-so-basic information about programming user interfaces, working with classes and objects, creating web applications, and dealing with files and databases.

You can (and probably should, eventually) buy separate books on each of these topics. It won't take long before your bookshelf is bulging with 10,000 or more pages of detailed information about every imaginable nuance of Java programming. But before you're ready to tackle each of those topics in depth, you need to get a bird's-eye picture. That's what this book is about.

And if you already *own* 10,000 pages or more of Java information, you may be overwhelmed by the amount of detail and wonder, "Do I really need to *read* 1,200 pages about JSP just to create a simple web page? And do I really *need* a six-pound book on JavaFX?" Truth is, most 1,200-page programming books have about 200 pages of really useful information — the kind you use every day — and about 1,000 pages of excruciating details that apply mostly if you're writing guidance-control programs for nuclear missiles or trading systems for the New York Stock Exchange.

The basic idea here is that I've tried to wring out the 100-or-so most useful pages of information on these different Java programming topics: setup and configuration, basic programming, object-oriented programming, advanced programming techniques, JavaFX, file and database programming, web programming, and animation and game programming. Thus you get a nice, trim book.

So whether you're just getting started with Java programming or you're a seasoned pro, you've found the right book.

About This Book

Java All-in-One For Dummies, 6th Edition, is a reference for all the great things (and maybe a few not-so-great things) that you may need to know when you're writing Java programs. You can, of course, buy a huge 1,200-page book on each of the programming topics covered in this book. But then, who would carry them home from the bookstore for you? And where would you find the shelf space to store them? And when will you find the time to read them?

In this book, all the information you need is conveniently packaged for you in-between one set of covers. And all of the information is current for the newest release of Java, known as JDK 14. This book doesn't pretend to be a comprehensive reference for every detail on every possible topic related to Java programming. Instead, it shows you how to get up and running fast so that you have more time to do the things you really want to do. Designed using the easy-to-follow *For Dummies* format, this book helps you get the information you need without laboring to find it.

Java All-in-One For Dummies, 6th Edition, is a big book made up of nine smaller books — minibooks, if you will. Each of these minibooks covers the basics of one key element of programming, such as installing Java and compiling and running programs, or using basic Java statements, or using JavaFX to write GUI applications.

Whenever one big thing is made up of several smaller things, confusion is always a possibility. That's why this book has multiple access points. At the beginning is a detailed table of contents that covers the entire book. Then each minibook begins with a minitable of contents that shows you at a miniglance what chapters are included in that minibook. Useful running heads appear at the top of each page to point out the topic discussed on that page. And handy thumbtabs run down the side of the pages to help you find each minibook quickly. Finally, a comprehensive index lets you find information anywhere in the entire book.

Foolish Assumptions

You and I have never met, so it is difficult for me to make any assumptions about why you are interested in this book. However, let's start with a few basic assumptions:

>> **You own or have access to a relatively modern computer.** The examples were created on a Windows computer, but you can learn to program in Java just as easily on a Mac or Linux computer.

>> **You're an experienced computer user.** In other words, I assume that you know the basics of using your computer, such as starting programs and working with the file system.

>> **You're interested in learning how to write programs in the Java language.** Since that's what this book teaches, it's a fair assumption.

I do *not* make any assumptions about any previous programming experience in Java or in any other programming language. Nor do I make any assumptions about *why* you want to learn about Java programming. There are all sorts of valid reasons for learning Java. Some want to learn Java for professional reasons; maybe you want to become a professional Java programmer, or maybe you are a C# or C++ programmer who occasionally needs to work in Java. On the other hand, maybe you think programming in Java would make an interesting hobby.

Regardless of your motivation, I *do* assume that you are a reasonably intelligent person. You don't have to have a degree in advanced physics, or a degree in anything at all for that matter, to master Java programming. All you have to be is someone who wants to learn and isn't afraid to try.

Icons Used in This Book

Like any *For Dummies* book, this book is chock-full of helpful icons that draw your attention to items of particular importance. You find the following icons throughout this book:

Danger, Will Robinson! This icon highlights information that may help you avert disaster.

WARNING

Did I tell you about the memory course I took?

REMEMBER

Pay special attention to this icon; it lets you know that some particularly useful tidbit is at hand.

TIP

**TECHNICAL
STUFF**

Hold it — overly technical stuff is just around the corner. Obviously, because this is a programming book, almost every paragraph of the next 900 or so pages could get this icon. So I reserve it for those paragraphs that go into greater depth, down into explaining how something works under the covers — probably deeper than you really need to know to use a feature, but often enlightening.

Beyond the Book

In addition to the material in the print or e-book you're reading right now, this product also comes with some access-anywhere goodies on the web. Check out the free Cheat Sheet for more on Java. To get this Cheat Sheet, simply go to www.dummies.com and type **Java All-in-One For Dummies Cheat Sheet** in the Search box.

Visit www.dummies.com/go/javaaiofd6e to dive even deeper into Java. You can find and download the code used in the book at that link. You can also download a bonus minibook covering how to use Java with files and databases.

Where to Go from Here

This isn't the kind of book you pick up and read from start to finish, as if it were a cheap novel. If I ever see you reading it at the beach, I'll kick sand in your face. Beaches are for reading romance novels or murder mysteries, not programming books. Although you could read straight through from start to finish, this book is a reference book, the kind you can pick up, open to just about any page, and start reading. You don't have to memorize anything in this book. It's a "need-to-know" book: You pick it up when you need to know something. Need a reminder on the constructors for the ArrayList class? Pick up the book. Can't remember the goofy syntax for anonymous inner classes? Pick up the book. After you find what you need, put the book down and get on with your life.

1

Java Basics

Contents at a Glance

Chapter **1**

Welcome to Java

This chapter is a gentle introduction to the world of Java. In the next few pages, you find out what Java is, where it came from, and where it's going. You also discover some of the unique strengths of Java, as well as some of its weaknesses. Also, you see how Java compares with other popular programming languages such as C, C++, and C#.

By the way, I assume in this chapter that you have at least enough background to know what computer programming is all about. That doesn't mean that I assume you're an expert or professional programmer. It just means that I don't take the time to explain such basics as what a computer program is, what a programming language is, and so on. If you have absolutely no programming experience, I suggest that you pick up a copy of *Java For Dummies,* 7th Edition, or *Beginning Programming with Java For Dummies,* 5th Edition, both by Barry Burd (Wiley).

Throughout this chapter, you find little snippets of Java program code, plus a few snippets of code written in other languages, including C, C++, and Basic. If you don't have a clue what this code means or does, don't panic. I just want to give you a feel for what Java programming looks like and how it compares with programming in other languages.

TIP

All the code listings used in this book are available for download at www.dummies.com/go/javaaiofd6e.

What Is Java, and Why Is It So Great?

Java is a programming language in the tradition of C and C++. As a result, if you have any experience with C or C++, you'll often find yourself in familiar territory as you discover the various features of Java. (For more information about the similarities and differences between Java and C or C++, see the section "Java versus Other Languages," later in this chapter.)

Java differs from other programming languages in a couple of significant ways, however. I point out the most important differences in the following sections.

Platform independence

One of the main reasons Java is so popular is its *platform independence,* which simply means that Java programs can be run on many types of computers.

Before Java, other programming languages promised platform independence by providing compatible compilers for different platforms. (A *compiler* is the program that translates programs written in a programming language into a form that can actually run on a computer.) The idea was that you could compile different versions of the programs for each platform. Unfortunately, this idea never really worked. The compilers were never identical on each platform; each had its own little nuances. As a result, you had to maintain a different version of your program for each platform you wanted to support.

Java's platform independence isn't based on providing compatible compilers for different platforms. Instead, Java is based on the concept of a *virtual machine* called the Java Virtual Machine (JVM). Think of the JVM as a hypothetical computer platform — a design for a computer that doesn't exist as actual hardware. Instead, the JVM simulates the operation of a hypothetical computer that is designed to run Java programs.

The Java compiler doesn't translate Java into the machine language of the computer that the program is running on. Instead, the compiler translates Java into the machine language of the JVM, which is called *bytecode.* Then the JVM runs the bytecode in the JVM.

When you compile a Java program, the runtime environment that simulates the JVM for the targeted computer type (Windows, Linux, macOS, and so on) is included with your compiled Java programs.

That's how Java provides platform independence — and believe it or not, it works pretty well. The programs you write run just as well on a PC running any version of Windows, a Macintosh, a Unix or Linux machine, or any other computer that has a compatible JVM — including smartphones or tablet computers.

While you lie awake tonight pondering the significance of Java's platform independence, here are a few additional thoughts to ponder:

TECHNICAL STUFF

» Platform independence goes only so far. If you have some obscure type of computer system — such as an antique Olivetti Programma 101 — and a JVM runtime environment isn't available for it, you can't run Java programs on it.

I didn't make up the Olivetti Programma 101. It was a desktop computer made in the early 1960s, and it happened to be my introduction to computer programming. (My junior high school math teacher had one in the back of his classroom, and he let me play with it during lunch.) Do a Google search for "Olivetti Programma 101" and you can find several interesting websites about it.

» Java's platform independence isn't perfect. Although the bytecode runs identically on every computer that has a JVM, some parts of Java use services provided by the underlying operating system. As a result, minor variations sometimes crop up, especially in applications that use graphical interfaces.

» Because a runtime system that emulates a JVM executes Java bytecode, some people mistakenly compare Java with interpreted languages such as Basic or Perl. Those languages aren't compiled at all, however. Instead, the interpreter reads and interprets each statement as it is executed. Java is a true compiled language; it's just compiled to the machine language of JVM rather than to the machine language of an actual computer platform.

TECHNICAL STUFF

» If you're interested, the JVM is completely *stack-oriented;* it has no registers for storing local data. (I'm not going to explain what that term means, so if it doesn't make sense to you, skip it. It's not important. It's just interesting to nerds who know about stacks, registers, and things of that ilk.)

Object orientation

Java is inherently *object-oriented*, which means that Java programs are made up from programming elements called objects. Simply put (don't you love it when you read *that* in a computer book?), an *object* is a programming entity that represents either some real-world object or an abstract concept.

All objects have two basic characteristics:

» Objects have data, also known as *state*. An object that represents a book, for example, has data such as the book's title, author, and publisher.

» Objects also have *behavior,* which means that they can perform certain tasks. In Java, these tasks are called *methods*. An object that represents a car might have methods such as start, stop, drive, and crash. Some methods simply allow you to access the object's data. A book object might have a getTitle method that tells you the book's title.

Classes are closely related to objects. A *class* is the program code you write to create objects. The class describes the data and methods that define the object's state and behavior. When the program executes, classes are used to create objects.

Suppose you're writing a payroll program. This program probably needs objects to represent the company's employees. So the program includes a class (probably named Employee) that defines the data and methods for each Employee object. When your program runs, it uses this class to create an object for each of your company's employees.

The Java API

The Java language itself is very simple, but Java comes with a library of classes that provide commonly used utility functions that most Java programs can't do without. This class library, called the *Java API* (short for *application programming interface*), is as much a part of Java as the language itself. In fact, the real challenge of finding out how to use Java isn't mastering the language; it's mastering the API. The Java language has only about 50 keywords, but the Java API has several thousand classes, with tens of thousands of methods that you can use in your programs.

The Java API has classes that let you do trigonometry, write data to files, create windows onscreen, and retrieve information from a database, among other things. Many of the classes in the API are general purpose and commonly used. A whole series of classes stores collections of data, for example. But many are obscure, used only in special situations.

Fortunately, you don't have to learn anywhere near all of the Java API. Most programmers are fluent with only a small portion of it: the portion that applies most directly to the types of programs they write. If you find a need to use some class from the API that you aren't yet familiar with, you can look up what the class does in the Java API documentation at http://docs.oracle.com/en/java/javase/14.

The Internet

Java is often associated with the Internet, and rightfully so, because Al Gore invented Java just a few days after he invented the Internet. Okay, Java wasn't really invented by Al Gore. It was developed right at the time the World Wide Web was becoming a phenomenon, and Java was specifically designed to take advantage of the web. In particular, the whole concept behind the JVM is to enable any computer connected to the Internet to run Java programs, regardless of the type of computer or the operating system it runs.

Most Java programming on the Internet uses *servlets,* which are web-based Java programs that run on an Internet server computer rather than in an Internet user's web browser.

A servlet generates a page of HTML and JavaScript that is sent to a user's computer to be displayed in the user's web browser. If you request information about a product from an online store, the store's web server runs a servlet to generate the HTML page containing the product information you requested.

Java versus Other Languages

Superficially, Java looks a lot like many of the programming languages that preceded it, most notably C and C++. For example, here's the classic `Hello, World!` program, written in the C programming language:

```
main()
{
    printf("Hello, World!");
}
```

This program simply displays the text `"Hello, World!"` on the computer's console. Here's the classic `Hello, World!` program written in Java:

```
public class HelloApp
{
    public static void main(String[] args)
    {
        System.out.println("Hello, World!");
    }
}
```

Although the Java version is a bit more verbose, the two have several similarities:

>> Both require each executable statement to end with a semicolon (;).

>> Both use braces ({ }) to mark blocks of code.

>> Both use a routine called `main` as the main entry point for the program.

Many other similarities aren't evident in these simple examples, but the examples bring the major difference between C and Java front and center: Object-oriented programming rears its ugly head even in simple examples. Consider the following points:

>> In Java, even the simplest program is a class, so you have to provide a line that declares the name of the class. In this example, the class is named HelloApp. HelloApp has a method named main, which the JVM automatically calls when a program is run.

>> In the C example, printf is a library function you call to print information to the console. In Java, you use the PrintStream class to write information to the console.

PrintStream? There's no PrintStream in this program! Wait a minute — yes, there is. Every Java program has available to it a PrintStream object that writes information to the console. You can get this PrintStream object by calling the out method of another class, named System. Thus, System.out gets the PrintStream object that writes to the console. The PrintStream class in turn has a method named println that writes a line to the console. So System.out.println really does two things, in the following order:

1. It uses the out field of the System class to get a PrintStream object.

2. It calls the println method of that object to write a line to the console.

Confusing? You bet. Everything will make sense, however, when you read about object-oriented programming in Book 3, Chapter 1.

>> void looks familiar. Although it isn't shown in the C example, you could have coded void on the main function declaration to indicate that the main function doesn't return a value. void has the same meaning in Java. But static? What does that mean? That, too, is evidence of Java's object orientation. It's a bit early to explain what it means in this chapter, but you can find out in Book 2, Chapter 7.

Important Features of the Java Language

If you believe the marketing hype put out by Oracle and others, you think that Java is the best thing to happen to computers since the invention of memory. Java may not be *that* revolutionary, but it does have many built-in features that set it apart from other languages. The following sections describe just three of the many features that make Java so popular.

Type checking

All programming languages must deal in one way or the other with *type checking* — the way that a language handles variables that store different types of data. Numbers, strings, and dates, for example, are commonly used *data types* available in most programming languages. Most programming languages also have several types of numbers, such as integers and real numbers.

All languages must check data types, so make sure that you don't try to do things that don't make sense (such as multiplying the gross national product by your last name). The question is, does the language require you to declare every variable's type so you can do type checking when it compiles your programs, or does the language do type checking only after it runs your program?

Some languages, such as Perl, are not as rigid about type checking as Java. For example, Perl does not require that you indicate whether a variable will contain an integer, a floating point number, or a string. Thus, all the following statements are allowed for a single variable named $a:

```
$a = 5
$a = "Strategery"
$a = 3.14159
```

Here three different types of data — integer, string, and double — have been assigned to the same variable.

Java, on the other hand, *does* complete type checking when the program is compiled. As a result, you must declare all variables as a particular type so that the compiler can make sure you use the variables correctly. The following bit of Java code, for example, won't compile:

```
int a = 5;
String b = "Strategery";
String c = a * b;
```

If you try to compile these lines, you get an error message saying that Java can't multiply an integer and a string.

In Java, every class you define creates a new type of data for the language to work with. Thus, the data types you have available to you in Java aren't just simple predefined types, such as numbers and strings. You can create your own types. If you're writing a payroll system, you might create an Employee type. Then you can declare variables of type Employee that can hold only Employee objects. This capability prevents a lot of programming errors. Consider this code snippet:

```
Employee newHire;
newHire = 21;
```

This code creates a variable (newHire) that can hold only Employee objects. Then it tries to assign the number 21 to it. The Java compiler won't let you run this program because 21 is a number, not an employee.

TECHNICAL STUFF

An important object-oriented programming feature of Java called *inheritance* adds an interesting — and incredibly useful — twist to type checking. Inheritance is way too complicated to dive into just yet, so I'll be brief here: In Java, you can create your own data types that are derived from other data types. Employees are people, for example, and customers are people too, so you might create a Person class and then create Employee and Customer classes that both inherit the Person class. Then you can write code like this:

```
Person p;
Employee e;
Customer c;
p = e;   // This is allowed because an Employee is also a Person.
c = e;   // This is not allowed because an Employee is not a Customer.
```

Confused yet? If so, that's my fault. Inheritance is a pretty heady topic for Chapter 1 of a Java book. Don't panic if it makes no sense just yet. It will all be clear by the time you finish reading Book 3, Chapter 4, which covers all the subtle nuances of using inheritance.

Exception handling

As Robert Burns said, "The best-laid schemes o' mice an' men gang oft agley, an' lea'e us nought but grief an' pain, for promis'd joy!" When you tinker with computer programming, you'll quickly discover what he meant. No matter how carefully you plan and test your programs, errors happen, and when they do, they threaten to bring your whole program to a crashing halt.

Java has a unique approach to error handling that's superior (in my opinion) to that of any other language (except C#, which just copies Java's approach, as I mention earlier in the chapter). In Java, the JRE intercepts and folds errors of all types into a special type of object called an *exception object.* After all, Java is object-oriented through and through, so why shouldn't its exception-handling features be object-oriented?

Java requires any statements that can potentially cause an exception to be bracketed by code that can catch and handle the exception. In other words, you, as the programmer, must anticipate errors that can happen while your program is running and make sure that those errors are dealt with properly. Although this necessity can be annoying, it makes the resulting programs more reliable.

On the Downside: Java's Weaknesses

So far, I've been tooting Java's horn pretty loudly. Lest you think that figuring out how to use it is a walk in the park, the following paragraphs point out some of Java's shortcomings (many of which have to do with the API rather than the language itself):

» **The API is way too big.** It includes so many classes and methods that you'll likely never use even half of them. Also, the sheer size of the Java API doesn't allow you to wander through it on your own, hoping to discover the one class that's perfect for the problem you're working on.

» **The API is overdesigned.** In some cases, it seems as though the Java designers go out of their way to complicate things that should be simple to use. For example, the API class that defines a multiline text-input area doesn't have a scroll bar. Instead, a separate class defines a panel that has a scroll bar. To create a multiline text area with a scroll bar, you have to use both classes. It would be simpler if there were an option on the multi-line text area class that let you specify whether you wanted to include a scroll bar.

» **Some corners of the API are haphazardly designed.** Most of the problems can be traced back to the initial version of Java, which was rushed to market so that it could ride the crest of the World Wide Web wave in the late 1990s. Since then, many parts of the API have been retooled more thoughtfully, but the API is still riddled with remnants of Java's early days.

TECHNICAL STUFF

» **In my opinion, the biggest weakness of Java is that it doesn't have a decimal data type.** This issue is a little too complicated to get into right now, but the implication is this: Without special coding (which few Java books explain and few Java programmers realize), Java doesn't know how to add. Consider this bit of code:

```
double x = 5.02;
double y = 0.01;
double z = x + y;
System.out.println(z);
```

This little program should print 5.03, right? It doesn't. Instead, it prints 5.029999999999999. This little error may not seem like much, but it can add up. If you ever make a purchase from an online store and notice that the sales tax is a penny off, this is why. The explanation for why these errors happen — and how to prevent them — is pretty technical, but it's something that every Java programmer needs to understand.

For the scoop on how to avoid this problem, refer to Book 2, Chapter 3.

Java Version Insanity

Like most products, Java gets periodic upgrades and enhancements. Since its initial release in 1996, Java has undergone the following version updates:

>> **Java 1.0:** This version was the original version of Java, released in 1996. Most of the language is still pretty much the same as it was in version 1.0, but the API has changed a lot since this release.

>> **Java 1.1:** This version was the first upgrade to Java, released in 1997. This release is important because most Internet browsers include built-in support for applets based on Java 1.1. To run applets based on later versions of Java, in most cases you must download and install a current JRE.

>> **Java 1.2:** This version, released in late 1998, was a huge improvement over the previous version — so much so, in fact, that Sun called it Java 2. It included an entirely new API called Swing for creating graphical user interfaces, as well as other major features.

>> **Java 1.3:** This version, released in 2000, was mostly about improving performance by changing the way the runtime system works. Oddly, though this version is technically Java 1.3, it's also called Java 2 version 1.3. Go figure.

>> **Java 1.4:** Released in 2001, this version offered a slew of improvements. As you might guess, it's called Java 2 version 1.4. Keep figuring. . ..

>> **Java 5.0:** Released in 2004, this version included more changes and improvements than any other version. To add to Sun's apparent unpredictability in its version numbering, this version officially has *two* version numbers. Sun's official Java website explains it like this:

- Both version numbers "1.5.0" and "5.0" are used to identify this release of the Java 2 Platform Standard Edition. Version "5.0" is the *product version*, while "1.5.0" is the *developer version*.

That clears everything right up, doesn't it?

» **Java 6.0:** Released in December 2006 (just in time for the holidays!), this version of Java offered minor improvements and better efficiency.

For Java 1.6, the product version is 6 (not 6.0). Remember the extra 2 that appeared magically in 1998? Well, the 2 is gone in Java 1.6. So unlike the versions between 1998 and 2006, Java 1.6 is officially named the Java Platform (not the Java 2 Platform). Personally, I think someone at Sun has been talking to George Lucas. I fully expect the next version of Java to be a prequel called Java 0 Episode 1.

» **Java 7.0:** Released in mid-2011, this was a relatively minor upgrade that added a few enhancements to the API and a few tweaks to the language itself.

» **Java 8.0:** Released in February 2014, Java 8 (as it is known) adds some significant and long-anticipated new features to Java. One of the most important is *lambda expressions,* a language feature that simplifies certain aspects of object-oriented programming. Other new features include a completely revamped API for working with dates and times, and a new framework for working with large collections of data in a way that can easily take advantage of multicore processors.

» **Java 9.0:** Released in 2017, Java 9 adds a significant and long-awaited feature called the *Java Module System,* which provides a new and improved way of managing the collections of Java code that make up a complete Java application and dramatically changes how Java applications are packaged. You learn about this new feature in Book 3, Chapter 8.

» **Java 10.0:** Released in March 2018, this version is the first to be released under the a new development schedule, in which new releases will come out twice per year, first in March and then in September.

» **Java 11.0:** Released in September 2018, this version introduced a dozen new features. The most popular is the *local type inference,* which introduces the `var` keyword (well, technically not a keyword but it sure looks and acts like a keyword), which allows the compiler to infer the type of a variable from its context.

» **Java 12.0:** Released in March 2019, this version introduced a dozen or so minor features, including a new standardized web client, which is covered in Book 7, Chapter 5.

» **Java 13.0:** Released in September 2019, this version introduced, among other things, a long-needed improvement to the `switch` statement, which I cover in Book 2, Chapter 6.

» **Java 14.0:** Released in March 2020, this is the current version as of the writing of this book. It includes a new type of data structure, called a `Record`, which I cover in Book 3, Chapter 2.

What's in a Name?

The final topic that I want to cover in this chapter is the names of the various pieces that make up Java's technology — specifically, the acronyms you constantly come across whenever you read or talk about Java, such as JVM, JRE, JDK, and J2EE. Here they are, in no particular order of importance:

» **JDK:** *Java Development Kit* — that is, the toolkit for developers that includes the Java compiler and the runtime environment. To write Java programs, you need the JDK. This term was used with the original versions of Java (1.0 and 1.1) and abandoned with version 1.2 in favor of SDK. But with versions 5.0, the term reappeared and is still in use today.

» **SDK:** *Software Development Kit* — what Sun called the JDK for versions 1.2, 1.3, and 1.4.

» **JRE:** *Java Runtime Environment* — the program that emulates the JVM so that users can run Java programs. To run Java programs, you need only download and install the JRE. Prior to Java 11, this was a separately installable program that had to be installed on computers that ran Java programs. This is no longer the case; the runtime environment is now packaged with executable Java applications.

» **JVM:** *Java Virtual Machine* — the platform-independent machine that's emulated by the JRE. All Java programs run in a JVM.

» **Java SE:** *Java Standard Edition* — a term that describes the Java language and the basic set of API libraries that are used to create Java programs that can run on Windows, Linux, and other platforms, such as Macintosh. Most of this book focuses on Java SE.

» **J2SE:** *Java 2 Standard Edition* — an older term for the Java language and basic libraries (for Java versions 1.2 through 1.5).

» **Java EE:** *Java Enterprise Edition,* also known as J2EE (*Java 2 Enterprise Edition*) — an expanded set of API libraries that provide special functions such as servlets.

Chapter **2**

Installing and Using Java Tools

Java development environments have two basic approaches. On the one hand, you can use a sophisticated integrated development environment (IDE) such as NetBeans or Eclipse. These tools combine a full-featured source editor that lets you edit your Java program files with integrated development tools, including visual development tools that let you create applications by dragging and dropping visual components onto a design surface.

At the other extreme, you can use just the basic command-line tools that are available free from Oracle's Java website (`http://java.oracle.com`). Then you can use any text editor you want to create the text files that contain your Java programs (called *source files*), and compile and run your programs by typing commands at a command prompt.

TIP

As a compromise, you may want to use a simple development environment, such as TextPad. TextPad is an inexpensive text editor that provides some nice features for editing Java programs (such as automatic indentation) and shortcuts for compiling and running programs. It doesn't generate any code for you or provide any type of visual design aids, however. TextPad is the tool I used to develop all the examples shown in this book. For information about downloading and using TextPad, see Book 1, Chapter 3.

TIP

If you prefer a free alternative, you can also investigate Notepad++ at `http://notepad-plus-plus.org`.

Downloading and Installing the Java Development Kit

Before you can start writing Java programs, you have to download and install the correct version of the Java Development Kit (JDK) for the computer system you're using. Oracle's Java website provides versions for Windows, Solaris, and Unix. The following sections show you how to download and install the JDK.

TIP

If you prefer, you can download and install the open-source version of Java from `http://openjdk.java.net`.

Downloading the JDK

To get to the download page, point your browser to `www.oracle.com/java/technologies`. Then follow the appropriate links to download the latest version of Java SE for your operating system. (At the time I wrote this, the latest version was 14.01.)

LEGAL MUMBO JUMBO

Before you can download the JDK, you have to approve of the Java license agreement — all 1,919 words of it, including all the *thereupons*, *whereases*, and *hithertos* so finely crafted by Oracle's legal department. I'm not a lawyer (and I don't play one on TV), but I'll try to summarize the license agreement for you:

- The party of the first part (that's Oracle) grants you the right to use Java as is and doesn't promise that Java will do anything at all.

- The party of the second part (that's you) in turn promises to use Java only to write programs. You're not allowed to try to figure out how Java works and sell your secrets to Microsoft.

- You can't use Java to make a nuclear bomb or a missile delivery system. (I'm not making that up. It's actually in the license agreement.)

When you get to the Java download page, you find a link to download the Java JDK. Click this link, and then select your operating system and click the JDK Download link to start the download.

The JDK download comes in two versions: an executable installer and a `.zip` file. Both are about the same size. I find it easier to download and run the `.exe` installer.

Installing the JDK

After you download the JDK file, you can install it by running the executable file you downloaded. The procedure varies slightly depending on your operating system, but basically, you just run the JDK installation program file after you download it, as follows:

>> On a Windows system, open the folder in which you saved the installation program and double-click the installation program's icon.

>> On a Linux or Solaris system, use console commands to change to the directory to which you downloaded the file and then run the program.

After you start the installation program, it prompts you for any information that it needs to install the JDK properly, such as which features you want to install and what folder you want to install the JDK in. You can safely choose the default answer for each option.

Perusing the JDK folders

When the JDK installs itself, it creates several folders on your hard drive. The locations of these folders vary depending on your system, but in all 32-bit versions of Windows, the JDK root folder is in the path `Program Files\Java` on your boot drive. On 64-bit versions of Windows, the root folder will be either `Program Files\Java` or `Program Files (x86)\Java`. The name of the JDK root folder also varies, depending on the Java version you've installed. For version 14.0.1, the root folder is named `jdk-14.0.1`.

Table 2-1 lists the subfolders created in the JDK root folder. As you work with Java, you'll refer to these folders frequently.

TABLE 2-1

Subfolders of the JDK Root Folder

Folder	Description
bin	The compiler and other Java development tools
conf	Configuration file
include	This library contains files needed to integrate Java with programs written in other languages
jmods	Modules for the Java Module System (new with Java 1.9)
legal	Copyright and license information for various Java components
lib	Library files, including the Java API class library

Setting the path

After you install the JDK, you need to configure your operating system so that it can find the JDK command-line tools. To do that, you must set the Path environment variable — a list of folders that the operating system uses to locate executable programs. Follow these steps:

1. **Open Windows Explorer, right-click This PC, and choose Properties.**

   This brings up the System Properties page.

2. **Click the Advanced System Settings link.**

3. **Click the Environment Variables button.**

   The Environment Variables dialog box appears, as shown in Figure 2-1.

4. **In the System Variables list, scroll to the Path variable, select it, and then click the Edit button.**

   This brings up a handy dialog box that lets you add or remove paths to the Path variable or change the order of the paths, shown in Figure 2-2.

5. **Add the** bin **folder to the beginning of the** Path **value.**

6. **Click New, key in the complete path to the** bin **folder, and press Enter.**

7. **Use the Move Up button to move the** bin **folder all the way to the top of the list.**

   Note: The name of the bin folder may vary on your system, as in this example:

   ```
   c:\Program Files\Java\jdk-14.0.1\bin;other directories...
   ```

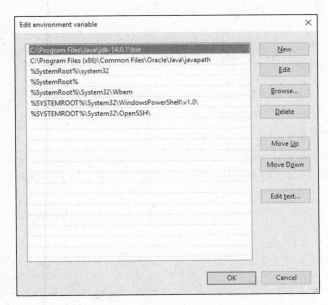

FIGURE 2-1:
The Environment
Variables
dialog box.

Installing and Using Java Tools

8. Click OK three times to exit.

The first OK gets you back to the Environment Variables dialog box; the second OK gets you back to the System Properties dialog box; and the third OK closes the System Properties dialog box.

FIGURE 2-2:
Editing the Path
variable.

For Linux or Solaris, the procedure depends on which shell you're using. For more information, consult the documentation for the shell you're using.

Using Java's Command-Line Tools

Java comes with several command-line tools that you can run directly from a command prompt. The two most important are `javac`, the Java compiler used to compile a program, and `java`, the command used to run a Java program. These tools work essentially the same way no matter what operating system you're using.

Compiling a program

You can compile a program from a command prompt by using the `javac` command. Before you can do that, however, you need a program to compile. Follow these steps:

1. **Using any text editor, type the following text in a file, and save it as** `HelloApp.java`:

    ```java
    public class HelloApp
    {
        public static void main(String[] args)
        {
            System.out.println("Hello, World!");
        }
    }
    ```

    WARNING

    Pay special attention to capitalization. If you type **Public** instead of **public**, for example, the program won't work. (If you don't want to bother with typing, you can download the sample programs from this book's website at www.dummies.com/go/javaaiofd6e.)

2. **Save the file in any directory you want.**

3. **Open a command prompt, use a** `cd` **command to change to the directory you saved the program file in, and then enter the command** `javac HelloApp.java`.

    This command compiles the program (`javac`) and creates a class file named `HelloApp.class`.

Assuming that you typed the program exactly right, the `javac` command doesn't display any messages at all. If the program contains any errors, however, you get one or more error messages onscreen. If you typed `Public` instead of `public` despite my warning earlier in this section, the compiler displays the following error message:

```
C:\java\samples>javac HelloApp.java
HelloApp.java:1: error: class, interface, or enum expected
Public class HelloApp
               ^
1 error
C:\java\samples>
```

The compiler error message indicates that an error is in line 1 of the HelloApp.java file. If the compiler reports an error message like this one, your program contains a coding mistake. You need to find the mistake, correct it, and compile the program again.

TIP

If a .java file contains just a single class, as most of the examples in this book do, you can skip the javac command altogether. The java command can both compile and run a .java file that contains a single class.

Compiling more than one file

Normally, the javac command compiles only the file that you specify on the command line, but you can coax javac into compiling more than one file at a time by using any of the techniques I describe in the following paragraphs:

» If the Java file you specify on the command line contains a reference to another Java class that's defined by a java file in the same folder, the Java compiler automatically compiles that class too.

Suppose you have a java program named TestProgram, which refers to a class called TestClass, and the TestClass.java file is located in the same folder as the TestProgram.java file. When you use the javac command to compile the TestProgram.java file, the compiler automatically compiles the TestClass.java file, too.

» You can list more than one filename in the javac command. The following command compiles three files:

```
javac TestProgram1.java TestProgram2.java TestProgram3.java
```

» You can use a wildcard to compile all the files in a folder, like this:

```
javac *.java
```

>> If you need to compile a lot of files at the same time but don't want to use a wildcard (perhaps you want to compile a large number of files but not all the files in a folder), you can create an *argument file,* which lists the files to compile. In the argument file, you can type as many filenames as you want, using spaces or line breaks to separate them. Here's an argument file named TestPrograms that lists three files to compile:

```
TestProgram1.java
TestProgram2.java
TestProgram3.java
```

You can compile all the programs in this file by using an @ character followed by the name of the argument file on the javac command line, like this:

```
javac @TestPrograms
```

Using Java compiler options

The javac command has a gaggle of options that you can use to influence the way it compiles your programs. For your reference, I list these options in Table 2-2.

To use one or more of these options, type the option before or after the source filename. Either of the following commands, for example, compiles the HelloApp.java file with the -verbose and -deprecation options enabled:

```
javac HelloWorld.java -verbose -deprecation
javac -verbose -deprecation HelloWorld.java
```

Don't get all discombobulated if you don't understand what all these options do. Most of them are useful only in unusual situations. The options you'll use the most are

>> -classpath or -cp: Use this option if your program makes use of class files that you've stored in a separate folder.

>> -deprecation: Use this option if you want the compiler to warn you whenever you use API methods that have been deprecated. (*Deprecated* methods are older methods that once were part of the Java standard API but are on the road to obsolescence. They still work but may not function in future versions of Java.)

>> -source: Use this option to limit the compiler to previous versions of Java. Note, however, that this option applies only to features of the Java language itself, not to the API class libraries. If you specify -source 1.4, for example, the compiler won't allow you to use new Java language features that were introduced in a version later than 1.4, such as generics, enhanced for loops,

or Lambda expressions. But you can still use the new API features that were added with version 1.5, such as the Scanner class.

» –help: Use this option to list the options that are available for the javac command.

TABLE 2-2 ## Java Compiler Options

Option	Description
–g	Generates all debugging info.
–g:none	Generates no debugging info.
–g:{lines,vars,source}	Generates only some debugging info.
–nowarn	Generates no warnings.
–verbose	Outputs messages about what the compiler is doing.
–deprecation	Outputs source locations where deprecated APIs are used.
–classpath <path>	Specifies where to find user class files.
–cp <path>	Specifies where to find user class files.
–sourcepath <path>	Specifies where to find input source files.
–bootclasspath <path>	Overrides locations of bootstrap class files.
–extdirs <dirs>	Overrides locations of installed extensions.
–endorseddirs <dirs>	Overrides location of endorsed standards path.
–d <directory>	Specifies where to place generated class files.
–encoding <encoding>	Specifies character encoding used by source files.
–source <release>	Provides source compatibility with specified release.
–target <release>	Generates class files for specific virtual-machine version.
–version	Provides version information.
–help	Prints a synopsis of standard options.
–X	Prints a synopsis of nonstandard options.
–J<flag>	Passes <flag> directly to the runtime system.
––enable–preview	Enables preview features — features that have been tentatively released in the current version but have not yet been adopted as standard. Use this option with caution, because a preview feature may be removed from the next version if it doesn't pan out.

Running a Java program

When you successfully compile a Java program, you can run the program by typing the java command followed by the name of the class that contains the program's main method. The JRE loads, along with the class you specify, and then runs the main method in that class. To run the HelloApp program, for example, type this command:

```
C:\java\samples>java HelloApp
```

The program responds by displaying the message "Hello, World!".

REMEMBER

The class must be contained in a file with the same name as the class, and its filename must have the extension .class. You usually don't have to worry about the name of the class file because it's created automatically when you compile the program with the javac command. Thus, if you compile a program in a file named HelloApp.java, the compiler creates a class named HelloApp and saves it in a file named HelloApp.class.

Understanding error messages

If Java can't find a filename that corresponds to the class, you get a simple error message indicating that the class can't be found. Here's what you get if you type JelloApp instead of HelloApp:

```
C:\java\samples>java JelloApp
Exception in thread "main"
    java.lang.NoClassDefFoundError: JelloApp
```

This error message simply means that Java couldn't find a class named JelloApp.

Specifying options

Like the Java compiler, the Java runtime command lets you specify options that can influence its behavior. Table 2-3 lists the most commonly used options.

Using the javap command

The javap command is called the Java *disassembler* because it takes class files apart and tells you what's inside them. You won't use this command often, but using it to find out how a particular Java statement works is fun sometimes. You can also use it to find out what methods are available for a class if you don't have the source code that was used to create the class.

TABLE 2-3 **Common Java Command Options**

Option	Description
–client	Runs the client virtual machine.
–server	Runs the server virtual, which is optimized for server systems.
–classpath *directories and archives*	Lists the directories or JAR or zip archive files used to search for class files.
–cp <search path>	Does the same thing as –classpath.
–D name=value	Sets a system property.
–verbose	Enables verbose output.
–version	Displays the JRE version number and then stops.
–showversion	Displays the JRE version number and then continues.
–? or –help	Lists standard options.
–X	Lists nonstandard options.
–ea or –enableassertions	Enables the assert command.
–ea *classes or packages*	Enables assertions for the specified classes or packages.
–esa or –enablesystemassertions	Enables system assertions.
–dsa or –disablesystemassertions	Disables system assertions.

Here's the information you get when you run the javap HelloApp command:

```
C:\java\samples>javap HelloApp
Compiled from "HelloApp.java"
public class HelloApp{
    public HelloApp();
    public static void main(java.lang.String[]);
}
```

As you can see, the javap command indicates that the HelloApp class was compiled from the HelloApp.java file and that it consists of a HelloApp public class and a main public method.

TECHNICAL STUFF

You may want to use two options with the javap command. If you use the –c option, the javap command displays the actual Java bytecodes created by the compiler for the class. And if you use the –verbose option, the bytecodes (plus a ton of other fascinating information about the innards of the class) are displayed.

USING OTHER COMMAND-LINE TOOLS

TIP

Java has many other command-line tools that come in handy from time to time. You can find a complete list of command-line tools at https://docs.oracle.com/en/java/javase/14/docs/specs/man.

I describe two of these additional tools in Book 3, Chapter 8:

- javadoc: Automatically creates HTML documentation for your Java classes.
- jar: Creates Java archive (JAR) files, which store classes in a compressed file that's similar to a zip file.

If you become a big-time Java guru, you can use this type of information to find out exactly how certain Java features work. Until then, you probably should leave the javap command alone except for those rare occasions when you want to impress your friends with your in-depth knowledge of Java. (Just hope that when you do, they don't ask you what the aload or invokevirtual instruction does.)

Using Java Documentation

Before you get too far into figuring out Java, don't be surprised if you find yourself wondering whether some class has some other method that I don't describe in this book — or whether some other class may be more appropriate for an application you're working on. When that time comes, you'll need to consult the Java help pages.

Complete documentation for Java is available on the Oracle Java website at https://docs.oracle.com/en/java/javase/14. Although this page contains many links to documentation pages, the one you'll use most is API Documentation. The API Documentation page (https://docs.oracle.com/en/java/javase/14/docs/api/index.html) provides complete documentation for all currently supported versions of the Java API. Figure 2-3 shows the home page for the Java SE 14 API documentation.

TIP

You can use this page to find complete information for any class in the API. You can drill down through the various modules and packages to find documentation for specific classes. For example, Figure 2-4 shows the documentation page for the String class. If you scroll down this page, you find complete information about everything you can do with this class, including an in-depth discussion of what the class does, a list of the various methods it provides, and a detailed description of what each method does. In addition, you find links to similar classes.

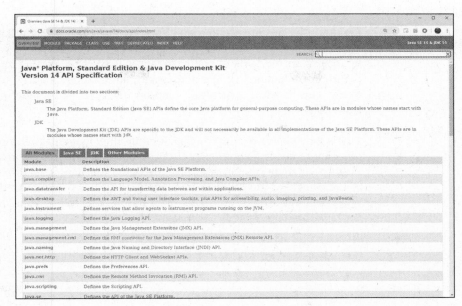

FIGURE 2-3:
A home page from the Java SE 14 API documentation.

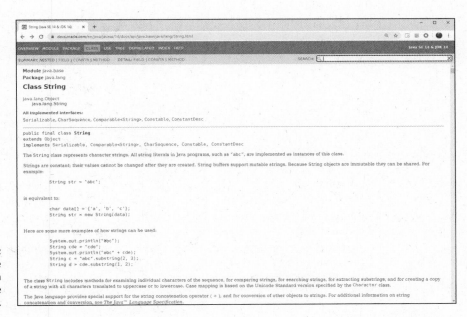

FIGURE 2-4:
The documentation page for the String class.

TECHNICAL STUFF

If you're interested in learning details about some element of the Java language itself (rather than the information about a class in the API class library), visit the Java Programming Language reference page at `http://docs.oracle.com/javase/specs/index.html`. That link takes you to a set of pages that describes — in sometimes excruciating and obscure detail — exactly how each element of the Java language works.

Frankly, this documentation isn't that much help for beginning programmers. It was written by computer scientists *for* computer scientists. You can tell just by looking at the table of contents that it isn't for novices. The first chapter is called "Introduction" (that's not so bad), but Chapters 2 and 3 are titled "Grammars" and "Lexical Structure," respectively, and matters just get more arcane from there.

That's why you're reading this book, after all. You won't even find a single sentence more about lexical structure in this book (other than this one, of course). Even so, at some time in your Java journeys, you may want to get to the bottom of the rules that govern such strange Java features as anonymous inner classes. When that day arrives, grab a six-pack of Jolt Cola, roll up your sleeves, and open the Java Language Specification pages.

Chapter **3**

Working with TextPad

TextPad is an inexpensive ($27) text editor that you can integrate with the Java Development Kit (JDK) to simplify the task of coding, compiling, and running Java programs. It isn't a true integrated development environment (IDE), as it lacks features such as integrated debugging, code generators, and drag-and-drop tools for creating graphical user interfaces.

TextPad is a popular tool for developing Java programs because of its simplicity and speed. It's ideal for learning Java because it doesn't generate any code for you. Writing every line of code yourself may seem like a bother, but the exercise pays off in the long run because you have a better understanding of how Java works.

Downloading and Installing TextPad

You can download a free evaluation version of TextPad from Helios Software Systems at www.textpad.com. You can use the evaluation version free of charge, but if you decide to keep the program, you must pay for it. (Helios accepts credit card payment online.)

If the Java JDK is already installed on your computer when you install TextPad, TextPad automatically configures itself to compile and run Java programs. If you install the JDK after you install TextPad, you need to configure TextPad for Java by following these steps:

1. **Choose Configure⇨Preferences to open the Preferences dialog box.**

2. **Click Tools in the tree that appears on the left side of the dialog box.**

3. **Click the Add button to reveal a drop-down list of options and then click Java SDK Commands.**

Figure 3-1 shows how the Preferences dialog box appears when the Java tools are installed. As you can see, the Tools item in the tree on the left side of the dialog box includes three Java tools: Compile Java, Run Java Application, and Run Java Applet. (The Run Java Applet tool is obsolete, so you can safely ignore it.)

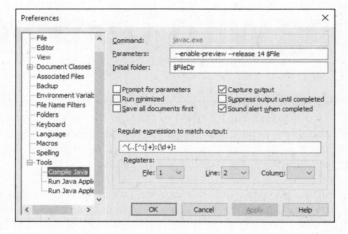

FIGURE 3-1:
Configuring tools
in TextPad.

4. **Click OK.**

The commands you need to compile and run Java programs are added to TextPad's Tools menu.

TIP

If you plan on using any preview features in Java, add the --enable-preview and --release 14 flags to the command line arguments for the Compile tool. The complete Compile tool arguments should look like this:

```
--enable-preview --release 14 $File
```

Then, add just the `--enable-preview` flag to the Run Java Application arguments list; the complete arguments should look like this:

```
--enable-preview $BaseName
```

WARNING

Do *not* add the `--release` flag to the Run Java Application tool. If you do, the tool won't be able to start the JVM, because `--release` is not a valid flag for the `java` command.

Editing Source Files

Figure 3-2 shows a Java source file being edited in TextPad. If you've worked with a Windows text editor before, you'll have no trouble mastering the basics of Text-Pad. I won't go over such basic procedures as opening and saving files because they're standard; instead, I describe some TextPad features that are useful for editing Java program files.

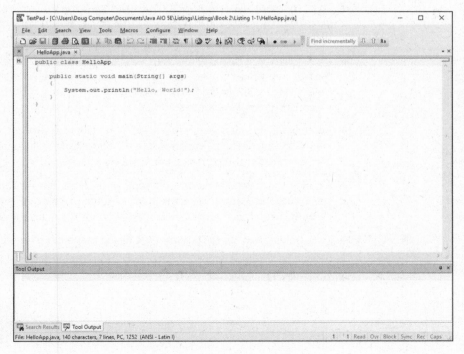

FIGURE 3-2:
Editing a Java file
in TextPad.

TIP

When you first create a file (by clicking the New button on the toolbar or by choosing File➪New), TextPad treats the file as a normal text file, not as a Java program file. After you save the file (by clicking the Save button or choosing File➪Save) and assign .java as the file extension, TextPad's Java-editing features kick in.

The following paragraphs describe some of TextPad's most noteworthy features for working with Java files:

>> You can't really tell from Figure 3-2, but TextPad uses different colors to indicate the function of each word or symbol in the program. Brackets are red so that you can spot them quickly and make sure that they're paired correctly. Keywords are blue. Comments and string literals are green. Other text, such as variable and method names, is black.

>> TextPad automatically indents whenever you type an opening bracket and then reverts to the previous indent when you type a closing bracket. This feature makes keeping your code lined up easy.

>> Line numbers display down the left edge of the editing window. You can turn these line numbers on or off by choosing View➪Line Numbers.

>> To go to a particular line, press Ctrl+G to bring up the Go To dialog box. Make sure that Line is selected in the Go to What box, enter the line number in the text box, and click OK.

>> If you have more than one file open, you can switch between the files by using the Document Selector — the pane on the left side of the TextPad window (refer to Figure 3-2). If the Document Selector isn't visible, choose View➪Document Selector to summon it.

>> Another way to switch between two (or more) files is to choose View➪Document Tabs. Tabs appear at the bottom of the document window, and you can click these tabs to switch documents.

>> A handy Match Bracket feature lets you pair brackets, braces, and parentheses. To use this feature, move the insertion point to a bracket, brace, or parenthesis and then press Ctrl+M. TextPad finds the matching element.

>> To search for text, press F5. In the Find dialog box, enter the text you're looking for, and click OK. To repeat the search, press Ctrl+F.

>> To replace text, press F8.

USING WORKSPACES

In TextPad, a *workspace* is a collection of files that you work on together. Workspaces are useful for projects that involve more than just one file. When you open a workspace, TextPad opens all the files in the workspace.

To create a workspace, first open all the files that you want to include in the workspace. Then choose File⇨Workspace⇨Save As and, in the Save As dialog box, give the workspace a name. (The files that make up the workspace are saved in a single file with the .tws extension.)

To open a workspace, choose File⇨Workspace⇨Open to display the Open dialog box, select the workspace file that you previously saved, and click Open. Alternatively, you can choose the workspace from the list of recently used workspaces at the bottom of the File⇨Workspace submenu.

To configure TextPad to open the most recently used workspace automatically whenever you start TextPad, choose Configure⇨Preferences to open the Preferences dialog box, click General in the tree on the left side of the dialog box, select the Reload Last Workspace at Startup check box, and click OK to close the dialog box.

Compiling a Program

To compile a Java program in TextPad, choose Tools⇨Compile Java or use the keyboard shortcut Ctrl+1. The javac command compiles the program, and the compiler output is displayed in the Tool Results pane of the TextPad window. If the program compiles successfully, the message Tool completed successfully appears in the Tool Results pane. If the compiler finds something wrong with your program, one or more error messages are displayed, as shown in Figure 3-3.

In this example, two compiler error messages are displayed:

```
C:\Users\Doug Computer\Documents\Java AIO 6E\Listings\Listings\Book 1\Listing
    1-1\HelloApp.java:5: error: unclosed string literal
        System.out.println("Hello, World!);
                           ^
C:\Users\Doug Computer\Documents\Java AIO 6E\Listings\Listings\Book 1\Listing
    1-1\HelloApp.java:7: error: reached end of file while parsing
}
 ^
2 errors

Tool completed with exit code 1
```

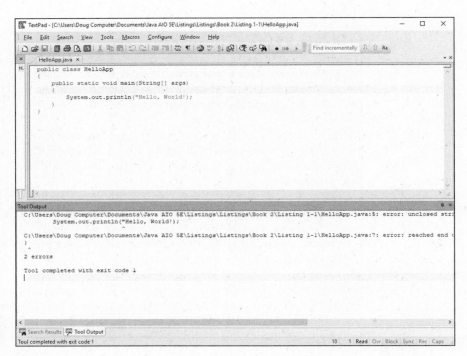

FIGURE 3-3:
Error messages
displayed by the
Java compiler.

TIP

If you double-click the first line of each error message, TextPad takes you to the spot where the error occurred. If you double-click the line with the `unclosed string literal` message, for example, you're taken to line 5, and the insertion point is positioned at the spot where the compiler found the error. Then you can correct the error and recompile the program.

REMEMBER

Often, a single error can cause more than one error message to display, as is the case in Figure 3-3. Here, a single mistake caused two errors.

Running a Java Program

After you compile a Java program with no errors, you can run it by choosing Tools⇨Run Java Application or pressing Ctrl+2. A command window opens, in which the program runs. Figure 3-4 shows the HelloApp program running in a separate window atop the TextPad window.

When the program finishes, the message Press any key to continue appears in the command window. When you press a key, the window closes, and TextPad comes back to life.

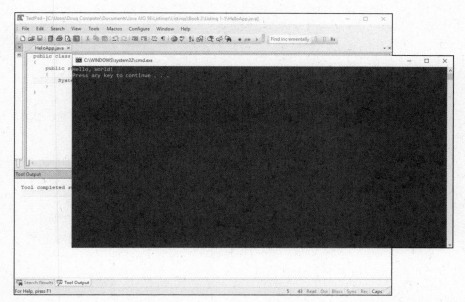

FIGURE 3-4:
Running a
program.

TECHNICAL
STUFF

In case you're wondering, TextPad actually runs your program by creating and running a *batch file* — a short text file that contains the commands necessary to run your program. This batch file is given a cryptic name, such as tp02a11c.BAT. Here's the batch file generated for the HelloApp program:

```
@ECHO OFF
C:
CD "\Users\Doug Computer\Documents\Java AIO 6E\Listings\Listings\Book 1\
    Listing 1-1"
"C:\Program Files\Java\jdk-14.0.1\bin\java.exe"  --enable-preview HelloApp
PAUSE
```

Here's a closer look at these commands:

>> The first command tells MS-DOS not to display the commands in the command window as the batch file executes.

>> The next two commands switch to the drive and directory that contain the java program.

>> Next, the java.exe program is called to run the HelloApp class.

>> Finally, a PAUSE command executes. That command is what displays the Press any key to continue message when the program finishes.

Working with TextPad

2

Programming Basics

Contents at a Glance

Chapter **1**

Java Programming Basics

I n this chapter, you find the basics of writing simple Java programs. The programs you see in this chapter are very simple: they just display simple information on a *console* (in Windows, that's a command-prompt window). You need to cover a few more chapters before you start writing programs that do anything worthwhile. But the simple programs you see in this chapter are sufficient to illustrate the basic structure of Java programs.

Be warned that in this chapter, I introduce you to several Java programming features that are explained in greater detail in later chapters. You see, for example, some variable declarations, a method, and even an `if` statement and a `for` loop. The goal of this chapter isn't to march you into instant proficiency with these programming elements, but just to introduce you to them.

TIP

You can find all the code listings used in this book at `www.dummies.com/go/javaaiofd6e`.

Looking at the Infamous Hello, World! Program

Many programming books begin with a simple example program that displays the text "Hello, World!" on the console. In Book 1, Chapter 1, I show you a Java program that does that to compare it with a similar program written in C. Now take a closer look at each element of this program, shown in Listing 1-1.

LISTING 1-1: **The HelloApp Program**

```
public class HelloApp                                             →1
{                                                                 →2
    public static void main(String[] args)                        →3
    {                                                             →4
        System.out.println("Hello, World!");                      →5
    }                                                             →6
}                                                                 →7
```

Later in this chapter, you discover in detail all the elements that make up this program. But first, I want to walk you through it word by word.

Lines 1 and 2 mark the declaration of a public class named HelloApp:

» →1 public: A *keyword* of the Java language that indicates that the element that follows should be made available to other Java elements. In this case, what follows is a class named HelloApp. As a result, this keyword indicates that the HelloApp class is a *public class,* which means other classes can use it. (In Book 3, Chapter 2, I cover the most common alternative to public: private. There are also other alternatives, but they're covered in later chapters.)

» class: Another Java keyword that indicates that the element being defined here is a class. All Java programs are made up of one or more *classes.* A class definition contains code that defines the behavior of the objects created and used by the program. Although most real-world programs consist of more than one class, the simple programs you see in this minibook have just one class.

» HelloApp: A *name* that identifies the class being defined here. Whereas keywords, such as public and class, are words that are defined by the Java programming language, names are words that you create to identify various elements you use in your program. In this case, the name HelloApp identifies the public class being defined here. (Although *name* is the technically correct term, sometimes names are called *identifiers.* Technically, a name is a type of identifier, but not all identifiers are names.)

» →2 {: The opening brace on line 2 marks the beginning of the *body* of the class. The end of the body is marked by the closing brace on line 7. Everything that appears within these braces belongs to the class. As you work with Java, you'll find that it uses these braces a lot. Pretty soon the third and fourth fingers on your right hand will know exactly where they are on the keyboard.

Lines 3 through 7 define a *method* of the HelloApp class named main:

» →3 public: The public keyword is used again, this time to indicate that a method being declared here should have public access. That means classes other than the HelloApp class can use it. All Java programs must have a class that declares a public method named main. The main method contains the statements that are executed when you run the program.

» static: You find all about the static keyword in Book 3, Chapter 3. For now, just take my word that the Java language requires that you specify static when you declare the main method.

» void: In Java, a *method* is a unit of code that can calculate and return a value. For example, you could create a method that calculates a sales total. Then the sales total would be the return value of the method. If a method doesn't need to return a value, you must use the void keyword to indicate that no value is returned. Because Java requires that the main method not return a value, you must specify void when you declare the main method.

» main: Finally, here's the *identifier* that provides the name for this method. As I've already mentioned, Java requires that this method be named main. Besides the main method, you can create additional methods with whatever names you want to use. You discover how to create additional methods in Book 2, Chapter 7. Until then, the programs consist of just one method named main.

» (String[] args): Oh, boy. This Java element is too advanced to thoroughly explain just yet. It's called a *parameter list,* and it's used to pass data to a method. Java requires that the main method must receive a single parameter that's an array of String objects. By convention, this parameter is named args. If you don't know what a parameter, a String, or an array is, don't worry about it. You can find out what a String is in the next chapter, and parameters are in Book 2, Chapter 7; arrays are in Book 4. In the meantime, realize that you have to code (String[] args) on the declaration for the main methods in all your programs.

» →4 Another {: Another set of braces begins at line 4 and ends at line 6. These braces mark the body of the main method. Notice that the closing brace in line 6 is paired with the opening brace in line 4, whereas the closing brace in line 7 is paired with the one in line 2. This type of pairing is commonplace in Java. In short, whenever you come to a closing brace, it's paired with the most recent opening brace that hasn't already been closed — that is, that hasn't already been paired with a closing brace.

» →5 `System.out.println("Hello, World!");`: This is the only statement in the entire program. It calls a method named `println` that belongs to the `System.out` object. The `println` method displays a line of text on the console. The text to be displayed is passed to the `println` method as a parameter in parentheses following the word `println`. In this case, the text is the string literal `Hello, World!` enclosed in a set of quotation marks. As a result, this statement displays the text `Hello, World!` on the console.

» Note that in Java, most (but not all) statements must end with a semicolon. Because this statement is the only one in the program, this line is the only one that requires a semicolon.

» →6 `}`: Line 6 contains the closing brace that marks the end of the `main` method body that was begun by the brace on line 4.

» →7 Another `}`: Line 7 contains the closing brace that marks the end of the `HelloApp` class body that was begun by the brace on line 2. Because this program consists of just one class, this line also marks the end of the program.

To run this program, you must first use a text editor such as Notepad or TextPad to enter it — exactly as it appears in Listing 1-1 — in a text file named `HelloApp.java`. Then you can compile it by running the following command at a command prompt:

```
javac HelloApp.java
```

This command creates a class file named `HelloApp.class` that contains the Java bytecode compiled for the `HelloApp` class.

You can run the program by entering this command:

```
java HelloApp
```

Now that you've seen what a Java program actually looks like, you're in a better position to understand exactly what this command does. First, it loads the Java Virtual Machine (JVM) into memory. Then it locates the `HelloApp` class, which must be contained in a file named `HelloApp.class`. Finally, it runs the `main` method of the `HelloApp` class. The `main` method, in turn, displays the message `"Hello, World!"` on the console.

TIP

Because the `HelloApp.java` file contains just one class, you can run it directly using the `java` command without first compiling. If you type **java HelloApp** and there is a `.java` file but no corresponding `.class` file, the `java` command will first compile the `HelloApp.java` file and then run the resulting class.

The rest of this chapter describes some of the basic elements of the Java programming language in greater detail.

Dealing with Keywords

A *keyword* is a word that has a special meaning defined by the Java programming language. The program shown earlier in Listing 1-1 uses four keywords: public, class, static, and void. In all, Java has 53 keywords. They're listed in alphabetical order in Table 1-1.

TABLE 1-1

Java's Keywords

abstract	default	goto	package	this
assert	do	if	private	throw
boolean	double	implements	protected	throws
break	else	import	public	transient
byte	enum	instanceof	return	true
case	extends	int	short	try
catch	false	interface	static	var
char	final	long	strictfp	void
class	finally	native	super	volatile
const	float	new	switch	while
continue	for	null	synchronized	_ (underscore)

TECHNICAL STUFF

Strangely enough, four keywords listed in Table 1-1 — true, false, null, and var — aren't technically considered to be keywords. Instead, true, false, and null are *literals* used to represent predefined values (or, in the case of null, the absence of a value), and var is a special type of identifier. Still, these words are reserved for use by the Java language in much the same way that keywords are, so I lumped them in with the keywords.

Stranger still, two keywords — const and goto — are reserved by Java but don't do anything. Both are carryovers from the C++ programming language. The const keyword defines a constant, which is handled in Java by the final keyword.

As for goto, it's a C++ statement that is considered anathema to object-oriented programming purists, so it isn't used in Java. Java reserves it as a keyword solely for the purpose of scolding you if you attempt to use it.

Even stranger yet is the underscore (_) character, which is reserved as a keyword but has no purpose — at least, not yet. Apparently, the underscore character is reserved just in case it may be used for some purpose in a future release of Java.

WARNING

Like everything else in Java, keywords are case-sensitive. Thus, if you type If instead of if or For instead of for, the compiler complains about your error. Because Visual Basic keywords begin with capital letters, you'll make this mistake frequently if you've programmed in Visual Basic.

Considering the Java community's disdain for Visual Basic, it's surprising that the error messages generated when you capitalize keywords aren't more insulting. Accidentally capitalizing a keyword in Visual Basic style can really throw the Java compiler for a loop. Consider this program, which contains the single error of capitalizing the word For:

```
public class CaseApp
{
    public static void main(String[] args)
    {
        For (int i = 0; i<5; i++)
        System.out.println("Hi");
    }
}
```

When you try to compile this program, the compiler generates a total of four error messages for this one mistake:

```
C:\Java AIO\CaseApp.java:5: '.class' expected
For (int i = 0; i<5; i++)
        ^
C:\Java AIO\CaseApp.java:5: illegal start of type
For (int i = 0; i<5; i++)
           ^
C:\Java AIO\CaseApp.java:5: not a statement
For (int i = 0; i<5; i++)
        ^
C:\Java AIO\CaseApp.java:5: ';' expected
For (int i = 0; i<5; i++)
                   ^
4 errors For (int i = 0; i<5; i++)
```

Even though this single mistake generates four error messages, not one of the messages actually points to the problem. The little arrow beneath the source line indicates what part of the line is in error, and none of these error messages has the arrow pointing anywhere near the word For! The compiler isn't smart enough to realize that you meant for instead of For. So it treats For as a legitimate identifier and then complains about everything else on the line that follows it. It would be much more helpful if the compiler generated an error message like this:

```
C:\Java AIO\CaseApp.java:5: 'For' is not a keyword
For (int i = 0; i<5; i++)
```

Better yet, for those of us old enough to remember *Get Smart* on TV:

```
C:\Java AIO\CaseApp.java:5: Thees ees Java! Vee do not capitalize keyverds here!
For (int i = 0; i<5; i++)
```

REMEMBER

The moral of the story is that Java is case-sensitive, and if your program won't compile and the error messages don't make any sense, check for keywords that you've mistakenly capitalized.

Working with Statements

Like most programming languages, Java uses statements to build programs. Unlike most programming languages, Java doesn't use statements as its fundamental unit of code. Instead, it gives that honor to the class. However, every class must have a body, and the body of a class is made up of one or more statements. In other words, you can't have a meaningful Java program without at least one statement. The following sections describe the ins and outs of working with Java statements.

Types of statements

Java has many types of statements. Some statements simply create variables that you can use to store data. These types of statements are often called *declaration statements* and tend to look like this:

```
int i;
String name;
```

Another common type of statement is an *assignment statement*, which assigns a value to a variable:

```
i = 42;
name = "Alexander Hamilton";
```

Declaration and assignment statements can be combined into a single statement, like this:

```
int i = 42;
String name = "Alexander Hamilton";
```

Another common type of statement is an *expression statement*, which performs calculations or other operations. For example:

```
System.out.println("Hello, World!");
```

Notice that this statement is the same as line 5 in Listing 1-1. Thus the single statement in the HelloApp program is an expression statement.

There are many kinds of statements besides these two. if statements, for example, execute other statements only if a particular condition has been met, and statements such as for, while, and do execute whole groups of statements one or more times.

TIP

It's often said that every Java statement must end with a semicolon. Actually, this isn't quite true. *Some* types of Java statements must end with semicolons — but others don't have to. The basic rule is that declaration and expression statements must end with a semicolon, but most other statement types do not. Where this rule gets tricky, however, is that most other types of statements include one or more declarations or expression statements that do use semicolons. Here's a typical if statement:

```
if (total > 100)
    discountPercent = 10;
```

Here, the variable named discountPercent is given a value of 10 if the value of the total variable is greater than 100. The assignment statement ends with a semicolon, but the if statement itself doesn't. (The Java compiler lets you know if you use a semicolon when you shouldn't.)

White space

In Java, the term *white space* refers to one or more consecutive space characters, tab characters, or line breaks. All white space is considered the same.

In other words, a single space is treated the same as a tab or line break or any combination of spaces, tabs, and line breaks.

If you've programmed in Visual Basic, white space is different from what you're used to. In Visual Basic, line breaks mark the end of statements unless special continuation characters are used. In Java, you don't have to do anything special to continue a statement onto a second line. Thus the statement

```
x = (y + 5) / z;
```

is identical to this statement:

```
x =
(y + 5) / z;
```

In fact, you could write the preceding statement like this if you wanted:

```
x
=
(
y
+
5
)
/
z
;
```

I wouldn't advise it, but the statement does compile and execute properly.

Be advised, however, that you can't put white space in the middle of a keyword or identifier. The following example won't work:

```
p u b l i c static v o i d main(String[] args)
```

Here the extra spaces between the letters in the words *public* and *void* will confuse the compiler.

TIP

Using white space liberally in your programs is a good idea. In particular, you should routinely use white space like this:

>> **Line breaks:** Place each statement on a separate line. In addition, you can break a longer statement into several lines for clarity.

>> **Tabs or spaces:** Use tabs or spaces to line up elements that belong together.

Java Programming Basics

The compiler ignores the extra white space, so it doesn't affect the bytecode that's created for your program. As a result, using extra white space in your program doesn't affect your program's performance in any way, but it does make the program's source code easier to read.

Working with Blocks

A *block* is a group of one or more statements that's enclosed in braces. A block begins with an opening brace ({) and ends with a closing brace (}). Between the opening and closing braces, you can code one or more statements. Here's a block that consists of three statements:

```
{
    int i, j;
    i = 100;
    j = 200;
}
```

TECHNICAL STUFF

A block is itself a type of statement. As a result, any time the Java language requires a statement, you can substitute a block to execute more than one statement. In Book 2, Chapter 4, you discover that the basic syntax of an if statement is this:

```
if ( expression ) statement
```

Here `statement` can be a single statement or a block. If you find this idea confusing, don't worry; it will make more sense when you turn to Book 2, Chapter 4.

TIP

You can code the braces that mark a block in two popular ways. One is to place both braces on separate lines and then indent the statements that make up the block. For example:

```
if ( i > 0)
{
    String s = "The value of i is " + i;
    System.out.print(s);
}
```

The other style is to place the opening brace for the block on the same line as the statement the block is associated with, like this:

```
if ( i > 0) {
    String s = "The value of i is " + i;
```

```
    System.out.print(s);
}
```

Which style you use is a matter of personal preference. I prefer the first style, and that's the style I use throughout this book. But either style works — and many programmers prefer the second style because it's more concise.

Note that even though a block can be treated as a single statement, you should *not* end a block with a semicolon. The statements within the block may require semicolons, but the block itself does not.

Creating Identifiers

An *identifier* is a word that you make up to refer to a Java programming element by name. Although you can assign identifiers to many types of Java elements, they're most commonly used for the following elements:

>> Classes, such as the HelloApp class in Listing 1-1

>> Methods, such as the main method in Listing 1-1

>> Variables and fields, which hold data used by your program

>> Parameters, which pass data values to methods

Identifiers are also sometimes called *names*. Strictly speaking, a name isn't quite the same thing as an identifier — all names are identifiers, but not all identifiers are names. But in practice, the terms *name* and *identifier* are used interchangeably.

You must follow a few simple rules when you create identifiers:

>> Identifiers are case-sensitive. As a result, SalesTax and salesTax are distinct identifiers.

>> Identifiers can be made up of upper- or lowercase letters, numerals, underscore characters (_), and dollar signs ($). Thus, identifier names such as Port1, SalesTax$, and Total_Sales.

>> All identifiers must begin with a letter. Thus, a15 is a valid identifier, but 13Unlucky isn't (because it begins with a numeral).

>> An identifier can't be the same as any of the Java keywords listed in Table 1-1. Thus, you can't create a variable named for or a class named public.

» The Java language specification recommends that you avoid using dollar signs in names you create, because code generators use dollar signs to create identifiers. Thus, avoiding dollar signs helps you avoid creating names that conflict with generated names.

Crafting Comments

A *comment* is a bit of text that provides explanations of your code. The compiler ignores comments, so you can place any text you want in a comment. Using plenty of comments in your program is a good way to explain what your program does and how it works.

Java has three basic types of comments: *end-of-line comments, traditional comments,* and *JavaDoc comments.* More about that is coming right up.

End-of-line comments

An *end-of-line comment* begins with the sequence // (a pair of consecutive slashes) and ends at the end of the line. You can place an end-of-line comment at the end of any line. Everything you type after the // is ignored by the compiler. For example:

```
total = total * discountPercent; // calculate the discounted total
```

If you want, you can also place end-of-line comments on separate lines, like this:

```
// calculate the discounted total
total = total * discountPercent;
```

You can place end-of-line comments in the middle of statements that span two or more lines. For example:

```
total = (total * discountPercent) // apply the discount first
+ salesTax;                        // then add the sales tax
```

Traditional comments

A *traditional comment* begins with the sequence /* , ends with the sequence */, and can span multiple lines. Here's an example:

```
/* HelloApp sample program.
This program demonstrates the basic structure
that all Java programs must follow. */
```

A traditional comment can begin and end anywhere on a line. If you want, you can even sandwich a comment between other Java programming elements, like this:

```
x = (y + /* a strange place for a comment */ 5) / z;
```

Usually, traditional comments appear on separate lines. One common use for traditional comments is to place a block of comment lines at the beginning of a class to indicate information about the class — such as what the class does, who wrote it, and so on. That type of comment, however, is usually better coded as a JavaDoc comment, as described in the next section.

WARNING

You may be tempted to temporarily comment out a range of lines by placing /* in front of the first line in the range and */ after the last line in the range. That practice can get you in trouble, however, if the range of lines you try to comment out includes a traditional comment, because traditional comments can't be *nested*. The following code won't compile, for example:

```
/*
int x, y, z;
y = 10;
z = 5;
x = (y + /* a strange place for a comment */ 5) / z;
*/
```

Here, I tried to comment out a range of lines that already included a traditional comment. Unfortunately, the */ sequence near the end of the fifth line is interpreted as the end of the traditional comment that begins in the first line, so when the compiler encounters the */ sequence in line 6, it generates an error message.

JavaDoc comments

JavaDoc comments are actually special types of traditional comments that you can use to create web-based documentation for your programs — automatically. Because you'll have a better appreciation of JavaDoc comments when you know more about object-oriented programming, I devote a section in Book 3, Chapter 8, to creating and using JavaDoc comments.

Introducing Object-Oriented Programming

Having presented some of the most basic elements of the Java programming language, most Java books would next turn to the important topics of variables and data types. Because Java is an inherently object-oriented programming language, however, and because classes are the heart of object-oriented programming, I look next at classes to explore the important role they play in creating objects. I get to variables and data types first thing in the next chapter.

Understanding classes and objects

As I've already mentioned, a *class* is code that defines the behavior of a Java programming element called an object. An *object* is an entity that has both state and behavior. The *state* of an object consists of any data that the object might be keeping track of, and the *behavior* consists of *actions* that the object can perform. The behaviors are represented in the class by one or more methods that can be called on to perform actions.

The difference between a class and an object is similar to the difference between a blueprint and a house. A blueprint is a plan for a house. A house is an implementation of a blueprint. One set of blueprints can be used to build many houses. Likewise, a class is a plan for an object, and an object is — in Java terms — an *instance* of a class. You can use a single class to create more than one object.

When an object is created, Java sets aside an area of computer memory that's sufficient to hold all the data that's stored by the object. As a result, each instance of a class has its own data, independent of the data used by other instances of the same class.

Understanding static methods

You don't necessarily have to create an instance of a class to use the methods of the class. If you declare a method with the static keyword, you can call the method without first creating an instance of the class, because static methods are called from classes, not from objects.

The main method of a Java application must be declared with the static keyword because when you start a Java program by using the java command from a command prompt, Java doesn't create an instance of the application class. Instead, it simply calls the program's static main method.

The difference between static and nonstatic methods will become more apparent when you look at object-oriented programming in more depth in Book 3. But for now, consider this analogy. The blueprints for a house include the details about systems that actually perform work in a finished house, such as electrical and plumbing systems. To use those systems, you have to actually build a house. In other words, you can't turn on the hot water by using the blueprint alone; you have to have an actual house with an actual device to heat the water.

The blueprints do include detailed measurements of the dimensions of the house, however. As a result, you *can* use the blueprints to determine the square footage of the living room.

Now imagine that the blueprints actually have a built-in calculator that displays the size of the living room if you push the Living Room button. That button would be like a static method in a class: You don't actually have to build a house to use the button; you can activate it from the blueprints alone.

Many Java programs — in fact, many of the programs in the rest of Book 2 — are entirely made up of static methods. Most realistic programs, however, require that you create one or more objects that the program uses as it executes. As a result, knowing how to create simple classes and how to create objects from those classes are basic skills in Java programming.

Creating an object from a class

In Java, you can create an object from a class in several ways. The most straightforward way is to create a variable that provides a name you can use to refer to the object, use the new keyword to create an instance of the class, and then assign the resulting object to the variable. The general form of a statement that does that bit of magic looks like this:

```
ClassName variableName = new ClassName();
```

To create an object instance of a class named Class1 and assign it to a variable named myClass1Object, you would write a statement like this:

```
Class1 myClass1Object = new Class1();
```

Why do you have to list the class name twice? The first time, you're providing a *type* for the variable. In other words, you're saying that the variable you're creating here can be used to hold objects created from the Class1 class. The second time you list the class name, you're creating an object from the class. The new keyword tells Java to create an object, and the class name provides the name of the class to use to create the object.

The equal sign (=) is an *assignment operator.* It simply says to take the object created by the new keyword and assign it to the variable. Thus, this statement actually does *three* things:

>> It creates a variable named myClass1Object that can be used to hold objects created from the Class1 class. At this point, no object has been created — just a variable that can be used to store objects.

>> It creates a new object in memory from the Class1 class.

>> It assigns this newly created object to the myClass1Object variable. That way, you can use the myClassObject variable to refer to the object that was created.

Viewing a program that uses an object

To give you an early look at what object-oriented programming really looks like, Listing 1-2 and Listing 1-3 show another version of the HelloApp application — this time using two classes, one of which is actually made into an object when the program is run. The first class, named HelloApp2, is shown in Listing 1-2. This class is similar to the HelloApp class shown in Listing 1-1 but uses an object created from the second class, named Greeter, to actually display the "Hello, World!" message on the console. The Greeter class is shown in Listing 1-3. It defines a method named sayHello that displays the message.

REMEMBER

Both the HelloApp and the Greeter classes are public classes. Java requires that each public class be stored in a separate file with the same name as the class; the filename ends with the extension .java. As a result, the HelloApp2 class is stored in a file named HelloApp2.java, and the Greeter class is stored in a file named Greeter.java.

The HelloApp2 class

The HelloApp2 class is shown in Listing 1-2.

LISTING 1-2: **The HelloApp2 Class**

```
// This application displays a hello message on        →1
// the console by creating an instance of the
// Greeter class and then calling the Greeter
// object's sayHello method.
public class HelloApp2                                  →5
{
    public static void main(String[] args)             →7
```

```
    {
        Greeter myGreeterObject = new Greeter();                    →8
        myGreeterObject.sayHello();                                 →9
    }
}
```

The following paragraphs describe the key points:

» →1 This class begins with a series of comment lines identifying the function of the program. For these comments, I used simple end-of-line comments rather than traditional comments. (For more on commenting, see the "Crafting Comments" section, earlier in this chapter.)

» →5 The HelloApp2 class begins on line 5 with the public class declaration. Because the public keyword is used, a file named HelloApp2.java must contain this class.

» →7 The main method is declared, using the same signature as the main method in the first version of this program (Listing 1-1). Get used to this form, because *all* Java applications must include a main method that's declared in this way.

» →8 The first line in the body of the main method creates a variable named myGreeterObject that can hold objects created from the Greeter class. Then it creates a new object using the Greeter class and assigns this object to the myGreeterObject variable.

» →9 The second line in the body of the main method calls the myGreeterObject object's sayHello method. As you'll see in a moment, this method simply displays the message "Hello, World!" on the console.

The Greeter class

The Greeter class is shown in Listing 1-3.

LISTING 1-3: **The Greeter Class**

```
// This class represents a Greeter object that displays           →1
// a hello message on the console.
public class Greeter                                              →3
{
    public void sayHello()                                        →5
    {
        System.out.println("Hello, World!");                      →7
    }
}
```

The following paragraphs describe the key points:

>> →1 This class also begins with a series of comment lines that identify the function of the program.

>> →3 The class declaration begins on this line. The class is declared as public so other classes can use it. Strictly speaking, the public declaration here isn't strictly required; the HelloApp2 class can access the Greeter class without it because they're in the same package.

>> →5 The sayHello method is declared using the public keyword so that it's available to other classes that use the Greeter class. The void keyword indicates that this method doesn't provide any data back to the statement that calls it, and sayHello simply provides the name of the method.

>> →7 The body of this method consists of just one line of code that displays the "Hello, World!" message on the console.

So what's the difference?

You may notice that the only line that actually does any real work in the Hello-App2 program is line 7 in the Greeter class (Listing 1-3), and this line happens to be identical to line 5 in the original HelloApp class (Listing 1-1). Other than the fact that the second version requires roughly twice as much code as the first version, what really *is* the difference between these two applications?

Simply put, the first version is procedural, and the second is object-oriented. In the first version of the program, the main method of the application class does all the work of the application by itself: It just says hello. The second version defines a class that knows how to say hello to the world and then creates an object from that class and asks that object to say hello. The application itself doesn't know (or even care) exactly how the Greeter object says hello. It doesn't know exactly what the greeting will be, what language the greeting will be in, or even how the greeting will be displayed.

To illustrate this point, consider what would happen if you used the Greeter class shown in Listing 1-4 rather than the one shown in Listing 1-3. This version of the Greeter class uses a Java library class called JOptionPane to display a message in a dialog box rather than in a console window. (I won't bother explaining in a list how this code works, but you can find out more about it in the next chapter.) If you were to run the HelloApp2 application using this version of the Greeter class, you'd get the dialog box shown in Figure 1-1.

FIGURE 1-1:
The class in
Listing 1-4
displays this
dialog box.

LISTING 1-4:   **Another Version of the Greeter Class**

```java
// This class creates a Greeter object that displays
// a hello message in a dialog box.

import javax.swing.JOptionPane;                                           →4

public class Greeter
{
    public void sayHello()
    {
        JOptionPane.showMessageDialog(null,
            "Hello, World!", "Greeter",
            JOptionPane.INFORMATION_MESSAGE);
    }
}
```

REMEMBER

The important point to realize here is that the HelloApp2 class doesn't have to be changed to use this new version of the Greeter class. Instead, all you have to do is replace the old Greeter class with the new one, recompile the Greeter class, and the HelloApp2 class won't know the difference. That's one of the main benefits of object-oriented programming.

Importing Java API Classes

You may have noticed that the Greeter class in Listing 1-4 includes this statement:

```java
import javax.swing.JOptionPane;
```

The purpose of the import statement is to let the compiler know that the program is using a class that's defined by the Java API called JOptionPane.

Because the Java API contains literally thousands of classes, some form of organization is needed to make the classes easier to access. Java does this by grouping classes into manageable groups called *packages.* In the previous example, the package that contains the JOptionPane class is named javax.swing.

Strictly speaking, import statements are never required. But if you don't use import statements to import the API classes your program uses, you must *fully qualify* the names of the classes when you use them by listing the package name in front of the class name. So if the class in Listing 1-4 didn't include the import statement in line 4, you'd have to code line 11 like this:

```
javax.swing.JOptionPane.showMessageDialog(null,
    "Hello, World!", "Greeter",
    javax.swing.JOptionPane.INFORMATION_MESSAGE);
```

In other words, you'd have to specify javax.swing.JOptionPane instead of just JOptionPane whenever you referred to this class.

TIP

Here are some additional rules for working with import statements:

>> import statements must appear at the beginning of the class file, before any class declarations.

>> You can include as many import statements as are necessary to import all the classes used by your program.

>> You can import all the classes in a particular package by listing the package name followed by an asterisk wildcard, like this:

```
import javax.swing.*;
```

>> Because many programs use the classes that are contained in the java.lang package, you don't have to import that package. Instead, those classes are automatically available to all programs. The System class is defined in the java.lang package. As a result, you don't have to provide an import statement to use this class.

>> JDK 9 introduced a new feature for managing packages called the *Java Platform Module System*. I cover this new feature in Book 3, Chapter 8.

Chapter **2**

Working with Variables and Data Types

I n this chapter, you find out the basics of working with variables in Java. Variables are the key to making Java programs general purpose. The `Hello, World!` programs in the previous chapter, for example, are pretty specific: The only thing they say is "Hello, World!" But with a variable, you can make this type of program more general. You could vary the greeting so that sometimes it would say "Hello, World!" and at other times it would say "Greetings, Foolish Mortals." Or you could personalize the greeting, so that it said "Hello, Bob!" or "Hello, Amanda!"

Variables are also the key to creating programs that can perform calculations. Suppose that you want to create a program that calculates the area of a circle, given the circle's radius. Such a program uses two variables: one to represent the radius of the circle and the other to represent the circle's area. The program asks the user to enter a value for the first variable. Then it calculates the value of the second variable.

Declaring Variables

In Java, you must explicitly declare all variables before using them. This rule is in contrast to some languages — most notably Python, which lets you use variables that haven't been explicitly declared.

WARNING

Allowing you to use variables that you haven't explicitly declared might seem a pretty good idea at first glance, but it's a common source of bugs that result from misspelled variable names. Java requires that you explicitly declare variables so that if you misspell a variable name, the compiler can detect your mistake and display a compiler error.

The basic form of a variable declaration is this:

```
type name;
```

Here are some examples:

```
int x;
String lastName;
double radius;
```

In these examples, variables named x, lastName, and radius are declared. The x variable holds integer values, the lastName variable holds String values, and the radius variable holds double values. For more information about what these types mean, see the section "Working with Primitive Data Types" later in this chapter. Until then, just realize that int variables can hold whole numbers (such as 5, 1,340, or −34), double variables can hold numbers with fractional parts (such as 0.5, 99.97, or 3.1415), and String variables can hold text values (such as "Hello, World!" or "Jason P. Finch").

TIP

Notice that variable declarations end with semicolons. That's because a variable declaration is itself a type of statement.

REMEMBER

Variable names follow the same rules as other Java identifiers, as I describe in Book 2, Chapter 1, so see that chapter for details. In short, a variable name can be any combination of letters, numerals, or underscores and dollar signs – but must start with a letter. Most programmers prefer to start variable names with lowercase letters and capitalize the first letter of individual words within the name. firstName and salesTaxRate, for example, are typical variable names.

Declaring two or more variables in one statement

You can declare two or more variables of the same type in a single statement by separating the variable names with commas. For example:

```
int x, y, z;
```

Here three variables of type `int` are declared, using the names x, y, and z.

TIP

As a rule, I suggest that you avoid declaring multiple variables in a single statement. Your code is easier to read and maintain if you give each variable a separate declaration.

Declaring class variables

A *class variable* is a variable that any method in a class can access, including static methods such as `main`. When declaring a class variable, you have two basic rules to follow:

>> You must place the declaration within the body of the class but not within any of the class methods.

>> You must include the word `static` in the declaration. The word `static` comes before the variable type.

TIP

Class variables are often called *static variables.* The key distinction between a static variable and an instance variable, which I cover in the next section, is that the value of a static variable is the same for all instances of the class. In contrast, each instance of a class has distinct values for its instance variables.

The following program shows the proper way to declare a class variable named `helloMessage`:

```
public class HelloApp
{
    static String helloMessage;

    public static void main(String[] args)
    {
        helloMessage = "Hello, World!";
        System.out.println(helloMessage);
    }
}
```

As you can see, the declaration includes the word static and is placed within the HelloApp class body but not within the body of the main method.

TIP

You don't have to place class variable declarations at the beginning of a class. Some programmers prefer to place them at the end of the class, as in this example:

```
public class HelloApp
{
    public static void main(String[] args)
    {
        helloMessage = "Hello, World!";
        System.out.println(helloMessage);
    }

    static String helloMessage;
}
```

Here the helloMessage variable is declared *after* the main method.

TIP

I think classes are easier to read if the variables are declared first, so that's where you see them in this book.

Declaring instance variables

An *instance variable* is similar to a class variable but doesn't specify the word static in its declaration. As the name suggests, instance variables are associated with instances of classes. As a result, you can use them only when you create an instance of a class. Because static methods aren't associated with an instance of the class, you can't use an instance variable in a static method — and that includes the main method.

The following example program won't compile:

```
public class HelloApp
{
    String helloMessage;                    // error -- should use static
    keyword

    public static void main(String[] args)
    {
        helloMessage = "Hello, World!";
        System.out.println(helloMessage);   // will not compile
    }
}
```

If you try to compile this program, you get the following error messages:

```
C:\Java\HelloApp.java:7: error: non-static variable helloMessage cannot be
    referenced from a static context
helloMessage = "Hello, World!";
^
C:\Java\HelloApp.java:8: non-static variable helloMessage cannot be referenced
    from a static context
System.out.println(helloMessage);
        ^
```

Both of these errors occur because the main method is static, so it can't access instance variables.

Instance variables are useful whenever you create your own classes, but because I don't cover that topic until Book 3, you won't see many examples of instance methods in the remainder of the chapters in Book 2.

Declaring local variables

A *local variable* is a variable that's declared within the body of a method. Then you can use the variable only within that method. Other methods in the class aren't even aware that the variable exists.

Here's a version of the HelloApp class in which the helloMessage variable is declared as a local variable:

```
public class HelloApp
{
    public static void main(String[] args)
    {
        String helloMessage;
        helloMessage = "Hello, World!";
        System.out.println(helloMessage);
    }
}
```

Note that you don't specify static on a declaration for a local variable. If you do, the compiler generates an error message and refuses to compile your program. Local variables always exist within the scope of a method, and they exist only while that method is executing. As a result, whether an instance of the class has been created is irrelevant. (For more information, see the section "Understanding Scope," later in this chapter.)

TIP

Unlike class and instance variables, a local variable is fussy about where you position the declaration for it. In particular, you must place the declaration before the first statement that actually uses the variable. Thus the following program won't compile:

```
public class HelloApp
{
    public static void main(String[] args)
    {
        helloMessage = "Hello, World!";      // error -- helloMessage
        System.out.println(helloMessage);    // is not yet declared
        String helloMessage;
    }
}
```

When it gets to the first line of the `main` method, the compiler generates two error messages complaining that it can't find the symbol `"helloMessage"`. That's because the symbol hasn't been declared.

Although most local variables are declared near the beginning of a method's body, you can also declare local variables within smaller blocks of code marked by braces. This will make more sense to you when you read about statements that use blocks, such as `if` and `for` statements. But here's an example:

```
if (taxRate > 0)
{
    double taxAmount;
    taxAmount = subTotal * taxRate;
    total = subTotal + total;
}
```

Here the variable `taxAmount` exists only within the set of braces that belongs to the `if` statement. (You can assume that the variables `taxRate`, `subtotal`, and `total` are defined outside of the code block defined by the braces.)

Initializing Variables

In Java, local variables are not given initial default values. The compiler checks to make sure that you have assigned a value before you use a local variable. The following example program won't compile:

```
public class testApp
{
```

```
    public static void main(String[] args)
    {
        int i;
        System.out.println("The value of i is " + i);
    }
}
```

If you try to compile this program, you get the following error message:

```
C:\Java\testApp.java:6: error: variable i might not have been initialized
System.out.println("The value of i is " + i);
                                          ^
```

To avoid this error message, you must initialize local variables before you use them. You can do that by using an assignment statement or an initializer, as I describe in the following sections.

Unlike local variables, class variables and instance variables are given default values. Numeric types are automatically initialized to zero, and String variables are initialized to empty strings. As a result, you don't have to initialize a class variable or an instance variable, although you can if you want them to have an initial value other than the default.

Initializing variables with assignment statements

One way to initialize a variable is to code an *assignment statement* following the variable declaration. Assignment statements have this general form:

```
variable = expression;
```

Here, the expression can be any Java expression that yields a value of the same type as the variable. Here's a version of the main method from the previous example that correctly initializes the i variable before using it:

```
public static void main(String[] args)
{
    int i;
    i = 0;
    System.out.println("i is " + i);
}
```

In this example, the variable is initialized to a value of zero before the println method is called to print the variable's value.

You find out a lot more about expressions in Book 2, Chapter 3. For now, you can just use simple literal values, such as 0 in this example.

Initializing variables with initializers

Java also allows you to initialize a variable on the same statement that declares the variable. To do that, you use an *initializer*, which has the following general form:

```
type name = expression;
```

In effect, the initializer lets you combine a declaration and an assignment statement into one concise statement. Here are some examples:

```
int x = 0;
String lastName = "Lowe";
double radius = 15.4;
```

In each case, the variable is both declared and initialized in a single statement.

When you declare more than one variable in a single statement, each variable can have its own initializer. The following code declares variables named x and y, and initializes x to 5 and y to 10:

```
int x = 5, y = 10;
```

WARNING

When you declare two class or instance variables in a single statement but use only one initializer, you can mistakenly think that the initializer applies to both variables. Consider this statement:

```
static int x, y = 5;
```

Here you might think that both x and y would initialize to 5. But the initializer applies only to y, so x is initialized to its default value, 0. (If you make this mistake with a local variable, the compiler displays an error message for the first statement that uses the x variable because it isn't properly initialized.)

Using Final Variables (Constants)

A *final variable* is a variable whose value you can't change after it's been initialized. To declare a final variable, you add the final keyword to the variable declaration, like this:

```
final int WEEKDAYS = 5;
```

A variable that is both final and static is called a *constant*. Constants are often used for values that are universally the same, such as the number of days in June or the atomic weight of iridium. To create a constant, add static final (not final static) to the declaration, as follows:

```
static final WEEKDAYS = 5;
```

TIP

Although it isn't required, using all capital letters for static final variable names is common. When you do so, you can easily spot the use of constants in your programs.

In addition to values that are universally the same, constants are useful for values that are used in several places throughout a program and that don't change during the course of the program. Suppose that you're writing a game that features bouncing balls, and you want the balls always to have a radius of 6 pixels. This program probably needs to use the ball diameter in several places — to draw the ball onscreen, to determine whether the ball has hit a wall, to determine whether the ball has hit another ball, and so on. Rather than just specify 6 whenever you need the ball's radius, you can set up a class constant named BALL_RADIUS, like this:

```
static final BALL_RADIUS = 6;
```

Using a constant has two advantages:

>> If you decide later that the radius of the balls should be 7, you make the change in just one place: the initializer for the BALL_RADIUS constant.

>> The constant helps document the inner workings of your program. The operation of a complicated calculation that uses the ball's radius is easier to understand if it specifies BALL_RADIUS rather than 6, for example.

Working with Primitive Data Types

The term *data type* refers to the type of data that can be stored in a variable. Java is sometimes called a *strongly typed* language because when you declare a variable, you must specify the variable's type. Then the compiler ensures that you don't try to assign data of the wrong type to the variable. The following example code generates a compiler error:

```
int x;
x = 3.1415;
```

Because x is declared as a variable of type int (which holds whole numbers), you can't assign the value 3.1415 to it.

Java makes an important distinction between primitive types and reference types.

>> **Primitive types** are the data types defined by the language itself.

>> **Reference types** are types defined by classes in the Java application programming interface (API) or by classes you create rather than by the language itself.

A key difference between a primitive type and a reference type is that the memory location associated with a primitive-type variable contains the actual value of the variable. As a result, primitive types are sometimes called *value types*. By contrast, the memory location associated with a reference-type variable contains an address (called a *pointer*) that indicates the memory location of the actual object. I explain reference types more fully in the section "Using reference types," later in this chapter, so don't worry if this explanation doesn't make sense just yet.

It isn't quite true that reference types are defined by the Java API and not by the Java language specification. A few reference types, such as Object and String, are defined by classes in the API, but those classes are specified in the Java Language API. Also, a special type of variable called an *array*, which can hold multiple occurrences of primitive- or reference-type variables, is considered to be a reference type.

Java defines a total of eight primitive types, listed in Table 2-1. Of the eight primitive types, six are for numbers, one is for characters, and one is for true/false values.

Integer types

An *integer* is a whole number — that is, a number with no fractional or decimal portion. Java has four integer types, which you can use to store numbers of varying sizes. The most commonly used integer type is int. This type uses 4 bytes to store an integer value that can range from about negative 2 billion to positive 2 billion.

TABLE 2-1

Java's Primitive Types

Type	Explanation
int	A 32-bit (4-byte) integer value
byte	An 8-bit (1-byte) integer value
short	A 16-bit (2-byte) integer value
long	A 64-bit (8-byte) integer value
float	A 32-bit (4-byte) floating-point value
double	A 64-bit (8-byte) floating-point value
char	A 16-bit character using the Unicode encoding scheme
boolean	A true or false value

If you're writing the application that counts how many hamburgers McDonald's has sold, an int variable may not be big enough. In that case, you can use a long integer instead. long is a 64-bit integer that can hold numbers ranging from about negative 9,000 trillion to positive 9,000 trillion. (That's a big number, even by federal deficit standards.)

In some cases, you may not need integers as large as the standard int type provides. For those cases, Java provides two smaller integer types. The short type represents a two-byte integer, which can hold numbers from −32,768 to +32,767, and the byte type defines a single-byte integer that can range from −128 to +127.

Although the short and byte types require less memory than the int and long types, there's usually little reason to use them for desktop applications, where memory is usually plentiful. A few bytes here or there won't make any difference in the performance of most programs — so you should stick to int and long most of the time. Also, use long only when you know that you're dealing with numbers too large for int.

TECHNICAL STUFF

In Java, the size of integer data types is specified by the language and is the same regardless of what computer a program runs on. This is a huge improvement over the C and C++ languages, which let compilers for different platforms determine the optimum size for integer data types. As a result, a C or C++ program written and tested on one type of computer may not execute identically on another computer.

TIP

Java allows you to *promote* an integer type to a larger integer type. Java allows the following, for example:

```
int xInt;
long yLong;
xInt = 32;
yLong = xInt;
```

Here you can assign the value of the xInt variable to the yLong variable because yLong is larger than xInt. Java does not allow the converse, however:

```
int xInt;
long yLong;
yLong = 32;
xInt = yLong;
```

The value of the yLong variable cannot be assigned to the xInt because xInt is smaller than yLong. Because this assignment may result in a loss of data, Java doesn't allow it. (If you need to assign a long to an int variable, you must use explicit casting, as described in the "Type casting" section, later in this chapter.)

You can include underscores to make longer numbers easier to read. Thus, the following statements both assign the same value to the variables xLong1 and xLong2:

```
long xLong1 = 58473882;
long xLong2 = 58_473_882;
```

Floating-point types

Floating-point numbers are numbers that have fractional parts (usually expressed with a decimal point), such as 19.95 or 3.1415.

Java has two primitive types for floating-point numbers: float, which uses 4 bytes, and double, which uses 8 bytes. In almost all cases, you should use the double type whenever you need numbers with fractional values.

The *precision* of a floating-point value indicates how many significant digits the value can have following its decimal point. The precision of a float type is only about six or seven decimal digits, which isn't sufficient for many types of calculations.

By contrast, double variables have a precision of about 15 digits, which is enough for most purposes.

TECHNICAL STUFF

Floating-point numbers actually use *exponential notation* (also called *scientific notation*) to store their values. That means that a floating-point number actually records two numbers: a base value (also called the *mantissa*) and an exponent. The actual value of the floating-point number is calculated by multiplying the mantissa by 2 raised to the power indicated by the exponent. For float types, the exponent can range from −127 to +128. For double types, the exponent can range from −1,023 to +1,024. Thus both float and double variables are capable of representing very large and very small numbers.

You can find more information about some of the nuances of working with floating-point values in Book 2, Chapter 3.

When you use a floating-point literal, I suggest you include a decimal point, like this:

```
double period = 99.0;
```

That avoids the confusion of assigning what looks like an integer to a floating-point variable.

GETTING SCIENTIFIC WITH FLOATS AND DOUBLES

If you have a scientific mind, you may want to use scientific notation when you write floating-point literals. The equation

```
double e = 5.10e+6;
```

is equivalent to

```
double e = 5100000D;
```

The sign is optional if the exponent is positive, so you can also write

```
double e = 5.10e6;
```

Note that the exponent can be negative to indicate values smaller than 1. The equation

```
double impulse = 23e-7;
```

is equivalent to

```
double impulse = 0.0000023;
```

To save that time, you can add an F or D suffix to a floating-point literal to indicate whether the literal itself is of type float or double. For example:

```
float value1 = 199.33F;
double value2 = 200495.995D;
```

If you omit the suffix, D is assumed. As a result, you can usually omit the D suffix for double literals.

Interestingly, floating-point numbers have two distinct zero values: a negative zero and a positive zero. You don't have to worry about these much, because Java treats them as equal. Still, they would make for a good question on *Jeopardy!* ("I'll take Weird Numbers for $200, Alex.")

The char type

The char type represents a single character from the Unicode character set. It's important to keep in mind that a character is not the same as a String; you find out about strings later in this chapter, in the section "Working with Strings." For now, just realize that a char variable can store just one character, not a sequence of characters, as a String can.

To assign a value to a char variable, you use a character literal, which is always enclosed in apostrophes rather than quotes. Here's an example:

```
char code = 'X';
```

Here the character X is assigned to the variable named code.

The following statement won't compile:

```
char code = "X"; // error -- should use apostrophes, not quotes
```

That's because quotation marks are used to mark Strings, not character constants.

You can also assign an integer value from 0 to 255 to a char variable, like this:

```
char cr = 013;
```

Here, the decimal value 13, which represents a carriage return, is assigned to the variable named cr.

Unicode is a 2-byte character code that can represent the characters used in most languages throughout the world. Currently, about 35,000 codes in the Unicode character set are defined, which leaves another 29,000 codes unused. The first 256 characters in the Unicode character set are the same as the characters of the ASCII character set, which is the most commonly used character set for computers with Western languages.

TIP

For more information about the Unicode character set, see the official Unicode website at www.unicode.org.

Character literals can also use special *escape sequences* to represent special characters. Table 2-2 lists the allowable escape sequences. These escape sequences let you create literals for characters that can't otherwise be typed within a character constant.

TABLE 2-2

Escape Sequences for Character Constants

Escape Sequence	Explanation
\b	Backspace
\t	Horizontal tab
\n	Line feed
\f	Form feed
\r	Carriage return
\"	Double quote
\'	Single quote
\\	Backslash

The Boolean type

A Boolean type can have one of two values: true or false. Booleans are used to perform logical operations, most commonly to determine whether some condition is true. For example:

```
boolean enrolled = true;
boolean credited = false;
```

Here a variable named enrolled of type boolean is declared and initialized to a value of true, and another boolean named credited is declared and initialized to false.

WARNING

In some languages, such as C or C++, integer values can be treated as Booleans, with 0 equal to `false` and any other value equal to `true`. Not so in Java. In Java, you can't convert between an `integer` type and a `boolean` type.

Using wrapper classes

Every primitive type has a corresponding class defined in the Java API class library. This class is sometimes called a *wrapper class* because it wraps a primitive value with the object-oriented equivalent of pretty wrapping paper and a bow to make the primitive type look and behave like an object. Table 2-3 lists the wrapper classes for each of the eight primitive types.

TABLE 2-3 **Wrapper Classes for the Primitive Types**

Primitive Type	Wrapper Class
int	Integer
short	Short
long	Long
byte	Byte
float	Float
double	Double
char	Character
boolean	Boolean

As you find out later in this chapter, you can use these wrapper classes to convert primitive values to strings, and vice versa.

Using reference types

In Book 3, Chapter 1, you're introduced to some of the basic concepts of object-oriented programming. In particular, you see how all Java programs are made up of one or more classes, and how to use classes to create objects. In this section, I show how you can create variables that work with objects created from classes.

To start, a *reference type* is a type that's based on a class rather than on one of the primitive types that are built into the Java language. A reference type can be based on a class that's provided as part of the Java API class library or on a class that you write yourself. Either way, when you create an object from a class, Java allocates however much memory the object requires to store the object. Then, if you assign the object to a variable, the variable is actually assigned a *reference* to the object, not the object itself. This reference is the address of the memory location where the object is stored.

Suppose that you're writing a game program that involves balls, and you create a class named Ball that defines the behavior of a ball. To declare a variable that can refer to a Ball object, you use a statement like this:

```
Ball b;
```

Here, the variable b is a variable of type Ball.

To create a new instance of an object from a class, use the new keyword along with the class name. This second reference to the class name is actually a call to a special routine of the class called a *constructor*. The constructor is responsible for initializing the new object. Here's a statement that declares a variable of type Ball, calls the Ball class constructor to create a new Ball object, and assigns a reference to the Ball object to the variable:

```
Ball b = new Ball();
```

REMEMBER

One of the key concepts in working with reference types is the fact that a variable of a particular type doesn't actually contain an object of that type. Instead, it contains a reference to an object of the correct type. An important side effect is that two variables can refer to the same object.

Consider these statements:

```
Ball b1 = new Ball();
Ball b2 = b1;
```

Here I've declared two Ball variables, named b1 and b2, but I've created only one Ball object. In the first statement, the Ball object is created, and b1 is assigned a reference to it. Then, in the second statement, the variable b2 is assigned a reference to the same object that's referenced by b1. As a result, both b1 and b2 refer to the same Ball object.

If you use one of these variables to change some aspect of the ball, the change is visible to the ball no matter which variable you use. Suppose that the Ball class has a method called setSpeed that lets you set the speed of the ball to any int value and a `getSpeed` method that returns an integer value that reflects the ball's current speed. Now consider these statements:

```
b1.setSpeed(50);
b2.setSpeed(100);
int speed = b1.getSpeed();
```

When these statements complete, is the value of the `speed` variable 50 or 100? The correct answer is 100. Because `b1` and `b2` refer to the same `Ball` object, changing the speed by using `b2` affects `b1` as well.

This is one of the most confusing aspects of programming with an object-oriented language such as Java, so don't feel bad if you get tripped up from time to time.

Using inferred variable types

Java 10 introduced a new featured called *local variable type inference,* which lets you substitute the generic word `var` for the variable type whenever you delclare and initialize a local variable in the same sentence.

Whew! That's a moutful. A simple example should make this more clear. Traditionally, you declare and initialize a integer variable like this:

```
int x = 5;
```

Note in this example that when I declare the variable x, I explicitly state the type of the variable as `int`. I then assign the integer value 5 to the variable.

Beginning with Java 10, you don't have to explicitly identify the type of a variable if the Java compiler can infer the variable type from the value you assign to it. Instead of stating the variable's type as `int`, you can use the generic word `var`, like this:

```
var x = 5;
```

Here, the Java compiler can infer that the variable type should be `int` because you assign an integer value (5) to the variable.

Here's another example, this time creating a variable of type `Ball`:

```
var b1 = new Ball();
```

In this example, the variable `b1` is given the type `Ball` because the variable is assigned an instance of the `Ball` class.

Here are a few important details to note about local variable type inference:

>> When you use the var type, you must declare the variable and assign a value in the same statement. So, the following code will not compile:

```
var b1;                    // Does not compile

b1 = new Ball();
```

>> The var type can be used only for local variables. You must still explicitly state the type for class variables.

>> After you've created a var variable, you can't change its type. For example, the following will not compile:

```
var x = 5;

x = "Hello!";      // Can't change the type
```

>> You can use var only for local variables. You can't use it for method parameters, method return types, or class fields.

REMEMBER

Note that in some programming languages, including JavaScript, the keyword var is used to indicate that a variable does not have a type and, therefore, can be used to store any type of data. That is not the case in Java. In Java, *all* variables have a specific type. When you use var as a variable type in Java, you're simply allowing the compiler to infer the variable's type based on the value that is assigned to it. Thus, using var in Java does not create an untyped variable.

Many programmers like to use the var keyword because it slightly reduces the verbosity of their programs. Personally, I avoid the var keyword because I want to be certain that I'm aware of the type of every variable I declare. So you won't see very many examples that use the var keyword in this book.

Working with Strings

A *String* is a sequence of text characters, such as the message "Hello, World!" displayed by the HelloApp program illustrated in this chapter and the preceding chapter. In Java, strings are an interesting breed. Java doesn't define strings as a primitive type. Instead, strings are a *reference type* defined by the Java API String class. The Java language does have some built-in features for working with strings. In some cases, these features make strings appear to be primitive types rather than reference types.

Java's string-handling features are advanced enough to merit an entire chapter, so for the full scoop on strings, I refer you to Book 4, Chapter 1. The following sections present just the bare essentials of working with strings so that you can incorporate simple strings into your programs.

Declaring and initializing strings

Strings are declared and initialized much like primitive types. In fact, the only difference you may notice at first is that the word String is capitalized, unlike the keywords for the primitive types, such as int and double. That's because String isn't a keyword. Instead, it's the name of the Java API class that provides for string objects.

The following statements define and initialize a string variable:

```
String s;
s = "Hello, World!";
```

Here a variable named s of type String is declared and initialized with the *string literal* "Hello, World!". Notice that string literals are enclosed in quotation marks, not apostrophes. Apostrophes are used for character literals, which are different from string literals.

Like any variable declaration, a string declaration can include an initializer. Thus you can declare and initialize a string variable in one statement, like this:

```
String s = "Hello, World!";
```

TIP

Class variables and instance variables are automatically initialized to empty strings, but local variables aren't. To initialize a local string variable to an empty string, use a statement like this:

```
String s = "";
```

Combining strings

Combine two strings by using the plus sign (+) as a *concatenation operator.* (In Java-speak, combining strings is called *concatenation.*) The following statement combines the value of two string variables to create a third string:

```
String hello = "Hello, ";
String world = "World!";
String greeting = hello + world;
```

The final value of the `greeting` variable is `"Hello, World!"`.

TIP

When Java concatenates strings, it doesn't insert any blank spaces between the strings. Thus, if you want to combine two strings and have a space appear between them, make sure that the first string ends with a space or the second string begins with a space. (In the preceding example, the first string ends with a space.)

Alternatively, you can concatenate a string literal along with the string variables. For example:

```
String hello = "Hello";
String world = "World!";
String greeting = hello + ", " + world;
```

Here the comma and the space that appear between the words `Hello` and `World` are inserted as a string literal.

Concatenation is one of the most commonly used string-handling techniques, so you see plenty of examples in this book. In fact, I've already used concatenation once in this chapter. Earlier, I showed you a program that included the following line:

```
System.out.println("The value of i is " + i);
```

Here the `println` method of the `System.out` object prints the string that's created when the literal `"The value of i is "` is concatenated with the value of the `i` variable.

Converting primitives to strings

Because string concatenation lets you combine two or more string values, and because primitive types such as `int` and `double` are *not* string types, you may be wondering how the last example in the preceding section can work. In other words, how can Java concatenate the string literal `"The value of i is "` with the integer value of `i` in this statement?:

```
System.out.println("The value of i is " + i);
```

The answer is that Java automatically converts primitive values to string values whenever you use a primitive value in a concatenation.

Be careful here: Java can confuse you about when the numbers are converted to strings in the course of evaluating the complete expression. Consider this admittedly far-fetched example:

```
int i = 2;
System.out.println(i + i + " equals four.");
```

This prints the following on the console:

```
4 equals four.
```

Here, the first plus sign indicates the addition of two int variables rather than concatenation. For the second plus sign, the resulting int answer is converted to a string and concatenated with " equals four."

You can explicitly convert a primitive value to a string by using the toString method of the primitive type's wrapper class. To convert the int variable x to a string, for example, you use this statement:

```
String s = Integer.toString(x);
```

In the next chapter, you discover how to use a special class called the NumberFormat class to convert primitive types to strings while applying various types of formatting to the value, such as adding commas, dollar signs, or percentage marks.

Converting strings to primitives

Converting a primitive value to a string value is pretty easy. Going the other way — converting a string value to a primitive — is a little more complex, because it doesn't always work. If a string contains the value 10, for example, you can easily convert it to an integer. But if the string contains thirty-two, you can't.

To convert a string to a primitive type, you use a parse method of the appropriate wrapper class, as listed in Table 2-4. To convert a string value to an integer, you use statements like this:

```
String s = "10";
int x = Integer.parseInt(s);
```

Note that there is no parse method to convert a String to a Character. If you need to do that, you can find out how in Book 4, Chapter 1.

TABLE 2-4 **Methods That Convert Strings to Numeric Primitive Types**

Wrapper	parse Method	Example
Integer	parseInt(String)	int x = Integer.parseInt("100");
Short	parseShort(String)	short x = Short.parseShort("100");
Long	parseLong(String)	long x = Long.parseLong("100");
Byte	parseByte(String)	byte x = Byte.parseByte("100");
Float	parseByte(String)	float x = Float.parseFloat("19.95");
Double	parseByte(String)	double x = Double.parseDouble("19.95");
Character	(none)	
Boolean	parseBoolean	boolean x = Boolean.parseBoolean
(String)	("true");	

Converting and Casting Numeric Data

From time to time, you need to convert numeric data of one type to another. You may need to convert a double value to an integer, or vice versa. Some conversions can be done automatically; others are done using a technique called *casting*. I describe automatic type conversions and casting in the following sections.

Automatic conversions

Java can automatically convert some primitive types to others and do so whenever necessary. Figure 2-1 shows which conversions Java allows. Note that the conversions shown with dotted arrows in the figure may cause some of the value's precision to be lost. An int can be converted to a float, for example, but large int values won't be converted exactly because int values can have more digits than can be represented by the float type.

Whenever you perform a mathematical operation on two values that aren't of the same type, Java automatically converts one of them to the type of the other. Here are the rules Java follows when doing this conversion:

>> If one of the values is a double, the other value is converted to a double.

>> If neither is a double but one is a float, the other is converted to a float.

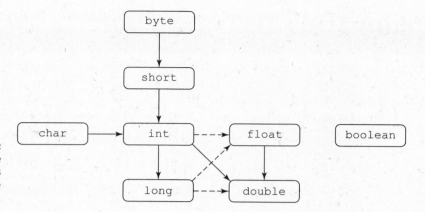

FIGURE 2-1:
Numeric type conversions that are done automatically.

>> If neither is a double nor a float but one is a long, the other is converted to a long.

>> If all else fails, both values are converted to int.

Type casting

Casting is similar to conversion but isn't done automatically. You use casting to perform a conversion that is *not* shown in Figure 2-1. If you want to convert a double to an int, for example, you must use casting.

WARNING

When you use casting, you run the risk of losing information. A double can hold larger numbers than an int, for example. In addition, an int can't hold the fractional part of a double. As a result, if you cast a double to an int, you run the risk of losing data or accuracy, so 3.1415 becomes 3, for example.

To cast a primitive value from one type to another, you use a *cast operator*, which is simply the name of a primitive type in parentheses placed before the value you want to cast. For example:

```
double pi = 3.1314;
int iPi;
iPi = (int) pi;
```

Note that the fractional part of a double is simply discarded when cast to an integer; it isn't rounded. For example:

```
double price = 9.99;
int iPrice = (int) price;
```

Here iPrice is assigned the value 9. If you want to round the double value when you convert it, use the round method of the Math class, as I show you in the next chapter.

Thinking Inside the Box

As I mention earlier in this chapter, in the section "Using wrapper classes," Java provides wrapper classes for each of its primitive data types. You may think that you would have to explicitly cast values from a primitive type to the corresponding wrapper type, like this:

```
int x = 1000;
Integer y = (Integer) x;
```

Because of a feature called *boxing*, you don't have to do that. Boxing occurs when Java automatically converts a primitive value to its corresponding wrapped object, and *unboxing* occurs when Java goes the other way (converts from a wrapped object to the corresponding primitive value). You don't have to explicitly cast the primitive value to its wrapper class, or vice versa.

Here's an example:

```
Integer wrap = 10;
System.out.println(wrap);
int prim = wrap;
System.out.println(prim);
```

The output of this code is 10 followed by another 10. In the first line, you assign a primitive value 10 to the wrapper object named wrap. In the third line of code, Java turns the big wrapped wrap object back into a primitive 10 (because the variable prim is of type int).

Think of *boxing* and *unboxing* as nicknames for *wrapping* and *unwrapping*. The bottom line is, Java can wrap and unwrap values automatically. That's very handy.

Understanding Scope

The *scope* of a variable refers to which parts of a class the variable exists in. In the simplest terms, every variable exists only within the block in which the variable is declared, as well as any blocks that are contained within that block. That's why class variables, which are declared in the class body, can be accessed by any methods defined by the class, but local variables defined within a method can be accessed only by the method in which they are defined.

REMEMBER

In Java, a *block* is marked by a matching pair of braces. Java has many kinds of blocks, including class bodies, method bodies, and block statements that belong to statements such as if or for statements. But in each case, a block marks the scope boundaries for the variables declared within it.

The program in Listing 2-1 can help clarify the scope of class and local variables.

LISTING 2-1:   A Program That Demonstrates Scope for Class and Local Variables

```
public class ScopeApp
{                                                                    →2
    static int x;
    public static void main(String[] args)
    {
        x = 5;
        System.out.println("main: x = " + x);
        myMethod();
    }
    public static void myMethod()
    {
        int y;
        y = 10;                                                      →13
        if (y == x + 5)
        {                                                            →15
            int z;
            z = 15;
            System.out.println("myMethod: z = " + z);
        }                                                            →19
        System.out.println("myMethod: x = " + x);
        System.out.println("myMethod: y = " + y);
    }                                                                →22
}
```

The following paragraphs explain the scope of each of the variables used in this class:

» →2 The variable x is a class variable. Its scope begins in line 2 and ends in line 23. As a result, both the main method and the myMethod method can access it.

» →13 The variable y is a local variable that's initialized in line 13. As a result, its scope begins in line 13 and ends in line 22, which marks the end of the body of the myMethod method.

» →15 The line markes the beginning of an if block that controls the scope of variable z.

>> →19 This line marks the end of the scope of variable z.

>> →22 This line marks the end of the myMethod method body and, therefore, the end of the scope of variable y.

When you run this program, you'll get the following output:

```
main: x = 5
myMethod: z = 15
myMethod: x = 5
myMethod: y = 10
```

TECHNICAL STUFF

Strictly speaking, the scope of a local variable begins when the variable is initialized and ends when the block that contains the variable's declaration ends. By contrast, the scope for a class or instance variable is the entire class in which the variable is declared. That means you can use a class or instance variable in a method that physically appears before the variable is declared, but you can't use a local variable before it's declared.

Shadowing Variables

A *shadowed variable* is a variable that would otherwise be accessible but is temporarily made unavailable because a variable with the same name has been declared in a more immediate scope. That's a mouthful, but the example in Listing 2-2 makes the concept clear. Here a class variable named x is declared. Then, in the main method, a local variable with the same name is declared.

LISTING 2-2: **A Class That Demonstrates Shadowing**

```
public class ShadowApp
{
    static int x;                                               →3
    public static void main(String[] args)
    {
        x = 5;                                                  →6
        System.out.println("x = " + x);
        int x;                                                  →8
        x = 10;                                                 →9
        System.out.println("x = " + x);
        System.out.println("ShadowApp.x = " + ShadowApp.x);     →11
    } →12

}                                                               →14
```

The following paragraphs explain the scoping issues in this program:

» →3 The class variable x is declared in line 3. Its scope is the entire class body, ending at line 14.

» →6 The class variable x is assigned a value in line 6. Then this value is printed to the console.

» →8 A local variable named x is declared. The local variable shadows the class variable x, so any reference to x through the end of this method in line 12 refers to the local variable rather than the class variable.

» →9 This line prints the value of x to the console. The local variable's value, not the class variable's value, is printed here.

» →11 While a class variable is shadowed, you can access it by specifying the class name as shown in line 11. Here ShadowApp.x refers to the class variable.

» →12 When the main method ends in line 12, the class variable x is no longer shadowed.

» →14 When the ShadowApp class body ends in line 14, the class variable x falls out of scope.

Here is the output you will get from this program:

```
x = 5
x = 10
ShadowApp.x = 5
```

Note that the scope and of the local variable begins when it's declared, and the shadowing also begins when the local variable is declared. So, if you try to access the variable between the declaration and the initialization, the Java compiler complains that the variable hasn't been initialized yet.

Because shadowing is a common source of errors, I suggest that you avoid it as much as possible.

Printing Data with System.out

You've already seen several programs that use System.out.println to display output on the console. In the following sections, I officially show you how this method works, along with a related method called just print.

Using standard input and output streams

Java applications are designed to work in a terminal input/output (I/O) environment. Every Java application has at its disposal three *I/O streams* that are designed for terminal-based input and output, which simply sends or receives data one character at a time. The three streams are

>> **Standard input:** A stream designed to receive input data. This stream is usually connected to the keyboard at the computer where the program is run. That way, the user can type characters directly into the standard input stream. In the section "Getting Input with the Scanner Class," later in this chapter, you connect this input stream to a class called Scanner, which makes it easy to read primitive data types from the standard input stream.

>> **Standard output:** A stream designed to display text output onscreen. When you run a Java program under Windows, a special console window opens, and the standard output stream is connected to it. Then any text you send to standard output is displayed in that window.

>> **Standard error:** Another stream designed for output. This stream is also connected to the console window. As a result, text written to the standard output stream is often intermixed with text written to the error stream.

TIP

Windows and other operating systems allow you to *redirect* standard output to some other destination — typically a file. When you do that, only the standard output data is redirected. Text written to standard error is still displayed in the console window.

To redirect standard output, you use a greater-than (>) sign on the command that runs the Java class, followed by the name of the file you want to save the standard output text to. Here's an example:

```
C:\Java>java TestApp >output.txt
```

Here the standard output created by the class TestApp is saved in a file named output.txt. Any text sent to the standard error stream still appears in the console window, however. As a result, the standard error stream is useful for programs that use output redirection to display status messages, error messages, or other information.

When you use a single > sign for redirection, Windows overwrites the redirected file if it already exists. If you prefer to add to the end of an existing file, use two greater-than signs, like this:

```
C:\Java>java TestApp >>output.txt
```

Then, if the output.txt file already exists, anything written to standard output by the TestApp class will be added to the end of the existing output.txt file.

All three standard streams are available to every Java program via the fields of the System class, described in Table 2-5.

TABLE 2-5

Static Fields of the System Object

Field	Description
System.in	Standard input
System.out	Standard output
System.err	Standard error

Using System.out and System.err

Both System.out and System.err represent instances of a class called Print-Stream, which defines the print and println methods used to write data to the console. You can use both methods with either a String argument or an argument of any primitive data type.

The only difference between the print and the println methods is that the println method adds a line-break character to the end of the output, so the output from the next call to print or println begins on a new line.

Because it doesn't start a new line, the print method is useful when you want to print two or more items on the same line. Here's an example:

```
int i = 64;
int j = 23;
System.out.print(i);
System.out.print(" and ");
System.out.println(j);
```

The console output produced by these lines is

```
64 and 23
```

Note that you could do the same thing with a single call to println by using string concatenation, like this:

```
int i = 64;
int j = 23;
System.out.println(i + " and " + j);
```

Getting Input with the Scanner Class

The Scanner class is used to get simple input values from the user in a console application. The techniques that I present here are used in many of the programs shown in the rest of this book.

Throughout the following sections, I refer to the program shown in Listing 2-3. This simple program uses the Scanner class to read an integer value from the user and then displays the value back to the console to verify that the program received the value entered by the user. Here's a sample of the console window for this program:

```
Enter an integer: 5
You entered 5.
```

The program begins by displaying the message Enter an integer: on the first line. Then it waits for you to enter a number. When you type a number (such as 5) and press the Enter key, it displays the confirmation message (You entered 5.) on the second line.

LISTING 2-3:	A Program That Uses the Scanner Class

```
import java.util.Scanner;                                          →1
public class ScannerApp
{
    static Scanner sc = new Scanner(System.in);                    →4
    public static void main(String[] args)
    {
        System.out.print("Enter an integer: ");                    →7
        int x = sc.nextInt();                                      →8
        System.out.println("You entered " + x + ".");              →9
    }
}
```

Importing the Scanner class

Before you can use the Scanner class in a program, you must import it. To do that, you code an import statement at the beginning of the program, before the class declaration, as shown in line 1 of Listing 2-3:

```
import java.util.Scanner;
```

Note that java and util aren't capitalized, but Scanner is.

TIP

If you're using other classes in the java.util package, you can import the entire package by coding the import statement like this:

```
import java.util.*;
```

Declaring and creating a Scanner object

Before you can use the Scanner class to read input from the console, you must declare a Scanner variable and create an instance of the Scanner class. I recommend that you create the Scanner variable as a class variable and create the Scanner object in the class variable initializer, as shown in line 4 of Listing 2-3:

```
static Scanner sc = new Scanner(System.in);
```

That way, you can use the sc variable in any method in the class.

To create a Scanner object, you use the new keyword followed by a call to the Scanner class constructor. Note that the Scanner class requires a parameter that indicates the *input stream* that the input comes from. You can use System.in here to specify standard keyboard console input.

Getting input

To read an input value from the user, you can use one of the methods of the Scanner class that are listed in Table 2-6. As you can see, the primitive data types have separate methods.

Notice in the first column of the table that each method listing begins with the type of the value that's returned by the method. The nextInt method, for example, returns an int value. Also, notice that each of the methods ends with an empty set of parentheses. That means that these methods don't require parameters. If a method does require parameters, the parameters are listed within these parentheses.

TABLE 2-6

Scanner Class Methods That Get Input Values

Method	Explanation
boolean nextBoolean()	Reads a boolean value from the user.
byte nextByte()	Reads a byte value from the user.
double nextDouble()	Reads a double value from the user.
float nextFloat()	Reads a float value from the user.
int nextInt()	Reads an int value from the user.
String nextLine()	Reads a String value from the user.
long nextLong()	Reads a long value from the user.
short nextShort()	Reads a short value from the user.

Because these methods read a value from the user and return the value, you most often use them in statements that assign the value to a variable. Line 8 in Listing 2-3, for example, reads an int and assigns it to a variable named x.

When the nextInt method is executed, the program waits for the user to enter a value in the console window. To let the user know what kind of input the program expects, usually you should call the System.out.print method before you call a Scanner method to get input. Line 7 in Listing 2-3 calls System.out.print to display the message Enter an integer: on the console. That way, the user knows that the program is waiting for input.

**TECHNICAL
STUFF**

If the user enters a value that can't be converted to the correct type, the program *crashes*, which means that it terminates abruptly. As the program crashes, it displays a cryptic error message that indicates what caused the failure. If you enter three instead of an actual number, for example, the console window looks something like this:

```
Enter an integer: three Exception in thread "main" java.util.
   InputMismatchException
at java.util.Scanner.throwFor(Scanner.java:819)
at java.util.Scanner.next(Scanner.java:1431)
at java.util.Scanner.nextInt(Scanner.java:2040)
at java.util.Scanner.nextInt(Scanner.java:2000)
at ScannerApp.main(ScannerApp.java:11)
```

This message indicates that an *exception* called InputMismatchException has occurred, which means that the program was expecting to see an integer but got something else instead. In Book 2, Chapter 8, you find out how to provide for exceptions like these so that the program can display a friendlier message and give the user another shot at entering a correct value. Until then, you have to put up with the fact that if the user enters incorrect data, your program crashes ungracefully.

TIP

You can prevent the nextInt and similar methods from crashing with incorrect input data by first using one of the methods listed in Table 2-7 to test the next input to make sure it's valid. I haven't yet covered the Java statements you need to perform this test, but don't worry; in Book 2, Chapter 8, I show you the solution.

TABLE 2-7 Scanner Class Methods That Check for Valid Input Values

Method	Explanation
boolean hasNextBoolean()	Returns true if the next value entered by the user is a valid boolean value.
boolean hasNextByte()	Returns true if the next value entered by the user is a valid byte value.
boolean hasNextDouble()	Returns true if the next value entered by the user is a valid double value.
boolean hasNextFloat()	Returns true if the next value entered by the user is a valid float value.
boolean hasNextInt()	Returns true if the next value entered by the user is a valid int value.
boolean hasNextLong()	Returns true if the next value entered by the user is a valid long value.
boolean hasNextShort()	Returns true if the next value entered by the user is a valid short value.

Getting Input with the JOptionPane Class

If you prefer to get your user input from a simple GUI rather than from a console, you can use the JOptionPane class. As shown in Figure 2-2, JOptionPane displays a simple dialog box to get text input from the user. Then you can use the parse methods of the primitive-type wrapper classes to convert the text entered by the user to the appropriate primitive type.

FIGURE 2-2:
A dialog box displayed by the JOptionPane class.

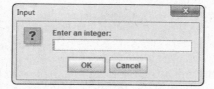

Although the JOptionPane class has many methods, the only one you need to use to get simple text input is the showInputDialog method. This method uses a single parameter that specifies the prompting message that's displayed in the dialog box. It returns a string value that you can then parse to the proper type.

The JOptionPane class is a part of the javax.swing package, so you need to add an import javax.swing.JOptionPane statement to the beginning of any program that uses this class.

Listing 2-4 shows a simple program that uses the JOPtionPane class to get an integer value and display it on the console.

LISTING 2-4: **A Program That Uses the JOptionPane Class to Get User Input**

```
import javax.swing.JOptionPane;                              →1
public class DialogApp
{
    public static void main(String[] args)
    {
        String s;
        s = JOptionPane.showInputDialog                      →7
            ("Enter an integer:");                           →8
        int x = Integer.parseInt(s);                         →9
        System.out.println("You entered " + x + ".");        →10
    }
}
```

The following paragraphs describe the important lines in this program:

- » →1 This line imports the JOptionPane class.

- » →7–8 This statement displays an input dialog box with the prompt Enter an integer: and assigns the string entered by the user to the variable named s.

- » →9 This statement uses the parseInt method of the Integer class to convert the string entered by the user to an integer.

- » →10 This statement displays the integer value to confirm that the data entered by the user was converted properly to an integer.

TECHNICAL STUFF

This program terminates abruptly if the user enters anything other than an integer in the input dialog box. If the user enters ten, for example, the program terminates, and a cryptic message indicating that a NumberFormatException has occurred is displayed. In Book 2, Chapter 8, you learn how to provide a more understandable message in this situation. Until then, just be careful to enter correct numbers when you use the JOptionPane class.

Using enum to Create Your Own Data Types

You will often find yourself using a `String` or `int` variable that you want to constrain to just a few different values. For example, suppose you're writing a program that plays a card game and you want a way to represent the suite of each card. You could do that with a `String` variable whose value should be `Hearts`, `Spades`, `Clubs`, or `Diamonds`. Or you might use an `int` and use the value 1 to represent Hearts, 2 for Spades, 3 for Clubs, or 4 for Diamonds. But such a scheme is error prone. What if you assign a value of 5 to the `int` variable?

A better way is to create an *enum*, which is basically a custom variable type which has a limited set of possible values. To define an enum, you use the `enum` keyword (usually modified by `public`) followed by a list of possible values enclosed in brackets:

```
public enum CardSuit {HEARTS, SPADES, CLUBS, DIAMONDS}
```

You can declare variables using the enum's name as the data type:

```
CardSuit suit;
```

Then, you can assign a value by using the enum name followed by a period and any of the enum values, as in this example:

```
suit = CardSuit.HEARTS;
```

Here's a complete program that defines an enum, creates a variable of the enum's type, assigns a value, and prints the result:

```
public class EnumTest
{
    public enum CardSuit {HEARTS, SPADES, CLUBS, DIAMONDS}
    public static void main(String[] args)
    {
    CardSuit suit;
    suit = CardSuit.HEARTS;
    System.out.println("The suit is " + suit);
    }
}
```

This program displays the following line on the console:

```
The suit is HEARTS
```

and /

» **Creating finely crafted expressions**

» **Incrementing and decrementing**

» **Accepting an assignment**

» **Using the Math class**

» **Formatting your numbers**

» **Seeing strange things that can happen with numbers**

Chapter **3**

Working with Numbers and Expressions

I n Book 2, Chapter 2, you discover the various primitive numeric types that are supported by Java. In this chapter, you build on that knowledge by doing basic operations with numbers. Much of this chapter focuses on the complex topic of *expressions*, which combine numbers with operators to perform calculations. This chapter also covers performing advanced calculations using the Math class and techniques for formatting numbers when you display them. In addition, you find out why Java's math operations sometimes produce results you may not expect.

Working with Arithmetic Operators

An *operator* is a special symbol or keyword that's used to designate a mathematical operation or some other type of operation that can be performed on one or more values, called *operands.* In all, Java has about 40 operators. This chapter focuses on the operators that do arithmetic. These *arithmetic operators* — seven of them in all, summarized in Table 3-1 — perform basic arithmetic operations, such as addition, subtraction, multiplication, and division.

TABLE 3-1

Java's Arithmetic Operators

Operator	Description
+	Addition
–	Subtraction
*	Multiplication
/	Division
%	Remainder (Modulus)
++	Increment
––	Decrement

The following section of code can help clarify how these operators work for `int` types:

```
int a = 32, b = 5;
int c = a + b; // c is 37
int d = a - b; // d is 27
int e = a * b; // e is 160
int f = a / b; // f is 5 (32 / 5 is 6 remainder 2)
int g = a % b; // g is 2 (32 / 5 is 6 remainder 2)
a++; // a is now 33
b--; // b is now 4
```

Notice that for division, the result is truncated. Thus `32 / 5` returns `6`, not `6.4`. For more information about integer division, see the section "Dividing Integers," later in this chapter.

Here's how the operators work for `double` values:

```
double x = 5.5, y = 2.0;
double m = x + y; // m is 7.5
double n = x - y; // n is 3.5
double o = x * y; // o is 11.0
double p = x / y; // p is 2.75
double q = x % y; // q is 1.5
x++; // x is now 6.5
y--; // y is now 1.0
```

When you divide two `int` values, the result is an integer value, even if you assign it to a `double` variable. For example:

```
int a = 21, b = 6;
double answer = a / b; // answer = 3.0
```

If that's not what you want, you can *cast* one of the operands to a `double` before performing the division. (*Casting an operand* means converting its value from one data type to another.) Here's how:

```
int a = 21, b = 6;
double answer = (double)a / b;        // answer = 3.5
```

CATEGORIZING OPERATORS BY THE NUMBER OF OPERANDS

A common way to categorize Java's operators is to use the number of operands the operator works on. When the operators are categorized this way, you find three types:

- **Unary operators:** Operators that work on just one operand. Examples of unary operators are negation (–x, which returns the negative of x) and increment (x++, which adds 1 to x).

  A unary operator can be a prefix operator or a postfix operator. A *prefix operator* is written before the operand, like this:

  ```
  operator operand
  ```

  A *postfix operator* is written after the operand, like this:

  ```
  operand operator
  ```

- **Binary operators:** Operators that work on two operands. Examples of binary operators are addition (x + y), multiplication (invoiceTotal * taxRate), and comparison operators (x < leftEdge). In Java, all binary operators are *infix operators*, which means they appear between the operands, like this:

  ```
  operand1 operator operand2
  ```

- **Ternary operators:** Operators that work on three operands. Java has only one ternary operator, called the *conditional operator* (?:). The conditional operator is also infix:

  ```
  operand1 ? operand2 : operand3
  ```

The moral of the story is that if you want to divide int values and get an accurate double result, you must cast at least one of the int values to a double.

TIP

Here are a few additional things to think about tonight as you lie awake pondering the wonder of Java's arithmetic operators:

>> In algebra, you can write a number right next to a variable to imply multiplication. In this case, *4x* means "four times *x*." Not so in Java. The following statement doesn't compile:

```
int x;

y = 4x;        // error, this line won't compile
```

>> The remainder operator (%) is also called a *modulus* operator. It returns the remainder when the first operand is divided by the second operand. The remainder operator is often used to determine whether one number is evenly divisible by another, in which case the result is zero. For more information, see the next section, "Dividing Integers."

>> All operators, including the arithmetic variety, are treated as separators in Java. As a result, any use of white space in an expression is optional. Thus the following two statements are equivalent:

```
a = ( (x + 4) * 7 ) / (y * x);
a=((x+4)*7)/(y*x);
```

Just remember that a little bit of white space never hurt anyone, and sometimes it helps make Java a little more readable.

Dividing Integers

When you divide one integer into another, the result is always another integer. Any remainder is simply discarded, and the answer is *not* rounded up. 5 / 4 gives the result 1, for example, and 3 / 4 gives the result 0. If you want to know that 5 / 4 is actually 1.25 or that 3 / 4 is actually 0.75, you need to use floats or doubles instead of integers.

If you need to know what the remainder is when you divide two integers, use the remainder operator (%). Suppose that you have a certain number of marbles to give away and a certain number of children to give them to. The program in Listing 3-1 lets you enter the number of marbles and the number of children. Then it calculates the number of marbles to give to each child and the number of marbles you have left over.

TIP

The remainder operator is also called the *modulus* operator.

Here's a sample of the console output for this program, where the number of marbles entered is 93 and the number of children is 5:

```
Welcome to the marble divvy-upper.
Number of marbles: 93
Number of children: 5
Give each child 18 marbles.
You will have 3 marbles left over.
```

LISTING 3-1: A Program That Divvies Up Marbles

```
import java.util.Scanner;                                          →1
public class MarblesApp
{
    static Scanner sc = new Scanner(System.in);                    →4
    public static void main(String[] args)
    {
        // declarations                                            →7
        int numberOfMarbles;
        int numberOfChildren;
        int marblesPerChild;
        int marblesLeftOver;                                       →11

        // get the input data                                      →13
        System.out.println("Welcome to the marble divvy-upper.");
        System.out.print("Number of marbles: ");
        numberOfMarbles = sc.nextInt();
        System.out.print("Number of children: ");
        numberOfChildren = sc.nextInt();                           →18

        // calculate the results
        marblesPerChild = numberOfMarbles / numberOfChildren;      →21
        marblesLeftOver = numberOfMarbles % numberOfChildren;      →22

        // print the results                                       →24
        System.out.println("Give each child " +
            marblesPerChild + " marbles.");
        System.out.println("You will have " +
            marblesLeftOver + " marbles left over.");
    }
}                                                                  →30
```

The following paragraphs describe the key lines in this program:

>> →1 Imports the java.util.Scanner class so that the program can use it to get input from the user.

>> →4 Creates the Scanner object and assigns it to a class variable so that it can be used in any method in the class.

>> →7–11 Declare the local variables used by the program.

>> →13-18 Get the input from the user.

>> →21 Calculates the number of marbles to give to each child by using integer division, which discards the remainder.

>> →22 Calculates the number of marbles left over.

>> →24–30 Print the results.

TIP

It's probably obvious if you think about it, but you should realize that if you use integer division to divide a by b, the result times b plus the remainder equals a. In other words:

```
int a = 29;          // any value will do
int b = 3;           // any value will do
int c = a / b;
int d = a % b;
int e = (c * b) + d;   // e will always equal a
```

Combining Operators

You can combine operators to form complicated expressions. When you do, the order in which the operations are carried out is determined by the *precedence* of each operator in the expression. The order of precedence for the arithmetic operators is

>> Increment (++) and decrement (−−) operators are evaluated first.

>> Next, sign operators (+ or −) are applied.

>> Then multiplication (∗), division (/), and remainder (%) operators are evaluated.

>> Finally, addition (+) and subtraction (−) operators are applied.

In the expression a + b * c, for example, multiplication has a higher precedence than addition. Thus b is multiplied by c first. Then the result of that multiplication is added to a.

If an expression includes two or more operators at the same order of precedence, the operators are evaluated left to right. Thus, in the expression a * b / c, a is multiplied by b and then the result is divided by c.

If you want, you can use parentheses to change the order in which operations are performed. Operations within parentheses are always performed before operations that aren't in parentheses. Thus, in the expression (a + b) * c, a is added to b first. Then the result is multiplied by c.

If an expression has two or more sets of parentheses, the operations in the innermost set are performed first. In the expression (a * (b + c)) / d, b is added to c. Then the result is multiplied by a. Finally, that result is divided by d.

TIP

Apart from the increment and decrement operators, these precedence rules and the use of parentheses are the same as they are for basic algebra. So if you were paying attention in the eighth grade, precedence should make sense.

WARNING

With double or float values, changing the left to right order for operators with the same precedence doesn't affect the result. With integer types, however, it can make a huge difference if division is involved. Consider these statements:

```
int a = 5, b = 6, c = 7;
int d1 = a * b / c;          // d1 is 4
int d2 = a * (b / c);        // d2 is 0
```

This difference occurs because integer division always returns an integer result, which is a truncated version of the actual result. Thus, in the first expression, a is first multiplied by b, giving a result of 30. Then this result is divided by c. Truncating the answer gives a result of 4. But in the second expression, b is first divided by c, which gives a truncated result of 0. Then this result is multiplied by a, giving a final answer of 0.

Using the Unary Plus and Minus Operators

The unary plus and minus operators let you change the sign of an operand. Note that the actual operator used for these operations is the same as the binary addition and subtraction operators. The compiler figures out whether you mean to use the binary or the unary version of these operators by examining the expression.

TIP

The unary minus operator doesn't necessarily make an operand have a negative value. Instead, it changes whatever sign the operand has to start with. Thus, if the operand starts with a positive value, the unary minus operator changes it to negative. But if the operand starts with a negative value, the unary minus operator makes it positive. The following examples illustrate this point:

```
int a = 5;          // a is 5
int b = -a;         // b is -5
int c = -b;         // c is +5
```

TECHNICAL STUFF

Interestingly enough, the unary plus operator doesn't actually do anything. For example:

```
int a = -5;         // a is -5
int b = +a;         // b is -5
a = 5;              // a is now 5
int c = +a;         // c is 5
```

Notice that if a starts out positive, +a is also positive. But if a starts out negative, +a is still negative. Thus the unary plus operator has no effect. I guess Java provides the unary plus operator out of a need for balance.

You can also use these operators with more complex expressions, like this:

```
int a = 3, b = 4, c = 5;
int d = a * -(b + c);      // d is -27
```

Here, b is added to c, giving a result of 9. Then the unary minus operator is applied, giving a result of -9. Finally, -9 is multiplied by a, giving a result of -27.

Using Increment and Decrement Operators

One of the most common operations in computer programming is adding or subtracting 1 from a variable. Adding 1 to a variable is called *incrementing* the variable. Subtracting 1 is called *decrementing*. The traditional way to increment a variable is this:

```
a = a + 1;
```

Here the expression a + 1 is calculated, and the result is assigned to the variable a.

Java provides an easier way to do this type of calculation: the increment (++) and decrement (−−) operators. These unary operators apply to a single variable. Thus, to increment the variable a, you can code just this:

```
a++;
```

Note that an expression that uses an increment or decrement operator is a statement by itself. That's because the increment or decrement operator is also a type of assignment operator, as it changes the value of the variable it applies to.

TIP

You can use the increment and decrement operators only on variables — not on numeric literals or other expressions. Java doesn't allow the following expressions, for example:

```
a = b * 5++;        // can't increment the number 5
a = (b * 5)++;      // can't increment the expression (b * 5)
```

Note that you can use an increment or decrement operator in an assignment statement. Here's an example:

```
int a = 5;
int b = a--;        // both a and b are set to 4
```

When the second statement is executed, the expression a−− is evaluated first, so a is set to 4. Then the new value of a is assigned to b. Thus both a and b are set to 4.

REMEMBER

The increment and decrement operators are unusual because they are unary operators that can be placed either before (*prefix*) or after (*postfix*) the variable they apply to. Whether you place the operator before or after the variable can have a major effect on how an expression is evaluated. If you place an increment or decrement operator before its variable, the operator is applied before the value of the variable is read. As a result, the incremented value of the variable is used in the expression. By contrast, if you place the operator after the variable, the operator is applied after the value has been read. Thus, the original value of the variable is used in its immediate context within the expression.

Confused yet? A simple example can clear things up. First, consider these statements with an expression that uses a postfix increment:

```
int a = 5;
int b = 3;
int c = a * b++;        // c is set to 15
```

Working with Numbers and Expressions

When the expression in the third statement is evaluated, the original value of b — 3 — is used in the multiplication. Thus c is set to 15. Then b is incremented to 4.

Now consider this version, with a prefix increment:

```
int a = 5;
int b = 3;
int c = a * ++b;          // c is set to 20
```

This time, b is incremented before the multiplication is performed, so c is set to 20. Either way, b ends up set to 4.

Similarly, consider this example:

```
int a = 5;
int b = a--;              // b is set to 5, a is set to 4.
```

This example is similar to an earlier example, but this time the postfix decrement operator is used. When the second statement is executed, the value of a is assigned to b. Then a is decremented. As a result, b is set to 5, and a is set to 4.

REMEMBER

Because the increment and decrement operators can be confusing when used with other operators in an expression, I suggest that you use them alone. Whenever you're tempted to incorporate an increment or decrement operator into a larger expression, pull the increment or decrement out of the expression, and make it a separate statement either before or after the expression. In other words, code

```
b++;
c = a * b;
```

instead of

```
c = a * ++b;
```

In the first version, it's crystal-clear that b is incremented before the multiplication is done.

Using the Assignment Operator

The standard assignment operator (=) is used to assign the result of an expression to a variable. In its simplest form, you code it like this:

```
variable = expression;
```

Here's an example:

```
int a = (b * c) / 4;
```

You've already seen plenty of examples of assignment statements like this one, so I won't belabor this point any further. I do want to point out — just for the record — that you *cannot* code an arithmetic expression on the left side of an equal sign. Thus the following statement doesn't compile:

```
int a;
a + 3 = (b * c);
```

WARNING

In the rest of this section, I point out some unusual ways in which you can use the assignment operator. I don't recommend that you actually use any of these techniques, as they're rarely necessary and almost always confusing, but knowing about them can shed light on how Java expressions work and sometimes can help you find sneaky problems in your code.

The key to understanding the rest of this section is realizing that in Java, assignments are expressions, not statements. In other words, a = 5 is an assignment expression, not an assignment statement. It becomes an assignment statement only when you add a semicolon to the end.

The result of an assignment expression is the value that's assigned to the variable. The result of the expression a = 5, for example, is 5. Likewise, the result of the expression a = (b + c) * d is the result of the expression (b + c) * d.

The implication is that you can use assignment expressions in the middle of other expressions. The following example is legal:

```
int a;
int b;
a = (b = 3) * 2;          // a is 6, b is 3
```

As in any expression, the part of the expression inside the parentheses is evaluated first. Thus, b is assigned the value 3. Then the multiplication is performed, and the result (6) is assigned to the variable a.

Now consider a more complicated case:

```
int a;
int b = 2;
a = (b = 3) * b;          // a is 9, b is 3
```

What's happening here is that the expression in the parentheses is evaluated first, which means that b is set to 3 before the multiplication is performed.

The parentheses are important in the previous example because without parentheses, the assignment operator is the last operator to be evaluated in Java's order of precedence. Consider one more example:

```
int a;
int b = 2;
a = b = 3 * b;        // a is 6, b is 6
```

This time, the multiplication 3 * b is performed first, giving a result of 6. Then this result is assigned to b. Finally, the result of that assignment expression (6) is assigned to a.

Incidentally, the following expression is also legal:

```
a = b = c = 3;
```

This expression assigns the value 3 to all three variables.

Using Compound Assignment Operators

A *compound assignment operator* is an operator that performs a calculation and an assignment at the same time. All of Java's binary arithmetic operators (that is, the ones that work on two operands) have equivalent compound assignment operators, which Table 3-2 lists.

TABLE 3-2   ## Compound Assignment Operators

Operator	Description
+=	Addition and assignment
−=	Subtraction and assignment
*=	Multiplication and assignment
/=	Division and assignment
%=	Remainder and assignment

The statement

```
a += 10;
```

is equivalent to

```
a = a + 10;
```

Also, the statement

```
z *=2;
```

is equivalent to

```
z = z * 2;
```

WARNING

To prevent confusion, use compound assignment expressions by themselves, not in combination with other expressions. Consider these statements:

```
int a = 2;
int b = 3;
a *= b + 1;
```

Is a set to 7 or 8?

In other words, is the third statement equivalent to

```
a = a * b + 1;        // This would give 7 as the result
```

or

```
a = a * (b + 1);      // This would give 8 as the result
```

At first glance, you might expect the answer to be 7, because multiplication has a higher precedence than addition. But assignment has the lowest precedence of all, and the multiplication here is performed as part of the assignment. As a result, the addition is performed before the multiplication — and the answer is 8. (Gotcha!)

Using the Math Class

Java's built-in operators are useful, but they don't come anywhere near providing all the mathematical needs of most Java programmers. That's where the Math class comes in. It includes a bevy of built-in methods that perform a wide variety of mathematical calculations, from basic functions such as calculating an absolute value or a square root to trigonometry functions such as sin and cos (sine and cosine), to practical functions such as rounding numbers or generating random numbers.

I was going to make a joke here about how you'd have to take a math class to fully appreciate the Math class; or how you'd better stay away from the Math class if you didn't do so well in math class; or how if you're on the football team, maybe you can get someone to do the Math class for you. But these jokes seemed too easy, so I decided not to make them.

TECHNICAL STUFF

All the methods of the Math class are declared as static methods, which means you can use them by specifying the class name Math followed by a period and a method name. Here's a statement that calculates the square root of a number stored in a variable named y:

```
double x = Math.sqrt(y);
```

The Math class is contained in the java.lang package, which is automatically available to all Java programs. As a result, you don't have to provide an import statement to use the Math class.

The following sections describe the most useful methods of the Math class.

Using constants of the Math class

The Math class defines two constants that are useful for many mathematical calculations. Table 3-3 lists these constants.

TABLE 3-3

Constants of the Math Class

Constant	What It Is	Value
PI	The constant pi (π), the ratio of a circle's radius and diameter	3.141592653589793
E	The base of natural logarithms	2.718281828459045

Note that these constants are only approximate values, because both π and e are irrational numbers.

The program shown in Listing 3-2 illustrates a typical use of the constant PI. Here, the user is asked to enter the radius of a circle. Then the program calculates the area of the circle in line 11. (The parentheses aren't really required in the expression in this statement, but they help clarify that the expression is the Java equivalent to the formula for the area of a circle, πr^2.)

Here's the console output for a typical execution of this program, in which the user entered 5 as the radius of the circle:

```
Welcome to the circle area calculator.
Enter the radius of your circle: 5
The area is 78.53981633974483
```

LISTING 3-2: **The Circle Area Calculator**

```java
import java.util.Scanner;
public class CircleAreaApp
{
    static Scanner sc = new Scanner(System.in);
    public static void main(String[] args)
    {
        System.out.println(
            "Welcome to the circle area calculator.");
        System.out.print("Enter the radius of your circle: ");
        double r = sc.nextDouble();
        double area = Math.PI * (r * r);                          →11
        System.out.println("The area is " + area);
    }
}
```

TIP

At the time I wrote this, the actual value of pi was known to a precision of more than 31 trillion digits. Unfortunately, the Math class's PI constant has a precision of just 15 digits. Fortunately, 15 digits is plenty of precision for most real-world applications. According to mathematician James Grime, you need just 39 digits of precision to calculate the circumference of the entire known universe to an accuracy of less than the size of a single hydrogen atom.

Working with mathematical functions

Table 3-4 lists the basic mathematical functions that are provided by the Math class. As you can see, you can use these functions to calculate such things as the absolute value of a number, the minimum and maximum of two values, square roots, powers, and logarithms.

TABLE 3-4   **Commonly Used Mathematical Functions Provided by the Math Class**

Method	Explanation
abs(argument)	Returns the absolute value of the argument. The argument can be an int, long, float, or double. The return value is the same type as the argument.
cbrt(argument)	Returns the cube root of the argument. The argument and return value are doubles.
exp(argument)	Returns e raised to the power of the argument. The argument and the return value are doubles.
hypot(arg1, arg2)	Returns the hypotenuse of a right triangle calculated according to the Pythagorean theorem — $\sqrt{x^2 + y^2}$. The argument and the return values are doubles.
log(argument)	Returns the natural logarithm (base e) of the argument. The argument and the return value are doubles.
log10(argument)	Returns the base 10 logarithm of the argument. The argument and the return value are doubles.
max(arg1, arg2)	Returns the larger of the two arguments. The arguments can be int, long, float, or double, but both must be of the same type. The return type is the same type as the arguments.
min(arg1, arg2)	Returns the smaller of the two arguments. The arguments can be int, long, float, or double, but both must be of the same type. The return type is the same type as the arguments.
pow(arg1, arg2)	Returns the value of the first argument raised to the power of the second argument. Both arguments and the return value are doubles.
random()	Returns a random number that's greater than or equal to 0.0 but less than 1.0. This method doesn't accept an argument, but the return value is a double.
signum(argument)	Returns a number that represents the sign of the argument: –1.0 if the argument is negative, 0.0 if the argument is zero, and 1.0 if the argument is positive. The argument can be a double or a float. The return value is the same type as the argument.
sqrt(argument)	Returns the square root of the argument. The argument and return value are doubles.

The program shown in Listing 3-3 demonstrates each of these methods except random. When run, it produces output similar to this:

```
abs(b) = 50
cbrt(x) = 2.924017738212866
exp(y) = 54.598150033144236
hypot(y,z)= 5.0
```

```
log(y) = 1.0986122886681096
log10(y) = 0.47712125471966244
max(a, b) = 100
min(a, b) = -50
pow(a, c) = 1000000.0
random() = 0.8536014557793756
signum(b) = -1.0
sqrt(x) = 1.7320508075688772
```

You can use this output to get an idea of the values returned by these Math class methods. You can see, for example, that the expression Math.sqrt(y) returns a value of 5.0 when y is 25.0.

The following paragraphs point out a few interesting tidbits concerning these methods:

>> You can use the abs and signnum methods to force the sign of one variable to match the sign of another, like this:

```
int a = 27;
int b = -32;

a = Math.abs(a) * Math.signum(b); // a is now -27;
```

>> You can use the pow method to square a number, like this:

```
double x = 4.0;
double y = Math.pow(x, 2); // a is now 16;
```

Simply multiplying the number by itself, however, is often just as easy and just as readable:

```
double x = 4.0;
double y = x * x; // a is now 16;
```

TECHNICAL STUFF

>> In the classic movie *The Wizard of Oz,* when the Wizard finally grants the Scarecrow his brains, the Scarecrow suddenly becomes intelligent and quotes the Pythagorean theorem, which is (coincidentally) used by the hypot method of the Math class. (Of course, he quotes it wrong. What the Scarecrow actually says in the movie is this: "The sum of the square root of any two sides of an isosceles triangle is equal to the square root of the remaining side." Silly Scarecrow. He didn't need to know this to be smart.)

>> Every time you run the program in Listing 3-3, you get a different result for the random method call. The random method is interesting enough that I describe it separately in the next section, "Creating random numbers."

```java
public class MathFunctionsApp
{
public static void main(String[] args)
{
    int a = 100;
    int b = -50;
    int c = 3;
    double x = 25.0;
    double y = 3.0;
    double z = 4.0;

    System.out.println("abs(b) = " + Math.abs(b));
    System.out.println("cbrt(x) = " + Math.cbrt(x));
    System.out.println("exp(y) = " + Math.exp(z));
    System.out.println("hypot(y,z)= " + Math.hypot(y,z));
    System.out.println("log(y) = " + Math.log(y));
    System.out.println("log10(y) = " + Math.log10(y));
    System.out.println("max(a, b) = " + Math.max(a, b));
    System.out.println("min(a, b) = " + Math.min(a, b));
    System.out.println("pow(a, c) = " + Math.pow(a, c));
    System.out.println("random() = " + Math.random());
    System.out.println("signum(b) = " + Math.signum(b));
    System.out.println("sqrt(x) = " + Math.sqrt(y));
    }
}
```

Creating random numbers

Sooner or later, you're going to want to write programs that play simple games. Almost all games have some element of chance built into them, so you need a way to create computer programs that don't work exactly the same every time you run them. The easiest way to do that is to use the random method of the Math class, which Table 3-4 lists later in this section, along with the other basic mathematical functions of the Math class.

The random method returns a double whose value is greater than or equal to 0.0 but less than 1.0. Within this range, the value returned by the random method is different every time you call it and is essentially random.

TECHNICAL STUFF

Strictly speaking, computers are not capable of generating *truly* random numbers, but over the years, clever computer scientists have developed ways to generate numbers that are random for all practical purposes. These numbers are called *pseudorandom numbers* because although they aren't completely random, they look random to most human beings.

TECHNICAL STUFF

Java has many methods and classes for generating random numbers. The `java.util.Random` class, for example, provides about ten specialized methods that generate random values. To generate a `double` with a value between 0.0 and 1.0, you can execute `new Random().nextDouble()`. In addition, the `java.security.SecureRandom` class provides random values for encrypting sensitive documents. And if size matters to you, the `java.math.BigInteger` class allows you to generate arbitrarily large random numbers (numbers with 1,000 digits, if that's what you need).

The `random` method generates a random `double` value between 0.0 (inclusive, meaning that it could be 0.0) and 1.0 (exclusive, meaning that it can't be 1.0). Most computer applications that need random values, however, need random integers between some arbitrary low value (usually 1, but not always) and some arbitrary high value. A program that plays dice needs random numbers between 1 and 6, whereas a program that deals cards needs random numbers between 1 and 52 (53 if a joker is used).

As a result, you need a Java expression that converts the `double` value returned by the `random` function to an `int` value within the range your program calls for. The following code shows how to do this, with the values set to 1 and 6 for a dice-playing game:

```
int low = 1; // the lowest value in the range
int high = 6; // the highest value in the range
int rnd = (int)(Math.random() * (high - low + 1)) + low;
```

This expression is a little complicated, so I show you how it's evaluated step by step:

1. The `Math.Random` method is called to get a random double value. This value is greater than 0.0 but less than 1.0.

2. The random value is multiplied by the high end of the range minus the low end, plus 1. In this example, the high end is 6 and the low end is 1, so you now have a random number that's greater than or equal to 0.0 but less than 6.0. (It could be 5.99999999999999, but it never is 6.0.)

3. This value is converted to an integer by the (`int`) cast. Now you have an integer that's 0, 1, 2, 3, 4, or 5. (Remember that when you cast a double to an `int`, any fractional part of the value is simply discarded. Because the number is less than 6.0, it never truncates to 6.0 when it is cast to an `int`.)

4. The `low` value in the range is added to the random number. Assuming that `low` is 1, the random number is now 1, 2, 3, 4, 5, or 6. That's just what you want: a random number between 1 and 6.

To give you an idea of how this random-number calculation works, Listing 3-4 shows a program that places this calculation in a method called randomInt and then calls it to simulate 100 dice rolls. The randomInt method accepts two parameters representing the low and high ends of the range, and it returns a random integer within the range. In the main method of this program, the randomInt method is called 100 times, and each random number is printed by a call to System.out.print.

The console output for this program looks something like this:

```
Here are 100 random rolls of the dice:
4 1 1 6 1 2 6 6 6 6 5 5 5 4 5 4 4 1 3 6 1 3 1 4 4 3 3 3 5 6 5 6 6 3 5 2 2 6 3 3
4 1 2 2 4 2 2 4 1 4 3 6 5 5 4 4 2 4 1 3 5 2 1 3 3 5 4 1 6 3 1 6 5 2 6 6 3 5 4 5
2 5 4 5 3 1 4 2 5 2 1 4 4 4 6 6 4 6 3 3
```

Every time you run this program, however, you see a different sequence of 100 numbers.

The program in Listing 3-4 uses several Java features that you haven't seen yet.

LISTING 3-4:   **Rolling the Dice**

```java
public class DiceApp
{
    public static void main(String[] args)
    {
        int roll;
        String msg = "Here are 100 random rolls of the dice:";
        System.out.println(msg);
        for (int i=0; i<100; i++)                              →8
        {
            roll = randomInt(1, 6);                            →10
            System.out.print(roll + " ");                      →11
        }
        System.out.println();
    }

    public static int randomInt(int low, int high)            →16
    {
        int result = (int)(Math.random()                      →18
            * (high - low + 1)) + low;
        return result;                                         →20
    }
}
```

The following paragraphs explain how the program works, but don't worry if you don't get all the elements in this program. The main thing to see is the expression that converts the random `double` value returned by the `Math.double` method to an integer.

» →8 The `for` statement causes the statements in its body (lines 10 and 11) to be executed 100 times. Don't worry about how this statement works for now; you find out about it in Book 2, Chapter 5.

» →10 This statement calls the `randomInt` method, specifying 1 and 6 as the range for the random integer to generate. The resulting random number is assigned to the `roll` variable.

» →11 The `System.out.print` method is used to print the random number followed by a space. Because this statement calls the `print` method rather than the `println` method, the random numbers are printed on the same line rather than on separate lines.

» →16 The declaration for the `randomInt` method indicates that the method returns an int value and accepts two int arguments: one named `low` and the other named `high`.

» →18 This expression converts the random `double` value to an integer between `low` and `high`.

» →20 The `return` statement sends the random number back to the statement that called the `randomInt` method.

Rounding functions

The `Math` class has four methods that round or truncate `float` or `double` values. Table 3-5 lists these methods. As you can see, each of these methods uses a different technique to calculate an `integer` value that's near the `double` or `float` value passed as an argument. Note that even though all four of these methods round a floating-point value to an `integer` value, only the round method actually returns an `integer` type (int or long, depending on whether the argument is a `float` or a `double`). The other methods return `double`s that happen to be `integer` values.

Listing 3-5 shows a program that uses each of the four methods to round three `double` values: `29.4`, `93.5`, and `-19.3`. Here's the output from this program:

```
round(x) = 29
round(y) = 94
round(z) = -19

ceil(x) = 30.0
```

```
ceil(y) = 94.0
ceil(z) = -19.0

floor(x) = 29.0
floor(y) = 93.0
floor(z) = -20.0

rint(x) = 29.0
rint(y) = 94.0
rint(z) = -19.0
```

TABLE 3-5   ## Rounding Functions Provided by the Math Class

Method	Explanation
ceil(argument)	Returns the smallest double value that is an integer and is greater than or equal to the value of the argument.
floor(argument)	Returns the largest double value that is an integer and is less than or equal to the value of the argument.
rint(argument)	Returns the double value that is an integer and is closest to the value of the argument. If two integer values are equally close, it returns the one that is even. If the argument is already an integer, it returns the argument value.
round(argument)	Returns the integer that is closest to the argument. If the argument is a double, it returns a long. If the argument is a float, it returns an int.

Note that each of the four methods produces a different result for at least one of the values:

>> All the methods except ceil return 29.0 (or 29) for the value 29.4. ceil returns 30.0, which is the smallest integer that's greater than 29.4.

>> All the methods except floor return 94.0 (or 94) for the value 93.5. floor returns 93.0 because that's the largest integer that's less than 93.5. rint returns 94.0 because it's an even number, and 93.5 is midway between 93.0 and 94.0.

>> All the methods except floor return -19.0 (or -19) for -19.3. floor returns -20 because -20 is the largest integer that's less than -19.3.

LISTING 3-5:   ## Program That Uses the Rounding Methods of the Math Class

```
public class RoundingApp
{
    public static void main(String[] args)
    {
```

```
        double x = 29.4;
        double y = 93.5;
        double z = -19.3;

        System.out.println("round(x) = " + Math.round(x));
        System.out.println("round(y) = " + Math.round(y));
        System.out.println("round(z) = " + Math.round(z));
        System.out.println();

        System.out.println("ceil(x) = " + Math.ceil(x));
        System.out.println("ceil(y) = " + Math.ceil(y));
        System.out.println("ceil(z) = " + Math.ceil(z));
        System.out.println();

        System.out.println("floor(x) = " + Math.floor(x));
        System.out.println("floor(y) = " + Math.floor(y));
        System.out.println("floor(z) = " + Math.floor(z));
        System.out.println();

        System.out.println("rint(x) = " + Math.rint(x));
        System.out.println("rint(y) = " + Math.rint(y));
        System.out.println("rint(z) = " + Math.rint(z));
    }
}
```

Formatting Numbers

Most of the programs you've seen so far have used the System.out.println or
System.out.print method to print the values of variables that contain numbers.
When you pass a numeric variable to one of these methods, the variable's value
is converted to a string before it's printed. The exact format used to represent the
value isn't very pretty: Large values are printed without any commas, and all the
decimal digits for double or float values are printed whether you want them to
be or not.

In many cases, you want to format your numbers before you print them — to add
commas to large values and limit the number of decimal places printed, for example. Or, if a number represents a monetary amount, you may want to add a dollar
sign (or whatever currency symbol is appropriate for your locale). To do that, you
can use the NumberFormat class. Table 3-6 lists the NumberFormat class methods.

TABLE 3-6	Methods of the NumberFormat Class
Method	Explanation
getCurrencyInstance()	A static method that returns a NumberFormat object that formats currency values
getPercentInstance()	A static method that returns a NumberFormat object that formats percentages
getNumberInstance()	A static method that returns a NumberFormat object that formats basic numbers
format(number)	Returns a string that contains the formatted number
setMinimumFractionDigits(int)	Sets the minimum number of digits to display to the right of the decimal point
setMaximumFractionDigits(int)	Sets the maximum number of digits to display to the right of the decimal point

REMEMBER

Like many aspects of Java, the procedure for using the NumberFormat class is a little awkward. It's designed to be efficient for applications that need to format a lot of numbers, but it's overkill for most applications.

The procedure for using the NumberFormat class to format numbers takes a little getting used to. First, you must call one of the static get*Xxx*Instance methods to create a NumberFormat object that can format numbers in a particular way. Then, if you want, you can call the setMinimumFractionDigits or setMaximumFrac-tionDigits method to set the number of decimal digits to be displayed. Finally, you call that object's format method to actually format a number.

Note that the NumberFormat class is in the java.text package, so you must include the following import statement at the beginning of any class that uses NumberFormat:

```
import java.text.NumberFormat;
```

Here's an example that uses the NumberFormat class to format a double value as currency:

```
double salesTax = 2.425;
NumberFormat cf = NumberFormat.getCurrencyInstance();
System.out.println(cf.format(salesTax));
```

When you run this code, the following line is printed to the console:

```
$2.43
```

Note that the currency format rounds the value from 2.425 to 2.43.

Here's an example that formats a number by using the general number format, with exactly three decimal places:

```
double x = 19923.3288;
NumberFormat nf = NumberFormat.getNumberInstance();
nf.setMinimumFractionDigits(3);
nf.setMaximumFractionDigits(3);
System.out.println(nf.format(x));
```

When you run this code, the following line is printed:

```
19,923.329
```

Here the number is formatted with a comma and the value is rounded to three places.

Here's an example that uses the percentage format:

```
double grade = .92;
NumberFormat pf = NumberFormat.getPercentInstance();
System.out.println(pf.format(grade));
```

When you run this code, the following line is printed:

```
92%
```

Recognizing Weird Things about Java Math

Believe it or not, computers — even the most powerful ones — have certain limitations when it comes to performing math calculations. These limitations are usually insignificant, but sometimes they sneak up and bite you. The following sections describe the things you need to watch out for when doing math in Java.

Integer overflow

WARNING

The basic problem with integer types is that they have a fixed size. As a result, there is a limit to the size of the numbers that can be stored in variables of type byte, short, int, or long. Although long variables can hold numbers that are huge, sooner or later you come across a number that's too big to fit in even a long variable.

Okay, consider this (admittedly contrived) example:

```
int a = 1000000000;
System.out.println(a);
a += 1000000000;
System.out.println(a);
a += 1000000000;
System.out.println(a);
a += 1000000000;
System.out.println(a);
```

Here you expect the value of a to get bigger after each addition. But here's the output that's displayed:

```
1000000000
2000000000
-1294967296
-294967296
```

The first addition seems to work, but after that, the number becomes negative! That's because the value has reached the size limit of the int data type. Unfortunately, Java doesn't tell you that this error has happened. It simply crams the int variable as full of bits as it can, discards whatever bits don't fit, and hopes that you don't notice. Because of the way int stores negative values, large positive values suddenly become large negative values. This effect is called *wrap around*.

The moral of the story is that if you're working with large integers, you should use long rather than int, because long can store much larger numbers than int. If your programs deal with numbers large enough to be a problem for long, consider using floating-point types instead. As you see in the next section, floating-point types can handle even larger values than long, and they let you know when you exceed their capacity.

Floating-point weirdness

WARNING

Floating-point numbers have problems of their own. For starters, floating-point numbers are stored using the binary number system (base 2), but humans work with numbers in the decimal number system (base 10). Unfortunately, accurately converting numbers between these two systems is sometimes impossible. That's because in any number base, certain fractions can't be represented exactly. One example: Base 10 has no way to exactly represent the fraction 1/3. You can approximate it as 0.3333333, but eventually you reach the limit of how many digits you can store, so you have to stop. In base 2, it happens that one of the fractions you can't accurately represent is the decimal value 1/10. In other words, a float or double variable can't accurately represent 0.1.

Don't believe me? Try running this code:

```
float x = 0.1f;
NumberFormat nf = NumberFormat.getNumberInstance();
nf.setMinimumFractionDigits(10);
System.out.println(nf.format(x));
```

The resulting output is this:

```
0.1000000015
```

Although 0.1000000015 is *close* to 0.1, it isn't exact.

WARNING

In most cases, Java's floating-point math is close enough not to matter. The margin of error is extremely small. If you're using Java to measure the size of your house, you'd need an electron microscope to notice the error. If you're writing applications that deal with financial transactions, however, normal rounding can sometimes magnify the errors to make them significant. You may charge a penny too much or too little sales tax. And in extreme cases, your invoices may actually have obvious addition errors.

REMEMBER

Integer types are stored in binary too, of course. But integers aren't subject to the same errors that floating-point types are — because integers don't represent fractions at all — so you don't have to worry about this type of error for integer types.

Division by zero

According to the basic rules of mathematics, you can't divide a number by zero. The reason is simple: Division is the inverse of multiplication — which means that if a * b = c, it is also true that a = c / b. If you were to allow b to be zero, division would be meaningless, because any number times zero is zero. Therefore, both a and c would also have to be zero. In short, mathematicians solved this dilemma centuries ago by saying that division by zero is simply not allowed.

So what happens if you *do* attempt to divide a number by zero in a Java program? The answer depends on whether you're dividing integers or floating-point numbers. If you're dividing integers, the statement that attempts the division by zero chokes up what is called an *exception*, which is an impolite way of crashing the program. In Book 2, Chapter 8, you find out how to intercept this exception to allow your program to continue. In the meantime, any program you write that attempts an integer division by zero crashes.

If you try to divide a floating-point type by zero, the results are not so abrupt. Instead, Java assigns to the floating-point result one of the special values listed in Table 3-7. The following paragraphs explain how these special values are determined:

>> If you divide a number by zero, and the sign of both numbers is the same, the result is positive infinity. 0.0 divided by 0.0 is positive infinity, as is –34.0 divided by –0.0.

>> If you divide a number by zero, and the signs of the numbers are different, the result is negative infinity. –40.0 divided by 0.0 is negative infinity, as is 34.0 divided by 0.0.

>> If you divide zero by zero, the result is not a number (NaN), regardless of the signs.

TABLE 3-7

Special Constants of the float and double Classes

Constant	Meaning
POSITIVE_INFINITY	Positive infinity
NEGATIVE_INFINITY	Negative infinity
NaN	Not a number

REMEMBER

Floating-point zeros can be positive or negative. Java considers positive and negative zeros to be equal numerically.

If you attempt to print a floating-point value that has one of these special values, Java converts the value to an appropriate string. Suppose that you execute the following statements:

```
double i = 50.0;
double j = 0.0;
double k = i / j;
System.out.println(k);
```

The resulting console output is

```
Infinity
```

If i were –50.0, the console would display –Infinity, and if i were zero, the console would display NaN.

TECHNICAL STUFF

The following paragraphs describe some final bits of weirdness I want to sneak in before closing this chapter:

» NaN is not equal to itself, which can have some strange consequences. For example:

```
double x = Math.sqrt(-50); // Not a number
double y = x;
if (x == y)
System.out.println("x equals y");
```

Okay, I know that I jumped the gun here on the `if` statement, because I don't cover `if` statements until Book 2, Chapter 4. So just assume, for the sake of argument, that the `if` statement tests whether the variable x is equal to the variable y. Because this test immediately follows an assignment statement that assigns the value of x to y, you can safely assume that x equals y, right?

Wrong. Because x is NaN, y also is NaN. NaN is never considered to be equal to any other value, including another NaN. Thus, the comparison in the `if` statement fails.

» Another strange consequence: You can't assume that a number minus itself is always zero. Consider this statement:

```
double z = x - x; // not necessarily zero
```

Shouldn't this statement always set z to zero? Not if x is NaN. In that case, not a number minus not a number is still not a number.

» One more weirdness and then I'll stop: Any mathematical operation involving infinity results in either another infinity or NaN. Infinity + 5, for example, still equals infinity, so Buzz Lightyear's call "To infinity and beyond!" just isn't going to happen. But infinity minus infinity gives you . . . NaN.

Working with Numbers and Expressions

Chapter **4**

Making Choices

S o far in this book, all the programs have run straight through from start to finish without making any decisions along the way. In this chapter, you discover two Java statements that let you create some variety in your programs. The if statement lets you execute a statement or a block of statements only if some conditional test turns out to be true. And the switch statement lets you execute one of several blocks of statements depending on the value of an integer variable.

The if statement relies heavily on the use of *Boolean expressions*, which are, in general, expressions that yield a simple true or false result. Because you can't do even the simplest if statement without a Boolean expression, this chapter begins by showing you how to code simple Java boolean expressions that test the value of a variable. Later, after looking at the details of how the if statement works, I revisit boolean expressions to show how to combine them to make complicated logical decisions. Then I get to the switch statement.

You're going to have to put your thinking cap on for much of this chapter, as most of it plays with logic puzzles. Find yourself a comfortable chair in a quiet part of the house, turn off the TV, and pour yourself a cup of coffee.

Using Simple Boolean Expressions

All if statements, as well as several of the other control statements that I describe in Book 2, Chapter 5 (while, do, and for), use boolean expressions to determine whether to execute or skip a statement (or a block of statements). A *boolean expression* is a Java expression that, when evaluated, returns a *boolean value:* true or false.

As you discover later in this chapter, boolean expressions can be very complicated. Most of the time, however, you use simple expressions that compare the value of a variable with the value of some other variable, a literal, or perhaps a simple arithmetic expression. This comparison uses one of the *relational operators* listed in Table 4-1. All these operators are *binary operators*, which means that they work on two operands.

TABLE 4-1

Relational Operators

Operator	Description
==	Returns true if the expression on the left evaluates to the same value as the expression on the right.
!=	Returns true if the expression on the left does not evaluate to the same value as the expression on the right.
<	Returns true if the expression on the left evaluates to a value that is less than the value of the expression on the right.
<=	Returns true if the expression on the left evaluates to a value that is less than or equal to the expression on the right.
>	Returns true if the expression on the left evaluates to a value that is greater than the value of the expression on the right.
>=	Returns true if the expression on the left evaluates to a value that is greater than or equal to the expression on the right.

A basic Java boolean expression has this form:

```
expression relational-operator expression
```

Java evaluates a `boolean` expression by first evaluating the expression on the left, then evaluating the expression on the right, and finally applying the relational operator to determine whether the entire expression evaluates to `true` or `false`.

Here are some simple examples of relational expressions. For each example, assume that the following statements were used to declare and initialize the variables:

```
int i = 5;
int j = 10;
int k = 15;
double x = 5.0;
double y = 7.5;
double z = 12.3;
```

Here are the sample expressions, along with their results (based on the values supplied):

Expression	Value	Explanation
i == 5	true	The value of i is 5.
i == 10	false	The value of i is not 10.
i == j	false	i is 5, and j is 10, so they are not equal.
i == j - 5	true	i is 5, and j - 5 is 5.
i > 1	true	i is 5, which is greater than 1.
j == i * 2	true	j is 10, and i is 5, so i * 2 is also 10.
x = i	true	Casting allows the comparison, and 5.0 is equal to 5.
k < z	false	Casting allows the comparison, and 15 is greater than 12.3.
i * 2 < y	false	i * 2 is 10, which is not less than 7.5.

WARNING

Note that the relational operator that tests for equality is two equal signs in a row (==). A single equal sign is the assignment operator. When you're first learning Java, you may find yourself typing the assignment operator when you mean the equals operator, like this:

```
if (i = 5)
```

Oops. But Java won't let you get away with this, so you have to correct your mistake and recompile the program. At first, doing so seems like a nuisance. The more you work with Java, the more you come to appreciate that comparison and assignment are two different things, and it's best that a single operator (=) isn't overloaded with both functions.

WARNING

Another important warning: Do *not* test strings by using any of the relational operators listed in Table 4-1, including the equals operator. You're probably going to feel tempted to test strings like this:

```
inputString == "Yes"
```

Note, however, that this is not the correct way to compare strings in Java. You find out the correct way in the section "Comparing Strings," later in this chapter.

Using if Statements

The if statement is one of the most important statements in any programming language, and Java is no exception. The following sections describe the ins and outs of using the various forms of Java's powerful if statement.

Simple if statements

In its most basic form, an if statement lets you execute a single statement or a block of statements only if a boolean expression evaluates to true. The basic form of the if statement looks like this:

```
if (boolean-expression)
    statement
```

Note that the boolean expression must be enclosed in parentheses. Also, if you use only a single statement, it must end with a semicolon. But the statement can also be a statement block enclosed by braces. In that case, each statement within the block needs a semicolon, but the block itself doesn't.

Here's an example of a typical if statement:

```
double commissionRate = 0.0;

if (salesTotal > 10000.0)
    commissionRate = 0.05;
```

In this example, a variable named commissionRate is initialized to 0.0 and then set to 0.05 if salesTotal is greater than 10000.0.

Some programmers find it helpful to visualize the operation of an if statement as a flowchart, as shown in Figure 4-1. In this flowchart, the diamond symbol represents the condition test: If the sales total is greater than $10,000, the statement in the rectangle is executed. If not, that statement is bypassed.

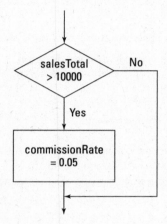

FIGURE 4-1:
The flowchart for
an if statement.

TIP

Indenting the statement under the if statement is customary because it makes the structure of your code more obvious. It isn't necessary, but it's always a good idea.

Here's an example that uses a block rather than a single statement:

```
double commissionRate = 0.0;

if (salesTotal > 10000.0)
{
    commissionRate = 0.05;
    commission = salesTotal * commissionRate;
}
```

In this example, the two statements within the braces are executed if sales Total is greater than $10,000. Otherwise neither statement is executed.

Here are a few additional points about simple `if` statements:

>> Some programmers prefer to code the opening brace for the statement block on the same line as the `if` statement itself, like this:

```
if (salesTotal > 10000.0) {
    commissionRate = 0.05;
    commission = salesTotal * commissionRate;

}
```

This method is simply a matter of style, so either technique is acceptable.

WARNING

>> Indentation by itself doesn't create a block. Consider this code:

```
if (salesTotal > 10000.0)
    commissionRate = 0.05;

commission = salesTotal * commissionRate;
```

Here I don't use the braces to mark a block but indent the last statement as though it were part of the `if` statement. Don't be fooled; the last statement is executed regardless of whether the expression in the `if` statement evaluates to `true`.

TIP

Some programmers like to code a statement block even for `if` statements that conditionally execute just one statement. Here's an example:

```
if (salesTotal > 10000.0)
{
    commissionRate = 0.05;

}
```

That's not a bad idea, because it makes the structure of your code a little more obvious by adding extra white space around the statement. Also, if you decide later that you need to add a few statements to the block, the braces are already there. (It's all too easy to later add extra lines to a conditional and forget to include the braces, which leads to a bug that can be hard to trace.)

>> If only one statement needs to be conditionally executed, some programmers use just one line for the whole thing, like this:

```
if (salesTotal > 10000.0) commissionRate = 0.05;
```

This method works, but I'd avoid it. Your classes are easier to follow if you use line breaks and indentation to highlight their structure.

if-else statements

An if-else statement adds an additional element to a basic if statement: a statement or block that's executed if the boolean expression is not true. Its basic format is

```
if (boolean-expression)
    statement
else
    statement
```

Here's an example:

```
double commissionRate;

if (salesTotal <= 10000.0)
    commissionRate = 0.02;
else
    commissionRate = 0.05;
```

In this example, the commission rate is set to 2 percent if the sales total is less than or equal to $10,000. If the sales total is greater than $10,000, the commission rate is set to 5 percent. Figure 4-2 shows a flowchart for this if-else statement.

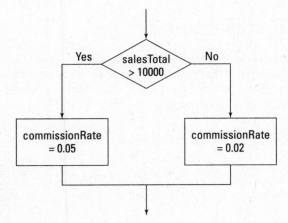

FIGURE 4-2:
The flowchart for an if-else statement.

In some cases, you can avoid using the else part of an if-else statement by cleverly rearranging your code. This code has the same effect as the preceding if-else statement:

```
double commissionRate = 0.05;

if (salesTotal <= 10000.0)
    commissionRate = 0.02;
```

You can use blocks for either or both of the statements in an if-else statement. Here's an if-else statement in which both statements are blocks:

```
double commissionRate;

if (salesTotal <= 10000.0)
{
    commissionRate = 0.02;
    level1Count++;
}
else
{
    commissionRate = 0.05;
    level2Count++;
}
```

Nested if statements

The statement that goes in the if or else part of an if-else statement can be any kind of Java statement, including another if or if-else statement. This arrangement is called *nesting*, and an if or if-else statement that includes another if or if-else statement is called a *nested if statement.*

The general form of a nested if statement is this:

```
if (expression-1)
    if (expression-2)
        statement-1
    else
        statement-2
else
    if (expression-3)
        statement-3
    else
        statement-4
```

In this example, *expression-1* is first to be evaluated. If it evaluates to true, *expression-2* is evaluated. If that expression is true, *statement-1* is executed; otherwise *statement-2* is executed. But if *expression-1* is false, *expression-3* is evaluated. If *expression-3* is true, *statement-3* is executed; otherwise *statement-4* is executed.

An if statement that's contained within another if statement is called an *inner if statement,* and an if statement that contains another if statement is called an *outer if statement.* Thus, in the preceding example, the if statement that tests *expression-1* is an outer if statement, and the if statements that test *expression-2* and *expression-3* are inner if statements.

TIP

Nesting can be as complex as you want, but try to keep it as simple as possible. Also, be sure to use braces and indentation to indicate the structure of the nested statements.

Suppose that your company has two classes of sales representatives (Class 1 and Class 2) and that these reps get different commissions for sales below $10,000 and sales above $10,000, according to this table:

Sales	Class 1	Class 2
$0 to $9,999	2%	2.5%
$10,000 and over	4%	5%

You could implement this commission structure with a nested if statement:

```
if (salesClass == 1)
    if (salesTotal < 10000.0)
        commissionRate = 0.02;
    else
        commissionRate = 0.04;
else
    if (salesTotal < 10000.0)
        commissionRate = 0.025;
    else
        commissionRate = 0.05;
```

This example assumes that if the salesClass variable isn't 1, it must be 2. If that's not the case, you have to use an additional if statement for Class 2 sales reps:

```
if (salesClass == 1)
    if (salesTotal < 10000.0)
        commissionRate = 0.02;
```

```
    else
        commissionRate = 0.04;
else if (salesClass == 2)
    if (salesTotal < 10000.0)
        commissionRate = 0.025;
    else
        commissionRate = 0.05;
```

Notice that I place this extra if statement on the same line as the else keyword. That's a common practice for a special form of nested if statements called *else-if statements*. You find more about this type of nesting in the next section.

You could just use a pair of separate if statements, of course, like this:

```
if (salesClass == 1)
    if (salesTotal < 10000.0)
        commissionRate = 0.02;
    else
        commissionRate = 0.04;
if (salesClass == 2)
    if (salesTotal < 10000.0)
        commissionRate = 0.025;
    else
        commissionRate = 0.05;
```

The result is the same.

Note that you could also have implemented the commission structure by testing the sales total in the outer if statement and the sales representative's class in the inner statements:

```
if (salesTotal < 10000)
    if (salesClass == 1)
        commissionRate = 0.02;
    else
        commissionRate = 0.04;
else
    if (salesClass == 1)
        commissionRate = 0.025;
    else
        commissionRate = 0.05;
```

WARNING

Be careful when you use nested if and else statements, as it is all too easy to end up with statements that don't work the way you expect them to. The key is knowing how Java pairs else keywords with if statements. The rule is actually very simple: Each else keyword is matched with the most previous if statement that hasn't already been paired with an else keyword.

The whole problem of knowing how else keywords are paired to if statements is called the *dangling else problem*. Whenever you use nested if statements with else clauses, you need to make sure you understand which else pairs to which if. Again, the rule is simple: Each else is matched with the most previous unmatched if.

Indentation is your friend here, but you must make sure that your indentation correctly matches the actual structure of your nested if and else statements.

But remember that Java doesn't care about your indentation. You can't coax Java into pairing the if and else keywords differently by using indentation.

Suppose that Class 2 sales reps don't get any commission, so the inner if statements in the preceding example don't need else statements. You may be tempted to calculate the commission rate by using this code:

```
if (salesTotal < 10000)
    if (salesClass == 1)
        commissionRate = 0.02;
else
    if (salesClass == 1)
        commissionRate = 0.025;
```

That won't work. The indentation creates the impression that the else keyword is paired with the first if statement, but in reality, it's paired with the second if statement. As a result, no sales commission rate is set for sales of $10,000 or more.

This problem has two solutions. The first, and preferred, solution is to use braces to clarify the structure:

```
if (salesTotal < 10000)
{
    if (salesClass == 1)
        commissionRate = 0.02;
}
else
{
    if (salesClass == 1)
        commissionRate = 0.025;
}
```

The other solution is to add an else statement that specifies an *empty statement* (a semicolon by itself) to the first inner if statement:

```
if (salesTotal < 10000)
    if (salesClass == 1)
        commissionRate = 0.02;
    else ;
else
    if (salesClass == 1)
        commissionRate = 0.025;
```

The empty else statement is paired with the inner if statement, so the second else keyword is properly paired with the outer if statement.

else-if statements

A common pattern for nested if statements is to have a series of if-else statements with another if-else statement in each else part:

```
if (expression-1)
    statement-1
else if (expression-2)
    statement-2
else if (expression-3)
    statement-3
```

These statements are sometimes called *else-if statements,* although that term is unofficial. Officially, all that's going on is that the statement in the else part happens to be another if statement — so this statement is just a type of a nested if statement. It's an especially useful form of nesting, however.

Suppose that you want to assign four commission rates based on the sales total, according to this table:

Sales	Commission
Over $10,000	5%
$5,000 to $9,999	3.5%
$1,000 to $4,999	2%
Under $1,000	0%

You can easily implement a series of else-if statements:

```
if (salesTotal >= 10000.0)
    commissionRate = 0.05;
else if (salesTotal >= 5000.0)
    commissionRate = 0.035;
else if (salesTotal >= 1000.0)
    commissionRate = 0.02;
else
    commissionRate = 0.0;
```

Figure 4-3 shows a flowchart for this sequence of else-if statements.

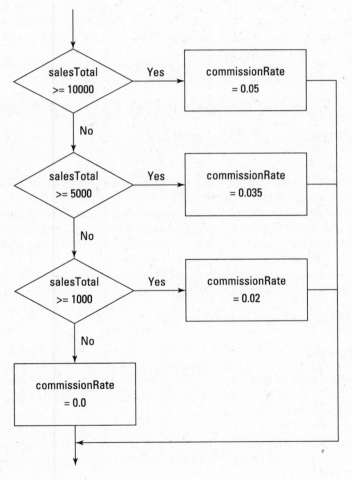

FIGURE 4-3:
The flowchart
for a sequence
of else-if
statements.

WARNING

You have to think through carefully how you set up these else-if statements. At first glance, for example, this sequence looks as though it might work:

```
if (salesTotal > 0.0)
    commissionRate = 0.0;
else if (salesTotal >= 1000.0)
    commissionRate = 0.02;
else if (salesTotal >= 5000.0)
    commissionRate = 0.035;
else if (salesTotal >= 10000.0)
    commissionRate = 0.05;
```

Nice try, but this scenario won't work. These if statements always set the commission rate to 0 percent because the boolean expression in the first if statement always tests true (assuming that the salesTotal isn't zero or negative — and if it is, none of the other if statements matter). As a result, none of the other if statements are ever evaluated.

Using Mr. Spock's Favorite Operators (Logical Ones, of Course)

A *logical operator* (sometimes called a *Boolean operator*) is an operator that returns a boolean result that's based on the boolean result of one or two other expressions. Expressions that use logical operators are sometimes called *compound expressions* because the effect of the logical operators is to let you combine two or more condition tests into a single expression. Table 4-2 lists the logical operators.

The following sections describe these operators in excruciating detail.

Using the ! operator

The simplest of the logical operators is *Not* (!). Technically, it's a *unary* prefix operator, which means that you use it with one operand, and you code it immediately in front of that operand. (Technically, this operator is called the *complement operator*, not the *Not operator*. But in real life, most people call it *Not*. And many programmers call it *bang*.)

The Not operator reverses the value of a boolean expression. Thus, if the expression is true, Not changes it to false. If the expression is false, Not changes it to true.

TABLE 4-2   **Logical Operators**

Operator	Name	Type	Description
!	Not	Unary	Returns `true` if the operand to the right evaluates to `false`. Returns `false` if the operand to the right is `true`.
&	And	Binary	Returns `true` if both of the operands evaluate to `true`. Both operands are evaluated before the And operator is applied.
|	Or	Binary	Returns `true` if at least one of the operands evaluates to `true`. Both operands are evaluated before the Or operator is applied.
^	Xor	Binary	Returns `true` if one — and only one — of the operands evaluates to `true`. Returns `false` if both operands evaluate to `true` or if both operands evaluate to `false`.
&&	Conditional And	Binary	Same as &, but if the operand on the left returns `false`, returns `false` without evaluating the operand on the right.
||	Conditional Or	Binary	Same as |, but if the operand on the left returns `true`, returns `true` without evaluating the operand on the right.

Here's an example:

```
!(i == 4)
```

This expression evaluates to `true` if i is any value other than 4. If i is 4, it evaluates to `false`. It works by first evaluating the expression (i == 4). Then it reverses the result of that evaluation.

TIP

Don't confuse the Not logical operator (!) with the Not Equals relational operator (!=). Although these operators are sometimes used in similar ways, the Not operator is more general. I could have written the preceding example like this:

```
i != 4
```

The result is the same. The Not operator can be applied to any expression that returns a `true`–`false` result, however, not just to an equality test.

REMEMBER

You must almost always enclose the expression that the ! operator is applied to in parentheses. Consider this expression:

```
! i == 4
```

Assuming that i is an integer variable, the compiler doesn't allow this expression because it looks like you're trying to apply the ! operator to the variable, not to the result of the comparison. A quick set of parentheses solves the problem:

```
!(i == 4)
```

Using the & and && operators

The & and && operators combine two boolean expressions and return true only if both expressions are true. This type of operation is called an *And operation*, because the first expression and the second expression must be true for the And operator to return true.

Suppose that the sales commission rate should be 2.5% if the sales class is 1 and the sales total is $10,000 or more. You could perform this test with two separate if statements (as I did earlier in this chapter), or you could combine the tests into one if statement:

```
if ((salesClass == 1) & (salesTotal >= 10000.0))
    commissionRate = 0.025;
```

Here the expressions (salesClass == 1) and (salesTotal >= 10000.0) are evaluated separately. Then the & operator compares the results. If they're both true, the & operator returns true. If one is false or both are false, the & operator returns false.

TIP

Notice that I use parentheses liberally to clarify where one expression ends and another begins. Using parentheses isn't always necessary, but when you use logical operators, I suggest that you always use parentheses to clearly identify the expressions being compared.

The && operator is similar to the & operator, but it leverages your knowledge of logic a bit more. Because both expressions compared by the & operator must be true for the entire expression to be true, there's no reason to evaluate the second expression if the first one returns false. The & operator isn't aware of this fact, so it blindly evaluates both expressions before determining the results. The && operator is smart enough to stop when it knows what the outcome is.

As a result, almost always use && instead of &. Here's the preceding example, and this time it's coded smartly with &&:

```
if ((salesClass == 1) && (salesTotal >= 10000.0))
    commissionRate = 0.025;
```

TIP

Why do I say you should *almost* always use &&? Because sometimes the expressions themselves have side effects that are important. The second expression might involve a method call that updates a database, for example, and you want the database to be updated whether the first expression evaluates to true or to false. In that case, you want to use & instead of && to ensure that both expressions get evaluated.

WARNING

Relying on the side effects of expressions can be risky — and you can almost always find a better way to write your code to avert the side effects. In other words, placing an important call to a database-update method inside a compound expression that's buried in an if statement probably isn't a good idea.

Using the | and || operators

The | and || operators are called *Or operators* because they return true if the first expression is true or if the second expression is true. They also return true if both expressions are true. (You find the | symbol on your keyboard just above the Enter key.)

Suppose that sales representatives get no commission if total sales are less than $1,000 or if the sales class is 3. You could do that with two separate if statements:

```
if (salesTotal < 1000.0)
    commissionRate = 0.0;
if (salesClass == 3)
    commissionRate = 0.0;
```

With an Or operator, however, you can do the same thing with a compound condition:

```
if ((salesTotal < 1000.0) | (salesClass == 3))
commissionRate = 0.0;
```

To evaluate the expression for this if statement, Java first evaluates the expressions on either side of the | operator. Then, if at least one of these expressions is true, the whole expression is true. Otherwise the expression is false.

TIP

In most cases, you should use the Conditional Or operator (||) instead of the regular Or operator (|), like this:

```
if ((salesTotal < 1000.0) || (salesClass == 3))
    commissionRate = 0.0;
```

Making Choices

Like the Conditional And operator (&&), the Conditional Or operator stops evaluating as soon as it knows what the outcome is. Suppose that the sales total is $500. Then there's no need to evaluate the second expression. Because the first expression evaluates to `true` and only one of the expressions needs to be `true`, Java can skip the second expression. If the sales total is $5,000, of course, the second expression must be evaluated.

As with the And operators, you should use the regular Or operator only if your program depends on some side effect of the second expression, such as work done by a method call.

Using the ^ operator

The ^ operator performs what in the world of logic is known as an *Exclusive Or*, commonly abbreviated as *Xor*. It returns `true` if one — and only one — of the two subexpressions is `true`. If both expressions are `true`, or if both expressions are `false`, the ^ operator returns `false`.

Put another way, the ^ operator returns `true` if the two subexpressions have different results. If they have the same result, it returns `false`.

Most programmers don't bother with the ^ operator because it's pretty confusing. My feelings won't be hurt if you skip this section.

Suppose that you're writing software that controls your model railroad set, and you want to find out whether two switches are set in a dangerous position that might allow a collision. If the switches are represented by simple `integer` variables named `switch1` and `switch2`, and 1 means the track is switched to the left and 2 means the track is switched to the right, you could easily test them like this:

```
if ( switch1 == switch2 )
    System.out.println("Trouble! The switches are the same");
else
    System.out.println("OK, the switches are different.");
```

Now, suppose that (for some reason) one of the switches is represented by an `int` variable where 1 means the switch goes to the left and *any* other value means the switch goes to the right — but the other switch is represented by an `int` variable where –1 means the switch goes to the left and any other value means the switch goes to the right. (Who knows — maybe the switches were made by different manufacturers.) You could use a compound condition like this:

```
if (((switch1==1)&&(switch2==-1)) || ((switch1!=1)&&(switch2!=-1)))
    System.out.println("Trouble! The switches are the same");
```

```
else
    System.out.println("OK, the switches are different.");
```

But an XOR operator could do the job with a simpler expression:

```
if ((switch1==1)^(switch2==-1))
    System.out.println("OK, the switches are different.");
else
    System.out.println("Trouble! The switches are the same");
```

Combining logical operators

You can combine simple boolean expressions to create more complicated expressions. For example:

```
if ((salesTotal<1000.0)||((salesTotal<5000.0)&&
        (salesClass==1))||((salestotal < 10000.0)&&
        (salesClass == 2)))
    CommissionRate = 0.0;
```

Can you tell what the expression in this if statement does? It sets the commission to zero if any one of the following three conditions is true:

>> The sales total is less than $1,000.

>> The sales total is less than $5,000, and the sales class is 1.

>> The sales total is less than $10,000, and the sales class is 2.

In many cases, you can clarify how an expression works just by indenting its pieces differently and spacing out its subexpressions. This version of the preceding if statement is a little easier to follow:

```
if (
        (salesTotal < 1000.0)
    || ( (salesTotal < 5000.0) && (salesClass == 1) )
    || ( (salestotal < 10000.0) && (salesClass == 2) )
   )
    commissionRate = 0.0;
```

Figuring out exactly what this if statement does, however, is still tough. In many cases, the better thing to do is skip the complicated expression and code separate if statements:

```
if (salesTotal < 1000.0)
    commissionRate = 0.0;
```

```
if ((salesTotal < 5000.0) && (salesClass == 1))
    commissionRate = 0.0;
if ((salestotal < 10000.0) && (salesClass == 2))
    commissionRate = 0.0;
```

WARNING

In Java, `boolean` expressions can get a little complicated when you use more than one logical operator, especially if you mix And and Or operators. Consider this expression:

```
if ( a==1 && b==2 || c==3 )
    System.out.println("It's true!");
else
    System.out.println("No it isn't!");
```

What do you suppose this `if` statement does if a is 5, b is 7, and c = 3? The answer is that the expression evaluates to `true`, and `"It's true!"` is printed. That's because Java applies the operators from left to right. So the && operator is applied to a==1 (which is `false`) and b==2 (which is also `false`, but that doesn't matter because this evaluation is skipped). Thus, the && operator returns `false`. Then the || operator is applied to that `false` result and the result of c==3, which is `true`. Thus the entire expression returns `true`.

TIP

Wouldn't this expression have been clearer if you had used a set of parentheses to clarify what the expression does? Consider this example:

```
if ( ( a==1 && b==2 ) || c==3 )
    System.out.println("It's true!");
else
    System.out.println("No it isn't!");
```

Here you can clearly see that the && operator is evaluated first.

Using the Conditional Operator

Java has a special operator called the *conditional operator* that's designed to eliminate the need for `if` statements in certain situations. It's a *ternary operator*, which means that it works with three operands. The general form for using the conditional operator is this:

```
boolean-expression ? expression-1 : expression-2
```

The `boolean` expression is evaluated first. If it evaluates to `true`, *expression-1* is evaluated, and the result of this expression becomes the result of the whole expression. If the expression is `false`, *expression-2* is evaluated, and its results are used instead.

Suppose that you want to assign a value of 0 to an integer variable named `sales-Tier` if total sales are less than $10,000 and a value of 1 if the sales are $10,000 or more. You could do that with this statement:

```
int tier = salesTotal > 10000.0 ? 1 : 0;
```

Although not required, a set of parentheses helps make this statement easier to follow:

```
int tier = (salesTotal > 10000.0) ? 1 : 0;
```

TIP

One common use for the conditional operator is when you're using concatenation to build a text string, and you have a word that may need to be plural based on the value of an integer variable. Suppose that you want to create a string that says `"You have x apples"`, with the value of a variable named `appleCount` substituted for x. But if `apples` is 1, the string should be `"You have 1 apple"`, not `"You have 1 apples"`.

The following statement does the trick:

```
String msg = "You have " + appleCount + " apple"
    + ((appleCount>1) ? "s." : ".");
```

When Java encounters the ? operator, it evaluates the expression (`appleCount>1`). If `true`, it uses the first string (`s.`). If `false`, it uses the second string (`"."`).

Comparing Strings

Comparing strings in Java takes a little extra care, because the == operator really doesn't work the way it should. Suppose that you want to know whether a String variable named `answer` contains the value `"Yes"`. You may be tempted to code an `if` statement like this:

```
if (answer == "Yes")
    System.out.println("The answer is Yes.");
```

Unfortunately, that's not correct. The problem is that in Java, strings are reference types, not primitive types; when you use the == operator with reference types, Java compares the references to the objects, not the objects themselves. As a result, the expression answer == "Yes" doesn't test whether the value of the string referenced by the answer variable is "Yes". Instead, it tests whether the answer string and the literal string "Yes" point to the same string object in memory. In many cases, they do — but sometimes they don't, and the results are difficult to predict.

The correct way to test a string for a given value is to use the equals method of the String class:

```
if (answer.equals("Yes"))
    System.out.println("The answer is Yes.");
```

This method actually compares the value of the string object referenced by the variable with the string you pass as a parameter and returns a Boolean result to indicate whether the strings have the same value.

The String class has another method, equalsIgnoreCase, that's also useful for comparing strings. It compares strings but ignores case, which is especially useful when you're testing string values entered by users. Suppose that you're writing a program that ends only when the user enters the word end. You could use the equals method to test the string:

```
if (input.equals("end"))
    // end the program
```

In this case, however, the user would have to enter end exactly. If the user enters End or END, the program won't end. It's better to code the if statement like this:

```
if (input.equalsIgnoreCase("end"))
    // end the program
```

Then the user could end the program by entering the word *end* spelled with any variation of upper- and lowercase letters, including end, End, END, or even eNd.

You can find much more about working with strings in Book 4, Chapter 1. For now, just remember that to test for string equality in an if statement (or in one of the other control statements presented in the next chapter), you must use the equals or equalsIgnoreCase method instead of the == operator.

Chapter **5**

Going Around in Circles (Or, Using Loops)

So far, all the programs in this book have started, run quickly through their main method, and then ended. If Dorothy from *The Wizard of Oz* were using these programs, she'd probably say, "My, programs come and go quickly around here!"

In this chapter, you find out how to write programs that don't come and go so quickly. They hang around by using *loops*, which let them execute the same statements more than once.

Loops are the key to writing one of the most common types of programs: programs that get input from the user, do something with it, get more input from the user and do something with that, and keep going this way until the user has had enough.

Put another way, loops are like the instructions on your shampoo: Lather. Rinse. *Repeat.*

TIP

Like if statements, loops rely on conditional expressions to tell them when to stop looping. Without conditional expressions, loops would go on forever, and your users would grow old watching them run. So if you haven't yet read Book 2, Chapter 4, I suggest that you do so before continuing much further.

Using Your Basic while Loop

The most basic of all looping statements in Java is while. The while statement creates a type of loop that's called a *while loop,* which is simply a loop that executes continuously as long as some conditional expression evaluates to true. while loops are useful in all sorts of programming situations, so you use while loops a lot. (I tell you about other kinds of loops later in this chapter.)

The while statement

The basic format of the while statement is this:

```
while (expression)
    statement
```

The while statement begins by evaluating the expression. If the expression is true, *statement* is executed. Then the expression is evaluated again, and the whole process repeats. If the expression is false, *statement* is not executed, and the while loop ends.

Note that the statement part of the while loop can either be a single statement or a block of statements contained in a pair of braces. Loops that have just one statement aren't very useful, so nearly all the while loops you code use a block of statements. (Well, okay, sometimes loops with a single statement are useful. It isn't unheard of — just not all that common.)

A counting loop

Here's a simple program that uses a while loop to print the even numbers from 2 through 20 on the console:

```
public class EvenCounter
{
    public static void main(String[] args)
    {
        int number = 2;

        while (number <= 20)
        {
            System.out.print(number + " ");
            number += 2;
        }
```

```
            System.out.println();
        }
    }
```

If you run this program, the following output is displayed in the console window:

```
2 4 6 8 10 12 14 16 18 20
```

The conditional expression in this program's `while` statement is `number <= 20`. That means the loop repeats as long as the value of `number` is less than or equal to 20. The body of the loop consists of two statements. The first prints the value of `number` followed by a space to separate this number from the next one. Then the second statement adds 2 to `number`.

Figure 5-1 shows a flowchart for this program. This flowchart can help you visualize the basic decision-making process of a loop.

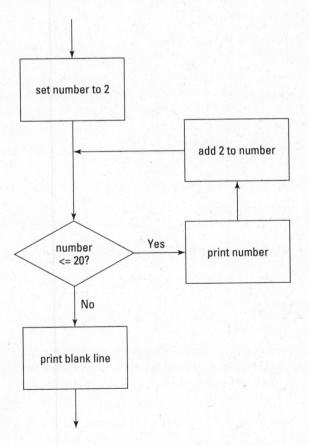

FIGURE 5-1:
The flowchart for
a while loop.

Going Around in Circles (Or, Using Loops)

CHAPTER 5 **Going Around in Circles (Or, Using Loops)** 153

Breaking Out of a Loop

In many programs, you need to set up a loop that has some kind of escape clause. Java's escape clause is the break statement. When a break statement is executed in a while loop, the loop ends immediately. Any remaining statements in the loop are ignored, and the next statement executed is the statement that follows the loop.

Suppose that you're afraid of the number 12. (I'm not a doctor, and I don't play one on TV, but I think the scientific name for this condition would be *dodecaphobia*.) You could modify the counting program shown in the preceding section so that when it gets to the number 12, it panics and aborts the loop:

```java
public class Dodecaphobia
{
    public static void main(String[] args)
    {
        int number = 2;

        while (number <= 20)
        {
            if (number == 12)
                break;
            System.out.print(number + " ");
            number += 2;
        }
        System.out.println();
    }
}
```

When you run this program, the following line is displayed on the console:

```
2 4 6 8 10
```

Whew! That was close. Almost printed the number 12 there.

Looping Forever

One common form of loop is called an *infinite loop*. That's a loop that goes on forever. You can create infinite loops many ways in Java (not all of them intentional), but the easiest is to just specify true for the while expression.

Here's an example:

```
public class CountForever
{
    public static void main(String[] args)
    {
        int number = 2;

        while (true)
        {
            System.out.print(number + " ");
            number += 2;
        }
    }
}
```

If you run this program, your console window quickly fills up with numbers and just keeps going. That's great if you *really like* even numbers, but eventually you'll tire of this loop and want it to stop. You can stop an infinite loop in any of three ways:

» Turn off your computer.

» Hit your computer with an ax or other heavy object.

» Close the console window.

The last one is probably the one you want to go with.

TIP

Obviously, infinite loops are something you want to avoid in your programs. So whenever you use a while expression that's always true, be sure to throw in a break statement to give your loop some way to terminate. You could use an infinite loop with a break statement in the Dodecaphobia program:

```
public class Dodecaphobia2
{
    public static void main(String[] args)
    {
        int number = 2;

        while (true)
        {
            if (number == 12)
                break;
            System.out.print(number + " ");
            number += 2;
        }
```

```
            System.out.println();
        }
    }
```

Here the loop looks as though it might go on forever, but the `break` statement panics out of the loop when it hits 12.

Letting the user decide when to quit

It turns out that infinite loops are also useful when you want to let the user be in charge of when to stop the loop. Suppose that you don't know what numbers a user is afraid of, so you want to count numbers until the user says to stop. Here's a program that does that:

```java
import java.util.Scanner;

public class NumberPhobia
{
    static Scanner sc = new Scanner(System.in);

    public static void main(String[] args)
    {
        int number = 2;
        String input;

        while (true)
        {
            System.out.println(number + " ");
            System.out.print
                ("Do you want to keep counting?"
                + " (Y or N)");
            input = sc.next();
            if (input.equalsIgnoreCase("N"))
                break;
            number += 2;
        }
        System.out.println("\nWhew! That was close.\n");
    }
}
```

Here's some typical console output from this program, for a user who has octophobia:

```
2
Do you want to keep counting? (Y or N)y
4
```

```
Do you want to keep counting? (Y or N)y
6
Do you want to keep counting? (Y or N)n
Whew! That was close.
```

Letting the user decide in another way

Another way to write a loop that a user can opt out of is to test the input string in the while condition. The only trick here is that you must first initialize the input string to the value that continues the loop. Otherwise, the loop doesn't execute at all!

Here's a variation of the NumberPhobia program named NumberPhobia2 that uses this technique:

```java
import java.util.Scanner;

public class NumberPhobia2
{
    static Scanner sc = new Scanner(System.in);

    public static void main(String[] args)
    {
        int number = 2;
        String input = "Y";

        while (input.equalsIgnoreCase("Y"))
        {
            System.out.println(number + " ");
            System.out.print
                ("Do you want to keep counting?"
                + " (Y or N)");
            input = sc.next();
            number += 2;
        }
        System.out.println("\nWhew! That was close.");
    }
}
```

This program works almost the same way as the preceding version, but with a subtle difference. In the previous version, if the user says N after the program displays 6, the value of the number variable after the loop is 6 because the break statement bails out of the loop before adding 2 to number. But in this version, the value of number is 8.

Using the continue Statement

The break statement is rather harsh: It completely bails out of the loop. Sometimes that's what you need — but just as often, you don't really need to quit the loop; you just need to skip a particular iteration of the loop. The Dodecaphobia program presented earlier in this chapter stops the loop when it gets to 12. What if you just want to skip the number 12, so you go straight from 10 to 14?

To do that, you can use the break statement's kinder, gentler relative, the continue statement. The continue statement sends control right back to the top of the loop, where the expression is immediately evaluated again. If the expression is still true, the loop's statement or block is executed again.

Here's a version of the Dodecaphobia program that uses a continue statement to skip the number 12 rather than stop counting altogether when it reaches 12:

```java
public class Dodecaphobia3
{
    public static void main(String[] args)
    {
        int number = 0;

        while (number < 20)
        {
            number += 2;
            if (number == 12)
                continue;
            System.out.print(number + " ");
        }
        System.out.println();
    }
}
```

Run this program, and you get the following output in the console window:

```
2 4 6 8 10 14 16 18 20
```

Notice that I had to make several changes in this program to get it to work with a continue statement instead of a break statement. If I had just replaced the word break with continue, the program wouldn't have worked, because the statement that added 2 to the number came after the break statement in the original version. As a result, if you just replace the break statement with a continue statement, you end up with an infinite loop when you reach 12, because the statement that adds 2 to number never gets executed.

To make this program work with a `continue` statement, I rearranged the statements in the loop body so that the statement that adds 2 to number comes before the `continue` statement. That way, the only statement skipped by the `continue` statement is the one that prints number to the console.

Unfortunately, this change affected other statements in the program. Because 2 is added to number before number is printed, I had to change the initial value of number from 2 to 0, and I had to change the `while` expression from number `<= 20` to number `< 20`.

Running do-while Loops

A *do-while loop* (sometimes just called a *do loop*) is similar to a `while` loop, but with a critical difference: In a `do-while` loop, the condition that stops the loop isn't tested until after the statements in the loop have executed at least once. The basic form of a `do-while` loop is this:

```
do
    statement
while (expression);
```

Note that the `while` keyword and the expression aren't coded until *after* the body of the loop. As with a `while` loop, the body for a `do-while` loop can be a single statement or a block of statements enclosed in braces.

Also, notice that the expression is followed by a semicolon. `do-while` is the only looping statement that ends with a semicolon.

Here's a version of the `EvenCounter` program that uses a `do-while` loop instead of a `while` loop:

```
public class EvenCounter2
{
    public static void main(String[] args)
    {
        int number = 2;

        do
        {
            System.out.print(number + " ");
            number += 2;
```

Going Around in Circles
(Or, Using Loops)

```
        } while (number <= 20);
        System.out.println();
    }
}
```

REMEMBER

Here's the most important thing to remember about do-while loops: The statement or statements in the body of a do-while loop *always* get executed at least once. By contrast, the statement or statements in the body of a while loop aren't executed at all if the while expression is false the first time it's evaluated.

Look at the flowchart in Figure 5-2 to see what I mean. You can see that execution starts at the top of the loop and flows through to the decision test after the loop's body has been executed once. Then, if the decision test is true, control flies back up to the top of the loop. Otherwise, it spills out the bottom of the flowchart.

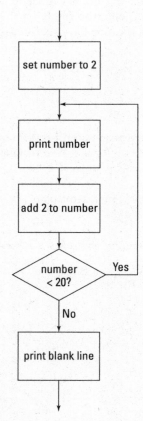

FIGURE 5-2:
The flowchart for
a do-while loop.

Here are a few other things to be aware of concerning do-while loops:

» You often can skip initializing the variables that appear in the expression before the loop, because the expression isn't evaluated until the statements in the loop body have been executed at least once. But remember that any variables mentiond in the while expression must be within scope of the do statement itself; variables declared *within* the do loop can't be used in the while expression because they're out of scope.

» You can use break and continue statements in a do-while loop, just as you can in a while loop.

» Some programmers like to place the brace that begins the loop body on the same line as the do statement and the while statement that ends the do-while loop on the same line as the brace that marks the end of the loop body. Whatever makes you happy is fine with me. Just remember that the compiler is agnostic when it comes to matters of indentation and spacing.

Validating Input from the User

do-while loops are especially useful for validating input by the user. Suppose you're writing a program that plays a betting game, and you want to get the amount of the user's bet from the console. The user can bet any dollar amount he wants (whole dollars only, though) but can't bet more than he has in the bank, and he can't bet a negative amount or zero. Here's a program that uses a do-while loop to get this input from the user:

```java
import java.util.Scanner;

public class GetABet
{
    static Scanner sc = new Scanner(System.in);

    public static void main(String[] args)
    {
        int bank = 1000;          // assume the user has $1,000
        int bet;                  // the bet entered by the user

        System.out.println("You can bet between 1 and " + bank);
        do
        {
            System.out.print("Enter your bet: ");
            bet = sc.nextInt();
```

```
        } while ( (bet <= 0) || (bet > bank) );
        System.out.println("Your money's good here.");
    }
}
```

Here the expression used by the `do-while` loop validates the data entered by the user, which means that it checks the data against some set of criteria to make sure the data is acceptable.

REMEMBER

The || operator performs an Or test. It returns `true` if at least one of the expressions on either side of the operator is true. So if the bet is less than or equal to zero (`bet <= 0`), or if the bet is greater than the money in the bank (`bet > bank`), this expression returns `true`.

WARNING

This type of validation testing checks only whether the user entered a valid number in an acceptable range. If the user entered something that isn't a valid number, such as the word `Buttercup` or `Humperdinck`, the program chokes badly and spews forth a bunch of vile exception messages upon the console. You find out how to clean up that mess in Book 2, Chapter 8. (Actually, you can avoid this problem by using either a `do` loop or a `while` loop and the `hasNextDouble` method of the `Scanner` class, which I describe in Book 2, Chapter 2.)

If you want to display an error message when the user enters incorrect input, you have to use an `if` statement inside the loop, and this `if` statement must duplicate the expression that validates the input data. Thus the expression that does the validation has to appear twice. For example:

```
import java.util.Scanner;

public class GetABet2
{
    static Scanner sc = new Scanner(System.in);

    public static void main(String[] args)
    {
        int bank = 1000;           // assume the user has $1,000
        int bet;                   // the bet entered by the user

        System.out.println ("You can bet between 1 and " + bank);
        do
        {
            System.out.print("Enter your bet: ");
            bet = sc.nextInt();
            if ( (bet <= 0) || (bet > bank) )
                System.out.println ("What, are you crazy?");
        } while ( (bet <= 0) || (bet > bank) );
```

```
            System.out.println("Your money's good here.");
        }
    }
```

Here, the `if` statement displays the message "What, are you crazy?" if the user tries to enter an inappropriate bet.

TIP

You can avoid duplicating the expression that does the data validation by adding a `boolean` variable that's set in the body of the `do-while` loop if the data is invalid, as in this example:

```
import java.util.Scanner;

public class GetABet3
{
    static Scanner sc = new Scanner(System.in);

    public static void main(String[] args)
    {
        int bank = 1000;            // assume the user has $1,000
        int bet;                    // the bet entered by the user
        boolean validBet;           // indicates if bet is valid

        System.out.println("You can bet between 1 and " + bank);
        do
        {
            System.out.print("Enter your bet: ");
            bet = sc.nextInt();
            validBet = true;
            if ( (bet <= 0) || (bet > bank) )
            {
                validBet = false;
                System.out.println("What, are you crazy?");
            }
        } while (!validBet);
        System.out.println("Your money's good here.");
    }
}
```

In this example, I use a `boolean` variable named `validBet` to indicate whether the user has entered a valid bet. After the user enters a bet, this variable is set to `true` before the `if` statement tests the validation criteria. Then, if the `if` statement finds that the bet is not valid, `validBet` is set to `false`.

Using the Famous for Loop

In addition to `while` and `do-while` loops, Java offers the *for loop*. You may have noticed that many of the loops presented so far in this minibook involve counting. It turns out that counting loops are quite common in computer programs, so the people who design computer programming languages (they're called computer programming language designers) long ago concocted a special kind of looping mechanism that's designed just for counting.

The basic principle behind a typical `for` loop is that the loop itself maintains a *counter variable* — that is, a variable whose value increases each time the body of the loop is executed. If you want a loop that counts from 1 to 10, you'd use a counter variable that starts with a value of 1 and is increased by 1 each time through the loop. Then you'd use a test to end the loop when the counter variable reaches 10. The `for` loop lets you set all this up in one convenient statement.

Understanding the formal format of the for loop

I would now like to inform you of the formal format of the `for` loop, so that you'll know how to form it from now on. The `for` loop follows this basic format:

```
for (initialization-expression; test-expression; count-expression)
    statement;
```

The three expressions in the parentheses following the keyword `for` control how the `for` loop works. The following paragraphs explain what these three expressions do:

>> The *initialization expression* is executed before the loop begins. Usually, you use this expression to initialize the counter variable. If you haven't declared the counter variable before the `for` statement, you can declare it here too.

>> The *test expression* is evaluated each time the loop is executed to determine whether the loop should keep looping. Usually, this expression tests the counter variable to make sure that it is still less than or equal to the value you want to count to. The loop keeps executing as long as this expression evaluates to `true`. When the test expression evaluates to `false`, the loop ends.

>> The *count expression* is evaluated each time the loop executes. Its job is usually to increment the counter variable.

Figure 5-3 shows a flowchart to help you visualize how a `for` loop works.

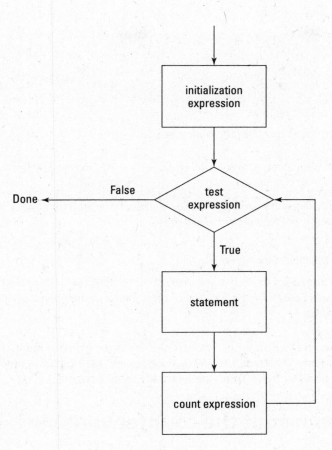

FIGURE 5-3:
The flowchart for
a for loop.

Here's a simple `for` loop that displays the numbers 1 to 10 on the console:

```java
public class CountToTen
{
    public static void main(String[] args)
    {
        for (int i = 1; i <= 10; i++)
            System.out.println(i);
    }
}
```

Run this program, and here's what you see on the console:

```
1
2
3
4
5
```

```
6
7
8
9
10
```

This for loop has the following pieces:

>> The initialization expression is int i = 1. This expression declares a variable named i of type int and assigns it an initial value of 1.

>> The test expression is i <= 10. As a result, the loop continues to execute as long as i is less than or equal to 10.

>> The count expression is i++. As a result, each time the loop executes, the variable i is incremented.

>> The body of the loop is the single statement System.out.println(i). As a result, each time the loop executes, the value of the i variable is printed to the console.

TECHNICAL STUFF

I made up the terms I use to describe the three expressions in a for loop. Officially, Java calls them the *ForInit expression*, the *expression*, and the *ForUpdate expression*. Don't you think my terms are more descriptive?

Scoping out the counter variable

If you declare the counter variable in the initialization statement, the scope of the counter variable is limited to the for statement itself. Thus, you can use the variable in the other expressions that appear within the parentheses and in the body of the loop, but you can't use it outside the loop. This example code causes a compiler error:

```
public class CountToTenError
{
    public static void main(String[] args)
    {
        for (int i = 1; i <=10; i++)
            System.out.println(i);
        System.out.println("The final value of i is " + i);
    }
}
```

That's because the last statement in the main method refers to the variable i, which has gone out of scope because it was declared within the for loop.

If you want to access the counter variable outside of the loop, you should declare the counter variable prior to the for statement, as in this example:

```
public class CountToTenErrorFixed
{
    public static void main(String[] args)
    {
        int i;
        for (i = 1; i <=10; i++)
            System.out.println(i);
        System.out.println("The final value of i is " + i);
    }
}
```

Note that because the i variable is declared before the for statement, the initialization expression doesn't name the variable's data type. When you run this program, the following appears in the console window:

```
1
2
3
4
5
6
7
8
9
10
The final value of i is 11
```

Counting even numbers

Earlier in this chapter, you saw a program that counts even numbers up to 20. You can do that with a for loop too. All you have to do is adjust the count expression. Here's a version of the CountEven program that uses a for loop:

```
public class ForEvenCounter
{
    public static void main(String[] args)
    {
        for (int number = 2; number <= 20; number += 2)
            System.out.print(number + " ");
        System.out.println();
    }
}
```

Run this program, and sure enough, the console window displays the following:

```
2 4 6 8 10 12 14 16 18 20
```

Counting backward

No rule says for loops can only count forward. To count backward, you simply have to adjust the three for loop expressions. As usual, the initialization expression specifies the starting value for the counter variable. The test expression uses a greater-than test instead of a less-than test, and the count expression subtracts from the counter variable rather than adding to it.

For example:

```
public class CountDown
{
    public static void main(String[] args)
    {
        for (int count = 10; count >= 1; count--)
            System.out.println(count);
    }
}
```

Run this program, and you see this result in the console window:

```
10
9
8
7
6
5
4
3
2
1
```

TIP

If you grew up in the 1960s watching NASA launches religiously, as I did, you'll appreciate this variation of the countdown program:

```
public class LaunchControl
{
    public static void main(String[] args)
    {
        System.out.print ("We are go for launch in T minus ");
        for (int count = 10; count >= 0; count--)
```

```
        {
            if (count == 8)
                System.out.println("Ignition sequence start!");
            else
                System.out.println(count + "...");
        }
        System.out.println("All engines running!");
        System.out.println("Liftoff! We have a liftoff!");
    }
}
```

When you run it, here's the output that's displayed:

```
We are go for launch in T minus 10...
9...
Ignition sequence start!
7...
6...
5...
4...
3...
2...
1...
0...
All engines running!
Liftoff! We have a liftoff!
```

Can't you just hear the voice of Paul Haney, the famous "Voice of Mission Control" for NASA in the 1960s? If you can't, you're not nearly as nerdy (or as old) as I am.

Using for loops without bodies

Some programmers get a kick out of writing code that is as terse as possible. I think *Seinfeld* did an episode about that. Jerry had a girlfriend who was a "terse-coder," and he had to dump her because he couldn't understand her code.

Anyway, terse-coders sometimes like to play with for statements in an effort to do away with the body of a for loop altogether. To do that, they take advantage of the fact that you can code any expression you want in the count expression part of a for statement, including method calls. Here's a program that prints the numbers 1 to 10 on the console, using a for statement that has no body:

```
public class TerseCoder
{
    public static void main(String[] args)
```

```
    {
        for (int i = 1; i <=10; System.out.println(i++));
    }
}
```

Here the count expression is a call to `System.out.println`. The parameter to the `println` method cleverly uses the increment operator, so the variable is both printed and incremented in the same expression.

Stay away from terse-coders! Seinfeld was right to dump her.

Ganging up your expressions

An obscure aspect of `for` loops is the fact that the initialization and count expressions can actually be a list of expressions separated by commas. This can be useful if you need to keep track of two counter variables at the same time. Here's a program that counts from 1 to 10 and 10 to 1 at the same time, using two counter variables:

```
public class CountBothWays
{
    public static void main(String[] args)
    {
        int a, b;
        for (a = 1, b = 10; a <= 10; a++, b--)
            System.out.println(a + " " + b);
    }
}
```

If you run this program, here's what you see in the console window:

```
1 10
2 9
3 8
4 7
5 6
6 5
7 4
8 3
9 2
10 1
```

Keep in mind these rules when you use more than one expression for the initialization and counter expressions:

>> In the initialization expression, you can't declare variables if you use more than one expression. That's why I declared the a and b variables before the for statement in the CountBothWays example.

>> The expressions in an expression list can be assignment statements, increment or decrement statements (such as a++), method calls, or object creation statements that use the new keyword to create an object from a class. Other types of statements, such as if statements or loops, are not allowed.

>> You can't list more than one expression in the test expression. You can use compound conditions created with boolean operators, however, so you don't need to use an expression list. Alternatively, you could craft a method that returns a boolean value and call that method as your test expression.

Here, just to prove that I can do it, is a version of the LaunchController program that uses a bodiless for loop:

```
public class ExpressionGanging
{
    public static void main(String[] args)
    {
        System.out.print ("We are go for launch in T minus ");
        for (int count = 10; count >= 0;
                System.out.println((count == 8) ?
                    "Ignition sequence start!" :
                count + "..."),
                count-- );
        System.out.println("All engines running!");
        System.out.println("Liftoff! We have a liftoff!");
    }
}
```

This program actually looks more complicated than it is. The count expression is a list of two expressions. First is a call to System.out.println that uses the ternary ?: operator to determine what to print. The ?: operator first evaluates the count variable to see if it equals 8. If so, the string "Ignition sequence start!" is sent to the println method. Otherwise, count + "..." is sent. The second expression simply increments the count variable.

I think you'll agree that coding the for statement like this example is way out of line. It's better to keep the expressions simple and do the real work in the loop's body.

Going Around in Circles
(Or, Using Loops)

Omitting expressions

Yet another oddity about for loops is that all three of the expressions are optional. If you omit one or more of the expressions, you just code the semicolon as a place-holder so that the compiler knows what's going on.

Omitting the test expression or the iteration expression is not common, but omitting the initialization expression is common. The variable you're incrementing in the for loop may already be declared and initialized before you get to the loop, for example. In that case, you can omit the initialization expression, like this:

```
Scanner sc = new Scanner(System.in);
System.out.print("Where should I start? ");
int a = sc.nextInt();
for ( ; a >= 0; a--)
    System.out.println(a);
```

This for loop simply counts down from whatever number the user enters to zero.

WARNING

If you omit the test expression, you'd better throw a break statement in the loop somewhere (as described earlier in the chapter in the "Breaking Out of a Loop" section). Otherwise you'll find yourself in an infinite loop.

You can omit all three of the expressions if you want to, as in this example:

```
for(;;)
    System.out.println("Oops");
```

This program also results in an infinite loop. There's little reason to do this, how-ever, because while(true) has the same effect and is more obvious.

Breaking and continuing your for loops

You can use a break in a for loop just as you can in a while or do-while loop. Here I revisit the Dodecaphobia program from earlier in the chapter, this time with a for loop:

```
public class ForDodecaphobia
{
    public static void main(String[] args)
    {
        for (int number = 2; number <=20; number += 2)
        {
            if (number == 12)
                break;
```

```
            System.out.print(number + " ");
        }
        System.out.println();
    }
}
```

As before, this version counts by 2 until it gets to 20. When it hits 12, however, it panics and aborts the loop, so it never actually gets to 14, 16, 18, or 20. The console output looks like this:

```
2 4 6 8 10
```

And here's a version that uses a `continue` statement to skip 12 rather than abort the loop:

```
public class ForDodecaphobia2
{
    public static void main(String[] args)
    {
        for (int number = 2; number <=20; number += 2)
        {
            if (number == 12)
                continue;
            System.out.print(number + " ");
        }
        System.out.println();
    }
}
```

The console output from this version looks like this:

```
2 4 6 8 10 14 16 18 20
```

Nesting Your Loops

Loops can contain loops. The technical term for this is *loop-de-loop*. Just kidding. Actually, the technical term is *nested loop*, which is simply a loop that is completely contained inside another loop. The loop that's inside is called the *inner loop*, and the loop that's outside is called the *outer loop*.

A simple nested for loop

To demonstrate the basics of nesting, here's a simple little program that uses a pair of nested for loops:

```
public class NestedLoop
{
    public static void main(String[] args)
    {
        for(int x = 1; x < 10; x++)
        {
            for (int y = 1; y < 10; y++)
                System.out.print(x + "-" + y + " ");
            System.out.println();
        }
    }
}
```

This program consists of two for loops. The outer loop uses x as its counter variable, and the inner loop uses y. For each execution of the outer loop, the inner loop executes 10 times and prints a line that shows the value of x and y for each pass through the inner loop. When the inner loop finishes, a call to System.out.println with no parameters forces a line break, thus starting a new line. Then the outer loop cycles so that the next line is printed.

When you run this program, the console displays this text:

```
1-1 1-2 1-3 1-4 1-5 1-6 1-7 1-8 1-9
2-1 2-2 2-3 2-4 2-5 2-6 2-7 2-8 2-9
3-1 3-2 3-3 3-4 3-5 3-6 3-7 3-8 3-9
4-1 4-2 4-3 4-4 4-5 4-6 4-7 4-8 4-9
5-1 5-2 5-3 5-4 5-5 5-6 5-7 5-8 5-9
6-1 6-2 6-3 6-4 6-5 6-6 6-7 6-8 6-9
7-1 7-2 7-3 7-4 7-5 7-6 7-7 7-8 7-9
8-1 8-2 8-3 8-4 8-5 8-6 8-7 8-8 8-9
9-1 9-2 9-3 9-4 9-5 9-6 9-7 9-8 9-9
```

A guessing game

Listing 5-1 shows a more complicated but realistic example of nesting. This program implements a simple guessing game in which the computer picks a number between 1 and 10, and you have to guess the number. After you guess, the computer tells you whether you're right or wrong and then asks whether you want to play again. If you enter Y or y, the game starts over.

The nesting comes into play because the entire game is written in a `while` loop that repeats as long as you say you want to play another game. Then — within that loop — each time the game asks for input from the user, it uses a `do-while` loop to validate the user's entry. Thus, when the game asks the user to guess a number between 1 and 10, it keeps looping until the number entered by the user is in that range. And when the game asks the user whether he or she wants to play again, it loops until the user enters Y, y, N, or n.

Here's a sample of the console output displayed by this program:

```
Let's play a guessing game!

I'm thinking of a number between 1 and 10.
What do you think it is? 5
You're wrong! The number was 8

Play again? (Y or N)y

I'm thinking of a number between 1 and 10.
What do you think it is? 32
I said, between 1 and 10. Try again: 5
You're wrong! The number was 6

Play again? (Y or N)maybe

Play again? (Y or N)ok

Play again? (Y or N)y
I'm thinking of a number between 1 and 10.
What do you think it is? 5
You're right!
Play again? (Y or N)n

Thank you for playing.
```

LISTING 5-1: **The Guessing Game**

```java
import java.util.Scanner;
public class GuessingGame
{
    static Scanner sc = new Scanner(System.in);
    public static void main(String[] args)
    {
        boolean keepPlaying = true;                              →7
        System.out.println("Let's play a guessing game!");
        while (keepPlaying)                                      →9
        {
```

(continued)

LISTING 5-1: *(continued)*

```java
    boolean validInput;                                        →11
    int number, guess;
    String answer;

    // Pick a random number
    number = (int)(Math.random() * 10) + 1;                    →16

    // Get the guess
    System.out.println("\nI'm thinking of a number "
        + "between 1 and 10.");
    System.out.print("What do you think it is? ");
    do                                                         →22
    {
        guess = sc.nextInt();
        validInput = true;
        if ( (guess < 1) || (guess > 10) )
        {
            System.out.print
                ("I said, between 1 and 10. "
                + "Try again: ");
            validInput = false;
        }
    } while (!validInput);                                     →33

    // Check the guess
    if (guess == number)                                       →36
        System.out.println("You're right!");
    else
        System.out.println("You're wrong! " +
            "The number was " + number);

    // Play again?
    do                                                         →43
    {
        System.out.print("\nPlay again? (Y or N)");
        answer = sc.next();
        validInput = true;
        if (answer.equalsIgnoreCase("Y"));
        else if (answer.equalsIgnoreCase("N"))
            keepPlaying = false;
        else
            validInput = false;
    } while (!validInput);                                     →53
    }                                                          →54
    System.out.println("\nThank you for playing!");            →55
    }
}
```

The following paragraphs describe some of the key lines in this program:

» →7 Defines a `boolean` variable named `keepPlaying` that's initialized to `true` and changed to `false` when the user indicates that he or she has had enough of this silly game.

» →9–54 Begins the main `while` loop for the game. The loop continues as long as `keepPlaying` is `true`. This loop ends on line 54.

» →11 Defines a `boolean` variable named `validInput` that's used to indicate whether the user's input is valid. The same variable is used for both the entry of the user's guess and the Y or N string at the end of each round.

» →16 Picks a random number between 1 and 10. For more information on random numbers, refer to Book 2, Chapter 3.

» →22–33 Begins the `do-while` loop that gets a valid guess from the user. This loop ends on line 33. The statements in this loop read the user's guess from the console and then test to make sure it is between 1 and 10. If so, `validInput` is set to `true`. Otherwise, `validInput` is set to `false`, an error message is displayed, and the loop repeats so that the user is forced to guess again. The loop continues as long as `validInput` is `false`.

» →36 The `if` statement compares the user's guess with the computer's number. A message is displayed to indicate whether the user guessed right or wrong.

» →43–53 Begins the `do-while` loop that asks whether the user wants to play again. This loop ends on line 53. The statements in this loop read a string from the user. If the user enters Y or y, `validInput` is set to `true`. (`keepPlaying` is already `true`, so it is left alone.) If the user enters N or n, `validInput` is set to `true`, and `keepPlaying` is set to `false`. And if the user enters anything else, `validInput` is set to `false`. The loop continues as long as `validInput` is `false`.

» →55 This statement is executed after the program's main `while` loop finishes; it thanks the user for playing the game.

Chapter **6**

Pulling a Switcheroo

I n Book 2, Chapter 4, you find out about the workhorses of Java decision-making: `boolean` expressions and the mighty `if` statement. In this chapter, you discover another Java tool for decision-making: the `switch` statement. The `switch` statement is a pretty limited beast, but it excels at making one particular type of decision: choosing one of several actions based on a value stored in an integer variable. As it turns out, the need to do just that comes up a lot. You want to keep the `switch` statement handy for use when such a need arises.

Battling else-if Monstrosities

Many applications call for a simple logical selection of things to be done depending on some value that controls everything. As I describe in Book 2, Chapter 4, such things are usually handled with big chains of `else-if` statements all strung together.

Unfortunately, these things can quickly get out of hand. `else-if` chains can end up looking like DNA double-helix structures or those things that dribble down from the tops of the computer screens in *The Matrix*, with hundreds of lines of code that string `else-if` after `else-if`. The `switch` statement provides a much more concise alternative.

Viewing an example else-if program

Listing 6-1 shows a bit of a program that might be used to decode error codes in a Florida or Ohio voting machine.

LISTING 6-1: **The else-if Version of a Voting Machine Error Decoder**

```java
import java.util.Scanner;

public class VoterApp
{
    static Scanner sc = new Scanner(System.in);

    public static void main(String[] args)
    {
        System.out.println
            ("Welcome to the voting machine "
            + "error code decoder.\n\n"
            + "If your voting machine generates "
            + "an error code,\n"
            + "you can use this program to determine "
            + "the exact\ncause of the error.\n");
        System.out.print("Enter the error code: ");

        int err = sc.nextInt();

        String msg;
        if (err==1)
            msg = "Voter marked more than one candidate.\n"
                + "Ballot rejected.";
        else if (err==2)
            msg = "Box checked and write-in candidate "
                + "entered.\nBallot rejected.";
        else if (err==3)
            msg = "Entire ballot was blank.\n"
                + "Ballot filled in according to "
                + "secret plan.";
        else if (err==4)
            msg = "Nothing unusual about the ballot.\n"
                + "Voter randomly selected for tax audit.";
        else if (err==5)
            msg = "Voter filled in every box.\n"
                + "Ballot counted twice.";
        else if (err==6)
            msg = "Voter drooled in voting machine.\n"
                + "Beginning spin cycle.";
        else if (err==7)
            msg = "Voter lied to pollster after voting.\n"
```

```
                + "Voter's ballot changed "
                + "to match polling data.";
        else
            msg = "Voter filled out ballot correctly.\n"
                + "Ballot discarded anyway.";
        System.out.println(msg);
    }
}
```

Wow! And this program has to decipher just 7 error codes. What if the machine had 500 codes?

Creating a better version of the example program

Fortunately, Java has a special statement that's designed just for the kind of task represented by the voting machine error decoder program: the `switch` statement. Specifically, the `switch` statement is useful when you need to select one of several alternatives based on the value of an `int`, `char`, `String`, or `enum` type variable.

TIP

An *enum* is a special kind of Java type whose value is one of several predefined constants. For example, you may have an enum named `TemperatureScale` with constant values `CELCIUS`, `FAHRENHEIT`, and `KELVIN`. A variable defined with the `TemperatureScale` type can have one of these three values. For more information, refer to Book 3, Chapter 2.

Listing 6-2 shows a version of the voting machine error decoder program that uses a `switch` statement instead of a big `else-if` structure. I think you'll agree that this version of the program is a bit easier to follow. The `switch` statement makes it clear that all the messages are selected based on the value of the `err` variable.

LISTING 6-2: **The switch Version of the Voting Machine Error Decoder**

```java
import java.util.Scanner;

public class VoterApp2
{
    static Scanner sc = new Scanner(System.in);

    public static void main(String[] args)
    {
        System.out.println
```

(continued)

LISTING 6-2: *(continued)*

```
                    ("Welcome to the voting machine "
                    + "error code decoder.\n\n"
                    + "If your voting machine generates "
                    + "an error code,\n"
                    + "you can use this program to determine "
                    + "the exact\ncause of the error.\n");
        System.out.print("Enter the error code: ");
        int err = sc.nextInt();

        String msg;

        switch (err)
        {
            case 1:
                msg = "Voter marked more than one "
                    + "candidate.\nBallot rejected.";
                break;
            case 2:
                msg = "Box checked and write-in candidate "
                    + "entered.\nBallot rejected.";
                break;
            case 3:
                msg = "Entire ballot was blank.\n"
                    + "Ballot filled in according to "
                    + "secret plan.";
                break;
            case 4:
                msg = "Nothing unusual about the ballot.\n"
                    + "Voter randomly selected for tax audit.";
                break;
            case 5:
                msg = "Voter filled in every box.\n"
                    + "Ballot counted twice.";
                break;
            case 6:
                msg = "Voter drooled in voting machine.\n"
                    + "Beginning spin cycle.";
                break;
            case 7:
                msg = "Voter lied to pollster after voting.\n"
                    + "Voter's ballot changed "
                    + "to match polling data.";
                break;
            default:
                msg = "Voter filled out ballot correctly.\n"
                    + "Ballot discarded anyway.";
                break;
```

```
            }
        System.out.println(msg);
    }
}
```

Using the switch Statement

The basic form of the switch statement is this:

```
switch (expression)
{
    case constant:
        statements;
        break;

    [ case constant-2:
        statements;
        break; ]...

    [ default:
        statements;
        break; ]...
}
```

The expression must evaluate to an int, short, byte, char, String, or enum. It can't be a long or a floating-point type.

You can code as many case groups as you want or need. Each group begins with the word case, followed by a constant (usually, a simple numeric or String literal) and a colon. Then you code one or more statements that you want executed if the value of the switch expression equals the constant. The last line of each case group is an optional break statement, which causes the entire switch statement to end.

The default group, which is optional, is like a catch-all case group. Its statements are executed only if none of the previous case constants match the switch expression.

TIP

Note that the case groups are not true blocks marked with braces. Instead, each case group begins with the case keyword and ends with the case keyword that starts the next case group. All the case groups together, however, are defined as a block marked with a set of braces.

The last statement in each case group usually is a break statement. A break statement causes control to skip to the end of the switch statement. If you omit the break statement, control falls through to the next case group. Accidentally leaving out break statements is the most common cause of trouble with the switch statement.

Viewing a boring switch example, complete with flowchart

Okay, the voting machine error decoder was kind of fun. Here's a more down-to-earth example. Suppose that you need to set a commission rate based on a sales class represented by an integer (1, 2, or 3) according to this table:

Class	Commission Rate
1	2%
2	3.5%
3	5%
Any other value	0%

You could do this with the following switch statement:

```
double commissionRate;

switch (salesClass)
{
    case 1:
        commissionRate = 0.02;
        break;

    case 2:
        commissionRate = 0.035;
        break;

    case 3:
        commissionRate = 0.05;
        break;

    default:
        commissionRate = 0.0;
        break;
}
```

Figure 6-1 shows a flowchart that describes the operation of this `switch` statement. As you can see, this flowchart is similar to the flowchart in Figure 4-3 (Book 2, Chapter 4), because the operation of the `switch` statement is similar to the operation of a series of `else-if` statements.

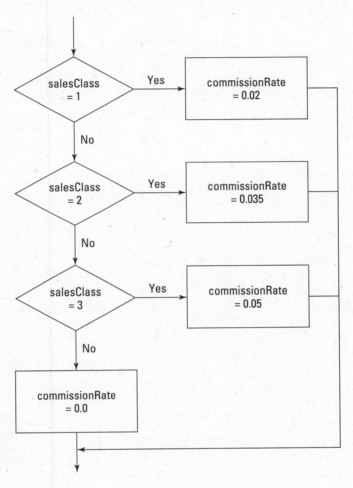

FIGURE 6-1: The flowchart for a switch statement.

TIP

Flowcharts remind me of the good old days, when many COBOL programming shops required their programmers to draw flowcharts for every program they wrote before they were allowed to write any code. The flowcharts didn't really help programmers write better programs, but they were fun to draw.

Putting if statements inside switch statements

You're free to include any type of statements you want in the case groups, including if statements. Suppose that your commission structure depends on total sales as well as sales class, as in this table:

Class	Sales < $10,000	Sales $10,000 and Above
1	1%	2%
2	2.5%	3.5%
3	4%	5%
Any other value	0%	0%

You can use the following switch statement:

```
double commissionRate;

switch (salesClass)
{
    case 1:
        if (salesTotal < 10000.0)
            commissionRate = 0.01;
        else
            commissionRate = 0.02;
        break;

    case 2:
        if (salesTotal < 10000.0)
            commissionRate = 0.025;
        else
            commissionRate = 0.035;
        break;

    case 3:
        if (salesTotal < 10000.0)
            commissionRate = 0.04;
        else
            commissionRate = 0.05;
        break;

    default:
        commissionRate = 0.0;
        break;
}
```

Here each case group includes an `if` statement. If necessary, these `if` statements could be complex nested `if` statements.

Other than the `if` statements within the case groups, there's nothing here to see, folks. Move along.

Creating Character Cases

Aside from having a nice alliterative title, this section shows how you can use a `char` variable rather than an integer in a `switch` statement. When you use a `char` type, providing two consecutive `case` constants for each case group is common, to allow for both lowercase and uppercase letters. Suppose that you need to set the commission rates for the sales class based on character codes rather than on integer values, according to this table:

Class	Commission Rate
A or a	2%
B or b	3.5%
C or c	5%
Any other value	0%

Here's a `switch` statement that can do the trick:

```
double commissionRate;

switch (salesClass)
{
    case 'A':
    case 'a':
        commissionRate = 0.02;
        break;

    case 'B':
    case 'b':
        commissionRate = 0.035;
        break;

    case 'C':
    case 'c':
        commissionRate = 0.05;
        break;
```

```
    default:
        commissionRate = 0.0;
        break;
}
```

The key to understanding this example is realizing that you don't have to code any statements at all for a case group — and that if you omit the `break` statement from a case group, control falls through to the next case group. Thus the `case 'A'` group doesn't contain any statements, but control falls through to the `case 'a'` group.

REMEMBER

You use apostrophes, not quotation marks, to create character literals.

Intentionally Leaving Out a Break Statement

Although the most common cause of problems with the `switch` statement is accidentally leaving out a `break` statement at the end of a case group, sometimes you need to do it on purpose. Many applications have features that are progressively added based on a control variable. Your local car wash, for example, may sell several packages with different services, as in this table:

Package	Services
A	Wash, vacuum, and hand-dry
B	Package A + wax
C	Package B + leather/vinyl treatment
D	Package C + tire treatment
E	Package D + new-car scent

Listing 6-3 shows an application that displays all the products you get when you order a specific package. It works by testing the package codes in a `switch` statement in reverse order (starting with package E) and adding the products that come with each package to the `details` variable. None of the case groups except the last includes a `break` statement. As a result, control falls through each case group to the next group. Thus, once a case group is matched, the rest of the case groups in the `switch` statement are executed.

LISTING 6-3: **The Car Wash Application**

```java
import java.util.Scanner;

public class CarWashApp
{
    static Scanner sc = new Scanner(System.in);

    public static void main(String[] args)
    {
        System.out.println("The car wash application!\n\n");
        System.out.print("Enter the package code: ");
        String s = sc.next();
        char p = s.charAt(0);

        String details = "";

        switch (p)
        {
            case 'E':
            case 'e':
                details += "\tNew Car Scent, plus ... \n";

            case 'D':
            case 'd':
                details += "\tTire Treatment, plus ... \n";

            case 'C':
            case 'c':
                details +=
                    "\tLeather/Vinyl Treatment, plus ... \n";

            case 'B':
            case 'b':
                details += "\tWax, plus ... \n";

            case 'A':
            case 'a':
                details += "\tWash, vacuum, and hand dry.\n";
                break;

            default:
                details = "That's not one of the codes.";
                break;
        }
        System.out.println("\nThat package includes:\n");
        System.out.println(details);
    }
}
```

TIP

Just between you and me, writing programs that depend on `switch` statements falling through the cracks (as in this example) isn't really a good idea. Instead, consider placing the statements for each case group in separate methods and then calling all the methods you need for each case group. Then you can use a `break` statement at the end of each group to prevent falling through. Listing 6-4 shows a version of the car wash application that uses this technique to prevent fall-throughs in the `switch` statement. (Using simple fall-throughs to treat uppercase and lowercase characters the same way isn't as confusing, so this program still uses that technique.)

LISTING 6-4: **A Version of the Car Wash Program That Prevents Nasty Falls**

```java
import java.util.Scanner;

public class CarWashApp2
{
    static Scanner sc = new Scanner(System.in);
    public static void main(String[] args)
    {
        System.out.println("The car wash application!\n\n");
        System.out.print("Enter the package code: ");
        String s = sc.next();
        char p = s.charAt(0);

        String details = "";
        switch (p)
        {
            case 'E':
            case 'e':
                details = packageE() + packageD() + packageC()
                    + packageB() + packageA();
                break;

            case 'D':
            case 'd':
                details = packageD() + packageC()
                    + packageB() + packageA();
                break;

            case 'C':
            case 'c':
                details = packageC() + packageB()
                    + packageA();
                break;

            case 'B':
            case 'b':
```

```java
                details = packageB() + packageA();
                break;

            case 'A':
            case 'a':
                details = packageA();
                break;

            default:
                details = "That's not one of the codes.";
                break;
        }
        System.out.println("\nThat package includes:\n");
        System.out.println(details);
    }

    public static String packageA()
    {
        return "\tWash, vacuum, and hand dry.\n";
    }

    public static String packageB()
    {
        return "\tWax, plus ... \n";
    }

    public static String packageC()
    {
        return "\tLeather/Vinyl Treatment, plus ... \n";
    }

    public static String packageD()
    {
        return "\tTire Treatment, plus ... \n";
    }

    public static String packageE()
    {
        return "\tNew Car Scent, plus ... \n";
    }
}
```

Switching with Strings

Listing 6-5 shows a version of the car wash program that uses the string codes PRESIDENTIAL, ELITE, DELUXE, SUPER, and STANDARD as the car wash types, instead of the letters A through E. Notice that to allow for variations in how a user might capitalize these codes, the user's input is converted to all capital letters before it is tested against the string constants in the switch statement.

LISTING 6-5: **A Version of the Car Wash Program That Uses a String**

```java
import java.util.Scanner;

public class CarWashStringApp
{
    static Scanner sc = new Scanner(System.in);

    public static void main(String[] args)
    {
        System.out.println("The car wash application\n\n");
        System.out.print("Enter the package code: ");
        String s = sc.next();

        String details = "";
        switch (s.toUpperCase())
        {
            case "PRESIDENTIAL":
                details += "\tNew Car Scent, plus ... \n";

            case "ELITE":
                details += "\tTire Treatment, plus ... \n";

            case "DELUXE":
                details += "\tLeather/Vinyl Treatment, plus ... \n";

            case "SUPER":
                details += "\tWax, plus ... \n";

            case "STANDARD":
                details += "\tWash, vacuum, and hand dry.\n";
                break;

            default:
                details = "That's not one of the codes.";
                break;
        }
```

```
        System.out.println("\nThat package includes:\n");
        System.out.println(details);
    }
}
```

Enhanced Switch Features with Java 13

With Java 13, several new features were added to the switch statement.

Technically, these features were introduced with Java 12 as *preview features,* which means that they weren't intended for use in production programs. With Java 13, these features have been upgraded to production status, so you can now use them at will.

The first improvement to the switch statement is that you can now now list more than one value in the case clause. The values must be separated by commas, as in this example:

```
case 'B', 'b':
```

In this example, either the value B or b will match the case. Prior to Java 13, it was necessary to code two case clauses in a row to accomplish this.

The other major new addition to the switch statement is that the switch statement itself can now return a value. You can, therefore, use a switch statement on the right side of an assignment statement, like this:

```
String msg = switch (p)
```

In this example, the value returned by the switch statement is assigned to the variable msg.

There are two ways you can provide a return value for a switch statement. The first is to use the new yield statement, which has a function similar to the break statement but provides a return value. Here's an example:

```
String msg = switch (p)
{
    case 'A', 'a'
        yield "Wash, vacuum, and hand dry.";
};
```

Here, the value "Wash, vacuum, and hand dry." is provided as the return value for the switch statement.

Note that when a switch statement returns a value and is used in an assignment statement, you must add a semicolon to the end of the statement. Strictly speaking, it isn't the switch statement that requires the semicolon; it's the assignment statement that requires the semicolon.

The second way to return a value in a switch statement is to use an arrow operator directly in the case clause, like this:

```
case 'A', 'a' -> "Wash, vacuum, and hand dry.";
```

The arrow operator is more concise than the yield statement, but there are few additional considerations for its use:

>> When you use the arrow operator to provide a return value, you canot simply list multiple statements as part of the case clause.

>> When you use the arrow operator, the case clause itself becomes a statement, which must be terminated with a semicolon.

>> If you do need to use more than one statement in a case clause, you must enclose the statements in a block, and the block must include a yield statement to provide the yielded value. For example:

```
case 'A', 'a' -> {
    System.out.println("Package A");
    yield "Wash, vacuum, and hand dry.";
}
```

>> You can also use the arrow operator in a default clause to provide a default return value.

Note that you can't mix and match the two methods for returning a value within a single switch statement. In other words, if you use the arrow operator for case clause, you must use it for all the case clauses.

Listing 6-6 shows a version of the Car Wash program that uses these new switch features.

LISTING 6-6: **A Version of the Car Wash Program That Uses Java 13 Switch Statement Features**

```java
import java.util.Scanner;

public class CarWashApp
{
    static Scanner sc = new Scanner(System.in);
    public static void main(String[] args)
    {
        System.out.println
            ("The car wash application!\n\n");
        System.out.print("Enter the package code: ");
        String s = sc.next();
        char p = s.charAt(0);

        String details = switch (p)
        {
            case 'E','e' -> packageE() + packageD() + packageC()
                    + packageB() + packageA();

            case 'D','d' -> packageD() + packageC()
                    + packageB() + packageA();

            case 'C','c' -> packageC() + packageB() + packageA();

            case 'B','b' -> packageB() + packageA();

            case 'A','a' -> packageA();

            default -> "That's not one of the codes.";
        };

        System.out.println("\nThat package includes:\n");
        System.out.println(details);
    }

    public static String packageA()
    {
        return "\tWash, vacuum, and hand dry.\n";
    }

    public static String packageB()
    {
        return "\tWax, plus ... \n";
    }

    public static String packageC()
```

(continued)

LISTING 6-6: **(continued)**

```
    {
        return "\tLeather/Vinyl Treatment, plus ... \n";
    }

    public static String packageD()
    {
        return "\tTire Treatment, plus ... \n";
    }

    public static String packageE()
    {
        return "\tNew Car Scent, plus ... \n";
    }
}
```

Chapter **7**

Adding Some Methods to Your Madness

I n Java, a *method* is a block of statements that has a name and can be executed by *calling* (also called *invoking*) it from some other place in your program. You may not realize it, but you're already very experienced in using methods. To print text to the console, for example, you use the `println` or `print` method. To get an integer from the user, you use the `nextInt` method. To compare string values, you use the `equals` or `equalsIgnoreCase` method. Finally, the granddaddy of all methods — `main` — contains the statements that are executed when you run your program.

All the methods you've used so far (with the exception of `main`) have been defined by the Java API and belong to a particular Java class. The `nextInt` method belongs to the `Scanner` class, for example, and the `equalsIgnoreCase` method belongs to the `String` class. By contrast, the `main` method belongs to the class defined by your application.

In this chapter, you find out how to create additional methods that are part of your application's class. Then you can call these methods from your `main` method. As you'll see, this technique turns out to be very useful for all but the shortest Java programs.

The Joy of Methods

The use of methods can dramatically improve the quality of your programming life. Suppose that the problem your program is supposed to solve is complicated, and you need at least 1,000 Java statements to get 'er done. You could put all those 1,000 statements in the main method, but it would go on for pages and pages. It's better to break your program into a few well-defined sections of code and place each of those sections in a separate method. Then your main method can simply call the other methods in the right sequence.

Or suppose that your program needs to perform some calculation, such as how long to let the main rockets burn to make a midcourse correction on a moon flight, and the program needs to perform this calculation in several places. Without methods, you'd have to duplicate the statements that do this calculation. That approach is not only error-prone, but also makes your programs more difficult to test and debug. But if you put the calculation in a method, you can simply call the method whenever you need to perform the calculation. Thus methods help you cut down on repetitive code.

Another good use for methods is to simplify the structure of your code that uses long loops. Suppose you have a while loop that has 500 statements in its body. That structure makes it pretty hard to track down the brace that marks the end of the body. By the time you find it, you probably will have forgotten what the while loop does. You can simplify this while loop by placing the code from its body in a separate method. Then all the while loop has to do is call the new method.

The Basics of Making Methods

All methods — including the main method — must begin with a *method declaration*. Here's the basic form of a method declaration, at least for the types of methods I talk about in this chapter:

```
public static return-type method-name (parameter-list)
{
    statements...
}
```

The following paragraphs describe the method declaration piece by piece:

>> **public:** This keyword indicates that the method's existence should be publicized to the world and that any Java program that knows about your

program (or, more accurately, the class defined for your Java program) should be able to use your method. That's not very meaningful for the types of programs you're dealing with at this point in the book, but it becomes more meaningful in Book 3: There you find out more about what `public` means and see some alternatives to `public` that are useful in various and sundry situations.

» **static:** This keyword declares that the method is a *static method,* which means that you can call it without first creating an instance of the class in which it's defined. The `main` method must always be static.

» **return-type:** After the word `static` comes the *return type,* which indicates whether the method returns a value when it is called and, if so, what type the value is. If the method doesn't return a value, specify `void`. (I talk more about methods that return values later in this chapter, in the section "Methods That Return Values.")

» **method-name:** Now comes the name of your method. The rules for making up method names are the same as the rules for creating variable names: You can use any combination of letters and numbers, but the name has to start with a letter. Also, it can include the dollar sign ($) and underscore character (_). No other special characters are allowed.

When picking a name for your method, try to pick a name that's relatively short but descriptive. A method name such as `calculateTheTotalAmountOfTheInvoice` is a little long, but just `calc` is pretty ambiguous. Something along the lines of `calculateInvoiceTotal` seems more reasonable to me.

» **parameter list:** You can pass one or more values to a method by listing the values in parentheses following the method name. The parameter list in the method declaration lets Java know what types of parameters a method should expect to receive and provides names so that the statements in the method's body can access the parameters as local variables. You discover more about parameters in the section "Methods That Take Parameters," later in this chapter.

If the method doesn't accept parameters, you must still code the parentheses that surround the parameter list. You just leave the parentheses empty.

» **Method body:** The method body consists of one or more Java statements enclosed in a set of braces. Unlike Java statements such as `if`, `while`, and `for`, the method body requires you to use the braces even if the body consists of only one statement.

An example

Okay, all that was a little abstract. Now, for a concrete example, I offer a version of the `Hello, World!` program in which the message is displayed not by the `main` method, but by a method named `sayHello` that's called by the `main` method:

```
public class HelloWorldMethod
{
    public static void main(String[] args)
    {
        sayHello();
    }

    public static void sayHello()
    {
        System.out.println("Hello, World!");
    }
}
```

This program is admittedly trivial, but it illustrates the basics of creating and using methods in Java. Here, the statement in the `main` method calls the `sayHello` method, which in turn displays a message on the console.

TIP

The order in which methods appear in your Java source file doesn't matter. The only rule is that all the methods must be declared within the body of the class — that is, between the first left brace and the last right brace. Here's a version of the `HelloWorldMethod` program in which I reverse the order of the methods:

```
public class HelloWorldMethod
{
    public static void sayHello()
    {
        System.out.println("Hello, World!");
    }

    public static void main(String[] args)
    {
        sayHello();
    }
}
```

This version of the program works exactly like the preceding version.

Another example

Okay, the last example was kind of dumb. No one in his (or her) right mind would create a method that has just one line of code and then call it from another method that *also* has just one line of code. The `Hello, World!` program is too trivial to illustrate anything remotely realistic.

A program in Book 2, Chapter 5, plays a guessing game. Most of this program's `main` method is a large `while` loop that repeats the game as long as the user wants to keep playing. This loop has 41 statements in its body. That's not so bad, but what if the game were 100 times more complicated, so that the `while` loop needed 4,100 statements to play a single cycle of the game? Do you really want a `while` loop that has 4,100 statements in its body? I should think not.

Listing 7-1 shows how you can simplify this game a bit just by placing the body of the main `while` loop in a separate method. I called this method `playARound`, because its job is to play one round of the guessing game. Now, instead of actually playing a round of the game, the `main` method of this program delegates that task to the `playARound` method.

LISTING 7-1: **A Version of the Guessing-Game Program That Uses a playARound Method**

```java
import java.util.Scanner;

public class GuessingGameMethod
{
    static Scanner sc = new Scanner(System.in);
    static boolean keepPlaying = true;                          →6

    public static void main(String[] args)
    {
        System.out.println("Let's play a guessing game!");
        while (keepPlaying)                                      →11
        {
            playARound();                                       →13
        }
        System.out.println("\nThank you for playing!");
    }

    public static void playARound()                             →18
    {
        boolean validInput;
        int number, guess;
        String answer;
```

(continued)

LISTING 7-1: *(continued)*

```
// Pick a random number
number = (int)(Math.random() * 10) + 1;
System.out.println("\nI'm thinking of a number "
    + "between 1 and 10.");

// Get the guess
System.out.print("What do you think it is? ");
do
{
    guess = sc.nextInt();
    validInput = true;
    if ((guess < 1) || (guess > 10))
    {
        System.out.print("I said, between 1 "
            + "and 10. Try again: ");
        validInput = false;
    }
} while (!validInput);

// Check the guess
if (guess == number)
    System.out.println("You're right!");
else
    System.out.println("You're wrong!"
        + " The number was " + number);

// Play again?
do
{
    System.out.print("\nPlay again? (Y or N)");
    answer = sc.next();
    validInput = true;
    if (answer.equalsIgnoreCase("Y"));
    else if (answer.equalsIgnoreCase("N"))
        keepPlaying = false;                              →58
    else
        validInput = false;
} while (!validInput);
    }
}
```

Here are a few important details to notice about this method:

➤ →6 Because both the main method (in line 11) and the playARound method (in line 58) must access the keepPlaying variable, I declare it as a class variable rather than as a local variable in the main method.

>> →13 The body of the while loop in the main method is just one line: a call to the playARound method. Thus, each time the loop repeats, the program plays one round of the game with the user.

>> →18 The declaration for the playARound method marks the method as static so that the static main method can call it.

TIP

The body of the playARound method is identical to the body of the while loop used in the single-method version of this program shown in Book 2, Chapter 5. If you want a refresher on how this code works, I politely refer you to Listing 5-1, near the end of that chapter.

Methods That Return Values

Some methods do some work, and then simply return when they're finished. But many methods need to return a value when they complete their work. For example, if a method's purpose is to perform a calculation, the method will likely return the result of the calculation to the calling method so that the calling method can do something with the value. You find out how to do that in the following sections.

Declaring the method's return type

To create a method that returns a value, you simply indicate the type of the value returned by the method on the method declaration in place of the void keyword. Here's a method declaration that creates a method that returns an int value:

```
public static int getRandomNumber()
```

Here the getRandomNumber method calculates a random number and then returns the number to the caller.

REMEMBER

The return type of a method can be any of Java's primitive return types (described in Book 2, Chapter 2):

```
int
long
float
char
short
byte
double
boolean
```

Alternatively, the return type can be a *reference type*, including a class defined by the API such as String or a class you create yourself.

Using the return statement to return the value

When you specify a return type other than void in a method declaration, the body of the method must include a return statement that specifies the value to be returned. The return statement has this form:

```
return expression;
```

The expression must evaluate to a value that's the same type as the type listed in the method declaration. In other words, if the method returns an int, the expression in the return statement must evaluate to an int.

Here's a program that uses a method that determines a random number between 1 and 10:

```java
public class RandomNumber
{
    public static void main(String[] args)
    {
        int number = getRandomNumber();
        System.out.println("The number is " + number);
    }

    public static int getRandomNumber()
    {
        int num = (int)(Math.random() * 10) + 1;
        return num;
    }
}
```

In this program, the getRandomNumber method uses the Math.random method to calculate a random number from 1 to 10. (For more information about the Math. random method, see Book 2, Chapter 3.) The return statement returns the random number that was calculated.

Because the return statement can specify an expression as well as a simple variable, I could just as easily have written the getRandomNumber method like this:

```java
public static int getRandomNumber()
{
```

```
    return (int)(Math.random() * 10) + 1;
}
```

Here the `return` statement includes the expression that calculates the random number.

Using a method that returns a type

You can use a method that returns a value in an assignment statement, like this:

```
int number = getRandomNumber();
```

Here the `getRandomNumber` method is called, and the value it returns is assigned to the variable `number`.

You can also use methods that return values in expressions — such as

```
number = getRandomNumber() * 10;
```

Here the value returned by the `getRandomNumber` method is multiplied by 10, and the result is assigned to `number`.

You gotta have a proper return statement

If a method declares a return type other than void, it *must* use a `return` statement to return a value. The compiler doesn't let you get away with a method that doesn't have a correct `return` statement.

Things can get complicated if your `return` statements are inside `if` statements. Sometimes, the compiler gets fooled and refuses to compile your program. To explain this situation, I offer the following tale of multiple attempts to solve what should be a simple programming problem.

Suppose that you want to create a random-number method that returns random numbers between 1 and 20 but never returns 12 (because you have the condition known as dodecaphobia, which — as Lucy from *Peanuts* would tell you — is the fear of the number 12). Your first thought is to just ignore the 12s, like this:

```
public static int getRandomNumber()
{
    int num = (int)(Math.random() * 20) + 1;
```

```
    if (num != 12)
        return num;
}
```

The compiler isn't fooled by your trickery here, however. It knows that if the number is 12, the return statement won't get executed, so it issues the message missing return statement and refuses to compile your program.

Your next thought is to simply substitute 11 whenever 12 comes up:

public static int getRandomNumber()

```
public static int getRandomNumber()
{
    int num = (int)(Math.random() * 20) + 1;
    if (num != 12)
        return num;
    else
        return 11;
}
```

Later that day, you realize that this solution isn't a good one because the number isn't really random anymore. One of the requirements of a good random-number generator is that any number should be as likely as any other number to come up next. But because you're changing all 12s to 11s, you've made 11 twice as likely to come up as any other number.

To fix this error, you decide to put the random-number generator in a loop that ends only when the random number is not 12:

```
public static int getRandomNumber()
{
    int num;
    do
    {
        num = (int)(Math.random() * 20) + 1;
        if (num != 12)
            return num;
    } while (num == 12);
}
```

But the compiler refuses to compile the method again. It turns out that the compiler is smart, but not very smart. It doesn't catch the fact that the condition in the do-while loop is the opposite of the condition in the if statement, meaning that the only way out of this loop is through the return statement in the if statement. So the compiler whines missing return statement again.

After thinking about it for a while, you come up with this solution:

```
public static int getRandomNumber()
{
    int num;
    while (true)
    {
        num = (int)(Math.random() * 20) + 1;
        if (num != 12)
            return num;
    }
}
```

Now everyone's happy. The compiler knows that the only way out of the loop is through the return statement, your dodecaphobic user doesn't have to worry about seeing the number 12, and you know that the random number isn't twice as likely to be 11 as any other number. Life is good, and you can move on to the next topic.

Trying another version of the guessing-game program

To illustrate the benefits of using methods that return values, Listing 7-2 presents another version of the guessing-game program that uses four methods in addition to main:

» **playARound:** This method plays one round of the guessing game. It doesn't return a value.

» **getRandomNumber:** This method returns a random number between 1 and 10.

» **getGuess:** This method gets the user's guess, makes sure that it is between 1 and 10, and returns the guess if it's within the acceptable range.

» **askForAnotherRound:** This method asks the user to play another round and returns a boolean value to indicate whether the user wants to continue playing.

LISTING 7-2: **Another Version of the Guessing-Game Program**

```java
import java.util.Scanner;

public class GuessingGameMethod2
{
    static Scanner sc = new Scanner(System.in);

    public static void main(String[] args)
    {
        System.out.println("Let's play a guessing game!");
        do                                                    →10
        {
            playARound();                                     →12
        } while (askForAnotherRound());                       →13
        System.out.println("\nThank you for playing!");
    }

    public static void playARound()                           →17
    {
        boolean validInput;
        int number, guess;
        String answer;

        // Pick a random number
        number = getRandomNumber();                           →24

        // Get the guess
        System.out.println("\nI'm thinking of a number "
            + "between 1 and 10.");
        System.out.print("What do you think it is? ");
        guess = getGuess();                                   →30

        // Check the guess
        if (guess == number)
            System.out.println("You're right!");
        else
            System.out.println("You're wrong!"
                + " The number was " + number);
    }

    public static int getRandomNumber()                       →40
    {
        return (int)(Math.random() * 10) + 1;                 →42
    }

    public static int getGuess()                              →45
    {
        while (true)                                          →47
```

```
        {
            int guess = sc.nextInt();
            if ((guess < 1) || (guess > 10))
            {
                System.out.print("I said, between 1 and 10. "
                    + "Try again: ");
            }
            else
                return guess;                                            →56
        }
    }

    public static boolean askForAnotherRound()                          →60
    {
        while (true)                                                    →62
        {
            String answer;
            System.out.print("\nPlay again? (Y or N) ");
            answer = sc.next();
            if (answer.equalsIgnoreCase("Y"))
                return true;                                            →68
            else if (answer.equalsIgnoreCase("N"))
                return false;                                           →70
        }
    }
}
```

The following paragraphs point out the key lines of this program:

» →10 The start of the do loop in the main method. Each cycle of this do loop plays one round of the game. The do loop continues until the user indicates that he or she wants to stop playing.

» →12 Calls the playARound method to play one round of the game.

» →13 Calls the askForAnotherRound method to determine whether the user wants to play another round. The boolean return value from this method is used as the expression for the do loop. Thus, the do loop repeats if the askForAnotherRound method returns true.

» →17 The start of the playARound method.

» →24 Calls the getRandomNumber method to get a random number between 1 and 10. The value returned by this method is stored in the number variable.

» →30 Calls the getGuess method to get the user's guess. This method returns a number between 1 and 10, which is stored in the guess variable.

>> →40 The start of the getRandomNumber method, which indicates that this method returns an int value.

>> →42 The return statement for the getRandomNumber method. The random number is calculated using the Math.random method, and the result of this calculation is returned as the value of the getRandomNumber method.

>> →45 The start of the getGuess method, which indicates that this method returns an int value.

>> →47 The getGuess method uses a while loop, which exits only when the user enters a number between 1 and 10.

>> →56 The return statement for the getGuess method. Note that this return statement is in the else part of an if statement that checks whether the number is less than 1 or greater than 10. If the number is outside the acceptable range, the return statement isn't executed. Instead, the program displays an error message, and the while loop repeats.

>> →60 The start of the askForAnotherRound method, which returns a boolean value.

>> →62 The askForAnotherRound method, which uses a while loop that exits only when the user enters a valid Y or N response.

>> →68 The askForAnotherRound method, which returns true if the user enters Y or y.

>> →70 The askForAnotherRound method, which returns false if the user enters N or n.

Methods That Take Parameters

A *parameter* is a value that you can pass to a method. Then the method can use the parameter as though it were a local variable initialized with the value of the variable passed to it by the calling method.

The guessing-game application shown in Listing 7-2 has a method named get-RandomNumber that returns a random number between 1 and 10:

```
public static int getRandomNumber()
{
    return (int)(Math.random() * 10) + 1;
}
```

This method is useful, but it would be even more useful if you could tell it the range of numbers you want the random number to fall in. It would be nice to call the method like this to get a random number between 1 and 10:

```
int number = getRandomNumber(1, 10);
```

Then, if your program needs to roll dice, you could call the same method:

```
int number = getRandomNumber(1, 6);
```

Or, to pick a random card from a deck of 52 cards, you could call it like this:

```
int number = getRandomNumber(1, 52);
```

You wouldn't have to start with 1, either. To get a random number between 50 and 100, you'd call the method like this:

```
int number = getRandomNumber(50, 100);
```

In the following sections, you write methods that accept parameters.

Declaring parameters

A method that accepts parameters must list the parameters in the method declaration. The parameters are placed in a *parameter list* inside the parentheses that follow the method name. For each parameter used by the method, you list the parameter type followed by the parameter name. If you need more than one parameter, you separate the parameters with commas.

Here's a version of the getRandomNumber method that accepts parameters:

```
public static int getRandomNumber(int min, int max)
{
    return (int)(Math.random() * (max - min + 1)) + min;
}
```

Here the method uses two parameters, both of type int, named min and max. Then, within the body of the method, these parameters can be used as though they were local variables.

TIP

The names you use for parameters can be the same as the names you use for the variables you pass to the method when you call it, but they don't have to be. You could call the getRandomNumber method like this:

```
int min = 1;
int max = 10;
int number = getRandomNumber(min, max);
```

Or you could call it like this:

```
int low = 1;
int high = 10;
int number = getRandomNumber(low, high);
```

Or you could dispense with the variables altogether and just pass literal values to the method:

```
int number = getRandomNumber(1, 10);
```

You can also specify expressions as the parameter values:

```
int min = 1;
int max = 10;
int number = getRandomNumber(min * 10, max * 10);
```

Here number is assigned a value between 10 and 100.

Scoping out parameters

The scope of a parameter is the method for which the parameter is declared. As a result, a parameter can have the same name as local variables used in other methods without causing any conflict. Consider this program:

```
public class ParameterScope
{
    public static void main(String[] args)
    {
        int min = 1;
        int max = 10;
        int number = getRandomNumber(min, max);

        System.out.println(number);
    }

    public static int getRandomNumber(int min, int max)
```

```
    {
        return (int)(Math.random() * (max - min + 1)) + min;
    }
}
```

Here the `main` method declares variables named `min` and `max`, and the `getRandom-Number` method uses `min` and `max` for its parameter names. This doesn't cause any conflict, because in each case the scope is limited to a single method.

Understanding pass-by-value

When Java passes a variable to a method via a parameter, the method itself receives a copy of the variable's value, not the variable itself. This copy is called a *pass-by-value*, and it has an important consequence: If a method changes the value it receives as a parameter, that change is *not* reflected in the original variable that was passed to the method. The following program can help clear this up:

```
public class ChangeParameters
{
    public static void main(String[] args)
    {
        int number = 1;
        tryToChangeNumber(number);
        System.out.println(number);
    }

    public static void tryToChangeNumber(int i)
    {
        i = 2;
    }
}
```

Here a variable named `number` is set to 1 and then passed to the method named `tryToChangeNumber`. This method receives the variable as a parameter named `i` and then sets the value of `i` to 2. Meanwhile, back in the `main` method, `println` is used to print the value of `number` after the `tryToChangeNumber` method returns.

Because `tryToChangeNumber` gets only a copy of `number`, not the `number` variable itself, this program displays the following on the console (drum roll, please): 1.

The key point is this: Even though the `tryToChangeNumber` method changes the value of its parameter, that change has no effect on the original variable that was passed to the method.

Trying yet another version of the guessing-game program

To show off the benefits of methods that accept parameters, Listing 7-3 shows one more version of the guessing-game program. This version uses the following methods in addition to main:

>> playARound: This method plays one round of the guessing game. It doesn't return a value, but it accepts two arguments, min and max, that indicate the minimum and maximum values for the number to be guessed.

>> getRandomNumber: This method returns a random number between min and max values passed as parameters.

>> getGuess: This method also accepts two parameters, min and max, to limit the range within which the user must guess.

>> askForAnotherRound: This method asks the user to play another round and returns a boolean value to indicate whether the user wants to continue playing. It accepts a String value as a parameter; this string is displayed on the console to prompt the user for a reply.

LISTING 7-3: **Yet Another Version of the Guessing-Game Program**

```
import java.util.Scanner;

public class GuessingGameMethod3
{
    static Scanner sc = new Scanner(System.in);

    public static void main(String[] args)
    {
        System.out.println("Let's play a guessing game!");
        do
        {
            playARound(1, getRandomNumber(7, 12));      →12
        } while (askForAnotherRound("Try again?"));
        System.out.println("\nThank you for playing!");
    }

    public static void playARound(int min, int max)
    {
        boolean validInput;
        int number, guess;
        String answer;
```

```
    // Pick a random number
    number = getRandomNumber(min, max);                            →24

    // Get the guess
    System.out.println("\nI'm thinking of a number "
        + "between " + min + " and " + max + ".");                 →28
    System.out.print("What do you think it is? ");
    guess = getGuess(min, max);                                    →30

    // Check the guess
    if (guess == number)
        System.out.println("You're right!");
    else
        System.out.println("You're wrong!"
            + " The number was " + number);
}

public static int getRandomNumber(int min, int max)                →40
{
    return (int)(Math.random()                                     →42
        * (max - min + 1)) + min;
}

public static int getGuess(int min, int max)                       →46
{
    while (true)
    {
        int guess = sc.nextInt();
        if ( (guess < min) || (guess > max) )                      →51
        {
            System.out.print("I said, between "
                + min + " and " + max
                + ". Try again: ");
        }
        else
            return guess;                                          →58
    }
}

public static boolean askForAnotherRound(String prompt)            →62
{
    while (true)
    {
        String answer;
        System.out.print("\n" + prompt + " (Y or N) ");
        answer = sc.next();
        if (answer.equalsIgnoreCase("Y"))
```

(continued)

LISTING 7-3: *(continued)*

```
            return true;
        else if (answer.equalsIgnoreCase("N"))
            return false;
    }
  }
}
```

The following paragraphs point out the key lines of this program:

» →12 This line calls the playARound method to play one round of the game. The values for min and max are passed as literals. To add a small amount of variety to the game, the getRandomNumber method is called here to set the value for the max to a random number from 7 to 12.

» →24 The call to the getRandomNumber method passes the values of min and max as parameters to set the range for the random numbers.

» →28 The message that announces to the user that the computer has chosen a random number uses the min and max parameters to indicate the range.

» →30 The call to the getGuess method now passes the range of acceptable guesses to the getGuess method.

» →40 The declaration for the getRandomNumber method specifies the min and max parameters.

» →42 The calculation for the random number is complicated a bit by the fact that min may not be 1.

» →46 The declaration for the getGuess method accepts the min and max parameters.

» →51 The if statement in the getGuess method uses the min and max values to validate the user's input.

» →58 This line is the return statement for the getGuess method. Note that this return statement is in the else part of an if statement that checks whether the number is less than 1 or greater than 10. If the number is outside the acceptable range, the return statement isn't executed. Instead, the program displays an error message, and the while loop repeats.

» →62 The askForAnotherRound method accepts a string variable to use as a prompt.

Chapter **8**

Handling Exceptions

This chapter is about what happens when Java encounters an error situation that it can't deal with. Over the years, computer programming languages have devised many ways to deal with these types of errors. The earliest programming languages dealt with them rudely, by abruptly terminating the program and printing out the entire contents of the computer's memory in hexadecimal. This output was called a *dump*.

Later programming languages tried various ways to keep the program running when serious errors occurred. In some languages, the statements that could potentially cause an error had elements added to them that would provide feedback about errors. A statement that read data from a disk file, for example, might return an error code if an I/O error occurred. Still other languages let you create a special error processing section of the program to which control would be transferred if an error occurred.

Being an object-oriented programming language, Java handles errors by using special *exception objects* that are created when an error occurs. In addition, Java has a special statement called the `try` statement that you must use to deal with exception objects. In this chapter, you find all the gory details of working with exception objects and `try` statements.

Understanding Exceptions

An *exception* is an object that's created when an error occurs in a Java program and Java can't automatically fix the error. The exception object contains information about the type of error that occurred. The most important information — the cause of the error — is indicated by the name of the exception class used to create the exception. You usually don't have to do anything with an exception object other than figure out which one you have.

Each type of exception that can occur is represented by a different exception class. Here are some typical exceptions:

>> `IllegalArgumentException`: You passed an incorrect argument to a method.

>> `InputMismatchException`: The console input doesn't match the data type expected by a method of the `Scanner` class.

>> `ArithmeticException`: You tried an illegal type of arithmetic operation, such as dividing an integer by zero.

>> `IOException`: A method that performs I/O encountered an unrecoverable I/O error.

>> `ClassNotFoundException`: A necessary class couldn't be found.

There are many other types of exceptions, and you find out about many of them later in this book.

You need to know a few other things about exceptions:

>> When an error occurs and an exception object is created, Java is said to have *thrown an exception.* Java has a pretty good throwing arm, so the exception is always thrown right back to the statement that caused it to be created.

>> The statement that caused the exception can *catch* the exception if it wants it, but it doesn't have to catch the exception if it doesn't want it. Instead, it can duck and let someone else catch the exception. That someone else is the statement that called the method that's currently executing.

>> If everyone ducks and the exception is never caught by the program, the program ends abruptly and displays a nasty-looking exception message on the console (more on that in the next section).

>> Two basic types of exceptions in Java are checked exceptions and unchecked exceptions:

- A *checked exception* is an exception that the compiler requires you to provide for it one way or another. If you don't, your program doesn't compile.

- An *unchecked exception* is an exception that you can provide for, but you don't have to.

>> So far in this book, I've avoided using any Java API methods that throw checked exceptions, but I have used methods that can throw unchecked exceptions. The nextInt method of the Scanner class, for example, throws an unchecked exception if the user enters something other than a valid integer value. For more information, read on.

Witnessing an exception

Submitted for your approval is a tale of a hastily written Java program, quickly put together to illustrate certain Java programming details while ignoring others. Out of sight, out of mind, as they say. Said program played a guessing game with the user, accepting numeric input via a class called Scanner. Yet this same program ignored the very real possibility that the user may enter strange and unexpected data — data that could hardly be considered numeric, at least not in the conventional sense. The time: Now. The place: Here. This program is about to cross over into . . . the Exception Zone.

The program I'm talking about here is, of course, the guessing-game program that's appeared in several forms in recent chapters. (You can find the most recent version at the end of Book 2, Chapter 7.) This program includes a validation routine that prevents the user from making a guess that's not between 1 and 10. That validation routine, however, assumes that the user entered a valid integer number. If the user enters something other than an integer value, the nextInt method of the Scanner class fails badly.

Figure 8-1 shows an example of what the console looks like if the user enters text (such as five) instead of a number. The first line after the user enters the incorrect data says the program has encountered an exception named Input-MismatchException. In short, this exception means that the data entered by the user couldn't be matched with the type of data that the Scanner class expected. The nextInt method expected to find an integer, and instead, it found the word five.

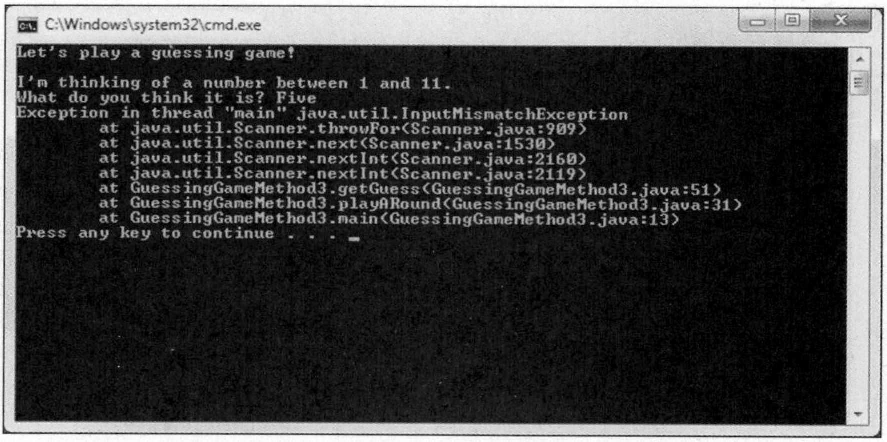

FIGURE 8-1:
This program has
slipped into the
Exception Zone.

Finding the culprit

You can find the exact statement in your program that caused the exception to occur by examining the lines that are displayed right after the line that indicates which exception was encountered. These lines, called the *stack trace,* list the methods that the exception passed through before your program was aborted. Often, the first method listed is deep in the bowels of the Java API, and the last method listed is your application's main method. Somewhere in the middle, you find the switch from methods in the Java API to a method in your program. That's usually where you find the statement in your program that caused the error.

In Figure 8-1, the stack trace lines look something like this:

```
at java.util.Scanner.throwFor(Scanner.java:909)
at java.util.Scanner.next(Scanner.java:1530)
at java.util.Scanner.nextInt(Scanner.java:2160)
at java.util.Scanner.nextInt(Scanner.java:2119)
at GuessingGameMethod3.getGuess(GuessingGameMethod3.java:51)
at GuessingGameMethod3.playARound(GuessingGameMethod3.java:31)
at GuessingGameMethod3.main(GuessingGameMethod3.java:13)
```

Each line lists not only a class and method name, but also the name of the source file that contains the class and the line number where the exception occurred. Thus, the first line in this stack trace indicates that the exception is handled in the throwFor method of the Scanner class at line 909 of the Scanner.java file. The next three lines also indicate methods in the Scanner class. The first line to mention the GuessingGame class (GuessingGameMethod3) is the fifth line. It shows that the exception happened at line 51 in the GuessingGameMethod3.java file. Sure enough, that's the line that calls the nextInt method of the Scanner class to get input from the user.

Catching Exceptions

Whenever you use a statement that might throw an exception, you should write special code to anticipate and catch the exception. That way, your program won't crash as shown in Figure 8-1 if the exception occurs.

You catch an exception by using a `try` statement, which usually follows this general form:

```
try
{
    statements that can throw exceptions
}
catch (exception-type identifier)
{
    statements executed when exception is thrown
}
```

Here, you place the statements that might throw an exception within a *try block*. Then you catch the exception with a *catch block*.

Here are a few things to note about `try` statements:

>> You can code more than one catch block. That way, if the statements in the try block might throw more than one type of exception, you can catch each type of exception in a separate catch block.

>> You can catch more than one exception in a single catch block.

>> For scoping purposes, the try block is its own self-contained block, separate from the catch block. As a result, any variables you declare in the try block are not visible to the catch block. If you want them to be, declare them immediately before the try statement.

>> You can also code a special block (called a *finally block*) after all the catch blocks. For more information about coding `finally` blocks, see the section "Using a `finally` Block," later in this chapter.

>> The various exception classes in the Java API are defined in different packages. If you use an exception class that isn't defined in the standard `java.lang` package that's always available, you need to provide an `import` statement for the package that defines the exception class.

A simple example

To illustrate how to provide for an exception, here's a program that divides two numbers and uses a `try/catch` statement to catch an exception if the second number turns out to be zero:

```
public class DivideByZero
{
    public static void main(String[] args)
    {
        int a = 5;
        int b = 0; // you know this won't work

        try
        {
            int c = a / b; // but you try it anyway
        }
        catch (ArithmeticException e)
        {
            System.out.println("Oops, you can't divide by zero.");
        }
    }
}
```

Here, the division occurs within a `try` block, and a `catch` block handles `ArithmeticException`. `ArithmeticException` is defined by `java.lang`, so an `import` statement for it isn't necessary.

When you run this program, the following is displayed on the console:

```
Oops, you can't divide by zero.
```

There's nothing else to see here. The next section shows a more complicated example, though.

Another example

Listing 8-1 shows a simple example of a program that uses a method to get a valid integer from the user. If the user enters a value that isn't a valid integer, the `catch` block catches the error and forces the loop to repeat.

LISTING 8-1: **Getting a Valid Integer**

```
import java.util.*;

public class GetInteger
    {
```

```
static Scanner sc = new Scanner(System.in);

public static void main(String[] args)
{
    System.out.print("Enter an integer: ");
    int i = GetAnInteger();
    System.out.println("You entered " + i);
}

public static int GetAnInteger()
{
    while (true)
    {
        try
        {
            return sc.nextInt();
        }
        catch (InputMismatchException e)
        {
            sc.next();
            System.out.print("That's not "
                + "an integer. Try again: ");
        }
    }
}
}
```

Here the statement that gets the input from the user and returns it to the calling method is coded within the `try` block. If the user enters a valid integer, this statement is the only one in this method that gets executed.

If the user enters data that can't be converted to an integer, however, the `nextInt` method throws an `InputMismatchException`. Then this exception is intercepted by the `catch` block — which disposes of the user's incorrect input by calling the `next` method and then displays an error message. Then the `while` loop repeats.

Here's what the console might look like for a typical execution of this program:

```
Enter an integer: three
That's not an integer. Try again: 3.001
That's not an integer. Try again: 3
You entered 3
```

Here are a couple other things to note about this program:

TECHNICAL STUFF

» The import statement specifies java.util.* to import all the classes from the java.util package. That way, the InputMismatchException class is imported.

» The next method must be called in the catch block to dispose of the user's invalid input because the nextInt method leaves the input value in the Scanner's input stream if an InputMismatchException is thrown. If you omit the statement that calls next, the while loop keeps reading it, throws an exception, and displays an error message in an infinite loop. If you don't believe me, look at Figure 8-2. I found this error out the hard way. (The only way to make it stop is to close the console window.)

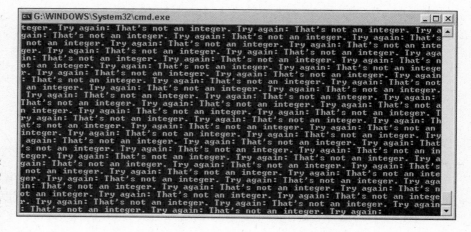

FIGURE 8-2: Why you have to call next to discard the invalid input.

Handling Exceptions with a Preemptive Strike

The try statement is a useful and necessary tool in any Java programmer's arsenal. The best way to handle exceptions, however, is to prevent them from happening in the first place. That's not possible all the time, but in many cases it is. The key is to test your data before performing the operation that can lead to an exception and then skipping or bypassing the operation of the data that is problematic. (One thing I really hate is problematic data.)

You can usually avoid the `ArithmethicException` that results from dividing integer data by zero by checking the data before performing the division:

```
if (b != 0)
    c = a / b;
```

This eliminates the need to enclose the division in a `try` block, because you know that the division by zero won't happen.

You can apply this same technique to input validation by using the `hasNextInt` method of the `Scanner` class. This method checks the next input value to make sure it's a valid integer. (The `Scanner` class calls the next input value a *token*, but that won't be on the test.) You can do this technique in several ways, and I've been encouraging you to ponder the problem since Book 2, Chapter 2. Now behold the long-awaited answer: Listing 8-2 shows a version of the `GetInteger` method that uses a `while` loop to avoid the exception.

LISTING 8-2: **Another Version of the GetInteger Method**

```java
import java.util.*;

public class GetInteger2
{
    static Scanner sc = new Scanner(System.in);

    public static void main(String[] args)
    {
        System.out.print("Enter an integer: ");
        int i = GetAnInteger();
        System.out.println("You entered " + i);
    }

    public static int GetAnInteger()
    {
        while (!sc.hasNextInt())
        {
            sc.nextLine();
            System.out.print("That's not an integer. Try again: ");
        }
        return sc.nextInt();
    }
}
```

This is a clever little bit of programming, don't you think? The conditional expression in the `while` statement calls the `hasNextInt` method of the `Scanner` to see whether the next value is an integer. The `while` loop repeats as long as this call returns `false`, indicating that the next value is not a valid integer. The body of the loop calls `nextLine` to discard the bad data and then displays an error message. The loop ends only when you know that you have good data in the input stream, so the `return` statement calls `nextInt` to parse the data to an integer and return the resulting value.

Catching All Exceptions at Once

Java provides a catch-all exception class called `Exception` that all other types of exceptions are based on. (Don't worry about the details of what I mean by that. When you read Book 3, Chapter 4, it makes more sense.)

If you don't want to be too specific in a `catch` block, you can specify `Exception` instead of a more specific exception class. For example:

```
try
{
    int c = a / b;
}
catch (Exception e)
{
    System.out.println("Oops, you can't divide by zero.");
}
```

In this example, the `catch` block specifies `Exception` rather than `ArithmeticException`.

If you have some code that might throw several types of exceptions, and you want to provide specific processing for some types but general processing for all the others, code the `try` statement this way:

```
try
{
    // statements that might throw several types of exceptions
}
catch (InputMismatchException e)
{
    // statements that process InputMismatchException
}
catch (IOException e)
```

```
{
    // statements that process IOException
}
catch (Exception e)
{
    // statements that process all other exception types
}
```

In this example, imagine that the code in the `try` block could throw an `InputMismatchException`, an `IOException`, and perhaps some other type of unanticipated exception. Here the three `catch` blocks provide for each of these possibilities.

If some of the exceptions to be caught require the same processing, you can combine them in a single `catch` clause. Just separate the exception types with a vertical bar, like this:

```
try
{
    // statements that might throw several types of exceptions
}
catch (InputMismatchException | IOException e)
{
    // statements that process InputMismatchException or IOException
}
catch (Exception e)
{
    // statements that process all other exception types
}
```

TIP

When you code more than one `catch` block on a `try` statement, always list the more specific exceptions first. If you include a `catch` block to catch `Exception`, list it last.

Displaying the Exception Message

In most cases, the `catch block` of a `try` statement won't do anything at all with the exception object passed to it. You may want to display an error message occasionally, however; exception objects have a few interesting methods that can come in handy from time to time. These methods are listed in Table 8-1.

TABLE 8-1   **Methods of the Exception Class**

Method	Description
`String getMessage()`	Describes the error in a text message.
`void printStackTrace()`	Prints the stack trace to the standard error stream.
`String toString()`	Returns a description of the exception. This description includes the name of the exception class followed by a colon and the `getMessage` message.

The following example shows how you could print the message for an exception in a `catch` block:

```
try
{
    int c = a / b;
}
catch (Exception e)
{
    System.out.println(e.getMessage());
}
```

This code displays the text / by zero on the console if b has a value of 0. You can get even more interesting output by using this line in the `catch` clause:

```
e.printStackTrace(System.out);
```

Using a finally Block

A `finally` block is a block that appears after any of the `catch` blocks for a statement. It's executed whether or not any exceptions are thrown by the `try` block or caught by any `catch` blocks. Its purpose is to let you clean up any mess that might be left behind by the exception, such as open files or database connections.

The basic framework for a `try` statement with a `finally` block is this:

```
try
{
    statements that can throw exceptions
}
catch (exception-type identifier)
{
```

```
    statements executed when exception is thrown
}
finally
{
    statements that are executed whether or not
    exceptions occur
}
```

Listing 8-3 shows a contrived but helpful example that demonstrates how to use the `finally` clause. In this example, a method called `divideTheseNumbers` tries to divide the numbers twice. If the division fails the first time (due to a divide-by-zero exception), it tries the division again. Completely irrational, I know. But persistent, like a teenager.

LISTING 8-3: **A Program That Uses a finally Clause**

```
public class CrazyWithZeros
{
    public static void main(String[] args)
    {
        try
        {
            int answer = divideTheseNumbers(5, 0);              →7
        }
        catch (Exception e)                                     →9
        {
            System.out.println("Tried twice, "
                + "still didn't work!");
        }
    }

    public static int divideTheseNumbers(int a, int b)          →16
        throws Exception
    {
        int c;
        try
        {
            c = a / b;     →22
            System.out.println("It worked!");                   →23
        }
        catch (Exception e)
        {
            System.out.println("Didn't work the first time.");  →27
            c = a / b;     →28
            System.out.println("It worked the second time!");   →29
        }
```

(continued)

LISTING 8-3: **(continued)**

```
        finally
        {
            System.out.println("Better clean up my mess.");              →33
        }
        System.out.println("It worked after all.");                     →35
        return c;                                                        →36
    }
}
```

Here's the console output for the program:

```
Didn't work the first time.
Better clean up my mess.
Tried twice, still didn't work!
```

The following paragraphs explain what's going on, step by step:

» →7 The main method calls the divideTheseNumbers method, passing 5 and 0 as the parameters. You already know that this method isn't going to work.

» →9 The catch clause catches any exceptions thrown by line 7.

» →16 The divideTheseNumbers method declares that it throws Exception.

» →22 This line is the first attempt to divide the numbers.

» →23 If the first attempt succeeds, this line is executed, and the message "It worked!" is printed. Alas, the division throws an exception, so this line never gets executed.

» →27 Instead, the catch clause catches the exception, and the message "Didn't work the first time." is displayed. That's the first line in the console output.

» →28 The divideTheseNumbers method stubbornly tries to divide the same two numbers again. This time, no try statement is there to catch the error.

» →29 Because another exception is thrown for the second division, however, this line is never executed. Thus you don't see the message "It worked the second time!" on the console. (If you do, you're in an episode of *The Twilight Zone*.)

» →33 This statement in the finally clause is always executed, no matter what happens. That's where the second line in the console output came from. After the finally clause executes, the ArithmeticException is thrown back up to the calling method, where it is caught by line 9. That's where the last line of the console output came from.

» →35 If the division did work, this line would be executed after the `try` block ends, and you'd see the message `"It worked after all."` on the console.

» →36 Then the `return` statement would return the result of the division.

Handling Checked Exceptions

Checked exceptions are exceptions that the designers of Java feel that your programs absolutely must provide for, one way or another. Whenever you code a statement that could throw a checked exception, your program must do one of two things:

» Catch the exception by placing the statement within a `try` statement that has a `catch` block for the exception.

» Specify a `throws` clause on the method that contains the statement to indicate that your method doesn't want to handle the exception, so it's passing the exception up the stack.

This is known as the *catch-or-throw* rule. In short, any method that includes a statement that might throw a checked exception must acknowledge that it knows the exception might be thrown. The method does this by handling it directly or by passing the exception up to its caller.

TIP

Don't be confused by the similar keywords `throw` and `throws`. They have related but distinct meanings:

» `throw` is a statement that throws an exception.

» `throws` is a keyword affixed to a method declaration to indicate that the method throws a checked exception that isn't captured by a `try-catch` statement.

WARNING

To illustrate the use of checked exceptions, I have to use some classes with methods that throw them. Up to now, I've avoided introducing classes that throw checked exceptions, so the following illustrations use some classes you aren't yet familiar with. Don't worry about what those classes do or how they work. The point is to see how to handle the checked exceptions they throw.

Viewing the catch-or-throw compiler error

Here's a program that uses a class called FileInputStream. To create an object from this class, you must pass the constructor a string that contains the path and name of a file that exists on your computer. If the file can't be found, the FileInputStream throws a FileNotFoundException that you must either catch or throw. This class is in the java.io package, so any program that uses it must include an import java.io statement.

Consider the following program:

```java
import java.io.*;

public class FileException1
{
    public static void main(String[] args)
    {
        openFile("C:\test.txt");
    }

    public static void openFile(String name)
    {
        FileInputStream f = new FileInputStream(name);
    }
}
```

This program won't compile. The compiler issues the following error message:

```
unreported exception java.io.FileNotFoundException; must be caught or declared
    to be thrown
```

This message simply means that you have to deal with the FileNotFoundException.

Catching FileNotFoundException

One way to deal with the FileNotFoundException is to catch it by using an ordinary try statement:

```java
import java.io.*;

public class FileException2
{
    public static void main(String[] args)
    {
        openFile("C:\test.txt");
```

```
        }

    public static void openFile(String name)
    {
        try
        {
            FileInputStream f = new FileInputStream(name);
        }
        catch (FileNotFoundException e)
        {
            System.out.println("File not found.");
        }
    }
}
```

In this example, the message "File not found." is displayed if the C:\test.txt file doesn't exist.

Throwing the FileNotFoundException

Suppose that you don't want to deal with this error condition in the openFile method, but would rather just pass the exception up to the method that calls the openFile method.

To do that, you omit the try statement. Instead, you add a throws clause to the openFile method's declaration. That clause indicates that the openFile method knows it contains a statement that might throw a FileNotFoundException but doesn't want to deal with that exception here. Instead, the exception is passed up to the caller.

Here's the openFile method with the throws clause added:

```
public static void openFile(String name)
    throws FileNotFoundException
{
    FileInputStream f = new FileInputStream(name);
}
```

As you can see, the throws clause simply lists the exception or exceptions that the method might throw. If more than one exception is in the list, separate the exceptions with commas:

```
public static void readFile(String name)
    throws FileNotFoundException, IOException
```

Adding a `throws` clause to the `openFile` method means that when the `FileNot-FoundException` occurs, it is simply passed up to the method that called the `open-File` method. That means the calling method (in this illustration, `main`) must catch or throw the exception. To catch the exception, the `main` method would have to be coded like this:

```
public static void main(String[] args)
{
    try
    {
        openFile("C:\test.txt");
    }
    catch (FileNotFoundException e)
    {
        System.out.println("File not found.");
    }
}
```

Then, if the file doesn't exist, the `catch` block catches the exception, and the error message is displayed.

Throwing an exception from main

If you don't want the program to handle the `FileNotFound` exception at all, you can add a `throws` clause to the `main` method, like this:

```
public static void main(String[] args)
    throws FileNotFoundException
{
    openFile("C:\test.txt");
}
```

Then the program abruptly terminates with an exception message and stack trace if the exception occurs.

Swallowing exceptions

What if you don't want to do anything if a checked exception occurs? In other words, you want to simply ignore the exception. You can do that by catching the exception in the `catch` block of a `try` statement but leaving the body of the `catch` block empty. Here's an example:

```
public static void openFile(String name)
{
```

```
    try
    {
        FileInputStream f = new FileInputStream(name);
    }
    catch (FileNotFoundException e)
    {
    }
}
```

Here the `FileNotFoundException` is caught and ignored. This technique is called *swallowing the exception.*

WARNING

Swallowing an exception is considered to be bad programming practice. Simply swallowing exceptions that you know you should handle when working on a complicated program is tempting. Because you plan to get back to that exception handler after you iron out the basic functions of the program, a little exception-swallowing doesn't seem like that bad an idea. The problem is that inevitably, you never get back to the exception handler, so your program gets rushed into production with swallowed exceptions.

If you must swallow exceptions, at least write a message to the console indicating that the exception occurred. That way, you have a constant reminder that the program has some unfinished details you must attend to.

Note that it is possible to intentionally swallow exceptions in a way that lets you manage the flow of your program. For example, suppose you want the `openFile` method to return a `boolean` value to indicate whether the file exists, rather than throw an exception. Then you could code the method something like this:

```
public static boolean openFile(String name)
{
    boolean fileOpened = false;
    try
    {
        FileInputStream f = new FileInputStream(name);
        fileOpened = true;
    }
    catch (FileNotFoundException e)
    {
    }
    return fileOpened;
}
```

Here the exception isn't really swallowed. Instead, its meaning is converted to a `boolean` result that's returned from the method. As a result, the error condition indicated by the `FileNotFoundException` isn't lost.

Swallowing exceptions in this way isn't generally considered to be a good practice, however. You'd be better off just testing to see if the file exists before you attempt to open it. If the file doesn't exist, you can simply return `false`.

Throwing Your Own Exceptions

Although such methods are uncommon, you may want to write methods that throw exceptions all on their own. To do that, you use a throw statement. The throw statement has the following basic format:

```
throw new exception-class ();
```

The *exception-class* can be Exception or a class that's derived from Exception. You find out how to create your own classes — including exception classes — in Book 3. For now, I just focus on writing a method that throws a general Exception.

Here's a program that demonstrates the basic structure of a method that throws an exception:

```
public class MyException
{
    public static void main(String[] args)
    {
        try
        {
            doSomething(true);
        }
        catch (Exception e)
        {
            System.out.println("Exception!");
        }
    }

    public static void doSomething(boolean t)
        throws Exception
    {
        if (t)
            throw new Exception();
    }
}
```

Here the doSomething method accepts a boolean value as a parameter. If this value is true, it throws an exception; otherwise it doesn't do anything.

Here are the essential points to glean from this admittedly trivial example:

>> You throw an exception by executing a throw statement. The throw statement specifies the exception object to be thrown.

>> If a method intentionally throws a checked exception, it must include a throws clause in its declaration.

>> A method that calls a method that throws an unchecked exception must either catch or throw the exception.

TIP

>> Yup, this example is pretty trivial. But it illustrates the essential points.

3

Object-Oriented Programming

Contents at a Glance

Chapter **1**

Understanding Object-Oriented Programming

This chapter is a basic introduction to object-oriented programming. It introduces some of the basic concepts and terms you need to know as you get a handle on the specific details of how object-oriented programming works in Java.

TIP

If you're more of a hands-on type, you may want to skip this chapter and go straight to Book 3, Chapter 2, where you find out how to create your own classes in Java. You can always return to this chapter later to review the basic concepts that drive object-oriented programming. Either way is okay by me. I get paid the same whether you read this chapter now or skip it and come back to it later.

What Is Object-Oriented Programming?

The term *object-oriented programming* means many different things. But at its heart, object-oriented programming is a type of computer programming based on the premise that all programs are essentially computer-based simulations of real-world objects or abstract concepts. For example:

>> Flight-simulator programs attempt to mimic the behavior of real airplanes. Some do an amazingly good job; military and commercial pilots train on them. In the 1960s, the Apollo astronauts used a computer-controlled simulator to practice for their moon landings.

>> Many computer games are simulations of actual games that humans play, such as baseball, NASCAR racing, and chess. But even abstract games such as Pac-Man and Final Fantasy IV attempt to model the behavior of creatures and objects that *could* exist somewhere. Those programs simulate a conceptual game — one that can't actually be played anywhere in the real world but *can* be simulated by a computer.

>> Business programs can be thought of as simulations of business processes — such as order taking, customer service, shipping, and billing. An invoice, for example, isn't just a piece of paper; it's a paper that represents a transaction that has occurred between a company and one of its customers. Thus, a computer-based invoice is really just a simulation of that transaction.

TECHNICAL STUFF

The notion of a programming language having a premise of this sort isn't new. Traditional programming languages such as C (and its predecessors, including even COBOL) are based on the premise that computer programs are computerized implementations of actual procedures — the electronic equivalent of "Step 1: Insert Tab A into Slot B." The LISP programming language is based on the idea that all programming problems can be looked at as different ways of manipulating lists. And the ever-popular database-manipulation language SQL views programming problems as ways to manipulate mathematical sets.

Here are some additional thoughts about the notion of computer programs being simulations of real-world objects or abstract concepts:

>> Sometimes the simulation is better than the real thing. Word-processing programs started out as simulations of typewriters, but a modern word-processing program is far superior to any typewriter.

>> The idea that all computer programs are simulations of one type or another isn't a new one. In fact, the first object-oriented programming language (Simula) was developed in the 1960s. By 1967, this language had many of the features we now consider fundamental to object-oriented programming — including classes, objects, inheritance, and virtual methods.

>> Come to think of it, manual business recordkeeping systems are simulations too. A file cabinet full of printed invoices doesn't hold actual orders; it holds written *representations* of those orders. A computer is a better simulation device than a file cabinet, but both are simulations.

Understanding Objects

All this talk of simulations is getting a little existential for me, so now I'm turning to the nature of the objects that make up object-oriented programming. *Objects* — both in the real world and in the world of programming — are entities that have certain basic characteristics. The following sections describe some of the most important of these characteristics: identity, type, state, and behavior.

Objects have identity

Every object in an object-oriented program has an *identity*. In other words, every occurrence of a particular type of object — called an *instance* — can be distinguished from every other occurrence of the same type of object, as well as from objects of other types.

In the real world, object identity is a pretty intuitive and obvious concept. Pick up two apples, and you know that although both of them are apples (that's the object type, described in the next section), you know that they aren't the same apple. Each has a distinct identity. Both are roughly the same color but not exactly. They're both roundish but have minor variations in shape. Either one (or both) could have a worm inside.

Open a file cabinet that's full of invoices, and you find page after page of papers that look almost identical, but each one has an invoice number printed somewhere near the top of the page. This number isn't what actually gives each of these invoices a unique identity, but it gives you an easy way to identify each individual invoice, just as your name gives other people an easy way to identify you.

In object-oriented programming, each object has its own location in the computer's memory. Thus, two objects, even though they may be of the same type, have their own distinct memory locations. The address of the starting location for an object provides a way of distinguishing one object from another, because no two objects can occupy the same location in memory.

Here are a few other important thoughts about object identity in Java:

» Java keeps each object's identity pretty much to itself. In other words, there's no easy way to get the memory address of an object; Java figures that it's none of your business, and rightfully so. If Java made that information readily available to you, you'd be tempted to tinker with it, which could cause all sorts of problems, as any C or C++ programmer can tell you.

» When used with objects, the equality operator (==) actually tests the object identity of two variables or expressions. If they refer to the same object instance, the two variables or expressions are considered equal.

» Java objects have something called a *hash code,* which is an int value that's automatically generated for every object and *almost* represents the object's identity. Two objects that are equal will always have the same hash code, but two objects that are not equal are not guaranteed to have different hash codes.

Objects have type

I remember studying "Naming of Parts," a fine poem written by Henry Reed in 1942, back when I was an English major in college:

To-day we have naming of parts. Yesterday,

We had daily cleaning. And to-morrow morning,

We shall have what to do after firing. But to-day,

To-day we have naming of parts. Japonica

Glistens like coral in all of the neighboring gardens,

And today we have naming of parts.

Sure, it's a fine antiwar poem and all that, but it's also a little instructive about object-oriented programming. After the first stanza, the poem goes on to name the parts of a rifle:

This is the lower sling swivel. And this

Is the upper sling swivel, whose use you will see,

When you are given your slings. And this is the piling swivel,

Which in your case you have not got.

Imagine a whole room of new soldiers taking apart their rifles, while the drill sergeant tells them, "This is the lower sling swivel. And this is the upper sling swivel. . . ." Each soldier's rifle has one of these parts — in object-oriented terms, an object of a particular type. The lower sling swivels in the soldiers' rifles are different objects, but all are of the type LowerSlingSwivel.

Like the drill sergeant in this poem, object-oriented programming lets you assign names to the different kinds of objects in a program. In Java, types are defined by classes. So when you create an object from a type, you're saying that the object is of the type specified by the class. The following example statement creates an object of type Invoice:

```
Invoice i = new Invoice();
```

In this case, a reference to the newly created Invoice object is assigned to the variable i, which the compiler knows can hold references to objects of type Invoice.

Objects have state

Now switch gears to another literary genius:

One fish, two fish,

Red fish, blue fish

In object-oriented terms, Dr. Seuss here is enumerating a pair of objects of type Fish. The Fish type apparently has two attributes; call them Number and Color. These two objects have differing values for these attributes:

Attribute	Object 1	Object 2
Number	One	Two
Color	Red	Blue

The type of an object determines what attributes the object has. Thus all objects of a particular type have the same attributes. They don't necessarily have the same values for those attributes, however. In this example, all Fish have attributes named Number and Color, but the two Fish objects have different values for these attributes.

The combination of the values for all the attributes of an object is called the object's *state*. Unlike its identity, an object's state can — and usually does — change over its lifetime. Some fish can change colors, for example. The total sales for a particular customer changes each time the customer buys another product. The grade-point average for a student changes each time a new class grade is recorded. The address and phone number of an employee change if the employee moves.

Here are a few more interesting details about object state:

>> Some of the attributes of an object are publicly known, but others can be private. The private attributes may be vital to the internal operation of the object, but no one outside the object knows that they exist. They're like your private thoughts: They affect what you say and do, but nobody knows them but you.

>> In Java, the state of an object is represented by instance variables, which are called *fields*. A *public field* is a field that's declared with the public keyword so that the variable can be visible to the outside world.

Objects have behavior

Another characteristic of objects is that they have *behavior,* which means that they can do things. Like state, the specific behavior of an object depends on its type. But unlike state, behavior isn't different for each instance of a type. Suppose that all the students in a classroom have calculators of the same type. Ask them all to pull out the calculators and add two numbers — any two numbers of their choosing. All the calculators display a different number, but they all add in the same way — that is, they all have a different state but the same behavior.

Another way to say that objects have behavior is to say that they provide services that can be used by other objects. You've already seen plenty of examples of objects that provide services to other objects. Objects created from the NumberFormat class, for example, provide formatting services that turn numeric values into nicely formatted strings such as $32.95.

In Java, the behavior of an object is provided by its methods. Thus the format method of the NumberFormat class is what provides the formatting behavior for NumberFormat objects.

Here are a few other notable points about object behavior:

>> The *interface* of a class is the set of methods and fields that the class makes public so that other objects can access them. (Note that I use the term

interface here generically; Java has a specific feature that goes by the name interface, which I introduce in Chapter 5 of this minibook.)

>> The *implementation* of a class refers to the Java code that determines exactly how an object does what it does. The implementation can and should be hidden within the object. Someone who uses the object needs to know what the object does but doesn't need to know how it works. If you later find a better way for the object to do its job, you can swap in the new improved version without anyone knowing the difference.

Understanding the Life Cycle of an Object

As you work with objects in Java, understanding how objects are born, live their lives, and die is important. This topic is called the *life cycle* of an object, and it goes something like this:

1. Before an object can be created from a class, the class must be *loaded*. To do that, the Java runtime locates the class on disk (in a .class file or, in the case of a file that contains just one class, a .java file) and reads it into memory. Then Java looks for any *static initializers* that initialize static fields — fields that don't belong to any particular instance of the class, but belong to the class itself and are shared by all objects created from the class.

A class is loaded the first time you create an object from the class or the first time you access a static field or method of the class. When you run the main method of a class, for example, the class is initialized because the main method is static.

2. An object is created from a class when you use the new keyword. To initialize the class, Java allocates memory for the object and sets up a reference to the object so that the Java runtime can keep track of it. Then the class *constructor* is called. The constructor is like a method but is called only once, when the object is created. The constructor is responsible for doing any processing required to initialize the object — initializing variables, opening files or databases, and so on.

3. The object lives its life, providing access to its public methods and fields to whoever wants and needs them.

4. When it's time for the object to die, the object is removed from memory, and Java drops its internal reference to it. You don't have to destroy objects yourself. A special part of the Java runtime called the *garbage collector* takes care of destroying all objects when they are no longer in scope.

Working with Related Classes

So far, most of the classes you've seen in this book have created objects that stand on their own, each being a little island unto itself. The real power of object-oriented programming, however, lies in its ability to create classes that describe closely related objects.

Baseballs, for example, are similar to softballs. Both are specific types of balls. Each type has a diameter and a weight; both types can be thrown, caught, or hit. Baseballs and softballs, however, have different characteristics that cause them to behave differently when they're thrown, caught, or hit.

If you're creating a program that simulates the way baseballs and softballs work, you need a way to represent these two types of balls. One option is to create separate classes to represent each type of ball. These classes are similar, so you can just copy most of the code from one class to the other.

Another option is to use a single class to represent both types of balls. Then you pass a parameter to the constructor to indicate whether an instance of the class behaves like a baseball or like a softball.

Both of these methods will work, but neither is ideal. In the first, you end up creating two classes that are clearly related, but there's no formal relationship established between the two classes. In the second, you create a class that represents two things that are similar but aren't exactly the same, so you end up with a class that has more than one purpose.

Java has two object-oriented programming features that are designed specifically to handle classes that are related this way: inheritance and interfaces. I briefly describe these features in the following sections.

Inheritance

Inheritance is an object-oriented programming technique that lets you use one class as the basis for another. The existing class is called the *base class, superclass,* or *parent class*; the new class that's derived from it is called the *derived class, subclass,* or *child class.*

When you create a subclass, the subclass is automatically given all the visible (that is, `public` or `protected`) methods and fields of its superclass. You can use these methods and fields as is, or you can override them to alter their behavior. In addition, you can add methods and fields that define data and behavior that's unique to the subclass.

You could use inheritance to solve the baseball/softball problem from the preceding section by creating a class named `Ball` that provides the basic features of all types of balls and then using it as the base class for separate classes named `BaseBall` and `SoftBall`. Then these classes could override the methods that need to behave differently for each type of ball.

One way to think of inheritance is as a way to implement *is-a-type-of relationships*. A softball is a type of ball, as is a baseball. Thus inheritance is an appropriate way to implement these related classes. (For more information about inheritance, see Book 3, Chapter 4.)

Interfaces

An *interface* is a set of methods and fields that a class must provide to *implement* the interface. The interface itself is simply a set of public method and field declarations that are given a name. Note that the interface itself doesn't provide any code that implements those methods. Instead, it just provides the declarations. Then a class that *implements* the interface provides code for each of the methods the interface defines.

You could use an interface to solve the baseball/softball problem by creating an interface named `Ball` that specifies all the methods and fields that a ball should have. Then you could create the `SoftBall` and `BaseBall` classes so that they both implement the `Ball` interface.

Interfaces are closely related to inheritance but have two key differences:

>> The interface itself doesn't provide code that implements any of its methods. An interface is just a set of method and field signatures. By contrast, a base class can provide the implementation for some or all of its methods.

>> A class can have only one base class, but a class can implement as many interfaces as necessary.

TECHNICAL STUFF

Strictly speaking, an interface can provide implementations for some of its methods. Specifically, an interface can provide an implementation for static methods, and an interface can provide a *default method*, which has a default implementation that can be overridden by the class that implements the interface.

You find out more about interfaces in Book 3, Chapter 5.

Designing a Program with Objects

An object-oriented program usually isn't just a single object. Instead, it's a group of objects that work together to get a job done. The most important part of developing an object-oriented program is designing the classes that are used to create the program's objects. The basic idea is to break a large problem into a set of classes, each of which is manageable in size and complexity. Then you write the Java code that implements those classes.

So the task of designing an object-oriented application boils down to deciding what classes the application requires — and what the public interface to each of those classes should be. If you plan your classes well, implementing the application is easy. If you plan your classes poorly, you'll have a hard time getting your application to work.

One common way to design object-oriented applications is to divide the application into several distinct *layers* or *tiers* that provide distinct types of functions. Most common is a three-layered approach, as shown in Figure 1-1. Here the objects of an application are split into three basic layers:

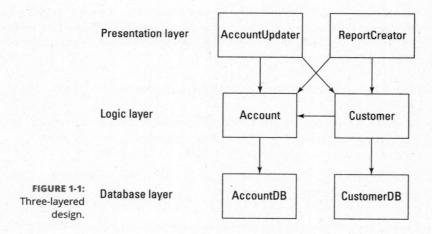

FIGURE 1-1: Three-layered design.

» **Presentation:** The objects in this layer handle all the direct interaction with users. The HTML pages in a web application go in this layer, as do the JavaFX classes in a GUI-based application. (I cover JavaFX in Book 6.)

» **Logic:** The objects in this layer represent the core objects of the application. For a typical business-type application, this layer includes objects that represent business entities such as customer, products, orders, suppliers, and the like. This layer is sometimes called the *business rules layer* because the objects in this layer are responsible for carrying out the rules that govern the application.

» **Database:** The objects in this layer handle all the details of interacting with whatever form of data storage is used by the application. If the data is stored in a SQL database, for example, the objects in this layer handle all the SQL.

Diagramming Classes with UML

Since the very beginning of computer programming, programmers have loved to create diagrams of their programs. Originally, they drew flowcharts that graphically represented a program's procedural logic.

Flowcharts were good at diagramming procedures, but they were way too detailed. When the structured programming craze hit in the 1970s, and programmers started thinking about the overall structure of their programs, they switched from flowcharts to *structure charts*, which illustrated the organizational relationships among the modules of a program or system.

Now that object-oriented programming is the thing, programmers draw *class diagrams* to illustrate the relationships among the classes that make up an application. Figure 1-2 shows a class diagram of a simple system that has four classes. The rectangles represent the classes themselves, and the arrows represent the relationships among the classes.

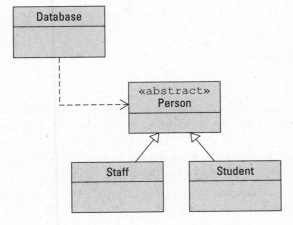

FIGURE 1-2:
A simple class diagram.

You can draw class diagrams in many ways. To add some consistency to their diagrams, most programmers use a standard called *UML*, which stands for *Unified Modeling Language*. The class diagram in Figure 1-2 is an example of a simple UML diagram, but UML diagrams can get much more complicated.

The following sections describe the details of creating UML class diagrams. Note that these sections don't even come close to explaining all the features of UML. I include just the basics of creating UML class diagrams so that you can make some sense of UML diagrams when you see them and so that you know how to draw simple class diagrams to design the class structure for your applications.

Drawing classes

The basic element in a class diagram is a class. In UML, each class is drawn as a rectangle. At minimum, the rectangle must include the class name. You can sub-divide the rectangle into two or three compartments that can contain additional information about the class, as shown in Figure 1-3.

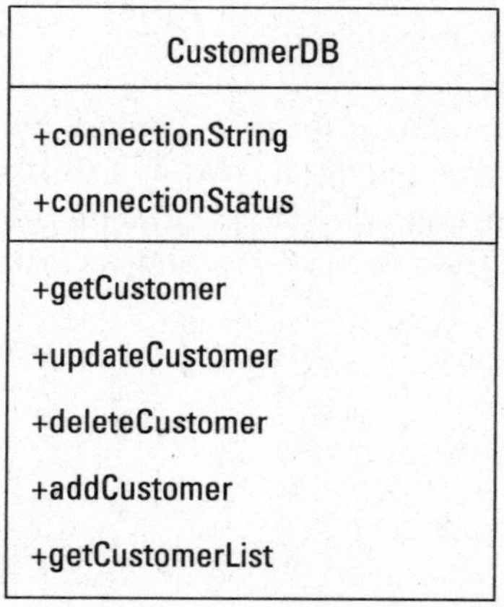

CustomerDB
+connectionString
+connectionStatus
+getCustomer
+updateCustomer
+deleteCustomer
+addCustomer
+getCustomerList

FIGURE 1-3:
A class.

The middle compartment of a class lists the instance variables, whereas the bottom compartment lists the class methods. The name of each variable or method can be preceded by a *visibility indicator*, which can be one of the symbols listed in Table 1-1. (In actual practice, it's common to omit the visibility indicator and list only those fields or methods that have public visibility.)

TABLE 1-1

Visibility Indicators for Class Variables and Methods

Indicator	Description
+	Public
–	Private
#	Protected

If you want, you can include type information for variables as well as for methods and parameters. The type of a variable is indicated by following the variable name with a colon and the type:

```
connectionString: String
```

A method's return type is indicated in the same way:

```
getCustomer(): Customer
```

Parameters are listed within the parentheses, and both the name and type are listed. For example:

```
getCustomer(custno: int): Customer
```

Note: Omitting the type and parameter information from UML diagrams is common.

TIP

Interfaces are drawn pretty much the same way as classes, but the class name is preceded by the word *interface*:

```
<<interface>>
ProductDB
```

Note: The word *interface* is enclosed within a set of double-left and double-right arrows. These arrows aren't just two less-than or greater-than symbols typed in a row; they're a special combination of symbols. Fortunately, the double-arrow symbol is a standard part of the ASCII character set. You can access it in Microsoft Word via the Insert Symbol command.

Drawing arrows

Besides using rectangles to represent classes, class diagrams include arrows to represent relationships among classes. UML uses a variety of types of arrows, as I describe in the following paragraphs.

A solid line with a hollow closed arrow at one end represents inheritance:

—————————————————▷

The arrow points to the base class. A dashed line with a hollow closed arrow at one end indicates that a class implements an interface:

– – – – – – – – – – – – – – –▷

The arrow points to the interface. A solid line with an open arrow indicates an *association*:

—————————————————→

An association simply indicates that two classes work together. It may be that one of the classes creates objects of the other class or that one class requires an object of the other class to perform its work. Or perhaps instances of one class contain instances of the other class.

You can add a name to an association arrow to indicate its purpose. If an association arrow indicates that instances of one class create objects of another class, you can place the word Creates next to the arrow.

Chapter **2**

Making Your Own Classes

O kay, class, it's time to learn how to create your own classes.

In this chapter, you discover the basics of creating classes in Java. All Java programs use or consist of classes, so you've already seen many examples of classes. You've seen class headers such as public class GuessingGame and static methods such as public static void main. Now, in this chapter, I show you how to create programs that have more than one class.

Declaring a Class

All classes must be defined by a *class declaration* — lines of code that provide the name for the class and the body of the class. Here's the most basic form of a class declaration:

```
[public] class ClassName {class-body}
```

The `public` keyword indicates that this class is available for use by other classes. Although it's optional, you usually include it in your class declarations. After all, the main reason you write class declarations is so other classes can create objects from the class you're defining. (Find out more about using the `public` keyword in the section "Seeing where classes go," later in this chapter.)

TECHNICAL STUFF

In later chapters of this book, you find out about some additional elements that can go in a class declaration. The format I'm describing here is just the basic format used to create basic classes.

Picking class names

The `ClassName` is a name that provides a name for your class. You can use any legal Java name you want to name a class, but the following three guidelines can simplify your life:

>> **Begin the class name with a capital letter.** If the class name consists of more than one word, capitalize each word: for example, `Ball`, `RetailCustomer`, and `GuessingGame`.

>> **Whenever possible, use nouns for your class names.** Classes create objects, and nouns are the words you use to identify objects. Thus most class names should be nouns.

>> **Avoid using the name of a Java API class.** No rule says that you absolutely have to, but if you create a class that has the same name as a Java API class, you have to use fully qualified names (such as `java.util.Scanner`) to tell your class apart from the API class with the same name.

TIP

There are thousands of Java API classes, so avoiding them all is pretty hard. But at the least, you should avoid commonly used Java class names, as well as any API classes that your application is likely to use. Creating a class named `String` or `Math`, for example, is just asking for trouble.

TECHNICAL STUFF

A Java class name must, of course, conform to Java's requirements for names in general. So, the name must begin with a letter, cannot contain white space or special characters other than underscores or dollar signs, and cannot be the same as any keyword or other reserved word such as `var`, `true`, `false`, or `null`.

Knowing what goes in the class body

The *class body* of a class is everything that goes within the braces at the end of the class declaration. The `public class ClassName` part of a class declaration takes just one line, but the body of the class declaration may take hundreds of lines (or thousands, if you get carried away).

The class body can contain the following elements:

>> **Fields:** Variable declarations define the public or private fields of a class.

>> **Methods:** Method declarations define the methods of a class.

>> **Constructors:** A *constructor* is a block of code that's similar to a method but is run to initialize an object when an instance is created. A constructor must have the same name as the class itself, and although it resembles a method, it doesn't have a return type.

>> **Initializers:** These stand-alone blocks of code are run only once, when the class is initialized. There are actually two types, called *static initializers* and *instance initializers*. I talk about instance initializers later in this chapter, in the section "Using Initializers." (For information about static initializers, refer to Book 3, Chapter 3.)

>> **Other classes and interfaces:** A class can include another class, which is then called an *inner class* or a *nested class*. Classes can also contain interfaces. (For more information about inner classes, see Book 3, Chapter 7. And for information about interfaces, refer to Book 3, Chapter 5.)

TIP

Unlike some programming languages, Java doesn't care about the order in which items appear in the class body. Still, being consistent about the order in which you place things in your classes is a good idea. That way you know where to find them. I usually code all the fields together at the start of the class, followed by constructors and then methods. If the class includes initializers, I place them near the fields they initialize. And if the class includes inner classes, I usually place them after the methods that use them.

Some programmers like to place the fields at the end of the class rather than at the beginning. Whatever brings you happiness is fine with me.

TECHNICAL STUFF

The fields, methods, classes, and interfaces contained within a class are called the *members* of the class. Constructors and initializers aren't considered to be members, for reasons that are too technical to explain at this point. (You can find the explanation in Book 3, Chapter 3.)

Seeing where classes go

A public class must be written in a source file that has the same name as the class, with the extension java. A public class named Greeter, for example, must be placed in a file named Greeter.java.

As a result, you can't place two public classes in the same file. The following source file (named DiceGame.java) won't compile:

```
public class DiceGame
{
    public static void main(String[] args)
    {
        Dice d = new Dice();
        d.roll();
    }
}

public class Dice
{
    public void roll()
    {
        // code that rolls the dice goes here
    }
}
```

The compiler coughs up a message indicating that Dice is a public class and must be declared in a file named Dice.java.

This problem has two solutions. The first is to remove the public keyword from the Dice class:

```
public class DiceGame
{
    public static void main(String[] args)
    {
        Dice d = new Dice();
        d.roll();
    }
}
```

```
class Dice
{
    public void roll()
    {
        // code that rolls the dice goes here
    }
}
```

The compiler gladly accepts this program.

TECHNICAL STUFF

When you code more than one class in a single source file, Java still creates a separate class file for each class. Thus, when you compile the DiceGame.java file, the Java compiler creates two class files: DiceGame.class and Dice.class.

Removing the public keyword from a class is acceptable for relatively small programs, but its limitation is that the Dice class is available only to the classes defined within the DiceGame.java file. If you want the Dice class to be more widely available, opt for the second solution: Place it, with the public keyword, in a separate file named Dice.java.

TIP

If you're going to create an application that has several public classes, create a separate folder for the application. Then save all the class files for the application to this folder. If you keep your class files together in the same folder, the Java compiler can find them. If you place them in separate folders, you may need to adjust your ClassPath environment variable to help the compiler find the classes.

Working with Members

The *members* of a class are the fields and methods defined in the class body. (Technically, classes and interfaces defined within a class are members too. I don't discuss them in this chapter, though, so you can ignore them for now.)

The following sections describe the basics of working with fields and methods in your classes.

Understanding fields

A *field* is a variable that's defined in the body of a class, outside any of the class's methods. Fields, which are also called *instance variables,* are available to all the methods of a class. In addition, if the field specifies the public keyword, the field is visible outside the class. If you don't want the field to be visible outside the class, use the private keyword instead.

A field is defined the same as any other Java variable, but it can also have a modifier that specifies the visibility of the field — that is, whether other classes can access the fields of the class you're defining. For now, I'll just use two basic forms of visibility: public and private. For a more complete discussion of visibility, see the section "Understanding Visibility," later in this chapter.

To create a public field that can be accessed by other classes, use the public modifier:

```
public int trajectory = 0;
public String name;
public Player player;
```

To create a private field, specify private instead of public:

```
private int x_position = 0;
private int y_position = 0;
private String error-message = "";
```

Fields can also be declared as final:

```
public final int MAX_SCORE = 1000;
```

The value of a final field can't be changed after it has been initialized. *Note:* Spelling static final field names with all capital letters is customary, but not required.

Understanding instance methods

You define methods for a class by using the same techniques that I describe in Book 2, Chapter 7. To declare a method that's available to users of your class, add the public keyword to the method declaration:

```
public boolean isActive()
{
    return this.isActive;
}
```

To create a private method that can be used within the class but isn't visible outside the class, use the private keyword:

```
private void calculateLunarTrajectory()
{
    // code to get the calculated lunar trajectory
}
```

Understanding visibility

In the preceding sections, I mention that both fields and methods can use the `public` or `private` keyword to indicate whether the field or method can be accessed from outside the class. This is called the *visibility* of the field or method.

There are actually four distinct levels of visibility you can use:

>> `private`: For fields that shouldn't be visible to any other classes — in other words, fields that are completely internal to the class.

>> `public`: For fields that should be visible to every other Java class, including classes that are outside of the current package.

>> `protected`: For fields that should be visible only to subclasses of the current class — that is, to subclasses or inner classes. (For more information, refer to Chapters 4 and 7 of this minibook.)

>> `package-private`: Use this visibility for fields that should be visible to any other class within the current package.

The combination of all the members that have `public` access is sometimes called the *public interface* of your class. These members are the only means that other objects have to communicate with objects created from your class. As a result, carefully consider which public fields and methods your class declares. (Again, I use the term *interface* here in a general sense, not to be confused with the specific Java feature called interface, which I cover in Chapter 5 of this minibook.)

The term *expose* is sometimes used to refer to the creation of public fields and methods. If a class has a public method named `isActive`, for example, you could say that the class exposes the `isActive` method. That simply means the method is available to other classes.

WARNING

You can use private fields and methods within a class — but not from other classes. Private fields and methods provide implementation details that may be crucial to the operation of your class but that shouldn't be exposed to the outside world. Private fields and methods are sometimes called *internal members* because they're available only from within the class.

Using Getters and Setters

One of the basic goals of object-oriented programming is to hide the implementation details of a class inside the class while carefully controlling what aspects of the class are exposed to the outside world. This is often referred to as *encapsulation*.

As a general rule, you hide as many of the details of your implementation from the outside world as you possibly can.

One way to do that is to avoid creating public fields. Instead, make your fields private. Then, selectively grant access to the data those fields contain by adding to the class special methods called *accessors*.

There are two types of accessors. A *get accessor* (also called a *getter*) is a method that retrieves a field value, whereas a *set accessor (setter)* is a method that sets a field value. These methods are usually named get*FieldName* and set*FieldName*, respectively. If the field is named count, for example, the getter and setter methods are named getCount and setCount.

TIP

For boolean values, it's common to use the name is*FieldName* for the getter method. For example, if a field is named Enabled, the corresponding getter method would be named isEnabled.

Here's a class that uses a private field named Health to indicate the health of a player in a game program:

```
public class Player
{
    private int health;

    public int getHealth()
    {
        return health;
    }

    public void setHealth(int h)
    {
        health = h;
    }
}
```

Here the health field itself is declared as private, so it can't be accessed directly. Instead, it can be accessed only through the methods getHealth and setHealth.

Creating classes with accessors rather than simple public fields offers several benefits:

» You can create a read-only property by providing a get accessor but not a set accessor. Then other classes can retrieve the property value — but can't change it.

» Instead of storing the property value in a private field, you can calculate it each time the get accessor method is called. Suppose you have a class named Order that includes fields named unitPrice and quantityOrdered. This class might also contain a getOrderTotal method that looks like this:

```
public double getOrderTotal()
{
    return unitPrice * quantityOrdered;
}
```

Here, instead of returning the value of a class field, the get accessor calculates the value to be returned.

» You can protect the class from bad data by validating data in a property set accessor and either ignoring invalid data or throwing an exception if invalid data is passed to the method. Suppose that you have a set accessor for an int property named Health whose value can range from 0 to 100. Here's a set accessor that prevents the Health property from being set to an incorrect value:

```
public void setHealth(int h)
{
    if (h < 0)
        health = 0;
    else if (h > 100)
        health = 100;
    else
        health = h;
}
```

Here, if the setHealth method is called with a value less than 0, health is set to 0. Likewise, if the value is greater than 100, health is set to 100.

For a little added insight on the use of accessors, see the nearby sidebar "The Accessor pattern."

Overloading Methods

A Java class can contain two or more methods with the same name, provided that those methods accept different parameters. This technique, called *overloading*, is one of the keys to building flexibility into your classes. With overloading, you can anticipate different ways that someone might want to invoke an object's functions and then provide overloaded methods for each alternative.

TIP

The term *overloading* is accurate but a little unfortunate. Normally, when you say that something is overloaded, there's a problem. I once saw a picture of a Volkswagen Jetta loaded down with 3,000 pounds of lumber. (You can find the picture courtesy of Snopes.com, the Urban Legend Reference Page website, at www.snopes.com/photos/automobiles/lumber.asp.) That's a classic example of overloading in the ordinary sense. Fortunately, you don't have to worry about Java collapsing under the weight of overloaded methods.

You're already familiar with several classes that have overloaded methods, though you may not realize it. The `PrintWriter` class, for example (which you access via `System.out`), defines 10 versions of the `println` method that allow you to print different types of data. The following lines show the method declaration for each of these overloads:

```
void println()
void println(boolean x)
void println(char x)
void println(char[] x)
void println(double x)
void println(float x)
void println(int x)
void println(long x)
void println(Object x)
void println(String x)
```

The basic rule in creating overloaded methods is that every method must have a unique signature. A method's *signature* is the combination of its name and the number and types of parameters it accepts. Thus, each of the `println` methods has a different signature, because although all the methods have the same name, each method accepts a different parameter type.

Two things that are *not* a part of a method's signature are

>> **The method's return type:** You can't code two methods with the same name and parameters but with different return types.

>> **The names of the parameters:** All that matters to the method signature are the types of the parameters and the order in which they appear. Thus the following two methods have the same signature:

```
double someMethodOfMine(double x, boolean y)
double someMethodOfMine(double param1, boolean param2)
```

Creating Constructors

A *constructor* is a block of code that's called when an instance of an object is created. In many ways, a constructor is similar to a method, but a few differences exist:

» A constructor doesn't have a return type.

» The name of the constructor must be the same as the name of the class.

» Unlike methods, constructors are not considered to be members of a class. (That's important only when it comes to inheritance, which is covered in Book 3, Chapter 4.)

» A constructor is called when a new instance of an object is created. In fact, it's the new keyword that calls the constructor. After creating the object, you can't call the constructor again.

Here's the basic format for coding a constructor:

```
public ClassName (parameter-list) [throws exception...]
{
    statements...
}
```

The public keyword indicates that other classes can access the constructor. That's usually what you want, although in the next chapter, you see why you might want to create a private constructor. ClassName must be the same as the name of the class that contains the constructor. You code the parameter list the same way that you code it for a method.

Notice also that a constructor can throw exceptions if it encounters situations that it can't recover from. (For more information about throwing exceptions, refer to Book 2, Chapter 8.)

Creating basic constructors

Probably the most common reason for coding a constructor is to provide initial values for class fields when you create the object. Suppose that you have a class named Actor that has fields named firstName and lastName. You can create a constructor for the Actor class:

```
public Actor(String first, String last)
{
    firstName = first;
    lastName = last;
}
```

Then you create an instance of the `Actor` class by calling this constructor:

```
Actor a = new Actor("Arnold", "Schwarzenegger");
```

A new `Actor` object for Arnold Schwarzenegger is created.

Like methods, constructors can be overloaded. In other words, you can provide more than one constructor for a class, provided that each constructor has a unique signature. Here's another constructor for the `Actor` class:

```
public Actor(String first, String last, boolean good)
{
    firstName = first;
    lastName = last;
    goodActor = good;
}
```

This constructor lets you create an `Actor` object with information besides the actor's name:

```
Actor a = new Actor("Arnold", "Schwarzenegger", false);
```

Creating default constructors

I grew up watching *Dragnet.* I can still hear Joe Friday reading some thug his rights: "You have the right to an attorney during questioning. If you desire an attorney and cannot afford one, an attorney will be appointed to you free of charge."

Java constructors are like that. Every class has a right to a constructor. If you don't provide a constructor, Java appoints one for you, free of charge. This free constructor is called the *default constructor.* It doesn't accept any parameters and doesn't do anything, but it does allow your class to be instantiated.

Thus, the following two classes are identical:

```
public Class1
{
    public Class1() { }
}
public Class1 { }
```

In the first example, the class explicitly declares a constructor that doesn't accept any parameters and has no statements in its body. In the second example, Java creates a default constructor that works just like the constructor shown in the first example.

The default constructor is *not* created if you declare any constructors for the class. As a result, if you declare a constructor that accepts parameters and still want to have an empty constructor (with no parameters and no body), you must explicitly declare an empty constructor for the class.

An example might clear this point up. The following code does *not* compile:

```
public class BadActorApp
{
    public static void main(String[] args)
    {
        Actor a = new Actor(); // error: won't compile
    }
}

class Actor
{
    private String lastName;
    private String firstName;
    private boolean goodActor;

    public Actor(String last, String first)
    {
        lastName = last;
        firstName = first;
    }

    public Actor(String last, String first, boolean good)
    {
        lastName = last;
        firstName = first;
        goodActor = good;
    }
}
```

This program won't compile because it doesn't explicitly provide a default constructor for the Actor class; because it does provide other constructors, the default constructor isn't generated automatically.

Calling other constructors

A constructor can call another constructor of the same class by using the special keyword this as a method call. This technique is commonly used when you have several constructors that build on one another.

Consider this class:

```
public class Actor
{
    private String lastName;
    private String firstName;
    private boolean goodActor;

    public Actor(String last, String first)
    {
        lastName = last;
        firstName = first;
    }

    public Actor(String last, String first, boolean good)
    {
        this(last, first);
        goodActor = good;
    }
}
```

Here the second constructor calls the first constructor to set the `lastName` and `firstName` fields. Then it sets the `goodActor` field.

You have a few restrictions in using the `this` keyword as a constructor call:

» You can call another constructor only in the very first statement of a constructor. Thus, the following code won't compile:

```
public Actor(String last, String first, boolean good)
{
    goodActor = good;
    this(last, first); // error: won't compile
}
```

If you try to compile a class with this constructor, you get a message saying `call to this must be first statement in constructor`.

» Each constructor can call only one other constructor, but you can chain constructors. If a class has three constructors, the first constructor can call the second one, which in turn calls the third one.

>> You can't create loops in which constructors call one another. Here's a class that won't compile:

```
class CrazyClass
{
    private String firstString;
    private String secondString;
    public CrazyClass(String first, String second)
    {
        this(first);
        secondString = second;
    }
    public CrazyClass(String first)
    {
        this(first, "DEFAULT"); // error: won't compile
    }
}
```

The first constructor starts by calling the second constructor, which calls the first constructor. The compiler complains that this error is a recursive constructor invocation and politely refuses to compile the class.

TECHNICAL STUFF

If you don't explicitly call a constructor in the first line of a constructor, Java inserts code that automatically calls the default constructor of the base class — that is, the class that this class inherits. (This little detail doesn't become too important until you get into inheritance, which is covered in Book 3, Chapter 4, so you can conveniently ignore it for now.)

Finding More Uses for the this Keyword

As I describe in the preceding section, you can use the this keyword in a constructor to call another constructor for the current class. You can also use this in the body of a class constructor or method to refer to the current object — that is, the class instance for which the constructor or method has been called.

The this keyword is usually used to qualify references to instance variables of the current object. For example:

```
public Actor(String last, String first)
{
    this.lastName = last;
    this.firstName = first;
}
```

Here this isn't really necessary because the compiler can tell that lastName and firstName refer to class variables. Suppose, however, that you use lastName and firstName as the parameter names for the constructor:

```
public Actor(String lastName, String firstName)
{
    this.lastName = lastName;
    this.firstName = firstName;
}
```

Here the this keywords are required to distinguish among the parameters named lastName and firstName and the instance variables with the same names.

You can also use this in a method body. For example:

```
public String getFullName()
{
    return this.firstName + " " + this.lastName;
}
```

Because this example has no ambiguity, this isn't really required. Many programmers like to use this even when it isn't necessary, however, because it clarifies that they're referring to an instance variable.

Sometimes you use the this keyword all by itself to pass a reference to the current object as a method parameter. You can print the current object to the console by using the following statement:

```
System.out.println(this);
```

TECHNICAL
STUFF

The println method calls the object's toString method to get a string representation of the object and then prints it to the console. By default, toString prints the name of the class that the object is an instance of the object's hash code. If you want the println method to print something more meaningful, provide a toString method of your own for the class.

Using Initializers

An *initializer* (sometimes called an *initializer block*) is a lonely block of code that's placed outside any method, constructor, or other block of code. Initializers are executed whenever an instance of a class is created, regardless of which constructor is used to create the instance.

Initializer blocks are similar to variable initializers used to initialize variables. The difference is that with an initializer block, you can code more than one statement. Here's a class that gets the value for a class field from the user when the class is initialized:

```
class PrimeClass
{
    private Scanner sc = new Scanner(System.in);

    public int x;
    {
        System.out.print(
            "Enter the starting value for x: ");
        x = sc.nextInt();
    }
}
```

You can almost always achieve the same effect by using other coding techniques, which usually are more direct. You could prompt the user for the value in the constructor, for example, or you could call a method in the field initializer, like this:

```
class PrimeClass
{
    private Scanner sc = new Scanner(System.in);
    public int x = getX();
    private int getX()
    {
        System.out.print("Enter the starting value "
            + "for x: ");
        return sc.nextInt();
    }
}
```

Either way, the effect is the same.

Here are a few other tidbits of information concerning initializers:

>> If a class contains more than one initializer, the initializers are executed in the order in which they appear in the program.

>> Initializers are executed before any class constructors.

>> A special kind of initializer block called a *static initializer* lets you initialize static fields. (For more information, see the next chapter.)

Using Records

Java 14 introduces a new type of class called a *record*. A record is designed to simplify the task of creating classes that consist of nothing more than a collection of data fields that — and here's the important part — cannot be changed after the record is created. (The Java term for an object that can't be changed after it has been created is *immutable*.)

You could create a class that implements this behavior using a traditional Java class as follows:

```java
public final class Person
{
    private final String firstName;
    private final String lastName;

    public ImmutablePersonClass(String f, String l)
    {
        firstName = f;
        lastName = l;
    }

    public String firstName()
    {
        return firstName;
    }

    public String lastName()
    {
        return lastName;
    }
}
```

Here, the Person class has two private fields named firstName and lastName, a constructor that accepts String arguments to initialize the private fields, and getter methods that retrieve the first and last name values. When a Person object has been created from this class, the class provides no mechanism for changing the first or last name values.

Here's a snippet of code that creates an instance of this class and then prints the full name on the console:

```java
Person p = new Person("William", "Shakespeare");
System.out.println(p.firstName() + " " + p.lastName());
```

With the new record feature, you could replace the entire Person class with the following single line of code:

```
public record Person(String firstName, String lastName){}
```

Here are the notable details you need to remember about creating records:

>> You specify the data components of the record in parentheses following the record name. In this example, there are two data components, firstName and lastName.

>> Java automatically creates private class fields for each of the data components.

>> Java automatically creates public getter methods for each of the data components, using the names you provide. (You can, if you wish, provide additional methods.)

>> Java automatically creates a constructor that accepts the data component values as arguments, in the order in which you list them when you define the record. (You can, if you wish, create additional constructors or override the default constructor to provide features such as data validation.)

TECHNICAL STUFF

In Java 14, the record feature is a *preview feature*, which means that it's being introduced into the language on a somewhat tentative basis. In future releases, the record feature may be promoted to a standard feature, or it may be removed altogether. Or the feature may be modified in future releases.

The record feature's status as a preview feature means two things:

>> You should use it with care, because it may not be supported in future releases.

>> You have to add special flags to the javac command line to compile any programs that use the record feature and also to the java command to run a compiled program that uses the feature.

Specifically, you should use this command to compile a program that uses the record feature:

```
javac --enable-preview --release 14 Person.java
```

And to run the program, you should add the --enable-preview switch to the java command, like this:

```
java --enable-preview Person.class
```

(Note that the --release flag is needed on the javac command but not on the java command.)

Chapter **3**

Working with Statics

A *static method* is a method that isn't associated with an instance of a class. (Unless you jumped straight to this chapter, you already knew that.) Instead, the method belongs to the class itself. As a result, you can call the method without first creating a class instance. In this chapter, you find out everything you need to know about creating and using static fields and methods.

Understanding Static Fields and Methods

According to my handy *Webster's* dictionary, the word *static* has several meanings, most of which relate to the idea of being stationary or unchanging. A *static display* is a display that doesn't move. *Static electricity* is an electrical charge that doesn't flow. A *static design* is a design that doesn't change.

As used by Java, however, the term *static* doesn't mean unchanging. You can create a static field, for example, and then assign values to it as a program executes. Thus, the value of the static field can change.

To confuse things further, the word *static* can also mean interference, as in radio static that prevents you from hearing music clearly on the radio. But in Java, the term *static* doesn't have anything to do with interference or bad reception.

So what does the term *static* mean in Java? It's used to describe a special type of field or method that isn't associated with a particular instance of a class. Instead, static fields and methods are associated with the class itself, which means that you don't have to create an instance of the class to access a static field or methods. Instead, you access a static field or method by specifying the class name, not a variable that references an object.

REMEMBER

The value of a static field is the same across all instances of the class. In other words, if a class has a static field named CompanyName, all objects created from the class will have the same value for CompanyName.

Static fields and methods have many common uses. Here are but a few:

» **To provide constants or other values that aren't related to class instances:** A Billing class might have a constant named SALES_TAX_RATE that provides the state sales tax rate.

» **To keep count of how many instances of a class have been created:** A Ball class used in a game might have a static field that counts how many balls currently exist. This count doesn't belong to any one instance of the Ball class.

» **In a business application, to keep track of a reference or serial number that's assigned to each new object instance:** An Invoice class might maintain a static field that holds the invoice number that is assigned to the next Invoice object created from the class.

» **To provide an alternative way to create instances of the class:** An excellent example is the NumberFormat class, which has static methods such as getCurrencyInstance and getNumberInstance that return object instances to format numbers in specific ways. One reason you might want to use this technique is to create classes that can have only one object instance. This type of class, called a *singleton class*, is described in the sidebar "The Singleton pattern," later in this chapter.

» **To provide utility functions that aren't associated with an object at all:** A good example in the Java API library is the Math class, which provides a bunch of static methods to do math calculations. Examples that you might code yourself are a data validation class with static methods that validate input data and a database class with static methods that perform database operations.

Working with Static Fields

A *static field* is a field that's declared with the static keyword, like this:

```
private static int ballCount;
```

Note that the position of the `static` keyword is interchangeable with the positions of the *visibility keywords* (`private`, `protected`, `public`, and `package-private`). As a result, the following statement works as well:

```
static private int ballCount;
```

As a convention, most programmers tend to put the visibility keyword first.

TIP

Note that you can't use the `static` keyword within a class method. Thus the following code won't compile:

```
private void someMethod()
{
    static int x;
}
```

In other words, fields can be static, but local variables can't.

You can provide an initial value for a static field. Here's an example:

```
private static String district = "Northwest";
```

If you don't provide an initial value for a static field, a default value appropriate to the type will be assigned automatically (0 for numeric types, empty string for Strings, and so on).

Static fields are created and initialized when the class is first loaded. That happens when a static member of the class is referred to or when an instance of the class is created, whichever comes first.

Another way to initialize a static field is to use a *static initializer,* which I cover later in this chapter, in the section "Using Static Initializers."

Using Static Methods

A *static method* is a method declared with the `static` keyword. Like static fields, static methods are associated with the class itself, not with any particular object created from the class. As a result, you don't have to create an object from a class before you can use static methods defined by the class.

The best-known static method is main, which is called by the Java runtime to start an application. The main method must be static — which means that applications are run in a static context by default.

One of the basic rules of working with static methods is that you can't access a nonstatic method or field from a static method, because the static method doesn't have an instance of the class to use to reference instance methods or fields. The following code won't compile:

```
public class TestClass
{
    private int x = 5; // an instance field

    public static void main(String[] args)
    {
        int y = x; // error: won't compile
    }
}
```

Here the main method is static, so it can't access the instance variable x.

Note: You *can* access static methods and fields from an instance method, however. The following code works fine:

```
public class Invoice
{
    private static double taxRate = 0.75;
    private double salesTotal;

    public double getTax()
    {
        return salesTotal * taxRate;
    }
}
```

Here the instance method named salesTotal has no trouble accessing the static field taxRate.

Counting Instances

One common use for static variables is to keep track of how many instances of a class have been created. To illustrate how you can do this, consider the program in Listing 3-1. This program includes two classes. The CountTest class is a simple class that keeps track of how many times its constructor has been called. Then the CountTestApp class uses a for loop to create 10 instances of the class, displaying the number of instances that have been created after it creates each instance.

Note that the instance count in this application is reset to zero each time the application is run. As a result, it doesn't keep track of how many instances of the CountTest class have ever been created — only of how many have been created *during a particular execution* of the program.

LISTING 3-1: **The CountTest Application**

```
public class CountTestApp                                              →1
{
    public static void main(String[] args)
    {
        printCount();
        for (int i = 0; i < 10; i++)
        {
            CountTest c1 = new CountTest();                            →8
            printCount();                                             →9
        }
    }

    private static void printCount()
    {
        System.out.println("There are now "                          →15
            + CountTest.getInstanceCount()
            + " instances of the CountTest class.");
    }
}

class CountTest                                                        →21
{
    private static int instanceCount = 0;                            →23

    public CountTest()                                               →25
    {
        instanceCount++;
    }

    public static int getInstanceCount()                             →30
    {
        return instanceCount;
    }
}
```

The following paragraphs describe some of the highlights of this program:

» →1 This line is the start of the CountTestApp class, which tests the CountTest class.

» →8 This line creates an instance of the CountTest class. Because this code is contained in a for loop, 10 instances are created.

» →9 This line calls the printCount method, which prints the number of CountTest objects that have been created so far.

» →15 This line prints a message indicating how many CountTest objects have been created so far. It calls the static getInstanceCount method of the CountTest class to get the instance count.

» →21 This line is the start of the CountTest class.

» →23 The static instanceCount variable stores the instance count.

» →25 This line is the constructor for the CountTest class. Notice that the instanceCount variable is incremented within the constructor. That way, each time a new instance of the class is created, the instance count is incremented.

» →30 The static getInstanceCount method simply returns the value of the static instanceCount field.

THE SINGLETON PATTERN

A *singleton* is a class that you can use to create only one instance. When you try to create an instance, the class first checks to see whether an instance already exists. If so, the existing instance is used; if not, a new instance is created.

You can't achieve this effect by using Java constructors, because a class instance has already been created by the time the constructor is executed. (That's why you can use the this keyword from within a constructor.) As a result, the normal way to implement a singleton class is to declare all the constructors for the class as private. That way, the constructors aren't available to other classes. Then you provide a static method that returns an instance. This method either creates a new instance or returns an existing instance.

Here's a bare-bones example of a singleton class:

```
class SingletonClass
{
    private static SingletonClass instance;

    private SingletonClass() {}

    public static SingletonClass getInstance()
    {
        if (instance == null)
            instance = new SingletonClass();
        return instance;
    }
}
```

Here the SingletonClass contains a private instance variable that maintains a reference to an instance of the class. Then a default constructor is declared with private visibility to prevent the constructor from being used outside the class. Finally, the static getInstance method calls the constructor to create an instance if the instance variable is null. Then it returns the instance to the caller.

Here's a bit of code that calls the getInstance method twice and then compares the resulting objects:

```
SingletonClass s1 = SingletonClass.getInstance();
SingletonClass s2 = SingletonClass.getInstance();
if (s1 == s2)
    System.out.println("The objects are the same");
else
    System.out.println("The objects are not the same");
```

When this code is run, the first call to getInstance creates a new instance of the SingletonClass class. The second call to getInstance simply returns a reference to the instance that was created in the first call. As a result, the comparison in the if statement is true, and the first message is printed to the console.

Preventing Instances

Sometimes you want to create a class that can't be instantiated at all. Such a class consists entirely of static fields and methods. A good example in the Java API is the Math class. Its methods provide utility-type functions that aren't really associated with a particular object. You may need to create similar classes yourself occasionally. You might create a class with static methods for validating input data, for example, or a database access class that has static methods to retrieve data from a database. You don't need to create instances of either of these classes.

You can use a simple trick to prevent anyone from instantiating a class. To create a class instance, you have to have at least one public constructor. If you don't provide a constructor in your class, Java automatically inserts a default constructor, which happens to be public.

All you have to do to prevent a class instance from being created, then, is provide a single private constructor, like this:

```
public final class Validation
{
    private Validation() {} // prevents instances
    // static methods and fields go here
}
```

Now, because the constructor is private, the class can't be instantiated. And because it's final, the class can't be extended, thwarting anyone who would try to foil your plans by simply inheriting your class and then instantiating it.

TECHNICAL STUFF

Incidentally, the Math class uses this technique to prevent you from creating instances from it. Here's an actual snippet of code from the Math class:

```
public final class Math {

/**
 * Don't let anyone instantiate this class.
 */

    private Math() {}
```

I figure that if this trick is good enough for the folks who wrote the Math class, it's good enough for me.

Using Static Initializers

In the preceding chapter, you discover *initializer blocks* that you can use to initialize instance variables. Initializer blocks aren't executed until an instance of a class is created, so you can't count on them to initialize static fields. After all, you might access a static field before you create an instance of a class.

Java provides a feature called a *static initializer* that's designed specifically to let you initialize static fields. The general form of a static initializer looks like this:

```
static
{
    statements...
}
```

As you can see, a static initializer is similar to an initializer block but begins with the word `static`. As with an initializer block, you code static initializers in the class body but outside any other block, such as the body of a method or constructor.

The first time you access a static member such as a static field or a static method, any static initializers in the class are executed — provided that you haven't already created an instance of the class. That's because the static initializers are also executed the first time you create an instance. In that case, the static initializers are executed *before* the constructor is executed.

TIP

If a class has more than one static initializer, the initializers are executed in the order in which they appear in the program.

Here's an example of a class that contains a static initializer:

```
class StaticInit
{
    public static int x;

    static
    {
        x = 32;
    }

    // other class members such as constructors and methods go here...

}
```

This example is pretty trivial. In fact, you can achieve the same effect just by assigning the value 32 to the variable when it is declared. If, however, you had to perform a complicated calculation to determine the value of x, a static initializer could be very useful.

Chapter 4

Using Subclasses and Inheritance

s you find out in Book 3, Chapter 1, a Java class can be based on another class. Then the class becomes like a child to the parent class: It inherits all the characteristics of the parent class, good and bad. All the visible fields and methods of the parent class are passed on to the child class. The child class can use these fields or methods as they are, or it can override them to provide its own versions. In addition, the child class can add fields or methods of its own.

In this chapter, you discover how this magic works, along with the basics of creating and using Java classes that inherit other classes. You also find out a few fancy tricks that help you get the most out of inheritance.

Introducing Inheritance

The word *inheritance* conjures up several noncomputer meanings:

» Children inherit certain characteristics from the parents. Two of my three children have red hair, for example. (Ideally, they won't be half bald by the time they're 30.)

» Children can also inherit behavior from their parents. As they say, the apple doesn't fall far from the tree.

» When someone dies, his heirs get his stuff. Some of it is good stuff, but some of it may not be. My kids are going to have a great time rummaging through my garage, deciding who gets what.

» You can inherit rights as well as possessions. You may be a citizen of a country by virtue of being born to parents who are citizens of that country.

In Java, *inheritance* refers to a feature of object-oriented programming that lets you create classes that are derived from other classes. A class that's based on another class is said to *inherit* the other class. The class that is inherited is called the *parent class,* the *base class,* or the *superclass.* The class that does the inheriting is called the *child class,* the *derived class,* or the *subclass.*

TIP

The terms *subclass* and *superclass* seem to be the preferred terms among Java gurus. So if you want to look like you know what you're talking about, use these terms. Also, be aware that the term *subclass* can be used as a verb. When you create a subclass that inherits a base class, for example, you are *subclassing* the base class.

You need to know a few important things about inheritance:

» A derived class automatically takes on all the behavior and attributes of its base class. Thus, if you need to create several classes to describe types that aren't identical but have many features in common, you can create a base class that defines all the common features. Then you can create several derived classes that inherit the common features.

» A derived class can add features to the base class it inherits by defining its own methods and fields. This is one way that a derived class distinguishes itself from its base class.

» A derived class can also change the behavior provided by the base class. A base class may provide that all classes derived from it have a method named play, for example, but each class is free to provide its own implementation of the play method. In this case, any class that extends the base class can provide its own implementation of the play method.

TIP

» Inheritance is best used to implement *is-a-type-of* relationships. Here are a few examples: Solitaire is a type of game; a truck is a type of vehicle; an invoice is a type of transaction. In each case, a particular kind of object is a specific type of a more general category of objects.

The following sections provide more examples that help illustrate these points.

Motorcycles, trains, and automobiles

Inheritance is often explained in terms of real-world objects such as cars and motorcycles or birds and reptiles. Consider various types of vehicles. Cars and motorcycles are two distinct types of vehicles. If you're writing software that represents vehicles, you could start by creating a class called Vehicle that would describe the features that are common to all types of vehicles, such as wheels; a driver; the ability to carry passengers; and the ability to perform actions such as driving, stopping, turning, and crashing.

A motorcycle is a type of vehicle that further refines the Vehicle class. The Motorcycle class would inherit the Vehicle class, so it would have wheels; a driver; possibly passengers; and the ability to drive, stop, turn, and crash. In addition, it would have features that differentiate it from other types of vehicles, such as two wheels and handlebars used for steering control.

A car is also a type of vehicle. The Car class would inherit the Vehicle class, so it too would have wheels; a driver; possibly some passengers; and the ability to drive, stop, turn, and crash. Also, it would have some features of its own, such as four wheels, a steering wheel, seat belts and air bags, and an optional automatic transmission.

Game play

Because you're unlikely ever to write a program that simulates cars, motorcycles, and other vehicles, take a look at a more common example: games. Suppose that you want to develop a series of board games such as Life, Sorry!, and Monopoly. Most board games have certain features in common:

>> They have a playing board with locations that players can occupy.

>> They have players that are represented by objects.

>> The game is played by each player taking a turn, one after the other. When the game starts, it keeps going until someone wins. (If you don't believe me, ask the kids who tried to stop a game of Jumanji before someone won.)

Each specific type of game has these basic characteristics but adds features of its own. The game Life adds features such as money, insurance policies, spouses, children, and a fancy spinner in the middle of the board. Sorry! has cards that you draw to determine each move and safety zones within which other players can't attack you. Monopoly has Chance and Community Chest cards, properties, houses, hotels, and money.

If you were designing classes for these games, you might create a generic BoardGame class that defines the basic features common to all board games and then use it as the base class for classes that represent specific board games, such as LifeGame, SorryGame, and MonopolyGame.

A businesslike example

If vehicles or games don't make the point clear enough, here's an example from the world of business. Suppose that you're designing a payroll system, and you're working on the classes that represent the employees. You realize that the payroll includes two types of employees: salaried employees and hourly employees. So you decide to create two classes, sensibly named SalariedEmployee and HourlyEmployee.

You quickly discover that most of the work done by these two classes is identical. Both types of employees have names, addresses, Social Security numbers, totals for how much they've been paid for the year, how much tax has been withheld, and so on.

The employee types also have important differences. The most obvious one is that the salaried employees have an annual salary, and the hourly employees have an hourly pay rate. Also, hourly employees have a schedule that changes week to week, and salaried employees may have a benefit plan that isn't offered to hourly employees.

Thus you decide to create three classes instead of just two. A class named Employee handles all the features that are common to both types of employees; then this class is the base class for the SalariedEmployee and Hourly Employee classes. These classes provide the additional features that distinguish salaried employees from hourly employees.

Inheritance hierarchies

One of the most important aspects of inheritance is that a class derived from a base class can in turn be used as the base class for another derived class. Thus you can use inheritance to form a hierarchy of classes.

You've already seen how an Employee class can be used as a base class to create two types of subclasses: a SalariedEmployee class for salaried employees and an HourlyEmployee class for hourly employees. Suppose that salaried employees fall into two categories: management and sales. Then you could use the SalariedEmployee class as the base class for two more classes: Manager and SalesPerson.

Thus, a Manager is a type of SalariedEmployee. Because a SalariedEmployee is a type of Employee, a Manager is also a type of Employee.

TECHNICAL STUFF

All classes ultimately derive from a Java class named Object. Any class that doesn't specifically state what class it is derived from is assumed to derive from the Object class. This class provides some of the basic features that are common to all Java classes, such as the toString method. For more information, see Book 3, Chapter 6.

Creating Subclasses

The basic procedure for creating a subclass is simple: You just use the extends keyword on the declaration for the subclass. The basic format of a class declaration for a class that inherits a base class is this:

```
public class ClassName extends BaseClass
{
    // class body goes here
}
```

Suppose that you have a class named Ball that defines a basic ball, and you want to create a subclass named BouncingBall that adds the ability to bounce:

```
public class BouncingBall extends Ball
{
    // methods and fields that add the ability to bounce to a basic Ball object:

    public void bounce()
    {
        // the bounce method
    }
}
```

Here I'm creating a class named BouncingBall that extends the Ball class. (*Extends* is Java's word for *inherits*.)

The subclass automatically has all the methods and fields of the class it extends. Thus, if the Ball class has fields named size and weight, the BouncingBall class has those fields too. Likewise, if the Ball class has a method named throw, the BouncingBall class gets that method too.

THE DELEGATION PATTERN

Inheritance is one of the great features of object-oriented programming languages such as Java, but it isn't the answer to every programming problem.

Suppose that you need to create a class named EmployeeCollection that represents a group of employees. One way to create this class would be to extend one of the collection classes supplied by the Java API, such as the ArrayList class. Then your EmployeeCollection class would be a specialized version of the ArrayList class and would have all the methods that are available to the ArrayList class.

A simpler alternative, however, would be to declare a class field of type ArrayList within your EmployeeCollection class. Then you could provide methods that use this ArrayList object to add or retrieve employees from the collection. In effect, your EmployeeCollection class is simply a wrapper of the ArrayList class, perhaps with a few bells and whistles added that pertain specifically to employees.

This technique is an application of a common design pattern called *Deletagion*. Why is it called that? Because instead of writing code that implements the functions of the collection, you *delegate* that task to an ArrayList object, which already knows how to perform these functions. (For more information about the ArrayList class, see Book 4, Chapter 3.)

You need to know some important details to use inheritance properly:

REMEMBER

>> A subclass inherits all the visible members from its base class. Constructors are *not* considered to be members, however. As a result, a subclass does *not* inherit constructors from its base class. And a subclass does not inherit members that are not visible to it (that is, private members).

>> You can *override* a method by declaring a new member with the same signature in the subclass. For more information, see the next section.

>> A special type of visibility called protected hides fields and methods from classes outside of the current package but makes them available to subclasses and other classes within the current package. For more information, see the section "Protecting Your Members," later in this chapter.

>> You can add more methods or fields with any level of visibility to a subclass. The BouncingBall class shown earlier in this section, for example, adds a public method named bounce.

Overriding Methods

If a subclass declares a method that has the same signature as a public method of the base class, the subclass version of the method *overrides* the base class version of the method. This technique lets you modify the behavior of a base class to suit the needs of the subclass.

Suppose you have a base class named Game that has a method named play. The base class, which doesn't represent any particular game, implements this method:

```
public class Game
{
    public void play()
    {
    }
}
```

Then you declare a class named Chess that extends the Game class but also provides an implementation for the play method:

```
public class Chess extends Game
{
    public void play()
    {
        System.out.println("I give up. You win.");
    }
}
```

Here, when you call the play method of a Chess object, the game announces that it gives up. (I was going to provide a complete implementation of an actual chess game program for this example, but it would have made this chapter about 600 pages long. So I opted for the simpler version here.)

Note that to override a method, several conditions have to be met:

>> The class must extend the class that defines the method you want to override.

>> The method must be visible to the subclass — you can't override a private method.

>> The method in the subclass must have the same signature as the method in the base class. In other words, the name of the method, the parameter types, and the return type must be the same. (Actually, the return type can be a more specific variant of the parent method's return type. For example, you can override a method that returns a Shape with a method that returns a Circle, because Circle is a subtype of Shape. This type of override is called a *covariant return type*.)

>> The overridden method can't reduce the visibility of the method it overrides. You can increase the visibility, but you can't decrease it.

Protecting Your Members

You're already familiar with the public and private keywords, which are used to indicate whether class members are visible outside the class or not. When you inherit a class, all the public members of the superclass are available to the subclass, but the private members aren't.

Java provides a third visibility option that's useful when you create subclasses: protected. A member with protected visibility is available to subclasses and classes in the same package, but not to classes outside of the package. Consider this example:

```java
public class Ball
{
    private double weight;

    protected double getWeight()
    {
        return this.weight;
    }

    protected void setWeight(double weight)
    {
        this.weight = weight;
    }
}

public class BaseBall extends Ball
{
    public BaseBall()
    {
        setWeight(5.125);
    }
}
```

Here, the getWeight and setWeight methods are declared with protected access, which means that they're visible in the subclass BaseBall. These methods aren't visible to classes that don't extend Ball, however.

Using this and super in Your Subclasses

You already know about the this keyword: It provides a way to refer to the current object instance. It's often used to distinguish between a local variable or a parameter and a class field with the same name. For example:

```
public class Ball
{
    private int velocity;
    public void setVelocity(int velocity)
    {
    this.velocity = velocity;
    }
}
```

Here the this keyword indicates that the velocity variable referred to on the left side of the assignment statement is the class field named velocity, not the parameter with the same name.

But what if you need to refer to a field or method that belongs to a base class? To do that, you use the super keyword. It works similarly to this but refers to the instance of the base class rather than the instance of the current class.

Consider these two classes:

```
public class Ball
{
    public void hit()
    {
    System.out.println("You hit it a mile!");
    }
}

class BaseBall extends Ball
{
    public void hit()
    {
        System.out.println("You tore the cover off!");
        super.hit();
    }
}
```

Here the `hit` method in the `BaseBall` class calls the `hit` method of its base class object. Thus, if you call the `hit` method of a `BaseBall` object, the following two lines are displayed on the console:

```
You tore the cover off!
You hit it a mile!
```

You can also use the `super` keyword in the constructor of a subclass to explicitly call a constructor of the superclass. For more information, see the next section.

Understanding Inheritance and Constructors

When you create an instance of a subclass, Java automatically calls the default constructor of the base class before it executes the subclass constructor. Consider the following classes:

```java
public class Ball
{
    public Ball()
    {
        System.out.println(
            "Hello from the Ball constructor");
    }
}

public class BaseBall extends Ball
{
    public BaseBall()
    {
        System.out.println(
            "Hello from the BaseBall constructor");
    }
}
```

If you create an instance of the `BaseBall` class, the following two lines are displayed on the console:

```
Hello from the Ball constructor
Hello from the BaseBall constructor
```

If you want, you can explicitly call a base class constructor from a subclass by using the `super` keyword. Because Java automatically calls the default constructor for you, the only reason to do this is to call a constructor of the base class that uses a parameter. Here's a version of the Ball and BaseBall classes in which the BaseBall constructor calls a Ball constructor that uses a parameter:

```java
public class Ball
{
    private double weight;
    public Ball(double weight)
    {
        this.weight = weight;
    }
}

public class BaseBall extends Ball
{
    public BaseBall()
    {
        super(5.125);
    }
}
```

Here the BaseBall constructor calls the Ball constructor to supply a default weight for the ball.

REMEMBER

You need to obey a few rules and regulations when working with superclass constructors:

>> If you use super to call the superclass constructor, you must do so in the very first statement in the constructor.

WARNING

>> If you don't explicitly call super, the compiler inserts a call to the default constructor of the base class. In that case, the base class must have a default constructor. If the base class doesn't have a default constructor, the compiler refuses to compile the program.

>> If the superclass is itself a subclass, the constructor for its superclass is called in the same way. This continues all the way up the inheritance hierarchy until you get to the Object class, which has no superclass.

Using final

Java has a final keyword that serves three purposes. When you use final with a variable, it creates a constant whose value can't be changed after it has been initialized. Constants are covered in Book 2, Chapter 2, so I won't describe this use of the final keyword more here. The other two uses of the final keyword are to create final methods and final classes. I describe these two uses of `final` in the following sections.

Final methods

A *final method* is a method that can't be overridden by a subclass. To create a final method, you simply add the keyword `final` to the method declaration. For example:

```
public class SpaceShip
{
    public final int getVelocity()
    {
        return this.velocity;
    }
}
```

Here the method `getVelocity` is declared as `final`. Thus, any class that uses the `SpaceShip` class as a base class can't override the `getVelocity` method. If it tries, the compiler issues the error message (`"Overridden method is final"`).

Here are some additional details about final methods:

TECHNICAL STUFF

>> Final methods execute more efficiently than nonfinal methods because the compiler knows at compile time that a call to a final method won't be overridden by some other method. The performance gain isn't huge, but for applications in which performance is crucial, it can be noticeable.

>> Private methods are automatically considered to be final because you can't override a method you can't see.

Final classes

A *final class* is a class that can't be used as a base class. To declare a class as final, just add the `final` keyword to the class declaration:

```
public final class BaseBall
{
    // members for the BaseBall class go here
}
```

Then no one can use the `BaseBall` class as the base class for another class.

When you declare a class to be final, all of its methods are considered to be final as well. That makes sense when you think about it. Because you can't use a final class as the base class for another class, no class can possibly be in a position to override any of the methods in the final class. Thus all the methods of a final class are final methods.

Casting Up and Down

An object of a derived type can be treated as though it were an object of its base type. If the BaseBall class extends the Ball class, for example, a BaseBall object can be treated as though it were a Ball object. This arrangement is called *upcasting*, and Java does it automatically, so you don't have to code a casting operator. Thus the following code is legal:

```
Ball b = new BaseBall();
```

Here an object of type BaseBall is created. Then a reference to this object is assigned to the variable b, whose type is Ball, not BaseBall.

Now suppose that you have a method in a ball-game application named hit that's declared like this:

```
public void hit(Ball b)
```

In other words, this method accepts a Ball type as a parameter. When you call this method, you can pass it either a Ball object or a BaseBall object, because BaseBall is a subclass of Ball. So the following code works:

```
BaseBall b1 = new BaseBall();
hit(b1);
Ball b2 = b1;
hit(b2);
```

WARNING

Automatic casting doesn't work the other way, however. Thus you can't use a Ball object where a BaseBall object is called for. Suppose your program has a method declared like this:

```
public void toss(BaseBall b)
```

Then the following code does *not* compile:

```
Ball b = new BaseBall();
toss(b);                    // error: won't compile
```

You can explicitly cast the b variable to a BaseBall object, however, like this:

```
Ball b = new BaseBall();
toss((BaseBall) b);
```

Note that the second statement throws an exception of type ClassCastException if the object referenced by the b variable isn't a BaseBall object. So the following code won't work:

```
Ball b = new SoftBall();
toss((BaseBall) b);      // error: b isn't a Softball
```

TIP

What if you want to call a method that's defined by a subclass from an object that's referenced by a variable of the superclass? Suppose that the SoftBall class has a method named riseBall that isn't defined by the Ball class. How can you call it from a Ball variable? One way to do that is to create a variable of the subclass and then use an assignment statement to cast the object:

```
Ball b = new SoftBall();
SoftBall s = (SoftBall)b;    // cast the Ball to a SoftBall
s.riseBall();
```

But there's a better way: Java lets you cast the Ball object to a SoftBall and call the riseBall method in the same statement. All you need is an extra set of parentheses, like this:

```
Ball b = new SoftBall();
((SoftBall) b).riseBall();
```

Here the expression ((SoftBall) b) returns the object referenced by the b variable, cast as a SoftBall. Then you can call any method of the SoftBall class by using the dot operator. (This operator throws a ClassCastException if b is not a SoftBall object.)

TIP

As a general rule, you should declare method parameters with types as far up in the class hierarchy as possible. Rather than create separate toss methods that accept BaseBall and SoftBall objects, for example, you can create a single toss method that accepts a Ball object. If necessary, the toss method can determine which type of ball it's throwing by using the instanceof operator, which is described in the next section.

Determining an Object's Type

As described in the preceding section, a variable of one type can possibly hold a reference to an object of another type. If SalariedEmployee is a subclass of the Employee class, the following statement is perfectly legal:

```
Employee emp = new SalariedEmployee();
```

Here the type of the emp variable is Employee, but the object it refers to is a SalariedEmployee.

Suppose you have a method named getEmployee whose return type is Employee but that actually returns either a SalariedEmployee or an HourlyEmployee object:

```
Employee emp = getEmployee();
```

In many cases, you don't need to worry about which type of employee this method returns, but sometimes you do. Suppose that the SalariedEmployee class extends the Employee class by adding a method named getFormattedSalary, which returns the employee's salary formatted as currency. Similarly, the HourlyEmployee class extends the Employee class with a getFormattedRate method that returns the employee's hourly pay rate formatted as currency. Then you'd need to know which type of employee a particular object is, to know whether you should call the getFormattedSalary method or the getFormattedRate method to get the employee's pay.

To tell what type of object has been assigned to the emp variable, you can use the instanceof operator, which is designed specifically for this purpose. Here's the preceding code rewritten with the instanceof operator:

```
Employee emp = getEmployee();
String msg;
if (emp instanceof SalariedEmployee)
{
    msg = "The employee's salary is ";
    msg += ((SalariedEmployee) emp).getFormattedSalary();
}
else
{
    msg = "The employee's hourly rate is ";
    msg += ((HourlyEmployee) emp).getFormattedRate();
}
System.out.println(msg);
```

Here the `instanceof` operator is used in an `if` statement to determine the type of the object returned by the `getEmployee` method. Then the `emp` can be cast without fear of `CastClassException`.

Poly What?

The term *polymorphism* refers to the ability of Java to use base class variables to refer to subclass objects; to keep track of which subclass an object belongs to; and to use overridden methods of the subclass, even though the subclass isn't known when the program is compiled.

This sounds like a mouthful, but it's not hard to understand when you see an example. Suppose that you're developing an application that can play the venerable game Tic-Tac-Toe. You start by creating a class named `Player` that represents one of the players. This class has a public method named `move` that returns an `int` to indicate which square of the board the player wants to mark:

```
class Player
{
    public int move()
    {
        for (int i = 0; i < 9; i++)
        {
            System.out.println(
                "\nThe basic player says:");
            System.out.println(
                "I'll take the first open square!");
            return firstOpenSquare();
        }
        return -1;
    }

    private int firstOpenSquare()
    {
        int square = 0;

        // code to find the first open square goes here

        return square;
    }
}
```

This basic version of the `Player` class uses a simple strategy to determine what its next move should be: It chooses the first open square on the board. This strategy stokes your ego by letting you think you can beat the computer every time. (To keep the illustration simple, I omit the code that actually chooses the move.)

Now you need to create a subclass of the Player class that uses a more intelligent method to choose its next move:

```
class BetterPlayer extends Player
{
    public int move()
    {
        System.out.println("\nThe better player says:");
        System.out.println(
            "I'm looking for a good move...");
        return findBestMove();
    }

    private int findBestMove()
    {
        int square = 0;

        // code to find the best move goes here

        return square;
    }
}
```

As you can see, this version of the Player class overrides the move method and uses a better algorithm to pick its move. (Again, to keep the illustration simple, I don't show the code that actually chooses the move.)

The next thing to do is write a short class that uses these two Player classes to play a game. This class contains a method named playTheGame that accepts two Player objects. It calls the move method of the first player and then calls the move method of the second player:

```
public class TicTacToeApp
{
    public static void main(String[] args)
    {
        Player p1 = new Player();
        Player p2 = new BetterPlayer();

        playTheGame(p1, p2);
    }

    public static void playTheGame(Player p1, Player p2)
    {
```

```
        p1.move();
        p2.move();
    }
}
```

Notice that the `playTheGame` method doesn't know which of the two players is the basic player and which is the better player. It simply calls the `move` method for each `Player` object.

When you run this program, the following output is displayed on the console:

```
Basic player says:
I'll take the first open square!
```

```
Better player says:
I'm looking for a good move...
```

When the `move` method for p1 is called, the `move` method of the `Player` class is executed. But when the `move` method for p2 is called, the `move` method of the `BetterPlayer` class is called.

TECHNICAL STUFF

Java knows to call the `move` method of the `BetterPlayer` subclass because it uses a technique called late binding. *Late binding* simply means that when the compiler can't tell for sure what type of object a variable references, it doesn't hard-wire the method calls when the program is compiled. Instead, it waits until the program is executing to determine exactly which method to call.

Creating Custom Exceptions

The last topic I want to cover in this chapter is how to use inheritance to create your own custom exceptions. I cover most of the details of working with exceptions in Book 2, Chapter 8, but I hadn't explored inheritance, so I couldn't discuss custom exception classes in that chapter. I promised that I'd get to it in this mini-book. The following sections deliver on that long-awaited promise.

Tracing the Throwable hierarchy

As you know, you use the try/catch statement to catch exceptions and the `throw` statement to throw exceptions. Each type of exception that can be caught or thrown is represented by a different exception class. What you may not have realized is that those exception classes use a fairly complex inheritance chain, as shown in Figure 4-1.

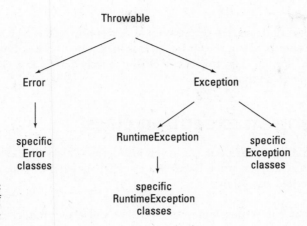

FIGURE 4-1:
The hierarchy of
exception classes.

The following paragraphs describe the classes in this hierarchy:

>> **Throwable:** The root of the exception hierarchy is the Throwable class. This class represents any object that can be thrown with a throw statement and caught with a catch clause.

>> **Error:** This subclass of Throwable represents serious error conditions that reasonable programs can't recover from. The subclasses of this class represent the specific types of errors that can occur. If the Java Virtual Machine runs out of memory, for example, a VirtualMachineError is thrown. You don't have to worry about catching these errors in your programs.

>> **Exception:** This subclass of Throwable represents an error condition that most programs should try to recover from. Thus, Exception is effectively the top of the hierarchy for the types of exceptions you catch with try/catch statements.

With the exception (sorry) of RuntimeException, the subclasses of Exception represent specific types of checked exceptions that you must catch or throw. Note that some of these subclasses have subclasses of their own. The exception class named IOException, for example, has more than 25 subclasses representing different kinds of I/O exceptions that can occur.

TIP

The "catch or throw" rule (see Book 2, Chapter 8) states that checked exceptions must either be caught in a try–catch statement or thrown upwards to the calling method by specifying throws in the method declaration.

>> **RuntimeException:** This subclass of Exception represents unchecked exceptions. You don't have to catch or throw unchecked exceptions, but you can if you want to. Subclasses of RuntimeException include NullPointerException and ArithmeticException.

If your application needs to throw a custom exception, you can create an exception class that inherits any of the classes in this hierarchy. Usually, however, you start with the Exception class to create a custom checked exception. The next section explains how to do that.

Creating an exception class

To create a custom exception class, you just define a class that extends one of the classes in the Java exception hierarchy. Usually you extend Exception to create a custom checked exception.

Suppose that you're developing a class that retrieves product data from a file or database, and you want methods that encounter I/O errors to throw a custom exception rather than the generic IOException that's provided in the Java API. You can do that by creating a class that extends the Exception class:

```
public class ProductDataException extends Exception
{
}
```

Unfortunately, constructors aren't considered to be class members, so they aren't inherited when you extend a class. As a result, the ProductDataException has only a default constructor. The Exception class itself and most other exception classes have a constructor that lets you pass a string message that's stored with the exception and can be retrieved via the getMessage method. Thus you want to add this constructor to your class, which means that you want to add an explicit default constructor too. So now the ProductDataException class looks like this:

```
public class ProductDataException extends Exception
{
    public ProductDataException
    {
    }

    public ProductDataException(String message)
    {
        super(message);
    }
}
```

Although it's possible to do so, adding fields or methods to a custom exception class is unusual.

Throwing a custom exception

As for any exception, you use a throw statement to throw a custom exception. You usually code this throw statement in the midst of a catch clause that catches some other, more generic exception. Here's a method that retrieves product data from a file and throws a ProductDataException if an IOException occurs:

```
public class ProductDDB
{
    public static Product getProduct(String code)
        throws ProductDataException
    {
        try
        {
            Product p;

            // code that gets the product from a file and might throw an
    IOException

            p = new Product();
            return p;
        }
        catch (IOException e)
        {
            throw new ProductDataException(
                "An IO error occurred.");
        }
    }
}
```

Here's some code that calls the getProduct method and catches the exception:

```
try
{
    Product p = ProductDB.getProduct(productCode);
}
catch (ProductDataException e)
{
    System.out.println(e.getMessage());
}
```

Here the message is simply displayed on the console if a ProductDataException is thrown. In an actual program, you want to log the error, inform the user, and figure out how to continue the program gracefully even though this data exception has occurred.

Chapter **5**

Using Abstract Classes and Interfaces

In this chapter, you find out how to use two similar but subtly distinct features: abstract classes and interfaces. Both let you declare the signatures of the methods and fields that a class implements separately from the class itself. Abstract classes accomplish this by way of inheritance. Interfaces do it without using inheritance, but the effect is similar. This chapter also covers a feature called *default methods*, which are designed to make interfaces easier to work with.

Using Abstract Classes

Java lets you declare that a method or an entire class is *abstract*. When a class is declared to be abstract, the class cannot be instantiated. An abstract begs to be extended; it serves as a superclass to one or more subclasses that *can* be instantiated. Classes that can be instantiated are sometimes called *concrete classes* to distinguish them from abstract classes.

Most abstract classes contain one or more *abstract methods*, which are prototypes for actual methods. An abstract method defines a return type, a name, a list of parameters, and (optionally) a throws clause, but has no body.

To create an abstract method, you specify the modifier abstract and replace the method body with a semicolon:

```
public abstract int hit(int batSpeed);
```

Here the method named hit is declared as an abstract method that returns an int value and accepts an int parameter.

To declare an abstract class, you use the abstract modifier on the class definition. Note that any class that contains one or more abstract methods *must* be declared as an abstract class. Here's an example:

```
public abstract class Ball
{
    public abstract int hit(int batSpeed);
}
```

WARNING

If you omit the abstract modifier from the class declaration, the Java compiler coughs up an error message to remind you that the class must be declared abstract.

Again, you can't create an instance of an abstract class. So, the following code won't compile:

```
Ball b = new Ball();
```

The problem here isn't with declaring the variable b as a Ball; it's using the new keyword with the Ball class in an attempt to create a Ball object. Because Ball is an abstract class, you can't use it to create an object instance.

You can create a subclass from an abstract class like this:

```
public class BaseBall extends Ball
{
    public int hit(int batSpeed)
    {
        // code that implements the hit method goes here
    }
}
```

THE FACTORY PATTERN

One common use for abstract classes is to provide a way to obtain an instance of one of several subclasses when you don't know which subclass you need in advance. To do this, you can create a *Factory class* that has one or more methods that return subclasses of the abstract class.

Suppose that you want to create a Ball object, but you want to let the user choose whether to create a SoftBall or a BaseBall. To use the Factory pattern, you create a class (I call it BallFactory) that has a method named getBallInstance. This method accepts a String parameter that's set to "BaseBall" if you want a BaseBall object or "SoftBall" if you want a SoftBall object.

Here's the factory class:

```
class BallFactoryInstance
{
    public static Ball getBall(String t)
    {
        if (s.equalsIgnoreCase("BaseBall"))
            return new BaseBall();
        if (s.equalsIgnoreCase("SoftBall"))
            return new SoftBall();
        return null;
    }
}
```

Then, assuming that the String variable userChoice has been set according to the user's choice, you can create the selected type of Ball object like this:

```
Ball b = BallFactory.getBallInstance(userChoice);
```

In an actual application, using an enum variable is better than using a String variable to indicate the type of object to be returned.

When you subclass an abstract class, the subclass must provide an implementation for each abstract method in the abstract class. In other words, it must override each abstract method with a nonabstract method. (If it doesn't, the subclass must also be declared as abstract, so it can't be instantiated either.)

TIP

Abstract classes are useful when you want to create a generic type that is used as the superclass for two or more subclasses, but the superclass itself doesn't represent an actual object. If all employees are either salaried or hourly, for example, it makes sense to create an abstract Employee class and then use it as the base class for the SalariedEmployee and HourlyEmployee subclasses.

Here are a few additional points to ponder concerning abstract classes:

TECHNICAL STUFF

>> Not all the methods in an abstract class have to be abstract. A class can provide an implementation for some of its methods but not others. In fact, even if a class doesn't have any abstract methods, you can still declare it as abstract. (In that case, the class can't be instantiated.)

A private method can't be abstract. That makes sense, because a subclass can't override a private method, and abstract methods must be overridden.

>> Although you can't create an instance of an abstract class, you can declare a variable by using an abstract class as its type. Then use the variable to refer to an instance of any of the subclasses of the abstract class.

>> A class or method can't specify both `abstract` and `final`. That would cause one of those logical paradoxes that result in the annihilation of the universe. Well, ideally, the effect would be localized. But the point is that because an `abstract` class can be used only if you subclass it, and because a `final` class can't be subclassed, letting you specify both `abstract` and `final` for the same class doesn't make sense.

TECHNICAL STUFF

Abstract classes are used extensively in the Java API. Many of the abstract classes have names that begin with `Abstract` — such as `AbstractBorder`, `AbstractCollection`, and `AbstractMap` — but most of the abstract classes don't. The `InputStream` class (used by `System.in`) is abstract, for example.

Using Interfaces

An *interface* is similar to an abstract class, but there's a crucial difference: A class can inherit only one other class, but a class can *implement* as many interfaces as it needs. A class implements an interface by using the `implements` keyword in the class definition and by providing implementations for any abstract methods defined by the interface.

In fact, in its original form, an interface could contain *only* abstract methods (plus final fields). In subsequent releases, Java has added additional types of methods that can be included in a method, such as default methods, static methods, and private methods. Still, the heart of an interface lies in its definition of abstract methods, which are implemented by classes that implement the interface.

TIP

Interfaces have two advantages over inheritance:

>> Interfaces are easier to work with than an abstract class, because you don't have to worry about providing any implementation details in the interface.

>> A class can extend only one other class, but it can implement as many interfaces as you need.

The following sections describe the details of creating and using interfaces.

Creating a basic interface

Here's a basic interface that defines a single method, named `Playable`, that includes a single method named `play`:

```
public interface Playable
{
    void play();
}
```

This interface declares that any class that implements the `Playable` interface must provide an implementation for a method named `play` that accepts no parameters and doesn't return a value.

This interface has a few interesting details:

>> The interface itself is declared as `public` so that it can be used by other classes. Like a public class, a public interface must be declared in a file with the same name. Thus this interface must be in a file named `Playable.java`.

TIP

>> The name of the interface (`Playable`) is an adjective. Most interfaces are named with adjectives rather than nouns because they describe some additional capability or quality of the classes that implement the interface. Thus classes that implement the `Playable` interface represent objects that can be played.

In case you haven't been to English class in a while, an *adjective* is a word that modifies a noun. You can convert many verbs to adjectives by adding *-able* to the end of the word — *playable, readable, drivable,* and *stoppable,* for example. This type of adjective is commonly used for interface names.

>> Another common way to name interfaces is to combine an adjective with a noun to indicate that the interface adds some capability to a particular type of object. You could call an interface that provides methods unique to card games `CardGame`, and this interface might have methods such as `deal`, `shuffle`, and `getHand`.

>> All the methods in an interface are assumed to be public and abstract. If you want, you can code the `public` and `abstract` keywords on interface methods. That kind of coding is considered to be bad form, however, because it might indicate that you think the default is private and not abstract.

Implementing an interface

To implement an interface, a class must do two things:

>> It must specify an `implements` clause on its class declaration.

>> It must provide an implementation for every method declared by the interface.

Here's a class that implements the `Playable` interface:

```
public class TicTacToe implements Playable
{
    // additional fields and methods go here

    public void play()
    {
        // code that plays the game goes here
    }

    // additional fields and methods go here
}
```

Here the declaration for the `TicTacToe` class specifies `implements Playable`. Then the body of the class includes an implementation of the `play` method.

A class can implement more than one interface:

```
public class Hearts implements Playable, CardGame
{
    // must implement methods of the Playable
    // and CardGame interfaces
}
```

Here, the `Hearts` class implements two interfaces: `Playable` and `CardGame`.

A class can possibly inherit a superclass and implement one or more interfaces. Here's an example:

```
public class Poker extends Game
    implements Playable, CardGame
{
    // inherits all members of the Game class
    // must implement methods of the Playable
    // and CardGame interfaces
}
```

Using an interface as a type

In Java, an interface is a kind of type, just like a class. As a result, you can use an interface as the type for a variable, parameter, or method return value.

Consider this snippet of code:

```
Playable game = getGame();
game.play();
```

Here I assume that the getGame method returns an object that implements the Playable interface. This object is assigned to a variable of type Playable in the first statement. Then the second statement calls the object's play method.

For another (slightly more complex) example, suppose that you have an interface named Dealable defining a method named deal that accepts the number of cards to deal as a parameter:

```
public interface Dealable
{
    void deal(int cards);
}
```

Now suppose that you have a method called startGame that accepts two parameters: a Dealable object and a String that indicates what game to play. This method might look something like this:

```
private void startGame(Dealable deck, String game)
{
    if (game.equals("Poker"))
        deck.deal(5);
    else if (game.equals("Hearts"))
        deck.deal(13);
    else if (game.equals("Gin"))
        deck.deal(10);
}
```

Assuming that you also have a class named CardDeck that implements the Dealable interface, you might use a statement like this example to start a game of Hearts:

```
Dealable d = new CardDeck();
startGame(d, "Hearts");
```

Notice that the variable d is declared as a Dealable. You could just as easily declare it as a CardDeck:

```
CardDeck d = new CardDeck();
startGame(d, "Hearts");
```

Because the CardDeck class implements the Dealable interface, it can be passed as a parameter to the startGame method.

More Things You Can Do with Interfaces

There's more to interfaces than just creating abstract methods. The following sections describe some additional interesting things you can do with interfaces. Read on.

Adding fields to an interface

Besides abstract methods, an interface can include final fields — that is, constants. Interface fields are used to provide constant values that are related to the interface. For example:

```
public interface GolfClub
{
    int DRIVER = 1;
    int SPOON = 2;
    int NIBLICK = 3;
    int MASHIE = 4;
}
```

Here any class that implements the GolfClub interface has these four fields (that is, constants) available.

TECHNICAL STUFF

Note that interface fields are automatically assumed to be static, final, and public. You can include these keywords when you create interface constants, but you don't have to.

Extending interfaces

You can extend interfaces by using the extends keyword. An interface that extends an existing interface is called a *subinterface*, and the interface being extended is called the *superinterface*.

When you use the extends keyword with interfaces, all the fields and methods of the superinterface are effectively copied into the subinterface. Thus the subinterface consists of a combination of the fields and methods in the superinterface *plus* the fields and methods defined for the subinterface.

Here's an example:

```
public interface ThrowableBall
{
    void throwBall();
    void catchBall();
}

public interface KickableBall
{
    void kickBall();
    void catchBall();
}

public interface PlayableBall
    extends ThrowableBall, KickableBall
{
    void dropBall();
}
```

Here three interfaces are declared. The first, named ThrowableBall, defines two methods: throwBall and catchBall. The second, named KickableBall, also defines two methods: kickBall and catchBall. The third, named PlayableBall, extends ThrowableBall and KickableBall, and adds a method of its own named dropBall.

Thus any class that implements the PlayableBall interface must provide an implementation for four methods: throwBall, catchBall, kickBall, and dropBall. Note that because the catchBall methods defined by the ThrowableBall and KickableBall interfaces have the same signature, only one version of the catchBall method is included in the PlayableBall interface.

Using interfaces for callbacks

In the theater, a callback is when you show up for an initial audition, they like what they see, and they tell you that they want you to come back so they can have another look.

In Java, a *callback* is sort of like that. It's a programming technique in which an object lets another object know that the second object should call one of the first object's methods whenever a certain event happens. The first object is called an *event listener* because it waits patiently until the other object calls it. The second object is called the *event source* because it's the source of events that result in calls to the listener.

Okay, my theater analogy was a bit of a stretch. Callbacks in Java aren't really that much like callbacks when you're auditioning for a big part. A callback is more like when you need to get hold of someone on the phone, and you call him when you know he isn't there and leave your phone number on his voicemail so he can call you back.

THE MARKER INTERFACE PATTERN

A *marker interface* is an interface that doesn't have any members. Its sole purpose in life is to identify a class as belonging to a set of classes that possess some capability or have some characteristic in common.

The best-known example of a marker interface is the Java API Cloneable interface, which marks classes that can be cloned. The Object class, which all classes ultimately inherit, provides a method named clone that can be used to create a copy of the object. You're allowed to call the clone method only if the object implements the Cloneable interface, however. If you try to call clone for an object that doesn't implement Cloneable, CloneNotSupportedException is thrown. (For more information about the clone method, refer to Book 3, Chapter 6.)

Here's the actual code for the Cloneable interface:

```
public interface Cloneable {
}
```

In some cases, you might find a use for marker interfaces in your own application. If you're working on a series of classes for creating games, you might create a marker interface named Winnable to distinguish games that have a winner from games that you just play for enjoyment.

Callbacks are handled in Java by a set of interfaces designed for this purpose. One handy use for callbacks is in the `Timer` class, which is part of the `javax.Swing` package. This class implements a basic timer that generates events at regular intervals — and lets you set up a *listener object* to handle these events. The listener object must implement the `ActionListener` interface, which defines a method named `actionPerformed` that's called for each timer event.

The `Timer` class constructor accepts two parameters:

>> The first parameter is an `int` value that represents how often the timer events occur.

>> The second parameter is an object that implements the `ActionListener` interface. This object's `actionPerformed` method is called when each timer event occurs.

The `ActionListener` interface is defined in the `java.awt.event` package. It includes the following code:

```
public interface ActionListener extends EventListener {

/**
* Invoked when an action occurs.
*/

    public void actionPerformed(ActionEvent e);
}
```

As you can see, the `ActionListener` interface consists of a single method named `actionPerformed`. It receives a parameter of type `ActionEvent`, but you don't use this parameter here. (You do use the `ActionEvent` class in Book 6.)

The `Timer` class has about 20 methods, but I talk about only one of them here: `start`, which sets the timer in motion. This method doesn't require any parameters and doesn't return a value.

Listing 5-1 shows a program that uses the `Timer` class to alternately display the messages `Tick...` and `Tock...` on the console at 1-second intervals. The `JOptionPane` class is used to display a dialog box; the program runs until the user clicks the OK button in this box. Figure 5-1 shows the Tick Tock program in action. (Actually it takes a while for the JOptionPane class to shut down the timer, so you'll see a few extra tick-tocks after clicking OK.)

LISTING 5-1: **The Tick Tock Program**

```java
import java.awt.event.*;                                          →1
import javax.swing.*;                                             →2

public class TickTock
{
    public static void main(String[] args)
    {
        // create a timer that calls the Ticker class
        // at one second intervals
        Timer t = new Timer(1000, new Ticker());                 →10
        t.start();                                               →11

        // display a message box to prevent the
        // program from ending immediately
        JOptionPane.showMessageDialog(null,                      →15
            "Click OK to exit program");
    }
}

class Ticker implements ActionListener                           →20
{
    private boolean tick = true;                                 →22

    public void actionPerformed(ActionEvent event)              →24
    {
        if (tick)
        {
            System.out.println("Tick...");                       →28
        }
        else
        {
            System.out.println("Tock...");                       →32
        }
        tick = !tick;                                            →34
    }
}
```

FIGURE 5-1:
The Tick Tock application in action.

The following paragraphs describe the important details of this program's operation:

» →1 The ActionListener interface is part of the java.awt.event package, so this import statement is required.

» →2 The Timer class is part of the javax.swing package, so this import statement is required.

» →10 This statement creates a new Timer object. The timer's interval is set to 1,000 milliseconds — which is equivalent to 1 second. A new instance of the Ticker class is passed as the second parameter. The timer calls this object's actionPerformed method at each timer tick — in other words, once per second.

» →11 This statement calls the start method to kick the timer into action.

» →15 The JOptionPane class is used to display a dialog box that tells the user to click the OK button to stop the application. You might think I include this dialog box to give the user a way to end the program. In reality, I use it to give the timer some time to run. If I just end the main method after starting the timer, the application ends, which kills the timer. Because I use JOptionPane here, the application continues to run as long as the dialog box is displayed. (For more information about JOptionPane, see Book 2, Chapter 2.)

» →20 This line is the declaration of the Ticker class, which implements the ActionListener interface.

» →22 This line is a private boolean class field that's used to keep track of whether the Ticker displays Tick... or Tock... Each time the actionPerformed method is called, this field is toggled.

>> →24 This line is the `actionPerformed` method, which is called at each timer interval.

>> →28 This line prints `Tick...` on the console if `tick` is true.

>> →32 This line prints `Tock...` on the console if `tick` is false.

>> →34 This line toggles the value of the `tick` variable. In other words, if `tick` is true, it's set to `false`. If `tick` is false, it's set to `true`.

Using Additional Interface Method Types

Although interfaces are an incredibly useful feature of Java, they have an inherent limitation: After you define an interface and then build classes that implement the interface, there's no easy way to modify the interface by adding additional methods to it.

For example, suppose you have created the following interface:

```
public interface Playable
{
    void play();
}
```

You then build several classes that implement this interface. Here's a simple example:

```
class Game implements Playable
{
    public void play()
    {
        System.out.println("Good luck!");
    }
}
```

This is a pretty pointless game, of course; it simply prints the message "Good luck!" whenever the `play` method is called.

Now suppose that you decide that the `Playable` interface should have an additional feature — specifically, you want to add the ability to end the game by calling a method named `quit`.

You'd be tempted to just add the new method to the existing interface, like this:

```
public interface Playable
{
    void play();
    void quit();
}
```

Unfortunately, however, doing so will break the Game class because it doesn't provide an implementation of the quit method.

You could, of course, modify the Game class by adding an implementation of the quit method. But what if you have written dozens, or even hundreds, of classes that implement Playable? As you can imagine, once an interface has become popular, it becomes nearly impossible to modify.

To alleviate this problem, Java lets you include several additional types of methods in your interfaces. In addition to abstract methods, you can include the following types of methods in an interface:

>> default **methods:** Provide a default implementation that is used if the method is not overridden.

>> static **methods:** Like static methods for a class, provide methods that are invoked apart from an instance of a class implementing the interface. Static methods cannot be overridden.

>> private **methods:** Provide for methods that can be called only by default methods or other private interface methods. A private method cannot be overridden or accessed by an implementing class.

>> private static **methods:** Similar to private methods but are static and can be called only by static interface methods or other private static interface methods. A private static method cannot be overridden or accessed by an implementing class.

You can use a default method to add a quit method to the Playable interface by specifying it as a default method, like this:

```
interface Playable
{
    void play();
    default void quit()
    {
    System.out.println("Sorry, quitting is not allowed.");
    }
}
```

Here the `Playable` interface specifies that if an implementing class does not provide an implementation of the `quit` method, the default method will be used. In this case, the default method simply prints the message `Sorry, quitting is not allowed`.

Here's a complete example that uses the `Playable` interface and its default method:

```
public class TestLambdaCollection
{
    public static void main(String[] args)
    {
        Game g = new Game();
        g.play();
        g.quit();
    }
}

interface Playable
{
    void play();
    default void quit()
    {
        System.out.println("Sorry, quitting is not allowed.");
    }
}

class Game implements Playable
{
    public void play()
    {
        System.out.println("Good luck!");
    }
}
```

When you run this program, the following will be displayed on the console:

```
Good luck!
Sorry, quitting is not allowed.
```

Chapter **6**

Using the Object and Class Classes

I n this chapter, you find out how to use two classes of the Java API that are important to object-oriented programming:

>> The Object class, which every other class inherits — including all the classes in the Java API and any classes you create yourself

>> The Class class, which is used to get information about an object's type

TECHNICAL STUFF

If I could, I'd plant a huge Technical Stuff icon on this entire chapter. All this stuff is a bit on the technical side, and many Java programmers get by for years without understanding or using it. Still, I recommend that you read this chapter carefully. Even if it all doesn't sink in, it may help explain why some things in Java don't work quite the way you think they should, and the information in this chapter may someday help you program your way out of a difficult corner.

The Mother of All Classes: Object

Object is the mother of all classes. In Java, every class ultimately inherits the Object class. This class provides a set of methods that is available to every Java object.

Every object is an Object

Any class that doesn't have an extends clause implicitly inherits Object. Thus you never have to code a class like this:

```
public class Product extends Object...
```

If a subclass has an extends clause that specifies a superclass other than Object, the class still inherits Object. That's because the inheritance hierarchy eventually gets to a superclass that doesn't have an extends clause, and that superclass inherits Object and passes it down to all its subclasses.

Suppose you have these classes:

```
public class Manager extends SalariedEmployee...
public class SalariedEmployee extends Employee...
public class Employee extends Person...
public class Person...
```

Here the Manager class inherits the Object class indirectly because it inherits SalariedEmployee, which inherits Employee, which inherits Person, which inherits Object.

REMEMBER

In Java, creating a class that *doesn't* inherit Object is not possible.

Object as a type

If you don't know or care about the type of an object referenced by a variable, you can specify its type as Object. The following example is perfectly legal:

```
Object emp = new Employee();
```

You can't do anything useful with the emp variable, however, because the compiler doesn't know that it's an Employee. If the Employee class has a method named setLastName, the following code doesn't work:

```
Object emp = new Employee();
emp.setLastName("Smith"); // error: won't compile
```

Because emp is an Object, not an Employee, the compiler doesn't know about the setLastName method.

Note that you could still cast the object to an Employee:

```
Object emp = new Employee();
((Employee)emp).setLastName("Smith"); // this works
```

But what's the point? You may as well make emp an Employee in the first place.

Declaring a variable, parameter, or return type as Object in certain situations, however, does make perfect sense. The Java API provides a set of classes designed to maintain collections of objects. One of the most commonly used of these classes is the ArrayList class, which has a method named add that accepts an Object as a parameter. This method adds the specified object to the collection. Because the parameter type is Object, you can use the ArrayList class to create collections of any type of object. (For more information about the ArrayList class and other collection classes, see Book 4.)

Methods of the Object class

Table 6-1 lists all the methods of the Object class. Ordinarily, I wouldn't list all the methods of a class; I'd just list the ones that I think are most useful. Because Object is such an important player in the game of object-oriented programming, however, I thought it best to show you all its capabilities, even though some of them are a bit obscure.

TABLE 6-1   **Methods of the Object Class**

Method	What It Does
protected Object clone()	Returns a copy of this object.
boolean equals(Object obj)	Indicates whether this object is equal to the obj object.
protected void finalize()	Is called by the garbage collector when the object is destroyed.
Class getClass()	Returns a Class object that represents this object's runtime class.
int hashCode()	Returns this object's hash code.
void notify()	Is used with threaded applications to wake up a thread that's waiting on this object.
void notifyAll()	Is used with threaded applications to wake up all threads that are waiting on this object.
String toString()	Returns a String representation of this object.
void wait()	Causes this object's thread to wait until another thread calls notify or notifyAll.
void wait(Long timeout)	Is a variation of the basic wait method.
void wait(Long timeout, int nanos)	Is yet another variation of the wait method.

TECHNICAL
STUFF

I warned you — this entire chapter should have a Technical Stuff icon.

Note: Almost half of these methods (`notify`, `notifyAll`, and the three `wait` methods) are related to threading. You find complete information about those five methods in Book 5, Chapter 1. Here's the rundown on the remaining six methods:

>> `clone`: This method is commonly used to make copies of objects, and overriding it in your own classes is not uncommon. I explain this method in detail later in this chapter, in the section "The `clone` Method."

>> `equals`: This method is commonly hashed to compare objects. Any class that represents an object that can be compared with another object should override this method. Turn to the section "The `equals` Method," later in this chapter, for more info.

>> `finalize`: This method is called when the garbage collector realizes that an object is no longer being used and can be discarded. The intent of this method is to let you create objects that clean up after themselves by closing open files and performing other cleanup tasks before being discarded. But because of the way the Java garbage collector works, there's no guarantee that the `finalize` method is ever actually called. As a result, this method isn't commonly used.

>> `getClass`: This method is sometimes used in conjunction with the `Class` class, which I describe later in this chapter, in the section "The `Class` Class."

>> `hashCode`: Every Java object has a *hash code,* which is an `int` representation of the class that's useful for certain operations.

>> `toString`: This method is one of the most commonly used methods in Java. I describe it in the section "The `toString` Method," later in this chapter.

Primitives aren't objects

I need to note that primitive types, such as `int` and `double`, are not objects. As a result, they do not inherit the `Object` class and don't have access to the methods listed in the preceding section.

As a result, the following code won't work:

```
int x = 50;
String s = x.toString();            // error: won't compile
```

If you really want to convert an `int` to a string in this way, you can use a wrapper class such as `Integer` to create an object from the value and then call its `toString` method:

```
String s = new Integer(x).toString();  // OK
```

Each of the wrapper classes also defines a static `toString` method, which you can use like this:

```
String s = Integer.toString(x);
```

TIP

Sometimes, using the compiler shortcut that lets you use primitive types in string concatenation expressions is easier:

```
String s = "" + x;
```

Here the `int` variable x is concatenated with an empty string.

REMEMBER

The point of all this is that primitive types aren't objects, so they don't inherit anything from `Object`. If you want to treat a primitive value as an object, you can use the primitive type's wrapper class (as I describe in Book 2, Chapter 2).

The toString Method

The `toString` method returns a `String` representation of an object. By default, the `toString` method returns the name of the object's class plus its hash code. In the sections that follow, I show you how to use the `toString` method and how to override it in your own classes to create more useful strings.

Using toString

Here's a simple program that puts the `toString` method to work:

```
public class TestToString
{
    public static void main(String[] args)
    {
        Employee emp = new Employee("Martinez",
            "Anthony");
        System.out.println(emp.toString());
    }
}
```

```
class Employee
{
    private String lastName;
    private String firstName;

    public Employee(String lastName, String firstName)
    {
        this.lastName = lastName;
        this.firstName = firstName;
    }
}
```

This code creates a new Employee object; then the result of its toString method is printed to the console. When you run this program, the following line is printed on the console:

```
Employee@82ba41
```

Note: The hash code — in this case, 82ba41 — will undoubtedly be different on your system.

It turns out that the explicit call to toString isn't really necessary in this example. I could just as easily have written the second line of the main method like this:

```
System.out.println(emp);
```

That's because the println method automatically calls the toString method of any object you pass it.

Overriding toString

The default implementation of toString isn't very useful in most situations. You don't really learn much about an Employee object by seeing its hash code, for example. Wouldn't it be better if the toString method returned some actual data from the object, such as the employee's name?

To do that, you must override the toString method in your classes. In fact, one of the basic guidelines of object-oriented programming in Java is to *always* override toString. Here's a simple program with an Employee class that overrides toString:

```
public class TestToString
{
    public static void main(String[] args)
    {
```

```
        Employee emp = new Employee("Martinez",
            "Anthony");
        System.out.println(emp.toString());
    }
}

class Employee
{
    private String lastName;
    private String firstName;

    public Employee(String lastName, String firstName)
    {
        this.lastName = lastName;
        this.firstName = firstName;
    }

    public String toString()
    {
        return "Employee["
            + this.firstName + " "
            + this.lastName + "]";
    }
}
```

When you run this program, the following line is displayed on the console:

```
Employee[Anthony Martinez]
```

Note that the output consists of the class name followed by some data from the object in brackets. This convention is common in Java programming.

TIP

The only problem with the preceding example is that the class name is hard-coded into the toString method. You can use the getClass method to retrieve the actual class name at runtime:

```
public String toString()
{
    return this.getClass().getName() + "["
        + this.firstName + " "
        + this.lastName + "]";
}
```

Here the getClass method returns a Class object that represents the class of the current object. Then the Class object's getName method is used to get the actual class name. (You discover more about the Class object later in this chapter.)

The equals Method

Testing objects to see whether they are equal is one of the basic tasks of any object-oriented programming language. Unfortunately, Java isn't very good at it. Consider this program:

```
public class TestEquality1
{
    public static void main(String[] args)
    {
        Employee emp1 = new Employee(
            "Martinez", "Anthony");
        Employee emp2 = new Employee(
            "Martinez", "Anthony");
        if (emp1 == emp2)
            System.out.println(
                "These employees are the same.");
        else
            System.out.println(
                "These are different employees.");
    }
}

class Employee
{
    private String lastName;
    private String firstName;

    public Employee(String lastName, String firstName)
    {
        this.lastName = lastName;
        this.firstName = firstName;
    }
}
```

Here the `main` method creates two `Employee` objects with identical data and then compares them. Alas, the comparison returns `false`. Even though the `Employee` objects have identical data, they're not considered to be equal because the equality operator (`==`) compares the object references, not the data contained by the objects. Thus the comparison returns `true` only if both `emp1` and `emp2` refer to the same instance of the `Employee` class.

If you want to create objects that are considered to be equal if they contain identical data, you have to do two things:

» Compare them with the equals method rather than the equality operator.

» Override the equals method in your class to compare objects based on their data.

The following sections describe both of these steps.

Using equals

To test objects using the equals method rather than the equality operator, you simply rewrite the comparison expression like this:

```
if (emp1.equals(emp2))
    System.out.println("These employees are equivalent.");
else
    System.out.println
        ("These are different employees.");
```

Here, the equals method of emp1 is used to compare emp1 with emp2.

By default, the equals operator (inherited from the Object class) returns the same result as the equality operator. So just replacing == with the equals method doesn't have any effect unless you also override the equals method, as explained in the next section.

TIP

Which object's equals method you use shouldn't matter. Thus the if statement shown here returns the same result:

```
if (emp2.equals(emp1))
    System.out.println("These employees are the same.");
else
    System.out.println
        ("These are different employees.");
```

TECHNICAL STUFF

Note that I said it *shouldn't* matter. Whenever you override the equals method, you're supposed to make sure that comparisons work in both directions. Sloppy programming, however, sometimes results in equals methods where a equals b but b doesn't equal a. Be on your toes.

Overriding the equals method

You can override the `equals` method so that objects can be compared based on their values. At the surface, you might think this is easy to do. You could be tempted to write the `equals` method for the `Employee` class like this:

```
// warning -- there are several errors in this code!
public boolean equals(Employee emp)
{
    if (this.getLastName().equals(emp.getLastName())
            && this.getFirstName().equals(emp.getFirstName()))
        return true;
    else
        return false;
}
```

The basic problem with this code — and the challenge of coding a good `equals` method — is that the parameter passed to the `equals` method must be an `Object`, not an `Employee`. That means that the `equals` method must be prepared to deal with anything that comes its way. Someone might try to compare an `Employee` object with a `Banana` object, for example, or with a null. The `equals` method must be prepared to deal with all possibilities.

Specifically, the Java API documentation says that whenever you override the `equals` method, you must ensure that the `equals` method meets five specific conditions. Here they are, quoted right out of the API documentation:

>> **It is *reflexive*.** For any non-null reference value x, x.equals(x) should return true.

>> **It is *symmetric*.** For any non-null reference values x and y, x.equals(y) should return true if — and only if — y.equals(x) returns true.

>> **It is *transitive*.** For any non-null reference values x, y, and z, if x.equals(y) returns true and y.equals(z) returns true, x.equals(z) should return true.

>> **It is *consistent*.** For any non-null reference values x and y, multiple invocations of x.equals(y) consistently return true or consistently return false, provided that no information used in equals comparisons on the objects is modified.

>> For any non-null reference value x, x.equals(null) should return false.

Sound confusing? Fortunately, it's not as complicated as it seems at first. You can safely ignore the transitive rule, because if you get the other rules right, this one happens automatically. The consistency rule basically means that you return consistent results. As long as you don't throw a call to Math.random into the comparison, that shouldn't be a problem.

Here's a general formula for creating a good equals method (assume that the parameter is named obj):

1. **Test the reflexive rule.**

Use a statement like this:

```
if (this == obj)
    return true;
```

In other words, if someone is silly enough to see whether an object is equal to itself, it returns true.

2. **Test the non-null rule.**

Use a statement like this:

```
if (obj == null)
    return false;
```

Null isn't equal to anything.

3. **Test that obj is of the same type as this.**

You can use the getClass method to do that, like this:

```
if (this.getClass() != obj.getClass())
    return false;
```

The two objects can't possibly be the same if they aren't of the same type. (It may not be apparent at first, but this test is required to fulfill the symmetry rule — that if x equals y, y must also equal x.)

4. **Cast obj to a variable of your class; then compare the fields you want to base the return value on, and return the result.**

Here's an example:

```
Employee emp = (Employee) obj;
return this.lastName.equals(emp.getLastName())
    && this.firstname.equals(emp.getFirstName());
```

Notice that the field comparisons for the String values use the equals method rather than ==. This is because you can't trust == to compare strings. If you need to compare primitive types, you can use ==. But you should use equals to compare strings and any other reference types.

Putting it all together, Listing 6-1 shows a program that compares two Employee objects by using a properly constructed equals method.

LISTING 6-1: **Comparing Objects**

```java
public class TestEquality2
{
public static void main(String[] args)
{
    Employee emp1 = new Employee(                           →5
        "Martinez", "Anthony");
    Employee emp2 = new Employee(                           →7
        "Martinez", "Anthony");
    if (emp1.equals(emp2))                                  →9
        System.out.println(
            "These employees are the same.");
    else
        System.out.println(
            "These are different employees.");
    }
}

class Employee                                              →18
{
    private String lastName;
    private String firstName;

    public Employee(String lastName, String firstName)
    {
        this.lastName = lastName;
        this.firstName = firstName;
    }

    public String getLastName()
    {
        return this.lastName;
    }

    public String getFirstName()
    {
        return this.firstName;
    }

    public boolean equals(Object obj)                       →39
    {
        // an object must equal itself
        if (this == obj)                                    →42
```

```
            return true;

        // no object equals null
        if (obj == null)                                    →46
            return false;

        // objects of different types are never equal
        if (this.getClass() != obj.getClass())              →50
            return false;

        // cast to an Employee, then compare the fields
        Employee emp = (Employee) obj;                      →54
        return this.lastName.equals(emp.getLastName())      →55
            && this.firstName.equals(emp.getFirstName());
    }
}
```

Following are some noteworthy points in this listing:

» →5 Creates an Employee object with the name Anthony Martinez.

» →7 Creates another Employee object with the name Anthony Martinez.

» →9 Compares the two Employee objects by using the equals method.

» →18 The Employee class.

» →39 The overridden equals method.

» →42 Returns true if the same object instances are being compared. This meets the first equality test: that an object must always be equal to itself.

» →46 Returns false if the object being compared is null. This meets the last equality test: that nothing is equal to null.

» →50 Returns false if the object being compared isn't of the correct type. This helps ensure the symmetry test: that if x equals y, y must equal x.

» →54 Having slid through the other tests, you can assume that you're comparing two different Employee objects, so the next step is to cast the other object to an Employee.

» →55 Having cast the other object to an Employee, the two fields (lastName and firstName) are compared, and the result of the compound comparison is returned.

The clone Method

Cloning refers to the process of making an exact duplicate of an object. Unfortunately, this process turns out to be a pretty difficult task in an object-oriented language such as Java. You'd think that cloning would be as easy as this:

```
Employee emp1 = new Employee("Stewart", "Martha");
Employee emp2 = emp1;
```

This code doesn't make a copy of the `Employee` object at all, however. Instead, you now have two variables that refer to the same object, which usually isn't what you want. Suppose that you execute these statements:

```
emp1.setLastName("Washington");
emp2.setLastName("Graham");
String lastName = emp1.getLastName();
```

After these statements execute, does `lastName` return `Washington` or `Graham`? The correct answer is `Graham`, because both `emp1` and `emp2` refer to the same `Employee` object.

By contrast, a *clone* is an altogether new object that has the same values as the original object. Often you can create a clone manually by using code like this:

```
Employee emp1 = new Employee("Stewart", "Martha");
Employee emp2 = new Employee();
emp2.setLastName(emp1.getLastName());
emp2.setFirstName(emp1.getFirstName());
emp2.setSalary(emp1.getSalary());
```

Here a new `Employee` object is created, and its fields are set to the same values as the original object.

TIP

Java provides a more elegant way to create object copies: the `clone` method, which is available to all classes because it's inherited from the `Object` class. As you discover in the following sections, however, the `clone` method can be difficult to create and use. For this reason, you want to implement it only for those classes that you think can really benefit from cloning.

Implementing the clone method

The `clone` method is defined by the `Object` class, so it's available to all Java classes, but `clone` is declared with `protected` access in the `Object` class. As a result, the `clone` method for a given class is available only within that class. If you want other objects to be able to clone your object, you must override the clone method and give it public access.

Note that the clone method defined by the Object class returns an Object type. That makes perfect sense, because the Object class doesn't know the type of the class in which you'll be overriding the clone method. An inconvenient side effect of returning an Object is that whenever you call the clone method for a class that overrides clone, you must cast the result to the desired object type.

Listing 6-2 gives a simple example of a program that clones Employee objects. In a nutshell, this program overrides the clone method for the Employee class: It creates an Employee object, clones it, changes the name of the original Employee object, and prints out both objects to the console.

LISTING 6-2: **A Cloning Example**

```
public class CloneTest
{
    public static void main(String[] args)
    {
        Employee emp1 = new Employee(                          →5
            "Martinez", "Anthony");
        emp1.setSalary(40000.0);                               →7
        Employee emp2 = (Employee)emp1.clone();                →8
        emp1.setLastName("Smith");                             →9
        System.out.println(emp1);                              →10
        System.out.println(emp2);                              →11
    }
}

class Employee                                                 →15
{
    private String lastName;
    private String firstName;
    private Double salary;
    public Employee(String lastName,

    String firstName)
    {
        this.lastName = lastName;
        this.firstName = firstName;
    }

    public String getLastName()
    {
        return this.lastName;
    }
```

(continued)

LISTING 6-2:   *(continued)*

```
public void setLastName(String lastName)
{
    this.lastName = lastName;
}

public String getFirstName()
{
    return this.firstName;
}

public void setFirstName(String firstName)
{
    this.firstName = firstName;
}

public Double getSalary()
{
    return this.salary;
}

public void setSalary(Double salary)
{
    this.salary = salary;
}

public Object clone()                                        →58
{
    Employee emp;
    emp = new Employee(                                      →61
    this.lastName, this.firstName);
    emp.setSalary(this.salary);                              →63
    return emp;                                              →64
}

public String toString()
{
    return this.getClass().getName() + "["
        + this.firstName + " "
        + this.lastName + ", "
        + this.salary + "]";
}
}
```

When you run this program, the following lines appear on the console:

```
Employee[Anthony Smith, 40000.0]
Employee[Anthony Martinez, 40000.0]
```

As you can see, the name of the second Employee object was successfully changed without affecting the name of the first Employee object.

The following paragraphs draw your attention to some of the highlights of this program:

» →5 Creates the first Employee object for an employee named Anthony Martinez.

» →7 Sets Mr. Martinez's salary.

» →8 Creates a clone of the Employee object for Mr. Martinez. Notice that the return value must be cast to an Employee, because the return value of the clone method is Object.

» →9 Changes the last name for the second Employee object.

» →10 Prints the first Employee object.

» →11 Prints the second Employee object.

» →15 The Employee class. This class defines private fields to store the last name, first name, and salary, as well as getter and setter methods for each field.

» →58 Overrides the clone method. Notice that its return type is Object, not Employee.

» →61 Creates a new Employee object, using the last name and first name from the current object.

» →63 Sets the new employee's salary to the current object's salary.

» →64 Returns the cloned Employee object.

Using clone to create a shallow copy

In the preceding example, the clone method manually creates a copy of the original object and returns it. In many cases, this is the easiest way to create a clone. But what if your class has a hundred or more fields that need to be duplicated? The chance of forgetting to copy one of the fields is high, and if you add a field to the class later on, you may forget to modify the clone method to include the new field.

Fortunately, you can solve this problem by using the `clone` method of the `Object` class directly in your own `clone` method. The `clone` method of the `Object` class can automatically create a copy of your object that contains duplicates of all the fields that are primitive types (such as `int` and `double`), as well as copies of immutable reference types — most notably, strings. So if all the fields in your class are either primitives or strings, you can use the `clone` method provided by the `Object` class to clone your class.

This type of clone is known as a *shallow copy*, for reasons that I explain in the next section.

To call the `clone` method from your own `clone` method, just specify `super.clone()`. Before you can do that, however, you must do two things:

>> Declare that the class supports the `Cloneable` interface. The `Cloneable` interface is a marker interface that doesn't provide any methods. It simply marks a class as being appropriate for cloning.

>> Enclose the call to `super.clone()` in a `try/catch` statement that catches the exception `CloneNotSupportedException`. This exception is thrown if you try to call `clone` on a class that doesn't implement the `Cloneable` interface. Provided that you implement `Cloneable`, this exception won't ever happen, but because `CloneNotSupportedException` is a checked exception, you must catch it.

Here's an example of an `Employee` class with a clone method that uses `super.clone()` to clone itself:

```
class Employee implements Cloneable
{
    // Fields and methods omitted...

    public Object clone()
    {
        Employee emp;
        try
        {
            emp = (Employee) super.clone();
        }
        catch (CloneNotSupportedException e)
        {
            return null; // will never happen
        }
```

```
        return emp;
    }
}
```

Notice that this method doesn't have to be aware of any of the fields declared in the Employee class. This clone method, however, works only for classes whose fields are all either primitive types or immutable objects such as strings.

Creating deep copies

TECHNICAL STUFF

It's not uncommon for some fields in a class actually to be other objects. The Employee class, for example, might have a field of type Address that's used to store each employee's address:

```
class Employee
{
public Address address;
// other fields and methods omitted
}
```

If that's the case, the super.clone() method won't make a complete copy of the object. The clone won't get a clone of the address field. Instead, it has a reference to the same address object as the original.

To solve this problem, you must do a deep copy of the Employee object. A *deep copy* is a clone in which any subobjects within the main object are also cloned or copied. To accomplish this feat, the clone method override first calls super.clone() to create a shallow copy of the object. Then it calls the clone method of each of the subobjects contained by the main object to create clones of those objects. (For a deep copy to work, of course, those objects must also support the clone methods or contain code to copy their values.)

Listing 6-3 shows an example. Here, an Employee class contains a public field named address, which holds an instance of the Address class. As you can see, the clone method of the Employee class creates a shallow copy of the Employee object and then sets the copy's address field to a clone of the original object's address field. To make this example work, the Address class also overrides the clone method. Its clone method calls super.clone() to create a shallow copy of the Address object.

LISTING 6-3: **Creating a Deep Copy**

```java
public class CloneTest2
{
    public static void main(String[] args)
    {
        Employee emp1 = new Employee(                          →5
            "Martinez", "Anthony");
        emp1.setSalary(40000.0);
        emp1.address = new Address(                            →8
            "1300 N. First Street",
            "Fresno", "CA", "93702");
        Employee emp2 = (Employee)emp1.clone();                →11

        System.out.println(                                    →13
            "**** after cloning ****\n");
        printEmployee(emp1);
        printEmployee(emp2);
        emp2.setLastName("Smith");                             →17
        emp2.address = new Address(                            →18
            "2503 N. 6th Street",
            "Fresno", "CA", "93722");

        System.out.println(                                    →22
            "**** after changing emp2 ****\n");
        printEmployee(emp1);
        printEmployee(emp2);
    }

    private static void printEmployee(Employee e)              →28
    {
        System.out.println(e.getFirstName()
            + " " + e.getLastName());
        System.out.println(e.address.getAddress());
        System.out.println("Salary: " + e.getSalary());
        System.out.println();
    }
}

class Employee implements Cloneable                            →38
{
    private String lastName;
    private String firstName;
    private Double salary;
    public Address address;                                    →43
    public Employee(String lastName, String firstName)
    {
        this.lastName = lastName;
        this.firstName = firstName;
```

```
        this.address = new Address();
    }

    public String getLastName()
    {
        return this.lastName;
    }

    public void setLastName(String lastName)
    {
        this.lastName = lastName;
    }

    public String getFirstName()
    {
        return this.firstName;
    }

    public void setFirstName(String firstName)
    {
        this.firstName = firstName;
    }

    public Double getSalary()
    {
        return this.salary;
    }

    public void setSalary(Double salary)
    {
        this.salary = salary;
    }

    public Object clone()                                    →81
    {
        Employee emp;
        try
        {
            emp = (Employee) super.clone();                  →86
            emp.address = (Address)address.clone();          →87
        }
        catch (CloneNotSupportedException e)                 →89
        {
            return null; // will never happen
        }
        return emp;                                          →93
    }
```

(continued)

LISTING 6-3:   *(continued)*

```java
    public String toString()
    {
        return this.getClass().getName() + "["
            + this.firstName + " "
            + this.lastName + ", "
            + this.salary + "]";
    }
}

class Address implements Cloneable                              →105
{
    private String street;
    private String city;
    private String state;
    private String zipCode;

    public Address()
    {
        this.street = "";
        this.city = "";
        this.state = "";
        this.zipCode = "";
    }

    public Address(String street, String city,
        String state, String zipCode)
    {
        this.street = street;
        this.city = city;
        this.state = state;
        this.zipCode = zipCode;
    }

    public Object clone()                                       →129
    {
        try
        {
            return super.clone();                               →133
        }
        catch (CloneNotSupportedException e)
        {
            return null; // will never happen
        }
    }

    public String getAddress()
```

```
        {
            return this.street + "\n"
                + this.city + ", "
                + this.state + " "
                + this.zipCode;
        }
    }
```

The `main` method in the `CloneTest2` class creates an `Employee` object and sets its name, salary, and address. Then it creates a clone of this object and prints the data contained in both objects. Next, it changes the last name and address of the second employee and prints the data again. Here's the output that's produced when this program is run:

```
**** after cloning ****

Anthony Martinez
1300 N. First Street
Fresno, CA 93702
Salary: 40000.0

Anthony Martinez
1300 N. First Street
Fresno, CA 93702
Salary: 40000.0

**** after changing emp2 ****

Anthony Martinez
1300 N. First Street
Fresno, CA 93702
Salary: 40000.0

Anthony Smith
2503 N. 6th Street
Fresno, CA 93722
Salary: 40000.0
```

As you can see, the two `Employee` objects have identical data after they are cloned, but they have different data after the fields for the second employee have been changed. Thus, you can safely change the data in one of the objects without affecting the other object.

The following paragraphs describe some of the highlights of this program:

» →5 Creates an employee named Anthony Martinez.

» →8 Sets the employee's address.

» →11 Clones the employee (okay, just the object, not the co-worker).

» →13 Prints the two Employee objects after cloning. They should have identical data.

» →17 Changes the second employee's name.

» →18 Changes the second employee's address.

» →22 Prints the two Employee objects after changing the data for the second employee. The objects should now have different data.

» →28 A utility method that prints the data for an Employee object.

» →38 The Employee class. Notice that this class implements Cloneable.

» →43 The address field, which holds an object of type Address.

» →81 The clone method in the Employee class.

» →86 Creates a shallow copy of the Employee object.

» →87 Creates a shallow copy of the Address object and assigns it to the address field of the cloned Employee object.

» →89 Catches CloneNotSupportedException, which won't ever happen because the class implements Cloneable. The compiler requires the try/catch statement here because CloneNotSupportedException is a checked exception.

» →93 Returns the cloned Employee object.

» →105 The Address class, which also implements Cloneable.

» →129 The clone method of the Address class.

» →133 Returns a shallow copy of the Address object.

The Class Class

Okay, class, it's time for one last class in this chapter: the Class class. The wording might get confusing, so put your thinking cap on.

Every class used by a Java application is represented in memory by an object of type Class. If your program uses Employee objects, for example, there's also a Class object for the Employee class. This Class object has information not about specific employees but about the Employee class itself.

You've already seen how you can get a Class object by using the getClass method. This method is defined by the Object class, so it's available to every object. Here's an example:

```
Employee emp = new Employee();
Class c = emp.getClass();
```

WARNING

Note that you have to initialize a variable with an object instance before you can call its getClass method. That's because the getClass method returns a Class object that corresponds to the type of object the variable refers to, not the type the variable is declared as.

Suppose that an HourlyEmployee class extends the Employee class. Then consider these statements:

```
HourlyEmployee emp = new Employee();
Class c = emp.getClass();
```

Here c refers to a Class object for the HourlyEmployee class, not the Employee class.

TECHNICAL STUFF

The Class class has more than 50 methods, but only 2 of them are worthy of your attention:

>> getName(): Returns a String representing the name of the class

>> getSuperclass(): Returns another Class object representing this Class object's superclass

If you're interested in the other capabilities of the Class class, you can always check it out in the Java API documentation.

TIP

One of the most common uses of the getClass method is to tell whether two objects are of the same type by comparing their Class objects. This works because Java guarantees that the Class object has only one instance for each different class used by the application. So even if your application instantiates 1,000 Employee objects, there is only one Class object for the Employee class.

As a result, the following code can determine whether two objects are both objects of the same type:

```
Object o1 = new Employee();
Object o2 = new Employee();

if (o1.getClass() == o2.getClass())
    System.out.println("They're the same.");
else
    System.out.println("They are not the same.");
```

In this case, the type of both objects is Employee, so the comparison is true.

To find out whether an object is of a particular type, use the object's getClass method to get the corresponding Class object. Then use the getName method to get the class name, and use a string comparison to check the class name. Here's an example:

```
if (emp.getClass().getName().equals("Employee"))
    System.out.println("This is an employee object.");
```

If all the strung-out method calls give you a headache, you can break the code apart:

```
Class c = emp.getClass();
String s = c.getName();

if (s.equals("Employee"))
    System.out.println("This is an employee object.");
```

The result is the same.

Chapter **7**

Using Inner Classes, Anonymous Classes, and Lambda Expressions

I n this chapter, you find out how to use three advanced types of classes: inner classes, static inner classes, and anonymous inner classes. All three types are useful in certain circumstances. In particular, inner classes and anonymous inner classes are commonly used with graphical applications created with JavaFX. For more information about JavaFX, refer to Book 6. In this chapter, I concentrate on the mechanics of creating these types of classes.

You'll also learn about a feature that was introduced with Java 8 called *lambda expressions*, which simplify the task of creating and using anonymous classes.

TECHNICAL STUFF

Once again, this chapter could have a Technical Stuff icon pasted next to every other paragraph. The immediate usefulness of some of the information I present in this chapter may seem questionable. But trust me — you need to know this stuff when you start writing Swing applications. If you want to skip this chapter for now, that's okay. You can always come back to it when you're learning Swing and need to know how inner classes and anonymous inner classes work.

Declaring Inner Classes

An *inner class* is a class that's declared inside another class. Thus the basic structure for creating an inner class is as follows:

```
class outerClassName
{
    private class innerClassName
    {
    // body of inner class
    }
}
```

The class that contains the inner class is called an *outer class.* You can use a visibility modifier with the inner class to specify whether the class should be `public`, `protected`, `private-package`, or `private`. This visibility determines whether other classes can see the inner class.

Understanding inner classes

At the surface, an inner class is simply a class that's contained inside another class, but there's more to it than that. Here are some key points about inner classes:

>> An inner class automatically has access to all the fields and methods of the outer class, even to private fields and methods. Thus an inner class has more access to its outer class than a subclass has to its superclass. (A subclass can access public and protected members of its superclass, but not private members.)

>> An inner class carries with it a reference to the current instance of the outer class that enables it to access instance data for the outer class.

>> Because of the outer-class instance reference, you can't create or refer to an inner class from a static method of the outer class. You can, however, create a *static inner class,* as I describe in the section "Using Static Inner Classes," later in this chapter.

TIP

>> One of the main reasons for creating an inner class is to create a class-that's of interest only to the outer class. As a result, you usually declare inner classes to be private so that other classes can't access them.

>> Occasionally, code in an inner class needs to refer to the instance of its outer class. To do that, you list the name of the outer class followed by the dot operator and `this`. If the outer class is named `MyOuterClass`, for example, you would use `MyOuterClass.this` to refer to the instance of the outer class.

Viewing an example

Book 3, Chapter 5 introduces an application that uses the Timer class in the Swing package (javax.swing.Timer) to display the lines Tick... and Tock... on the console at one-second intervals. It uses a class named Ticker that implements the ActionListener interface to handle the Timer object's clock events.

TIP

In this chapter, you see three versions of this application. You may want to quickly review Book 3, Chapter 5 if you're unclear on how this application uses the Timer class to display the Tick... and Tock... messages or why the JOptionPane dialog box is required.

Listing 7-1 shows a version of this application that implements the Ticker class as an inner class.

LISTING 7-1: **Tick Tock with an Inner Class**

```java
import java.awt.event.*;
import javax.swing.*;

public class TickTockInner
{
    private String tickMessage = "Tick...";            →6
    private String tockMessage = "Tock...";            →7

    public static void main(String[] args)
    {
        TickTockInner t = new TickTockInner();         →11
        t.go();                                        →12
    }

    private void go()                                  →15
    {
        // create a timer that calls the Ticker class
        // at one second intervals
        Timer t = new Timer(1000, new Ticker());       →19
        t.start();

        // display a message box to prevent the
        // program from ending immediately
        JOptionPane.showMessageDialog(null,            →24
            "Click OK to exit program");
```

(continued)

CHAPTER 7 **Using Inner Classes, Anonymous Classes, and Lambda Expressions**

LISTING 7-1: *(continued)*

```
        System.exit(0);                                         →26
    }

class Ticker implements ActionListener                          →29
{
    private boolean tick = true;

    public void actionPerformed(ActionEvent event)             →33
    {
        if (tick)
        {
            System.out.println(tickMessage);                   →37
        }
        else
        {
            System.out.println(tockMessage);                   →41
        }
        tick = !tick;
    }
}
}
```

THE OBSERVER PATTERN

Event listeners in Java are part of a Java model called the *Delegation Event Model,* which is an implementation of a more general design pattern called the Observer pattern. This pattern is useful when you need to create objects that interact with one another when a change in the status of one of the objects occurs. The object whose changes are being monitored is called the *observable object,* and the object that monitors those changes is called the *observer object.* The observer object registers itself with the observable object, which then notifies the observer object when its status changes.

You discover more about how Java implements this pattern for event handling in Book 6. But if you're interested, you may want to investigate the Observer and Observable interfaces that are part of the Java API. They provide a standard way to create simple implementations of the Observer pattern.

The following paragraphs describe some of the highlights of this program:

>> →6 The String variables named tickMessage and tockMessage (line 7) contain the messages to be printed on the console. Note that these variables are defined as fields of the outer class. As you'll see, the inner class Ticker is able to access these fields directly.

>> →11 Because an inner class can be used only by an instantiated object, you can't use it directly from the static main method. As a result, the main method in this program simply creates an instance of the application class (TickTockInner).

>> →12 This line executes the go method of the new instance of the TickTockInner class.

TIP

The technique used in lines 11 and 12 is a fairly common programming technique that lets an application get out of a static context quickly and into an object-oriented mode.

>> →15 This line is the go method, called from line 12.

>> →19 This line creates an instance of the Timer class with the timer interval set to 1,000 milliseconds (1 second) and the ActionListener set to a new instance of the inner class named Ticker.

>> →24 Here, the JOptionPane class is used to display a dialog box. This dialog box is necessary to give the timer a chance to run. The application ends when the user clicks OK.

>> →26 This line calls the exit method of the System class, which immediately shuts down the Java Virtual Machine. This method call isn't strictly required here, but if you leave it out, the timer continues to run for a few seconds after you click OK before the JVM figures out that it should kill the timer.

>> →29 This line is the declaration for the inner class named Ticker. Note that this class implements the ActionListener interface.

>> →33 The actionPerformed method is called by the Timer object every 1,000 milliseconds.

>> →37 In this line and in line 41, the inner class directly accesses a field of the outer class.

Using Static Inner Classes

A *static inner class* is similar to an inner class but doesn't require an instance of the outer class. Its basic form is the following:

```
class outerClassName
{
    private static class innerClassName
    {
        // body of inner class
    }
}
```

Like a static method, a static inner class can't access any nonstatic fields or methods in its outer class. It can access static fields or methods, however.

Listing 7-2 shows a version of the Tick Tock application that uses a static inner class rather than a regular inner class.

LISTING 7-2: **Tick Tock with a Static Inner Class**

```java
import java.awt.event.*;
import javax.swing.*;

public class TickTockStatic
{
    private static String tickMessage = "Tick...";        →6
    private static String tockMessage = "Tock...";        →7

    public static void main(String[] args)
    {
        TickTockStatic t = new TickTockStatic();
        t.go();
    }

    private void go()
    {
        // create a timer that calls the Ticker class
        // at one second intervals
        Timer t = new Timer(1000, new Ticker());
        t.start();

        // display a message box to prevent the
        // program from ending immediately
        JOptionPane.showMessageDialog(null,
            "Click OK to exit program");
```

```
        System.exit(0);
    }

    static class Ticker implements ActionListener                    →29
    {
        private boolean tick = true;
        public void actionPerformed(
            ActionEvent event)
        {
            if (tick)
            {
                System.out.println(tickMessage);
            }
            else
            {
                System.out.println(tockMessage);
            }
            tick = !tick;
        }
    }
}
```

This version of the application and the Listing 7-1 version have only three differences:

>> →6 The `tickMessage` field is declared as static. This is necessary so that the static class can access it.

>> →7 The `tockMessage` field is also declared as static.

>> →29 The `Ticker` class is declared as static.

Using Anonymous Inner Classes

Anonymous inner classes (usually just called *anonymous classes*) are probably the strangest feature of the Java programming language. The first time you see an anonymous class, you'll almost certainly think that someone made a mistake and that the code can't possibly compile. But compile it does, and it even works. When you get the hang of working with anonymous classes, you'll wonder how you got by without them.

An anonymous class is a class that's defined on the spot, right at the point where you want to instantiate it. Because you code the body of the class right where you need it, you don't have to give it a name. (That's why it's called an *anonymous* class.)

Creating an anonymous class

The basic form for declaring and instantiating an anonymous class is this:

```
new ClassOrInterface() { class-body }
```

As you can see, you specify the new keyword followed by the name of a class or interface that specifies the type of the object created from the anonymous class. This class or interface name is followed by parentheses, which may include a parameter list that's passed to the constructor of the anonymous class. Then you code a class body enclosed in braces. This class body can include anything that a regular class body can include: fields, methods, and even other classes or interfaces.

Here's an example of a simple anonymous class:

```java
public class AnonClass
{
    public static void main(String[] args)
    {
        Ball b = new Ball()
            {
                public void hit()
                {
                    System.out.println("You hit it!");
                }
            };
        b.hit();
    }

    interface Ball
    {
        void hit();
    }
}
```

In this example, I create an interface named Ball that has a single method named hit. Then, back in the main method, I declare a variable of type Ball and use an anonymous class to create an object. The body of the anonymous class consists of an implementation of the hit method that simply displays the message You hit it! on the console. After the anonymous class is instantiated and assigned to the b variable, the next statement calls the hit method.

When you run this program, the single line You hit it! is displayed on the console.

Here are some things to ponder when you work with anonymous classes:

>> You can't create a constructor for an anonymous class, because the anonymous class doesn't have a name. What would you call the constructor, anyway?

>> You can't pass parameters if the anonymous class is based on an interface. That makes sense; interfaces don't have constructors, so Java wouldn't have anything to pass the parameters to.

>> An assignment statement can use an anonymous class as shown in this example. In that case, the anonymous class body is followed by a semicolon that marks the end of the assignment statement. Note that this semicolon is part of the assignment statement, not the anonymous class. (In the next section, you see an example of an anonymous class that's passed as a method parameter. In that example, the body isn't followed by a semicolon.)

>> An anonymous class is a special type of inner class; like any inner class, it automatically has access to the fields and methods of its outer class.

>> An anonymous class can't be static.

Creating a program with an anonymous class

Listing 7-3 shows a more complex example of an anonymous class: a version of the Tick Tock application that uses an anonymous class as the action listener for the timer.

LISTING 7-3: **Tick Tock with an Anonymous Class**

```
import java.awt.event.*;
import javax.swing.*;

public class TickTockAnonymous
{
    private String tickMessage = "Tick...";
    private String tockMessage = "Tock...";

    public static void main(String[] args)                    →9
    {
        TickTockAnonymous t = new TickTockAnonymous();
        t.go();
    }
```

(continued)

CHAPTER 7 **Using Inner Classes, Anonymous Classes, and Lambda Expressions** 357

LISTING 7-3: *(continued)*

```
private void go()
{
    // create a timer that calls the Ticker class
    // at one second intervals
    Timer t = new Timer(1000,                                   →19
        new ActionListener()                                   →20
        {                                                      →21
            private boolean tick = true;
            public void actionPerformed(                       →23
                ActionEvent event)
            {
                if (tick)
                {
                    System.out.println(tickMessage);
                }
                else
                {
                    System.out.println(tockMessage);
                }
                tick = !tick;
            }
        } );                                                   →36

    t.start();

    // display a message box to prevent the
    // program from ending immediately
    JOptionPane.showMessageDialog(null,
        "Click OK to exit program");
    System.exit(0);
}
}
```

By now, you've seen enough versions of this program that you should understand how it works. The following paragraphs explain how this version uses an anonymous class as the ActionListener parameter supplied to the Timer constructor:

» →9 The main method creates an instance of the TickTockAnonymous class and executes the go method.

» →19 n the go method, an instance of the Timer class is created.

» →20 The second parameter of the TimerClass constructor is an object that implements the ActionListener interface. This object is created here via an anonymous class. ActionListener is specified as the type for this class.

» →21 This left brace marks the beginning of the body of the anonymous class.

>> →23 The `actionPerformed` method is called every 1,000 milliseconds by the timer. Note that this method can freely access fields defined in the outer class.

>> →36 The right brace on this line marks the end of the body of the anonymous class. Then the right parenthesis marks the end of the parameter list for the `Timer` constructor. The left parenthesis that's paired with this right parenthesis is on line 19. Finally, the semicolon marks the end of the assignment statement that started on line 19.

Using Lambda Expressions

A *lambda expression* is a new feature that in some ways is similar to anonymous classes, but with more concise syntax. More specifically, a lambda expression lets you create an anonymous class that implements a specific type of interface called a *functional interface* — which has one and only one abstract method.

The `Ball` interface that was presented in the previous section meets that definition:

```
interface Ball
{
    void hit();
}
```

Here the only abstract method is the `hit` method.

TECHNICAL STUFF

A functional interface can contain additional methods, provided they are not abstract.

A lambda expression is a concise way to create an anonymous class that implements a functional interface. Instead of providing a formal method declaration that includes the return type, method name, parameter types, and method body, you simply define the parameter types and the method body. The Java compiler infers the rest based on the context in which you use the lambda expression.

The parameter types are separated from the method body by a new operator, called the *arrow operator*, which consists of a hyphen followed by a greater-than symbol. Here's an example that implements the `Ball` interface:

```
() -> { System.out.println("You hit it!");}
```

Here the lambda expression implements a functional interface whose single method does not accept parameters. When the method is called, the text `"You hit it!"` is printed.

You can use a lambda expression anywhere you can use a normal Java expression. You'll use it most in assignment statements or as passed parameters. The only restriction is that you can use a lambda expression only in a context that requires an instance of a functional interface. For example, here's a complete program that uses a lambda expression to implement the `Ball` interface:

```
public class LambdaBall
{
    public static void main(String[] args)
    {
        Ball b = () -> { System.out.println("You hit it!"); };
        b.hit();
    }
    interface Ball
    {
        void hit();
    }
}
```

The general syntax for a lambda expression is this:

```
(parameters) -> expression
```

or this:

```
(parameters) -> { statement;...}
```

If you use an expression, a semicolon is not required. If you use one or more statements, the statements must be enclosed in curly braces and a semicolon is required at the end of each statement.

Note also that if the functional interface that the lambda is based on returns a value (not void), the value of *expression* in the first syntax is used as the return value. In the second syntax, in which one or more statements are enclosed in a block, you must use a `return` statement to return an appropriate value.

Don't forget that the statement in which you use the lambda expression must itself end with a semicolon. Thus, the lambda expression in the previous example has two semicolons in close proximity:

```
Ball b = () -> { System.out.println("You hit it!"); };
```

The first semicolon marks the end of the statement that calls `System.out.println`; the second semicolon marks the end of the assignment statement that assigns the lambda expression to the variable `b`.

Chapter **8**

Working with Packages and the Java Module System

This chapter shows you what to do with the classes you create. Specifically, I show you how to organize classes into neat packages. Packages enable you to keep your classes separate from classes in the Java API and allow you to reuse your classes in other applications, and even let you distribute your classes to others, assuming other people might be interested in your classes. If that's the case, you probably won't want to send those people all your separate class files. Instead, bundle them into a single file called a JAR file. That's covered in this chapter too.

I then show you how to use a feature called Javadoc that lets you add documentation comments to your classes. With Javadoc, you can build professional-looking documentation pages automatically. Your friends will think you're a real Java guru when you post your Javadoc pages to your website.

Finally, I show you how to use a feature introduced with Java 9 called the Java Module System, which provides an improved way of working with packages that avoids some of the more common pitfalls of the old Java package system.

Working with Packages

A *package* is a group of classes that belong together. Without packages, the entire universe of Java classes would be a huge, unorganized mess. Imagine the thousands of classes that are available in the Java API combined with millions of Java classes created by Java programmers throughout the world and all thrown into one big pot. Packages let you organize this mass into smaller, manageable collections of related classes.

Importing classes and packages

When you use `import` statements at the beginning of a Java source file, you make classes from the packages mentioned in the `import` statements available throughout the file. (I cover `import` statements in Book 2, Chapter 1, but it doesn't hurt to repeat it here.)

An `import` statement can import all the classes in a package by using an asterisk wildcard. Here all the classes in the `java.util` package are imported:

```
import java.util.*;
```

Alternatively, you can import classes one at a time. Here just the `ArrayList` class is imported:

```
import java.util.ArrayList;
```

Note: You don't have to use an `import` statement to use a class from a package. But if you don't use an `import` statement, you must *fully qualify* any references to the class. For example, you can use the `ArrayList` class without importing `java.util`:

```
java.util.ArrayList<String> list = new java.util.ArrayList<String>();
```

Because fully qualified names are a pain to always spell out, you should always use `import` statements to import the packages or individual classes your application uses.

You never have to explicitly import two packages:

>> `java.lang`: This package contains classes that are so commonly used that the Java compiler makes them available to every program. Examples of the classes in this package are `String`, `Exception`, and the various wrapper classes, such as `Integer` and `Boolean`. (For complete documentation on this package and all of the other Java packages described in this book, refer to https://docs.oracle.com/en/java/javase/14/docs/api/index.html.

>> **The default package:** This package contains classes that aren't specifically put in some other package. All the programs I show in this book up to this point are in the default package.

For simple program development and experimentation, using the default package is acceptable. However, if you start work on a serious Java application, create a separate package for it and place all of the application's classes there. You find out how to do that in the next section.

TECHNICAL
STUFF

You can't import the default package, even if you want to. Suppose you have two packages — the default package and the com.lowewriter.util package. The default package's code contains the statement import com.lowewriter.util.*. That's okay. But the default package doesn't have a name — at least it has no name that you can use inside a program. The com.lowewriter.util package's code can't contain a statement like this one:

```
import that_default_package.you_know.the_one_with_no_name
```

Creating your own packages

Creating your own packages to hold your classes is easy. Well, relatively easy, anyway. You must go through a few steps:

1. **Pick a name for your package.**

You can use any name you wish, but I recommend you follow the established convention of using your Internet domain name (if you have one), only backwards. I own a domain called LoweWriter.com, so I use the name com.lowewriter for all my packages. (Using your domain name backwards ensures that your package names are unique.)

Notice that package names are in all-lowercase letters. That's not an absolute requirement, but it's a Java convention that you ought to stick to. If you start using capital letters in your package names, you'll be branded a rebel for sure. And since Java is case-sensitive, a package named com.lowewriter is a different package from one named com.LoweWriter.

TIP

You can add additional levels beyond the domain name if you want. For example, I put my utility classes in a package named com.lowewriter.util.

If you don't have a domain all to yourself, try using your email address backwards. For example, if your email address is SomeBody@SomeCompany.com, use com.somecompany.somebody for your package names. That way they are still unique.

2. **Choose a directory on your hard drive to be the root of your class library.**

You need a place on your hard drive to store your classes. I suggest you create a directory such as c:\javaclasses.

This folder becomes the *root directory* for your Java packages.

3. **Create subdirectories within the root directory for your package name.**

For example, for the package named com.lowewriter.util, create a directory named com in the c:\javaclasses directory (assuming that's the name of your root). Then, in the com directory, create a directory named lowewriter. Then, in lowewriter, create a directory named util. Thus, the complete path to the directory that contains the classes for the com.lowewriter.util package is c:\javaclasses\com\lowewriter\util.

4. **Add the root directory for your package to the** ClassPath **environment variable.**

The exact procedure for doing this depends on your operating system. You can set the ClassPath by double-clicking System from the Control Panel. Click the Advanced tab, and then click Environment Variables.

Be careful not to disturb any directories already listed in the ClassPath. To add your root directory to the ClassPath, add a semicolon followed by the path to your root directory to the end of the ClassPath value. For example, suppose your ClassPath is already set to this:

```
.;c:\util\classes
```

Then you modify it to look like this:

```
.;c:\util\classes;c:\javaclasses
```

Here I added ;c:\javaclasses to the end of the ClassPath value.

5. **Save the files for any classes you want to be in a particular package in the directory for that package.**

For example, save the files for a class that belongs to the com.lowewriter.util package in c:\javaclasses\com\lowewriter\util.

6. **Add a package statement to the beginning of each source file that belongs in a package.**

The package statement simply provides the name for the package that any class in the file is placed in. For example:

```
package com.lowewriter.util;
```

The package statement must be the first non-comment statement in the file.

An example

Suppose you've developed a utility class named Console that has a bunch of handy static methods for getting user input from the console. For example, this class has a static method named askYorN that gets a Y or N from the user and returns a boolean value to indicate which value the user entered. You decide to make this class available in a package named com.lowewriter.util so you and other like-minded programmers can use it in their programs.

Here's the source file for the Console class:

```
package com.lowewriter.util;

import java.util.Scanner;

public class Console
{
    static Scanner sc = new Scanner(System.in);
    public static boolean askYorN(String prompt)
    {
        while (true)
        {
            String answer;
            System.out.print("\n" + prompt
                + " (Y or N) ");
            answer = sc.next();
            if (answer.equalsIgnoreCase("Y"))
                return true;
            else if (answer.equalsIgnoreCase("N"))
                return false;
        }
    }
}
```

Okay, so far this class has just the one method (askYorN), but one of these days you'll add a bunch of other useful methods to it. In the meantime, you want to get it set up in a package so you can start using it right away.

So you create a directory named c:\javaclasses\com\lowewriter\util (as described in the preceding section) and save the source file to this directory. Then you compile the program so the Console.class file is stored in that directory too. And you add c:\javaclasses to your ClassPath environment variable.

Now you can use the following program to test that your package is alive and well:

```
import com.lowewriter.util.*;

public class PackageTest
{
    public static void main(String[] args)
    {
        while (Console.askYorN("Keep going?"))
        {
            System.out.println("D'oh!");
        }
    }
}
```

Here the `import` statement imports all the classes in the `com.lowewriter.util` package. Then, the `while` loop in the `main` method repeatedly asks the user if he or she wants to keep going.

Putting Your Classes in a JAR File

A *JAR file* is a single file that can contain more than one class in a compressed format that the Java Runtime Environment can access quickly. (*JAR* stands for *Java archive.*) A JAR file can have just a few classes in it, or thousands.

JAR files are created by the `jar` utility, which you find in the JDK's `bin` directory along with the other Java command-line tools, such as `java` and `javac`. JAR files are similar in format to Zip files, a compressed format made popular by the PKZIP program. The main difference is that JAR files contain a special file, called the *manifest file,* that contains information about the files in the archive. This manifest is automatically created by the `jar` utility, but you can supply a manifest of your own to provide additional information about the archived files.

JAR files are a common way to distribute finished Java applications. After finishing your application, you run the `jar` command from a command prompt to prepare the JAR file. Then, another user can copy the JAR file to his or her computer. The user can then run the application directly from the JAR file.

jar command-line options

The `jar` command is an old-fashioned Unix-like command, complete with arcane command-line options that you have to get right if you expect to coax `jar` into doing something useful.

The basic format of the jar command is this:

```
jar options jar-file [manifest-file] class-files...
```

The options specify the basic action you want jar to perform and provide additional information about how you want the command to work. Table 8-1 lists the options.

TABLE 8-1   ## Options for the jar Command

Option	Description
c	Creates a new jar file.
u	Updates an existing jar file.
x	Extracts files from an existing jar file.
t	Lists the contents of a jar file.
f	Indicates that the jar file is specified as an argument. You almost always want to use this option.
v	Verbose output. This option tells the jar command to display extra information while it works.
0	Doesn't compress files when it adds them to the archive. This option isn't used much.
m	Specifies that a manifest file is provided. It's listed as the next argument following the jar file.
M	Specifies that a manifest file should not be added to the archive. This option is rarely used.

Note that you must specify at least the c, u, x, or t option to tell jar what action you want to perform.

Archiving a package

The most common use for the jar utility is to create an archive of an entire package. The procedure for doing that varies slightly depending on what operating system you're using. However, the jar command itself is the same regardless of your operating system. Here's the procedure for archiving a package on a PC running any version of Windows:

1. **Open a command window.**

The easiest way to do that is to choose Start➪Run, type **cmd** in the Open text box, and click OK. On Windows 10, click Start, type **cmd** and press Enter.

TIP

If you have trouble running the jar command in Step 3, you may need to open the command prompt in Administrator mode. To do so, click the Start menu, type **cmd**, right-click cmd.exe at the top of the Start menu, and choose Run as Administrator.

2. **Use a cd command to navigate to your package root.**

   For example, if your packages are stored in c:\javaclasses, use this command:

   ```
   cd \javaclasses
   ```

3. **Use a jar command that specifies the options cf, the name of the jar file, and the path to the class files you want to archive.**

   For example, to create an archive named utils.jar that contains all the class files in the com.lowewriter.util package, use this command:

   ```
   jar cf utils.jar com\lowewriter\util\*.class
   ```

4. **To verify that the jar file was created correctly, use the jar command that specifies the options tf and the name of the jar file.**

   For example, if the jar file is named utils.jar, use this command:

   ```
   jar tf utils.jar
   ```

   This lists the contents of the jar file so you can see what classes were added. Here's some typical output from this command:

   ```
   META-INF/
   META-INF/MANIFEST.MF
   com/lowewriter/util/Console.class
   ```

   As you can see, the utils.jar file contains the Console class, which is in my com.lowewriter.util package.

5. **That's all!**

   You're done. You can leave the jar file where it is, or you can give it to your friends so they can use the classes it contains.

Adding a jar to your classpath

To use the classes in an archive, you must add the jar file to your ClassPath environment variable. I describe the procedure for modifying the ClassPath variable in Windows 10 earlier in this chapter, in the section "Creating your own packages." So I won't repeat the details here.

To add an archive to the ClassPath variable, just add the complete path to the archive, making sure to separate it from any other paths already in the ClassPath with a semicolon. Here's an example:

```
.;c:\javaclasses\utils.jar;c:\javaclasses
```

Here I added the path c:\javaclasses\utils.jar to my ClassPath variable.

You can add all the jar files from a particular directory to the ClassPath in one fell swoop. For example, imagine that your c:\javaclasses directory contains two jar files — utils.jar and extras.jar. To add both jar files to the Class-Path, use a forward slash (/) followed by an asterisk:

```
.;c:\javaclasses/*
```

The forward slash looks strange, especially when combined with the back slash in c:\javaclasses. But that's the way you use a ClassPath wildcard.

REMEMBER

The first path in a ClassPath variable is always a single dot (.), which allows Java to find classes in the current directory.

TIP

Also, be aware that Java searches the various paths and archive files in the Class Path variable in the order in which you list them. Thus, with the ClassPath .;c:\javaclasses\utils.jar;c:\javaclasses, Java searches for classes first in the current directory, then in the utils archive, and finally in the c:\javaclasses directory.

Running a program directly from an archive

With just a little work, you can set up an archive so that a Java program can be run directly from it. All you have to do is create a *manifest file* before you create the archive. Then, when you run the jar utility to create the archive, you include the manifest file on the jar command line.

TIP

For this procedure to work, you must have the JDK deployed to the target computer. This technique is for testing programs in a development environment, not for deploying to end users.

A *manifest file* is a simple text file that contains information about the files in the archive. Although it can contain many lines of information, it needs just one line to make an executable jar file:

```
Main-Class: ClassName
```

The class name is the fully qualified name of the class that contains the main method that is executed to start the application. It isn't required, but it's typical to use the extension .mf for manifest files.

For example, suppose you have an application whose main class is GuessingGame, and all the class files for the application are in the package com.lowewriter.game. First, create a manifest file named game.mf in the com\lowewriter\game directory. This file contains the following line:

```
Main-Class: com.lowewriter.game.GuessingGame
```

Then run the jar command with the options cfm, the name of the archive to create, the name of the manifest file, and the path for the class files. Here's an example:

```
jar cfm game.jar com\lowewriter\game\game.mf com\lowewriter\game\*.class
```

Now you can run the application directly from a command prompt by using the java command with the -jar switch and the name of the archive file. Here's an example:

```
java -jar game.jar
```

This command starts the JRE and executes the main method of the class specified by the manifest file in the game.jar archive file.

TIP

If your operating system is configured properly, you can also run the application by double-clicking an icon for the jar file.

Using Javadoc to Document Your Classes

One last step remains before you can go public with your hot new class library or application: preparing the documentation for its classes. Fortunately, Java provides a tool called *Javadoc* that can automatically create fancy HTML-based documentation based on comments in your source files. All you have to do is add a comment for each public class, field, and method; then run the source files through the javadoc command. *Voilà!* You have professional-looking, web-based documentation for your classes.

The following sections show you how to add Javadoc comments to your source files, how to run the source files through the javadoc command, and how to view the resulting documentation pages.

Adding Javadoc comments

The basic rule for creating Javadoc comments is that they begin with /** and end with */. You can place Javadoc comments in any of three different locations in a source file:

>> Immediately before the declaration of a public class

>> Immediately before the declaration of a public field

>> Immediately before the declaration of a public method or constructor

A Javadoc comment can include text that describes the class, field, or method. Each subsequent line of a multiline Javadoc comment usually begins with an asterisk. Javadoc ignores this asterisk and any white space between it and the first word on the line.

The text in a Javadoc comment can include HTML markup if you want to apply fancy formatting. You should avoid using heading tags (<h1> and so on) because Javadoc creates those, and your heading tags just confuse things. But you can use tags for boldface and italics (<b> and <i>) or to format code examples (use the <pre> tag).

In addition, you can include special *doc tags* that provide specific information used by Javadoc to format the documentation pages. Table 8-2 summarizes the most commonly used tags.

TABLE 8-2   **Commonly Used Javadoc Tags**

Tag	Explanation
@author	Provides information about the author, typically the author's name, email address, website information, and so on.
@version	Indicates the version number.
@since	Used to indicate the version with which this class, field, or method was added.
@param	Provides the name and description of a method or constructor.
@return	Provides a description of a method's return value.
@throws	Indicates exceptions that are thrown by a method or constructor.
@deprecated	Indicates that the class, field, or method is deprecated and shouldn't be used.

To give you an idea of how Javadoc comments are typically used, Listing 8-1 shows a class named Employee with Javadoc comments included. (This Java file also includes a class named Address, which is required for the Employee class to work. For the sake of brevity, I do not provide Javadoc comments for the Address class.)

LISTING 8-1: **An Employee Class with Javadoc Comments**

```
package com.lowewriter.payroll;

/** Represents an employee.
 * @author Doug Lowe
 * @author www.LoweWriter.com
 * @version 1.5
 * @since 1.0
 */
public class Employee
{
    private String lastName;
    private String firstName;
    private Double salary;

/** Represents the employee's address.
 */
    public Address address;

/** Creates an employee with the specified name.
 * @param lastName The employee's last name.
 * @param firstName The employee's first name.
 */
    public Employee(String lastName, String firstName)
    {
        this.lastName = lastName;
        this.firstName = firstName;
        this.address = new Address();
    }

/** Gets the employee's last name.
 * @return A string representing the employee's last
 *                 name.
 */
    public String getLastName()
    {
        return this.lastName;
    }

/** Sets the employee's last name.
 * @param lastName A String containing the employee's
 *                 last name.
 */
    public void setLastName(String lastName)
    {
```

```java
        this.lastName = lastName;
    }
/** Gets the employee's first name.
 * @return A string representing the employee's first
 *                  name.
*/
    public String getFirstName()
    {
        return this.firstName;
    }
/** Sets the employee's first name.
 * @param firstName A String containing the
 *                  employee's first name.
*/
    public void setFirstName(String firstName)
    {
        this.firstName = firstName;
    }
/** Gets the employee's salary.
 * @return A double representing the employee's salary.
*/
    public double getSalary()
    {
        return this.salary;
    }


/** Sets the employee's salary.
 * @param salary A double containing the employee's
 *                  salary.
*/
    public void setSalary(double salary)
    {
        this.salary = salary;
    }
}
class Address implements Cloneable
{
    public String street;
    public String city;
    public String state;
    public String zipCode;
}
```

Using the javadoc command

The javadoc command has a few dozen options you can set, making it a complicated command to use. However, you can ignore all these options to create a basic

set of documentation pages. Just specify the complete path to all the Java files you want to create documentation for, like this:

```
javadoc com\lowewriter\payroll\*.java
```

The `javadoc` command creates the documentation pages in the current directory, so you may want to switch to the directory where you want the pages to reside first.

For more complete information about using this command, refer to the `javadoc` documentation at the Oracle website. You can find it here: `http://www.oracle.com/technetwork/articles/java/index-jsp-135444.html`.

Viewing Javadoc pages

After you run the `javadoc` command, you can access the documentation pages by starting with the `index.html` page. To quickly display this page, just type **index. html** at the command prompt after you run the `javadoc` command. Or you can start your browser, navigate to the directory where you created the documentation pages, and open the `index.html` page. Either way, Figure 8-1 shows an index page that lists two classes.

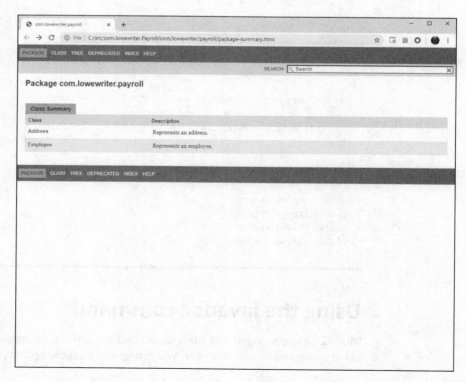

FIGURE 8-1:
A Javadoc index page.

If you think this page looks familiar, that's because the documentation for the Java API was created using Javadoc. So you should already know how to find your way around these pages.

To look at the documentation for a class, click the class name's link. A page with complete documentation for the class comes up. For example, Figure 8-2 shows part of the documentation page for the Employee class. Javadoc generated this page from the source file shown in Listing 8-1.

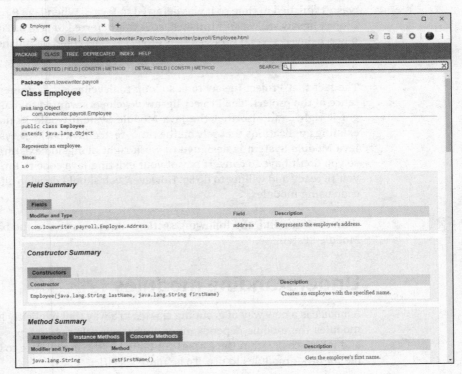

FIGURE 8-2: Documentation for the Employee class.

Using the Java Module System

Java packages as described so far in this chapter have served Java programmers well since the very first version of Java, introduced back in 1995. Packages were state-of-the-art when Java was first released, but they've been showing their age now for several years. For example, developers often have problems managing packages for large applications that use a large numbers of packages, especially when those packages require different versions of Java.

Another problem with packages is that they don't provide an easy way to create a lightweight application whose runtime contains only those portions of Java that are actually needed. This can limit Java's ability to run on devices with limited resources, such as embedded controllers or smartphones.

The Java Module System, new with Java 9, is designed to address these two issues and a few other more subtle problems with the old package system. The Java Module System, also known as Project Jigsaw, has been in development now for more than six years. It was originally planned to be released with Java 7 in 2011, but it wasn't finished in time so it was deferred to Java 8. When Java 8 rolled out in 2014, Project Jigsaw still wasn't finished, so it was deferred to Java 9. The release of Java 9 was delayed several times so that Project Jigsaw could be finished, and the good news is that the Java Module System is finally here!

The fact that Project Jigsaw took so long to develop is a testament to the importance of the project. The Project Jigsaw developers wanted to make sure that they got it right, and also that the Java Module System would in no way break any existing applications that rely on the time-tested Java package system. In fact, the Java Module System is designed to work right alongside existing Java packages, so you don't have to convert any of your existing Java code to Project Jigsaw until you're ready and willing to do so. However, you should develop all new production code using modules.

With that in mind, the following sections introduce you to the basics of the Java Module System.

Understanding modules

A *module* is a new way of grouping classes in a way that explicitly lists which other modules the module depends on and what specific public types (that is, classes and interfaces) within the classes contained in the module are to be made available for other modules to use. To be more specific:

>> A module must explicitly list its *dependencies* — that is, what other modules are required for the module to compile and run. For example, if one or more of the classes in the module require database access, the module must explicitly indicate that it requires the Java database module (known as java.sql).

>> A module must also explicitly list the *visibility* of any packages contained within the module. As you already know, you can create public types within a traditional package, and those public classes are available throughout the package and also externally to the package. With modules, public types in a package are visible outside of the module only if the package is explicitly listed as an exported type.

If this sounds complicated, don't worry — it isn't. Like traditional packages, modules are stored in JAR files. A JAR file that contains a module is called a *modular JAR file.* The only difference between a modular JAR file and a regular JAR file is that a modular JAR file contains a special class called `module-info.class`. The `module-info.class` class file identifies the module's dependencies (that is, what other modules are required) and the packages that are visible to other modules. All this is done by means of a source file called `module-info.java`, which is explained in the next section.

The module-info.java file

The `module-info.java` file is a Java source file that defines a module. The module is defined by using a combination of several new Java language elements that are introduced by Java 9.

Here is a simple `module-info.java` file that does creates a module but does not define any dependencies or exported packages:

```
module com.lowewriter.payroll {}
```

In this example, the name of the module is `com.lowewriter.payroll`. Note that modules are named just like packages, typically using the reverse-domain notation. In other words, for this example, I used my own personal domain (`lowewriter.com`) in reverse, followed by the name of the module.

To specify that the module is dependent on another module, you add a `requires` statement that specifies the name of the module. For example, if the module will require database access, it will need the standard `java.sql` module:

```
module com.lowewriter.payroll
{
    requires java.sql;
}
```

TIP As I explain in Book 6, the JavaFX modules used to create programs that have a graphical user interface (GUI) require that you name the necessary modules in the manifest file. For more information, refer to Book 6, Chapter 1.

If the module also depends on another module you've created named `com.lowewriter.util`, you would add a `requires` statement for that module as well:

```
module com.lowewriter.payroll
{
    requires java.sql;
    requires com.lowewriter.util;
}
```

You can also add `exports` statements to export packages that are contained in the module. When you export a package, any public classes or interfaces within the package are visible to other modules on the module path. For example:

```
module com.lowewriter.payroll
{
    requires java.SQL;
    requires com.lowewriter.util;
    exports com.lowewriter.payrolldb;
}
```

In this example, any public classes or interfaces defined by the package `com.lowewriter.payrolldb` are visible to other modules.

Setting up folders for a module

Getting the source folders set up for a module is similar to setting up the folders for a package, but with the added requirement that the `module-info.java` file must be in the root folder for the module. A common scheme is for the root folder to have the same name as the module, including the dots. For example, the root folder for the `com.lowewriter.payroll` module is `com.lowewriter.payroll`.

The `module-info.java` file lives within the root folder. In addition, the path to packages that are a part of the module typically follow the same convention as for normal packages. Thus, within the module root folder is a folder named `com`, within the `com` folder is a folder named `lowewriter`, and within the `lowewriter` folder is a folder named `payroll`. The `java` files for the payroll package are within the `payroll` folder.

The manifest file, if needed, can also live in the project's root folder. Or, you can create a separate folder within the root folder — commonly named `meta-inf` — and place the manifest file there. Because the manifest consists of just a single file, I think it's easy enough to place it in the root folder.

Assuming that there are two source files for the `payroll` package (named `Address.java` and `Employee.java`) and a manifest file named `payroll.mf`, the complete folder structure including the source files for the payroll example looks like this:

```
com.lowewriter.payroll
    module-info.java
    payroll.mf
    com
        lowewriter
            payroll
                Address.java
                Employee.java
```

Maybe now the `exports` command in the `module-info.java` file makes more sense:

```
exports com.lowewriter.payroll;
```

Notice that the path from the module root to the source files for the package correspond to the package name.

Compiling a module

To compile a module, you can use the `javac` command from a command prompt, specifying the name of all the source files you want to compile. First, navigate to the root folder for the module (in this case, `C:\java14\com.lowewriter.payroll`). Then use a `javac` command like this one:

```
javac module-info.java com\lowewriter\payroll\*.java
```

This command will create a `module-info.class` file in the module's root folder as well as `.class` files for all Java source files in `com\lowewriter\payroll`. For this example, two class files will be created in the `com\lowewriter\payroll` folder: `Address.class` and `Employee.class`.

If the `module.info.class` file calls out any modules that don't reside in the standard Java library folder, you'll need to provide both the location of the modules and the module names using command-line switches. You use the `--module-path` switch to provide the path to the required modules, and the `--add-modules` switch to name the modules. For example, if the payroll program requires the `javafx.controls` module (a requirement for JavaFX programs, as described in Book 6), use this line:

```
javac module-info.java com\lowewriter\payroll\*.java --module-path "c:\
    javafx14.0.1\lib" --add-modules javafx.controls
```

Again, refer to Book 6 for more information about compiling programs that use JavaFX.

Creating a modular JAR file

The final step for creating a Java module is to assemble the compiled class files (including `module-info.class`) into a modular JAR file. You can do that by using the `jar` command, like this:

```
jar cf com.lowewriter.payroll.jar *.class com\lowewriter\payroll\*.class
```

In this example, `cf` means to create a Jar file. The `cf` is followed by the class files to include. To keep the command simple, I used wildcards to include all the class files in the module root folder and all the class files in the `com\lowewriter\payroll` folder.

You can verify that the Jar file contains the correct contents by running the `jar` command with the `tf` option followed by the name of the Jar file. For example:

```
C:\java14\com.lowewriter.payroll>jar tf com.lowewriter.payroll.jar
META-INF/
META-INF/MANIFEST.MF
module-info.class
com/lowewriter/payroll/Address.class
com/lowewriter/payroll/Employee.class
```

As you can see, the `jar tf` command indicates that the Jar file contains three classes as expected: `module-info.class` and the two `com.lowewriter.payroll` classes.

There is a lot more to learn about the Java Module System that's beyond the scope of this book, but this should be enough to get you started with it. For more information, search the web for *Java Module System.* You'll find plenty of good information from Oracle and other sources about the advanced details of working with modules for larger applications.

Running a modular JAR file

After you've created a modular JAR file that specifies a main class (via a manifest file), you can run the file like a standard JAR file, using the java command:

```
java -jar com.lowewriter.payroll.jar
```

However, if the JAR file includes modules that aren't within the standard Java library, you'll need to use the `--module-path` and `--add-modules` switches just as you do on the `javac` command. So, for example, if the Payroll application required JavaFX, you'd run it with a command like this:

```
java -jar com.lowewriter.payroll.jar --module-path "c:\javafx14.0.1\lib" --add-
    modules javafx.controls
```

Once again, refer to Book 6 for more information about working with JavaFX.

4

Strings, Arrays, and Collections

Contents at a Glance

- » **Examining string class methods**

- » **Working with substrings**

- » **Splitting strings**

- » **Using the StringBuilder and StringBuffer classes**

- » **Using the CharSequence interface**

Chapter **1**

Working with Strings

S trings are among the most common types of objects in Java. Throughout this book are various techniques for working with strings. You've seen how to create string variables, how to concatenate strings, and how to compare strings. But so far, I've only scratched the surface of what you can do with strings. In this chapter, I dive deeper into what Java can do with strings.

I start with a brief review of what I've covered so far about strings, so that you don't have to go flipping back through the book to find basic information. Then I look at the `String` class itself and some of the methods it provides for working with strings. Finally, I examine two almost identical classes named `StringBuilder` and `StringBuffer` that offer features not available in the basic `String` class as well as an interface named `CharSequence` that is implemented by `String`, `StringBuilder`, and `StringBuffer`.

Reviewing Strings

To save you the hassle of flipping back through this book, the following paragraphs summarize what I present about strings in earlier chapters:

- >> Strings are reference types, not value types, such as `int` or `boolean`. As a result, a string variable holds a reference to an object created from the `String` class, not the value of the string itself.

- >> Even though strings aren't primitive types, the Java compiler has some features designed to let you work with strings almost as though they were primitive types. Java lets you assign string literals to string variables, for example, like this:

```
String line1 = "Oh what a beautiful morning!";
```

- >> Strings can include *escape sequences* that consist of a slash followed by another character. The most common escape sequences are \n for new line and \t for tab. If you want to include a back slash in a string, you must use the escape sequence \\. Here is a complete list of all the escape sequences you can use:

Escape Sequence	Explanation
\n	Newline
\t	Tab
\b	Backspace
\r	Carriage return
\f	Form feed
\"	Quotation mark
\\	Back slash

- >> Strings and characters are different. String literals are marked by quotation marks; character literals are marked by apostrophes. Thus, "a" is a string literal that happens to be one character long. By contrast, 'a' is a character literal.

- >> You can combine, or *concatenate,* strings by using the + operator, like this:

```
String line2 = line1 + "\nOh what a beautiful day!";
```

» You can also use the += operator with strings, like this:

```
line2 += "\nI've got a beautiful feeling";
```

» When a primitive type is used in a concatenation expression, Java automatically converts the primitive type to a string. Thus Java allows the following:

```
int empCount = 50;
String msg = "Number of employees: " + empCount;
```

» The various primitive wrapper classes (such as integer and double) have parse methods that can convert string values to numeric types. Here's an example:

```
String s = "50";
int i = Integer.parseInt(s);
```

» You shouldn't compare strings by using the equality operator (==). Instead, you should use the equals method. Here's an example:

```
if (lastName.equals("Lowe"))
    System.out.println("This is me!");
```

» The String class also has an equalsIgnoreCase method that compares strings without considering case. Here's an example:

```
if (lastName.equalsIgnoreCase("lowe"))
    System.out.println("This is me again!");
```

THE IMMUTABLE PATTERN

Many applications can benefit from classes that describe immutable objects. An *immutable object* is an object that, once created, can never be changed. The String class is the most common example of an immutable object. After you create a String object, you can't change it.

Suppose that you're designing a game in which the playing surface has fixed obstacles, such as trees. You can create the Tree class by using the Immutable pattern. The constructor for the Tree class could accept parameters that define the size, type, and location of the tree. But after you create the tree, you can't move it.

(continued)

Working with Strings

(continued)

Follow these three simple rules for creating an immutable object:

- Provide one or more constructors that accept parameters to set the initial state of the object.

- Don't allow any methods to modify any instance variables in the object. Set instance variables with constructors and then leave them alone.

- Any method that modifies the object should do so by creating a new object with the modified values. Then this method returns the new object as its return value.

Using the String Class

The `String` class is the class you use to create string objects. It has a whole gaggle of methods that are designed to let you find out information about the string that's represented by the `String` class. Table 1-1 lists the most useful of these methods.

TABLE 1-1 **String Class Methods**

Method	Description
`char charAt(int)`	Returns the character at the specified position in the string.
`int compareTo(String)`	Compares this string to another string, using alphanumeric order. Returns −1 if this string comes before the other string, 0 if the strings are the same, and 1 if this string comes after the other string. (Note that all uppercase numbers are considered less than any lowercase number, so, for example, *Merry* comes before *examine*.
`int compareTo IgnoreCase(String)`	Similar to compareTo but ignores case.
`boolean contains(CharSequence)`	Returns `true` if this string contains the parameter value. The parameter can be a `String`, `StringBuilder`, or `StringBuffer`.
`boolean endsWith(String)`	Returns `true` if this string ends with the parameter string.
`boolean equals(String)`	Returns `true` if this string has the same value as the parameter string.
`boolean equalsIgnore Case(String)`	Similar to `equals` but ignores case.
`String indent(int)`	Increases or decreases the indentation of each line within the original string. (Introduced with Java 12.)
`int indexOf(char)`	Returns the index of the first occurrence of the char parameter in this string. Returns −1 if the character isn't in the string.

Method	Description
`int indexOf(String)`	Returns the index of the first occurrence of the `String` parameter in this string. Returns –1 if the string isn't in this string.
`int indexOf(String, int start)`	Similar to `indexOf`, but starts the search at the specified position in the string.
`boolean isBlank()`	Returns `true` if the string is empty or contains only blanks. (Introduced in Java 11.)
`int lastIndexOf(char)`	Returns the index of the last occurrence of the `char` parameter in this string. Returns –1 if the character isn't in the string.
`int lastIndexOf(String)`	Returns the index of the last occurrence of the `String` parameter in this string. Returns –1 if the string isn't in this string.
`int lastIndexOf(String, int)`	Similar to `lastIndexOf`, but starts the search at the specified position in the string.
`int length()`	Returns the length of this string.
`String repeat(int)`	Returns a string that repeats the original string the indicated number of times. (Introduced in Java 11.)
`String replace(char, char)`	Returns a new string that's based on the original string, but with every occurrence of the first parameter replaced by the second parameter.
`String replaceAll(String old, String new)`	Returns a new string that's based on the original string, but with every occurrence of the first string replaced by the second parameter. Note that the first parameter can be a regular expression.
`String replaceFirst(String old, String new)`	Returns a new string that's based on the original string, but with the first occurrence of the first string replaced by the second parameter. Note that the first parameter can be a regular expression.
`String[] split(String)`	Splits the string into an array of strings, using the string parameter as a pattern to determine where to split the strings. The string parameter must be a valid regex expression.
`boolean startsWith(String)`	Returns `true` if this string starts with the parameter string.
`boolean startsWith (String, int)`	Returns `true` if this string contains the parameter string at the position indicated by the `int` parameter.
`String strip()`	Returns a copy of the string with all leading and trailing white spaces removed. (Introduced in Java 11.)
`String stripLeading()`	Returns a copy of the string with all leading white spaces removed. (Introduced in Java 11.)
`String stripTrailing()`	Returns a copy of the string with all trailing white spaces removed. (Introduced in Java 11.)

(continued)

Working with Strings

TABLE 1-1 *(continued)*

Method	Description
String substring(int)	Extracts a substring from this string, beginning at the position indicated by the int parameter and continuing to the end of the string.
String substring(int, int)	Extracts a substring from this string, beginning at the position indicated by the first parameter and ending at the position one character before the value of the second parameter.
char[] toCharArray()	Converts the string to an array of individual characters.
String toLowerCase()	Converts the string to lowercase.
String toString()	Returns the string as a String (pretty pointless, if you ask me, but all classes must have a toString method).
String toUpperCase()	Converts the string to uppercase.
String trim()	Returns a copy of the string with all leading and trailing white spaces removed.
String valueOf(primitiveType)	A static method that returns a string representation of any primitive type.

REMEMBER

The most important thing to remember about the String class is that in spite of the fact that it has a bazillion methods, none of those methods let you alter the string in any way. A String object is *immutable*, which means that it can't be changed.

Although you can't change a string after you create it, you can use methods of the String class to create new strings that are variations of the original string. The following sections describe some of the more interesting things you can do with these methods.

Finding the length of a string

One of the most basic string operations is determining the length of a string. You do that with the length method. For example:

```
String s = "A wonderful day for a neighbor.";
int len = s.length();
```

Here len is assigned a value of 31 because the string s consists of 31 characters.

Getting the length of a string usually isn't very useful by itself, but the length method often plays an important role in other string manipulations, as you see throughout the following sections.

Making simple string modifications

Several of the methods of the `String` class return modified versions of the original string. `toLowerCase`, for example, converts a string to all-lowercase letters:

```
String s1 = "Oompa Loompa";
String s2 = s1.toLowerCase();
```

Here `s2` is set to the string `oompa loompa`. The `toUpperCase` method works the same way but converts strings to all-uppercase letters.

The `trim` method removes white-space characters (spaces, tabs, newlines, and so on) from the start and end of a word. Here's an example:

```
String s = "   Oompa Loompa   ";
s = s.trim();
```

Here the spaces before and after `Oompa Loompa` are removed. Thus, the resulting string is 12 characters long.

TECHNICAL STUFF

Bear in mind that because strings are immutable, these methods don't actually change the `String` object. Instead, they create a new `String` with the modified value. A common mistake — especially for programmers who are new to Java but experienced in other languages — is to forget to assign the return value from one of these methods. The following statement has no effect on `s`:

```
s.trim();
```

Here the `trim` method trims the string — but then the program discards the result. The remedy is to assign the result of this expression back to `s`, like this:

```
s = s.trim();
```

Extracting characters from a string

You can use the `charAt` method to extract a character from a specific position in a string. When you do, keep in mind that the index number for the first character in a string is 0, not 1. Also, you should check the length of the string before extracting a character. If you specify an index value that's beyond the end of the string, the exception `StringIndexOutOfBoundsException` is thrown. (Fortunately, this is an unchecked exception, so you don't have to enclose the `charAt` method in a `try/catch` statement.)

Working with Strings

Here's an example of a program that uses the `charAt` method to count the number of vowels in a string entered by the user:

```java
import java.util.Scanner;

public class CountVowels
{
    static Scanner sc = new Scanner(System.in);

    public static void main(String[] args)
    {
        System.out.print("Enter a string: ");
        String s = sc.nextLine();

        int vowelCount = 0;

        for (int i = 0; i < s.length(); i++)
        {
            char c = s.charAt(i);
            if (    (c == 'A') || (c == 'a')
                 || (c == 'E') || (c == 'e')
                 || (c == 'I') || (c == 'i')
                 || (c == 'O') || (c == 'o')
                 || (c == 'U') || (c == 'u') )
                vowelCount++;
        }
        System.out.println("That string contains "
            + vowelCount + " vowels.");
    }
}
```

Here the `for` loop checks the length of the string to make sure that the index variable i doesn't exceed the string length. Then each character is extracted and checked with an `if` statement to see whether it is a vowel. The condition expression in this `if` statement is a little complicated because it must check for five different vowels, both uppercase and lowercase.

Following is an example that shows the output created by this program if you enter a string listing all of the letters of the alphabet:

```
Enter a string: abcdefghijklmnopqrstuvwxyz
That string contains 5 vowels.
```

Extracting substrings from a string

The `substring` method lets you extract a portion of a string. This method has two forms. The first version accepts a single integer parameter. It returns the substring that starts at the position indicated by this parameter and extends to the rest of the string. (Remember that string positions start with 0, not 1.) Here's an example:

```
String s = "Baseball";
String b = s.substring(4);        // "ball"
```

Here b is assigned the string ball.

The second version of the substring method accepts two parameters to indicate the start and end of the substring you want to extract. Note that the substring actually ends at the character that's immediately before the position indicated by the second parameter. So, to extract the characters at positions 2 through 5, specify 2 as the start position and 6 as the ending position. For example:

```
String s = "Baseball";
String b = s.substring(2, 6);     // "seba"
```

Here b is assigned the string seba.

The following program uses substrings to replace all the vowels in a string entered by the user with asterisks:

```
import java.util.Scanner;

public class MarkVowels
{
    static Scanner sc = new Scanner(System.in);

    public static void main(String[] args)
    {
        System.out.print("Enter a string: ");
        String s = sc.nextLine();
        String originalString = s;

        int vowelCount = 0;

        for (int i = 0; i < s.length(); i++)
        {
            char c = s.charAt(i);
            if (    (c == 'A') || (c == 'a')
                 || (c == 'E') || (c == 'e')
                 || (c == 'I') || (c == 'i')
                 || (c == 'O') || (c == 'o')
                 || (c == 'U') || (c == 'u') )
```

```
        {
            String front = s.substring(0, i);
            String back = s.substring(i+1);
            s = front + "*" + back;
        }
    }
}
System.out.println();
System.out.println(originalString);
System.out.println(s);
    }
}
```

This program uses a for loop and the charAt method to extract each character from the string. Then, if the character is a vowel, a string named front is created that consists of all the characters that appear before the vowel. Next, a second string named back is created with all the characters that appear after the vowel. Finally, the s string is replaced by a new string that's constructed from the front string, an asterisk, and the back string.

Here's some sample console output from this program so that you can see how it works:

```
Enter a string: Where have all the vowels gone?

Where have all the vowels gone?
Wh*r* h*v* *ll th* v*w*ls g*n*?
```

Splitting a string

The split command is especially useful for splitting a string into separate strings based on a delimiter character. Suppose you have a string with the parts of an address separated by colons, like this:

```
1500 N. Third Street:Fresno:CA:93722
```

With the split method, you can easily separate this string into four strings. In the process, the colons are discarded.

TECHNICAL STUFF

Unfortunately, the use of the split method requires that you use an array, and arrays are covered in the next chapter. I'm going to plow ahead with this section anyway on a hunch that you already know a few basic things about arrays. (If not, you can always come back to this section after you read the next chapter.)

The split method carves a string into an array of strings separated by the delimiter character passed via a string parameter. Here's a routine that splits an address into separate strings and then prints out all the strings:

```
String address =
    "1500 N. Third Street:Fresno:CA:93722";

String[] parts = address.split(":");

for (int i = 0; i < parts.length; i++)
    System.out.println(parts[i]);
```

If you run this code, the following lines are displayed on the console:

```
1500 N. Third Street
Fresno
CA
93722
```

The string passed to the split method is actually a special type of string used for pattern recognition, called a *regular expression*. You discover regular expressions in Book 5. For now, here are a few regular expressions that might be useful when you use the split method:

Regular Expression	Explanation
\\t	A tab character
\\n	A newline character
\\|	A vertical bar
\\s	Any white-space character
\\s+	One or more occurrences of any white-space character

The last regular expression in this table, \\s+, is especially useful for breaking a string into separate words. The following program accepts a string from the user, breaks it into separate words, and then displays the words on separate lines:

```
import java.util.Scanner;

public class ListWords
{
    static Scanner sc = new Scanner(System.in);

    public static void main(String[] args)
```

```
    {
        System.out.print("Enter a string: ");
        String s = sc.nextLine();
        String[] word = s.split("\\s+");
        for (String w : word)
            System.out.println(w);
    }
}
```

Here's a sample of the console output for a typical execution of this program:

```
Enter a string: This string    has    several    words
This
string
has
several
words
```

Notice that some of the words in the string entered by the user are preceded by more than one space character. The \\s+ pattern used by the split method treats any consecutive white-space character as a single delimiter when splitting the words.

Replacing parts of a string

You can use the replaceFirst or replaceAll method to replace a part of a string that matches a pattern you supply with some other text. Here's the main method of a program that gets a line of text from the user and then replaces all occurrences of the string cat with dog:

```
public static void main(String[] args)
{
    Scanner sc = new Scanner(System.in);
    System.out.print("Enter a string: ");
    String s = sc.nextLine();

    s = s.replaceAll("cat", "dog");

    System.out.println(s);
}
```

And here's the console for a typical execution of this program:

```
Enter a string: I love cats.
I love dogs.
```

As with the `split` methods, the first parameter of `replace` methods can be a regular expression that provides a complex matching string. (For more information, see Book 5, Chapter 3.)

REMEMBER

Once again, don't forget that strings are immutable. As a result, the `replace` methods don't actually modify the `String` object itself. Instead, they return a new `String` object with the modified value.

Using the StringBuilder and StringBuffer Classes

The `String` class is powerful, but it's not very efficient for programs that require heavy-duty string manipulation. Because `String` objects are immutable, any method of the `String` class that modifies the string in any way must create a new `String` object and copy the modified contents of the original string object to the new string. That's not so bad if it happens only occasionally, but it can be inefficient in programs that do it a lot.

Even string concatenation is inherently inefficient. Consider these statements:

```
int count = 5;
String msg = "There are ";
msg += count;
msg += " apples in the basket.";
```

These four statements may actually create five `String` objects:

>> `"There are "`: Created for the literal in the second statement. The `msg` variable is assigned a reference to this string.

>> `"5"`: Created to hold the result of `count.toString()`. The `toString` method is implicitly called by the third statement, so `count` is concatenated with `msg`.

>> `"There are 5"`: Created as a result of the concatenation in the third statement. A reference to this object is assigned to `msg`.

>> `"apples in the basket."`: Created to hold the literal in the fourth statement.

>> `"There are 5 apples in the basket."`: Created to hold the result of the concatenation in the fourth statement. A reference to this object is assigned to `msg`.

For programs that do only occasional string concatenation and simple string manipulations, these inefficiencies aren't a big deal. In fact, the compiler may optimize them to eliminate the inefficiency. But for programs that do extensive string manipulation, Java offers two alternatives to the String class: the String-Builder and StringBuffer classes.

Creating a StringBuilder object

You can't assign string literals directly to a StringBuilder object, as you can with a String object. The StringBuilder class, however, has a constructor that accepts a String as a parameter. So to create a StringBuilder object, you use a statement such as this:

```
StringBuilder sb = new StringBuilder("Today is the day!");
```

Internally, a StringBuilder object maintains a fixed area of memory where it stores a string value. This area of memory is called the *buffer*. The string held in this buffer doesn't have to use the entire buffer. As a result, a StringBuilder object has both a length and a capacity. The *length* represents the current length of the string maintained by the StringBuilder, and the *capacity* represents the size of the buffer itself. Note that the length can't exceed the capacity.

When you create a StringBuilder object, initially the capacity is set to the length of the string plus 16. The StringBuilder class automatically increases its capacity whenever necessary, so you don't have to worry about exceeding the capacity.

Using StringBuilder methods

Table 1-2 lists the most useful methods of the StringBuilder class. Note that the StringBuffer class uses the same methods. If you have to use StringBuffer instead of StringBuilder, just change the class name and use the same methods.

TABLE 1-2   ## StringBuilder Methods

Method	Description
append(*primitiveType*)	Appends the string representation of the primitive type to the end of the string.
append(Object)	Calls the object's toString method and appends the result to the end of the string.
append(CharSequence)	Appends the string to the end of the StringBuilder's string value. The parameter can be a String, StringBuilder, or StringBuffer.
char charAt(int)	Returns the character at the specified position in the string.
delete(int, int)	Deletes characters starting with the first int and ending with the character before the second int.

Method	Description
`deleteCharAt(int)`	Deletes the character at the specified position.
`ensureCapacity(int)`	Ensures that the capacity of `StringBuilder` is at least equal to the `int` value; increases the capacity if necessary.
`int capacity()`	Returns the capacity of this `StringBuilder`.
`int indexOf(String)`	Returns the index of the first occurrence of the specified string. If the string doesn't appear, returns –1.
`int indexOf(String, int)`	Returns the index of the first occurrence of the specified string, starting the search at the specified index position. If the string doesn't appear, returns –1.
`insert(int, primitiveType)`	Inserts the string representation of the primitive type at the point specified by the `int` argument.
`insert(int, Object)`	Calls the `toString` method of the `Object` parameter and then inserts the resulting string at the point specified by the `int` argument.
`insert(int, CharSequence)`	Inserts the string at the point specified by the `int` argument. The second parameter can be a `String`, `StringBuilder`, or `StringBuffer`.
`int lastIndexOf(String)`	Returns the index of the last occurrence of the specified string. If the string doesn't appear, returns –1.
`int lastIndexOf(String, int)`	Returns the index of the last occurrence of the specified string, starting the search at the specified index position. If the string doesn't appear, returns –1.
`int length()`	Returns the length of this string.
`replace(int, int, String)`	Replaces the substring indicated by the first two parameters with the string provided by the third parameter.
`reverse()`	Reverses the order of characters.
`setCharAt(int, char)`	Sets the character at the specified position to the specified character.
`setLength(int)`	Sets the length of the string. If that length is less than the current length, the string is truncated; if it's greater than the current length, new characters — hexadecimal zeros — are added.
`String substring(int)`	Extracts a substring, beginning at the position indicated by the `int` parameter and continuing to the end of the string.
`String substring(int, int)`	Extracts a substring, beginning at the position indicated by the first parameter and ending at the position one character before the value of the second parameter.
`String toString()`	Returns the current value as a `String`.
`String trimToSize()`	Reduces the capacity of the `StringBuffer` to match the size of the string.

Viewing a StringBuilder example

To illustrate how the StringBuilder class works, here's a StringBuilder version of the MarkVowels program from earlier in this chapter:

```
import java.util.Scanner;

public class StringBuilderApp
{
    static Scanner sc = new Scanner(System.in);

    public static void main(String[] args)
    {
        System.out.print("Enter a string: ");

        String s = sc.nextLine();
        StringBuilder sb = new StringBuilder(s);

        int vowelCount = 0;

        for (int i = 0; i < s.length(); i++)
        {
            char c = s.charAt(i);
            if (    (c == 'A') || (c == 'a')
                 || (c == 'E') || (c == 'e')
                 || (c == 'I') || (c == 'i')
                 || (c == 'O') || (c == 'o')
                 || (c == 'U') || (c == 'u') )
            {
                sb.setCharAt(i, '*');
}
        }
        System.out.println();
        System.out.println(s);
        System.out.println(sb.toString());
    }
}
```

This program uses the setCharAt method to directly replace any vowels it finds with asterisks. That's much more efficient than concatenating substrings (which is the way the String version of this program worked).

Using the CharSequence Interface

The Java API includes a useful interface called CharSequence. All three of the classes discussed in this chapter — String, StringBuilder, and StringBuffer — implement this interface. This method exists primarily to let you use String, StringBuilder, and StringBuffer interchangeably.

Toward that end, several of the methods of the String, StringBuilder, and StringBuffer classes use CharSequence as a parameter type. For those methods, you can pass a String, StringBuilder, or StringBuffer object. Note that a string literal is treated as a String object, so you can use a string literal anywhere a CharSequence is called for.

TECHNICAL STUFF

In case you're interested, the CharSequence interface defines four methods:

» char charAt(int): Returns the character at the specified position.

» int length(): Returns the length of the sequence.

» subSequence(int start, int end): Returns the substring indicated by the start and end parameters.

» toString(): Returns a String representation of the sequence.

If you're inclined to use CharSequence as a parameter type for a method so that the method works with a String, StringBuilder, or StringBuffer object, be advised that you can use only these four methods.

Chapter **2**

Using Arrays

I could use a raise. . . .

Oh, *arrays*. Sorry.

Arrays are an important aspect of any programming language, and Java is no exception. In this chapter, you discover just about everything you need to know about using arrays. I cover run-of-the-mill one-dimensional arrays; multidimensional arrays; and two classes that are used to work with arrays, named `Array` and `Arrays`.

Understanding Arrays

An *array* is a set of variables that is referenced by using a single variable name combined with an index number. Each item of an array is called an *element*. All the elements in an array must be of the same type. Thus the array itself has a type that specifies what kind of elements it can contain. An `int` array can contain `int` values, for example, and a `String` array can contain strings.

The index number is written after the variable name and enclosed in brackets. So if the variable name is x, you could access a specific element with an expression like x[5].

TIP

You might think that x[5] would refer to the fifth element in the array. But index numbers start with zero for the first element, so x[5] actually refers to the sixth element. This little detail is one of the chief causes of problems when working with arrays — especially if you cut your array-programming teeth in a language in which arrays are indexed from one instead of from zero. So, in Java, get used to counting from zero instead of from one.

The real power of arrays comes from the simple fact that you can use a variable or even a complete expression as an array index. So (for example) instead of coding x[5] to refer to a specific array element, you can code x[i] to refer to the element indicated by the index variable i. You see plenty of examples of index variables throughout this chapter.

Here are a few additional tidbits of array information to ponder before you get into the details of creating and using arrays:

>> Even though an array has no corresponding class file, an array is still an object. You can refer to the array object as a whole, rather than to a specific element of the array, by using the array's variable name without an index. Thus, if x[5] refers to an element of an array, x refers to the array itself.

>> An array has a fixed length that's set when the array is created. This length determines the number of elements that can be stored in the array. The maximum index value you can use with any array is one less than the array's length. Thus, if you create an array of ten elements, you can use index values from 0 to 9.

>> You can't change the length of an array after you create the array.

>> You can access the length of an array by using the length field of the array variable. x.length, for example, returns the length of the array x.

Creating Arrays

Before you can create an array, you must declare a variable that refers to the array. This *variable declaration* should indicate the type of elements that are stored by the array followed by a set of empty brackets, like this:

```
String[] names;
```

Here a variable named `names` is declared. Its type is an array of `String` objects.

TECHNICAL
STUFF

Just to make sure that you're confused as much as possible, Java also lets you put the brackets on the variable name rather than on the type. The following two statements both create arrays of `int` elements:

```
int[] array1;   // an array of int elements
int array2[];   // another array of int elements
```

TIP

Both of these statements have exactly the same effect. Most Java programmers prefer to put the brackets on the type rather than on the variable name.

By itself, that statement doesn't create an array; it merely declares a variable that can refer to an array. You can actually create the array in two ways:

» Use the new keyword followed by the array type, this time with the brackets filled in to indicate how many elements the array can hold. For example:

```
String[] names;

names = new String[10];
```

Here, an array of `String` objects that can hold ten strings is created. Each of the strings in this array is initialized to an empty string.

» As with any other variable, you can combine the declaration and the creation into one statement:

```
String[] names = new String[10];
```

Here the array variable is declared and an array is created in one statement.

» Use a special shortcut that lets you create an array and populate it with values in one swoop:

```
String[] names = {"One", "Two", "Three"};
```

TIP

If you don't know how many elements the array needs at compile time, you can use a variable or an expression for the array length. Here's a routine from a method that stores player names in an array of strings. It starts by asking the user how many players are on the team. Then it creates an array of the correct size:

```
System.out.print("How many players? ");
int count = sc.nextInt();          // sc is a Scanner
String[] players = new String[count];
```

Initializing an Array

One way to initialize the values in an array is to simply assign them one by one:

```
String[] days = new Array[7];
Days[0] = "Sunday";
Days[1] = "Monday";
Days[2] = "Tuesday";
Days[3] = "Wednesday";
Days[4] = "Thursday";
Days[5] = "Friday";
Days[6] = "Saturday";
```

Java has a shorthand way to create an array and initialize it with values:

```
String[] days = { "Sunday", "Monday", "Tuesday",
                  "Wednesday", "Thursday",
                  "Friday", "Saturday" };
```

Here each element to be assigned to the array is listed in an *array initializer.* Here's an example of an array initializer for an int array:

```
int[] primes = { 2, 3, 5, 7, 11, 13, 17 };
```

Note: The length of an array created with an initializer is determined by the number of values listed in the initializer.

An alternative way to code an initializer is this:

```
int[] primes = new int[] { 2, 3, 5, 7, 11, 13, 17 };
```

To use this type of initializer, you use the new keyword followed by the array type and a set of empty brackets. Then you code the initializer.

Using for Loops with Arrays

One of the most common ways to process an array is with a for loop. In fact, for loops were invented specifically to deal with arrays. Here's a for loop that creates an array of 100 random numbers, with values ranging from 1 to 100:

```
int[] numbers = new int[100];
for (int i = 0; i < 100; i++)
    numbers[i] = (int)(Math.random() * 100) + 1;
```

And here's a loop that fills an array of player names with strings entered by the user:

```
String[] players = new String[count];
for (int i = 0; i < count; i++)
{
    System.out.print("Enter player name: ");
    players[i] = sc.nextLine();       // sc is a Scanner
}
```

For this example, assume that count is an int variable that holds the number of players to enter.

You can also use a for loop to print the contents of an array. For example:

```
for (int i = 0; i < count; i++)
    System.out.println(players[i]);
```

Here the elements of a String array named players are printed to the console.

The previous example assumes that the length of the array was stored in a variable before the loop was executed. If you don't have the array length handy, you can get it from the array's length property:

```
for (int i = 0; i < players.length; i++)
    System.out.println(players[i]);
```

Solving Homework Problems with Arrays

Every once in a while, an array and a for loop or two can help you solve your kids' homework problems for them. I once helped my daughter solve a tough homework assignment for a seventh-grade math class. The problem was stated something like this:

> Bobo (these problems always had a character named Bobo in them) visits the local high school on a Saturday and finds that all the school's 1,000 lockers are neatly closed. So he starts at one end of the school and opens them all. Then he goes back to the start and closes every other locker (lockers 2, 4, 6, and so on). Then he goes back to the start and hits every third locker: If it's open, he closes it; if it's closed, he opens it. Then he hits every fourth locker, every fifth locker, and so on. He keeps doing this all weekend long, walking the hallways opening and closing lockers 1,000 times. Then he gets bored and goes home. How many of the school's 1,000 lockers are left open, and which ones are they?

Sheesh!

This problem presented a challenge, and being the computer-nerd father I am, I figured that this was the time to teach my daughter about for loops and arrays. So I wrote a little program that set up an array of 1,000 booleans. Each represented a locker: true meant open, and false meant closed. Then I wrote a pair of nested for loops to do the calculation.

My first attempt told me that 10,000 of the 1,000 lockers were opened, so I figured that I'd made a mistake somewhere. And while I was looking at the code, I realized that the lockers were numbered 1 to 1,000, but the elements in my array were numbered 0 to 999, and that was part of what led to the confusion that caused my first answer to be ridiculous.

So I decided to create the array with 1,001 elements and ignore the first one. That way, the indexes corresponded nicely to the locker numbers.

After a few hours of work, I came up with the program in Listing 2-1.

LISTING 2-1:	The Classic Locker Problem Solved

```
public class BoboAndTheLockers
{
    public static void main(String[] args)
    {
        // true = open; false = closed
        boolean[] lockers = new boolean[1001];          →6

        // close all the lockers
        for (int i = 1; i <= 1000; i++)                 →9
            lockers[i] = false;
        for (int skip = 1; skip <= 1000; skip++)        →11
        {
            System.out.println("Bobo is changing every "
                + skip + " lockers.");
            for (int locker = skip; locker < 1000;      →15
                    locker += skip)
                lockers[locker] = !lockers[locker];     →17
        }
        System.out.println("Bobo is bored"
            + " now so he's going home.");
        // count and list the open lockers
        String list = "";
        int openCount = 0;
        for (int i = 1; i <= 1000; i++)                 →24
            if (lockers[i])
```

```
        {
            openCount++;
            list += i + " ";
        }
    System.out.println("Bobo left " + openCount
        + " lockers open.");
    System.out.println("The open lockers are: "
        + list);
    }
}
```

Here are the highlights of how this program works:

» →6 This line sets up an array of booleans with 1,001 elements. I created one more element than I needed so I could ignore element 0.

» →9 This for loop closes all the lockers. This step isn't really necessary because booleans initialize to false, but being explicit about initialization is good.

» →11 Every iteration of this loop represents one complete trip through the hallways opening and closing lockers. The skip variable represents how many lockers Bobo skips on each trip. First he does every locker, then every second locker, and then every third locker. So this loop simply counts from 1 to 1,000.

» →15 Every iteration of this loop represents one stop at a locker on a trip through the hallways. This third expression in the for statement (on the next line) adds the skip variable to the index variable so that Bobo can access every *n*th locker on each trip through the hallways.

» →17 This statement uses the not operator (!) to reverse the setting of each locker. Thus, if the locker is open (true), it's set to closed (false), and vice versa.

» →24 Yet another for loop spins through all the lockers and counts the ones that are open. It also adds the locker number for each open locker to the end of a string so that all the open lockers can be printed.

This program produces more than 1,000 lines of output, but only the last few lines are important. Here they are:

```
Bobo is bored now so he's going home.
Bobo left 31 lockers open.
The open lockers are: 1 4 9 16 25 36 49 64 81 100 121 144 169 196 225 256 289
    324 361 400 441 484 529 576 625 676 729 784 841 900 961
```

So there's the answer: 31 lockers are left open. I got an A. (I mean, my daughter got an A.)

TIP

By the way, did you notice that the lockers that were left open were the ones whose numbers are perfect squares? Or that 31 is the largest number whose square is less than 1,000? I didn't, either — until my daughter told me after school the next day.

Using the Enhanced for Loop

You can often eliminate the tedium of working with indexes in `for` loops by using a special type of `for` loop called an *enhanced* `for` *loop*. Enhanced `for` loops are often called `for-each` loops because they automatically retrieve each element in a iterable object. Using a `for-each` loop eliminates the need to create and initialize an index, increment the index, test for the last item in the array, and access each array item using the index.

When it's used with an array, the enhanced `for` loop has this format:

```
for (type identifier : array)
{
Statements...
}
```

The *type* identifies the type of the elements in the array, and the *identifier* provides a name for a local variable that is used to access each element. The colon operator (often read as "in") is then followed by the name of the array you want to process.

Here's an example:

```
String[] days = { "Sunday", "Monday", "Tuesday",
                  "Wednesday", "Thursday",
                  "Friday", "Saturday" };
for (String day : days)
{
    System.out.println(day);
}
```

This loop prints the following lines to the console:

```
Sunday
Monday
Tuesday
Wednesday
Thursday
Friday
Saturday
```

In other words, it prints each of the strings in the array on a separate line.

It's important to note that the for-each loop gives you a copy of each item in the array, not the item itself. So, you can't alter the contents of an array by using a for-each loop. Consider this code snippet:

```
int[] nums = {1, 2, 3, 4, 5};

for (int n : nums)
    n = n * 2;

for (int n : nums)
    System.out.println(n);
```

You may expect this code to print the values 2, 4, 6, 8, and 10 because you multiplied n by 2 in the for-each loop. But instead, the code prints 1, 2, 3, 4, and 5. That's because n holds a copy of each array item, not the array item itself.

Using Arrays with Methods

You can write methods that accept arrays as parameters and return arrays as return values. You just use an empty set of brackets to indicate that the parameter type or return type is an array. You've already seen this in the familiar main method declaration:

```
public static void main(String[] args)
```

Here's a static method that creates and returns a String array with the names of the days of the week:

```
public static String[] getDaysOfWeek()
{
    String[] days = { "Sunday", "Monday", "Tuesday",
                      "Wednesday", "Thursday",
                      "Friday", "Saturday" };
    return days;
}
```

And here's a static method that prints the contents of any `String` array to the console, one string per line:

```
public static void printStringArray(String[] strings)
{
    for (String s : strings)
        System.out.println(s);
}
```

Finally, here are two lines of code that call these methods:

```
String[] days = getDaysOfWeek();
printStringArray(days);
```

The first statement declares a `String` array and then calls `getDaysOfWeek` to create the array. The second statement passes the array to the `printString Array` method as a parameter.

Using Varargs

Varargs provides a convenient way to create a method that accepts a variable number of arguments. When you use varargs, the last argument in the method signature uses ellipses to indicate that the caller can provide one or more arguments of the given type.

Here's an example:

```
public static void PrintSomeWords(String... words)
{
    for (String word : words)
        System.out.println(word);
}
```

Here, the `PrintSomeWords` method specifies that the caller can pass any number of String arguments to the method (including none at all) and that the arguments will be gathered up in an array named `words`.

Here's a snippet of code that calls the `PrintSomeWords` methods using various numbers of arguments to prove the point:

```
PrintSomeWords();
PrintSomeWords("I");
PrintSomeWords("Am", "Not");
PrintSomeWords("Throwing", "Away", "My", "Shot");
```

The resulting console output looks like this:

```
I
Am
Not
Throwing
Away
My
Shot
```

An important caveat about using varargs is that the variable argument must always be the last argument in the argument list. This makes sense when you consider that, otherwise, the compiler wouldn't be able to keep track of the arguments if any other than the last argument had a variable number. So, the following won't compile:

```
public static void PrintSomeWords(String head, String... words, string tail)
```

You'll get a warning that says, "Varargs parameter must be the last parameter."

Using Two-Dimensional Arrays

The elements of an array can be any type of object you want, including another array. In the latter case, you have a *two-dimensional array*, sometimes called an *array of arrays.*

Two-dimensional arrays are often used to track data in column-and-row format, much the way that a spreadsheet works. Suppose that you're working on a program that tracks five years' worth of sales (2016 through 2020) for a company, with the data broken down for each of four sales territories (North, South, East, and West). You could create 20 separate variables, with names such as `sales2020North`, `sales2020South`, `sales2020East`, and so on. But that gets a little tedious.

Alternatively, you could create an array with 20 elements, like this:

```
double[] sales = new sales[20];
```

But then how would you organize the data in this array so that you know the year and sales region for each element?

With a two-dimensional array, you can create an array with an element for each year. Each of those elements in turn is another array with an element for each region.

Thinking of a two-dimensional array as a table or spreadsheet is common, like this:

Year	North	South	East	West
2016	23,853	22,838	36,483	31,352
2017	25,483	22,943	38,274	33,294
2018	24,872	23,049	39,002	36,888
2019	28,492	23,784	42,374	39,573
2020	31,932	23,732	42,943	41,734

Creating a two-dimensional array

To declare a two-dimensional array for this sales data, you simply list two sets of empty brackets, like this:

```
double sales[][];
```

Here `sales` is a two-dimensional array of type `double`. To put it another way, `sales` is an array of `double` arrays.

To create the array, you use the `new` keyword and provide lengths for each set of brackets, as in this example:

```
sales = new double[5][4];
```

Here the first dimension specifies that the `sales` array has five elements. This array represents the rows in the table. The second dimension specifies that each of those elements has an array of type `double` with four elements. This array represents the columns in the table.

TIP

A key point to grasp here is that one instance is of the first array, but a separate instance of the second array for each element is in the first array. So this statement actually creates five `double` arrays with four elements each. Then those five arrays are used as the elements for the first array.

Note that as with a one-dimensional array, you can declare and create a two-dimensional array in one statement, like this:

```
double[][] sales = new double[5][4];
```

Here the sales array is declared and created all in one statement.

Accessing two-dimensional array elements

To access the elements of a two-dimensional array, you use two indexes. This statement sets the 2016 sales for the North region:

```
sales[0][0] = 23853.0;
```

As you might imagine, accessing the data in a two-dimensional array by hard-coding each index value can get tedious. No wonder for loops are normally used instead. The following bit of code uses a for loop to print the contents of the sales array to the console, separated by tabs. Each year is printed on a separate line, with the year at the beginning of the line. In addition, a line of headings for the sales regions is printed before the sales data. Here's the code:

```
NumberFormat cf = NumberFormat.getCurrencyInstance();
System.out.println("\tNorth\t\tSouth\t\tEast\t\tWest");
int year = 2016;
for (int y = 0; y < 5; y++)
{
    System.out.print(year + "\t");
    for (int region = 0; region < 4; region++)
    {
        System.out.print(cf.format(sales[y][region]));
        System.out.print("\t");
    }
    year++;
    System.out.println();
}
```

Assuming that the sales array has already been initialized, this code produces the following output on the console:

	North	South	East	West
2016	$23,853.00	$22,838.00	$36,483.00	$31,352.00
2017	$25,483.00	$22,943.00	$38,274.00	$33,294.00
2018	$24,872.00	$23,049.00	$39,002.00	$36,888.00
2019	$28,492.00	$23,784.00	$42,374.00	$39,573.00
2020	$31,932.00	$23,732.00	$42,943.00	$41,734.00

WARNING

The order in which you nest the `for` loops that access each index in a two-dimensional array is crucial! The preceding example lists the sales for each year on a separate line, with the sales regions arranged in columns. You can print a listing with the sales for each region on a separate line, with the years arranged in columns, by reversing the order in which the `for` loops that index the arrays are nested:

```
for (int region = 0; region < 4; region++)
{
    for (int y = 0; y < 5; y++)
    {
        System.out.print(cf.format(sales[y][region]));
        System.out.print("    ");
    }
    System.out.println();
}
```

Here the outer loop indexes the region and the inner loop indexes the year:

```
$23,853.00    $25,483.00    $24,872.00    $28,492.00    $31,932.00
$22,838.00    $22,943.00    $23,049.00    $23,784.00    $23,732.00
$36,483.00    $38,274.00    $39,002.00    $42,374.00    $42,943.00
$31,352.00    $33,294.00    $36,888.00    $39,573.00    $41,734.00
```

Initializing a two-dimensional array

The technique for initializing arrays by coding the array element values in curly braces works for two-dimensional arrays too. You just have to remember that each element of the main array is actually another array. So you have to nest the array initializers.

Here's an example that initializes the sales array:

```
double[][] sales =
    { {23853.0, 22838.0, 36483.0, 31352.0},    // 2004
      {25483.0, 22943.0, 38274.0, 33294.0},    // 2005
      {24872.0, 23049.0, 39002.0, 36888.0},    // 2006
      {28492.0, 23784.0, 42374.0, 39573.0},    // 2007
      {31932.0, 23732.0, 42943.0, 41734.0} };  // 2008
```

Here I added a comment to the end of each line to show the year that the line initializes. Notice that the left brace for the entire initializer is at the beginning of the second line, and the right brace that closes the entire initializer is at the end of the last line. Then the initializer for each year is contained in its own set of braces.

Using jagged arrays

When you create an array with an expression such as `new int[5][3]`, you're specifying that each element of the main array is actually an array of type `int` with three elements. Java, however, lets you create two-dimensional arrays in which the length of each element of the main array is different. This is sometimes called a *jagged array* because the array doesn't form a nice rectangle. Instead, its edges are jagged.

Suppose that you need to keep track of four teams, each consisting of two or three people. The teams are as follows:

Team	Members
A	Henry Blake, Johnny Mulcahy
B	Benjamin Pierce, John McIntyre, Jonathan Tuttle
C	Margaret Houlihan, Frank Burns
D	Max Klinger, Radar O'Reilly, Igor Straminsky

The following code creates a jagged array for these teams:

```
String[][] teams
    = { {"Henry Blake", "Johnny Mulcahy"},
        {"Benjamin Pierce", "John McIntyre",
            "Jonathan Tuttle"},
        {"Margaret Houlihan", "Frank Burns"},
        {"Max Klinger", "Radar O'Reilly",
            "Igor Straminsky"} };
```

Here each nested array initializer indicates the number of strings for each subarray. The first subarray has two strings, the second has three strings, and so on.

You can use nested `for` loops to access the individual elements in a jagged array. For each element of the main array, you can use the `length` property to determine how many entries are in that element's subarray. For example:

```
for (int i = 0; i < teams.length; i++)
{
    for (int j = 0; j < teams[i].length; j++)
        System.out.println(teams[i][j]);
    System.out.println();
}
```

Notice that the length of each subarray is determined with the expression `teams[i].length`. This `for` loop prints one name on each line, with a blank line between teams, like this:

```
Margaret Houlihan
Frank Burns

Max Klinger
Radar O'Reilly
Igor Straminsky

Henry Blake
Johnny Mulcahy

Benjamin Pierce
John McIntyre
Jonathan Tuttle
```

If you don't want to fuss with keeping track of the indexes yourself, you can use an enhanced `for` loop and let Java take care of the indexes. For example:

```java
for (String[] team : teams)
{
    for (String player : team)
        System.out.println(player);
    System.out.println();
}
```

Here the first enhanced `for` statement specifies that the type for the `team` variable is `String[]`. As a result, each cycle of this loop sets `team` to one of the subarrays in the main `teams` array. Then the second enhanced `for` loop accesses the individual strings in each subarray.

Going beyond two dimensions

TECHNICAL
STUFF

Java doesn't limit you to two-dimensional arrays. Arrays can be nested within arrays, to as many levels as your program needs. To declare an array with more than two dimensions, you just specify as many sets of empty brackets as you need. For example:

```java
int[][][] threeD = new int[3][3][3];
```

Here a three-dimensional array is created, with each dimension having three elements. You can think of this array as a cube. Each element requires three indexes to access.

You can access an element in a multidimensional array by specifying as many indexes as the array needs. For example:

```
threeD[0][1][2] = 100;
```

This statement sets element 2 in column 1 of row 0 to 100.

You can nest initializers as deep as necessary, too. For example:

```
int[][][] threeD =
    { { {1,   2,  3}, { 4,  5,  6}, { 7,  8,  9} },
      { {10, 11, 12}, {13, 14, 15}, {16, 17, 18} },
      { {19, 20, 21}, {22, 23, 24}, {25, 26, 27} } };
```

Here a three-dimensional array is initialized with the numbers 1 through 27.

You can also use multiple nested if statements to process an array with three or more dimensions. Here's another way to initialize a three-dimensional array with the numbers 1 to 27:

```
int[][][] threeD2 = new int[3][3][3];
int value = 1;
for (int i = 0; i < 3; i++)
    for (int j = 0; j < 3; j++)
        for (int k = 0; k < 3; k++)
            threeD2[i][j][k] = value++;
```

Working with a Fun but Complicated Example: A Chessboard

Okay, so much for the business examples. Here's an example that's more fun — assuming you think chess is fun. The program in Listing 2-2 uses a two-dimensional array to represent a chessboard. Its sole purpose is to figure out the possible moves for a knight (that's the horsey, for the non-chess players among us), given its starting position. The user is asked to enter a starting position (such as f1), and the program responds by displaying the possible squares. Then the program prints out a crude-but-recognizable representation of the board, with the knight's position indicated with an X and each possible move indicated with a question mark (?).

TIP

In case you're not familiar with chess, it's played on a board that's 8 by 8 squares, with alternating light and dark squares. The normal way to identify each square is to use a letter and a number, where the letter represents the column (called a *file*) and the number represents the row (called a *rank*), as shown in Figure 2-1. The knight has an interesting movement pattern: He moves two squares in one direction and then makes a 90-degree turn and moves one square to the left or right. The possible moves for the knight, given a starting position of e4, are shaded dark. As you can see, this knight has eight possible moves: c3, c5, d6, f6, g5, g3, f2, and d2.

a8	b8	c8	d8	e8	f8	g8	h8
a7	b7	c7	d7	e7	f7	g7	h7
a6	b6	c6	d6	e6	f6	g6	h6
a5	b5	c5	d5	e5	f5	g5	h5
a4	b4	c4	d4	♘	f4	g4	h4
a3	b3	c3	d3	e3	f3	g3	h3
a2	b2	c2	d2	e2	f2	g2	h2
a1	b1	c1	d1	e1	f1	g1	h1

FIGURE 2-1:
A classic chessboard.

Here's a sample of what the console looks like if you enter e4 for the knight's position:

```
Welcome to the Knight Move calculator.

Enter knight's position: e4

The knight is at square e4
From here the knight can move to:
c5
d6
f6
g5
g3
```

```
f2
d2
c3
  -  -  -  -  -  -  -  -
  -  -  -  -  -  -  -  -
  -  -  -  ?  -  ?  -  -
  -  -  ?  -  -  -  ?  -
  -  -  -  -  X  -  -  -
  -  -  ?  -  -  -  ?  -
  -  -  -  ?  -  ?  -  -
  -  -  -  -  -  -  -  -
Do it again? (Y or N) n
```

As you can see, the program indicates that the knight's legal moves from e4 are
c5, d6, f6, g5, g3, f2, d2, and c3. Also, the graphic representation of the board
indicates where the knight is and where he can go.

LISTING 2-2: **Playing Chess in a *For Dummies* Book**

```
import java.util.Scanner;

public class KnightMoves
{
    static Scanner sc = new Scanner(System.in);

    // the following static array represents the 8
    // possible moves a knight can make
    // this is an 8 x 2 array
    static int[][] moves = { {-2, +1},                                  →10
                             {-1, +2},
                             {+1, +2},
                             {+2, +1},
                             {+2, -1},
                             {+1, -2},
                             {-1, -2},
                             {-2, -1} };

    public static void main(String[] args)
    {
        System.out.println("Welcome to the "
            + "Knight Move calculator.\n");
        do
        {
            showKnightMoves();                                          →25
        }
        while (getYorN("Do it again?"));
    }
```

(continued)

LISTING 2-2: *(continued)*

```java
public static void showKnightMoves()                          →29
{
    // The first dimension is the file (a, b, c, etc.)
    // The second dimension is the rank (1, 2, 3, etc.)
    // Thus, board[3][4] is square d5.
    // A value of 0 means the square is empty
    // 1 means the knight is in the square
    // 2 means the knight could move to the square
    int[][] board = new int[8][8];                            →37

    String kSquare; // the knight's position as a square
    Pos kPos;       // the knight's position as a Pos

    // get the knight's initial position
    do                                                        →43
    {
        System.out.print("Enter knight's position: ");
        kSquare = sc.nextLine();
        kPos = convertSquareToPos(kSquare);
    } while (kPos == null);

    board[kPos.x][kPos.y] = 1;                                →50
    System.out.println("\nThe knight is at square "
        + convertPosToSquare(kPos));
    System.out.println(
        "From here the knight can move to:");
    for (int move = 0; move < moves.length; move ++)          →55
    {
        int x, y;
        x = moves[move][0];  // the x for this move
        y = moves[move][1];  // the y for this move
        Pos p = calculateNewPos(kPos, x, y);
        if (p != null)
        {
            System.out.println(convertPosToSquare(p));
            board[p.x][p.y] = 2;
        }
    }

    printBoard(board);                                        →68

}
// this method converts squares such as a1 or d5 to
// x, y coordinates such as [0][0] or [3][4]
public static Pos convertSquareToPos(String square)           →73
{
```

```
        int x = -1;
        int y = -1;
        char rank, file;

        file = square.charAt(0);
        if (file == 'a') x = 0;
        if (file == 'b') x = 1;
        if (file == 'c') x = 2;
        if (file == 'd') x = 3;
        if (file == 'e') x = 4;
        if (file == 'f') x = 5;
        if (file == 'g') x = 6;
        if (file == 'h') x = 7;

        rank = square.charAt(1);
        if (rank == '1') y = 0;
        if (rank == '2') y = 1;
        if (rank == '3') y = 2;
        if (rank == '4') y = 3;
        if (rank == '5') y = 4;
        if (rank == '6') y = 5;
        if (rank == '7') y = 6;
        if (rank == '8') y = 7;

        if (x == -1 || y == -1)
        {
            return null;
        }
        else
            return new Pos(x, y);
    }

// this method converts x, y coordinates such as
// [0][0] or [3][4] to squares such as a1 or d5.
public static String convertPosToSquare(Pos p)
{
    String file = "";

    if (p.x == 0) file = "a";
    if (p.x == 1) file = "b";
    if (p.x == 2) file = "c";
    if (p.x == 3) file = "d";
    if (p.x == 4) file = "e";
    if (p.x == 5) file = "f";
    if (p.x == 6) file = "g";
    if (p.x == 7) file = "h";
```

→109

(continued)

LISTING 2-2:   *(continued)*

```
            return file + (p.y + 1);
        }

        // this method calculates a new Pos given a
        // starting Pos, an x move, and a y move
        // it returns null if the resulting move would
        // be off the board.
        public static Pos calculateNewPos(Pos p, int x, int y)          →129
        {
            // rule out legal moves
            if (p.x + x < 0)
                return null;
            if (p.x + x > 7)
                return null;
            if (p.y + y < 0)
                return null;
            if (p.y + y > 7)
                return null;

            // return new position
            return new Pos(p.x + x, p.y + y);
        }

        public static void printBoard(int[][] b)                         →145
        {
            for (int y = 7; y >= 0; y--)
            {
                for (int x = 0; x < 8; x++)
                {
                    if (b[x][y] == 1)
                        System.out.print(" X ");
                    else if (b[x][y] == 2)
                        System.out.print(" ? ");
                    else
                        System.out.print(" - ");
                }
                System.out.println();
            }
        }
        public static boolean getYorN(String prompt)                     →161
        {
            while (true)
            {
                String answer;
                System.out.print("\n" + prompt + " (Y or N) ");
                answer = sc.nextLine();
                if (answer.equalsIgnoreCase("Y"))
```

```
            return true;
        else if (answer.equalsIgnoreCase("N"))
            return false;
    }
  }
}

// this class represents x, y coordinates on the board
class Pos                                                      →177
{
    public int x;
    public int y;

    public Pos(int x, int y)
    {
        this.x = x;
        this.y = y;
    }
}
```

TECHNICAL STUFF

You have to put your thinking cap on to make your way through this program, which is a bit on the complicated side. The following paragraphs can help clear up the more complicated lines:

>> →10 This line declares a two-dimensional array that's used to store the possible moves for a knight in terms of x and y. The knight's move of two squares left and one square up, for example, is represented as {–2, 1}. There are a total of eight possible moves, and each move has two values (x and y). So this two-dimensional array has eight rows and two columns.

>> →25 The code that gets the user's starting position for the knight and does all the calculations is complicated enough that I didn't want to include it in the main method, so I put it in a separate method named showKnightMoves. That way, the do loop in the main method is kept simple. It just keeps going until the user enters N when getYorN is called.

>> →29 The showKnightMoves method begins here.

>> →37 The board array represents the chessboard as a two-dimensional array with eight rows for the ranks and eight columns for the files. This array holds int values. A value of 0 indicates that the square is empty. The square where the knight resides gets a 1, and any square that the knight can move to gets a 2.

>> →43 This do loop prompts the user for a valid square to plant the knight in. The loop includes a call to the method convertSquareToPos, which converts the

user's entry (such as e4) to a Pos object. (The Pos class is defined later in the program; it represents a board position as an x, y pair.) This method returns null if the user enters an incorrect square, such as a9 or x4. So to get the user to enter a valid square, the loop just repeats if the converSquareToPos returns null.

» →50 The board position entered by the user is set to 1 to indicate the position of the knight.

» →55 A for loop is used to test all the possible moves for the knight to see whether they're valid from the knight's current position, using the moves array that was created way back in line 10. In the body of this loop, the calculateNewPos method is called. This method accepts a board position and x and y values to indicate where the knight can be moved. If the resulting move is legal, it returns a new Pos object that indicates the position the move leads to. If the move is not legal (that is, it takes the knight off the board), the calculateNewPos method returns null.

Assuming that calculateNewPos returns a non-null value, the body of this loop prints the square (it calls convertPosTosquare to convert the Pos object to a string, such as c3). Then it marks the board position represented by the move with 2 to indicate that the knight can move to this square.

» →68 After all the moves are calculated, the printBoard method is called to print the board array.

» →73 This is the convertSquareToPos method. It uses a pair of brute-force if statements to convert a string such as a1 or e4 to a Pos object representing the same position. I probably could have made this method a little more elegant by converting the first letter in the string to a Char and then subtracting the offset of the letter a to convert the value to a proper integer. But I think the brute-force method is clearer, and it requires only a few more lines of code.

Note that if the user enters an incorrect square (such as a9 or x2), null is returned.

» →109 This is the convertPosToSquare method, which does the opposite of the convertSquareToPos method. It accepts a Pos argument and returns a string that corresponds to the position. It uses a series of brute-force if statements to determine the letter that corresponds to the file but does a simple addition to calculate the rank. (The Pos object's y member is an array for the y position. Array indexes are numbered starting with 0, but chess rank numbers start with 1. That's why 1 is added to the y position to get the rank number.)

» →129 The calculateNewPos method accepts a starting position, an x offset, and a y offset. It returns a new position if the move is legal; otherwise it returns null. To find illegal moves, it adds the x and y offsets to the starting x and y position and checks to see whether the result is less than 0 or greater than 7. If the move is legal, it creates a new Pos object whose position is calculated by adding the x and y offsets to the x and y values of the starting position.

» → 145 The printBoard method uses a nested for loop to print the board. The outer loop prints each rank. Notice that it indexes the array backward, starting with 7 and going down to 0. That's necessary so that the first rank is printed at the bottom of the console output. An inner for loop is used to print the squares for each rank. In this loop, an if statement checks the value of the board array element that corresponds to the square to determine whether it prints an X, a question mark, or a hyphen.

» → 161 The getYorN method simply displays a prompt on-screen and asks the user to enter Y or N. It returns true if the user enters Y or false if the user enters N. If the user enters anything else, this method prompts the user again.

» → 177 The Pos class simply defines two public fields, x and y, to keep track of board positions. It also defines a constructor that accepts the x and y positions as parameters.

Using the Arrays Class

The final topic for this chapter is the Arrays class, which provides a collection of static methods that are useful for working with arrays. The Arrays class is in the java.util package, so you have to use an import statement for the java.util. Arrays class or the entire java.util.* package to use this class. Table 2-1 lists the most commonly used methods of the Arrays class.

TABLE 2-1   Handy Methods of the Arrays Class

Method	Description
static int binarySearch (*array, key*)	Searches for the specified key value in an array. The return value is the index of the element that matches the key. The method returns –1 if the key couldn't be found. The array and the key must be of the same type and can be any primitive type or an object.
static array copyOf (arrayOriginal, newLength)	Returns an array that's a copy of arrayOriginal. The newLength parameter need not equal the original array's length. If newLength is larger, the method pads the new array with zeros. If newLength is smaller, the method doesn't copy all of the original array's values.

(continued)

TABLE 2-1 *(continued)*

Method	Description
`static array copyOfRange (arrayOriginal,` *`from,`* `to)`	Does basically what the `copyOf` method does, but copies only a selected slice of values (from one index to another) of the original array.
`boolean deepEquals(`*`array1,`* *`array2`*`)`	Returns `true` if the two arrays have the same element values. This method works for arrays of two or more dimensions.
`boolean equals(`*`array1,`* *`array2`*`)`	Returns `true` if the two arrays have the same element values. This method checks equality only for one-dimensional arrays.
`static void fill(`*`array,`* *`value`*`)`	Fills the array with the specified value. The value and array must be of the same type and can be any primitive type or an object.
`static void fill(`*`array,`* *`from, to`*`, value)`	Fills the elements indicated by the *from* and *to* `int` parameters with the specified value. The value and array must be of the same type and can be any primitive type or an object.
`static void setAll(`*`array,`* *`generator function`*`)`	Sets the elements of an array using a function provided by *generator,* usually written with a lambda expression.
`static void sort(`*`array`*`)`	Sorts the array in ascending sequence.
`static void sort(`*`array,`* *`from, to`*`)`	Sorts the specified elements of the array in ascending sequence.
`static String toString (`*`array`*`)`	Formats the array values in a string. Each element value is enclosed in brackets, and the element values are separated with commas.

Filling an array

The `fill` method can be handy if you want to prefill an array with values other than the default values for the array type. Here's a routine that creates an array of integers and initializes each element to 100:

```
int[] startValues = new int[10];
Arrays.fill(startValues, 100);
```

WARNING

Although you can code a complicated expression as the second parameter, the `fill` method evaluates this expression only once. Then it assigns the result of this expression to each element in the array.

You might think that you could fill an array of 1,000 integers with random numbers from 1 to 100, like this:

```
int[] ran = new int[1000]
Arrays.fill(ran, (int)(Math.random() * 100) + 1);
```

Unfortunately, this code won't work. What happens is that the expression is evaluated once to get a random number; then all 1,000 elements in the array are set to that random number.

TIP

You can easily accomplish this task by using the setAll method along with a lambda expression, like this:

```
int[] a = new int[10];
Arrays.setAll(a, i -> (int)(Math.random() * 100) + 1);
```

Copying an array

In Java 1.6, the Arrays class has some useful new methods. Using the new copyOf and copyOfRange methods, you can copy a bunch of elements from an existing array into a brand-new array. If you start with something named arrayOriginal, for example, you can copy the arrayOriginal elements to something named arrayNew, as shown in Listing 2-3.

LISTING 2-3: **The Copycat**

```
import java.util.Arrays;

class CopyDemo
{
    public static void main(String args[])
    {
        int arrayOriginal[] = {42, 55, 21};
        int arrayNew[] =
            Arrays.copyOf(arrayOriginal, 3);                    →9
        printIntArray(arrayNew);
    }
    static void printIntArray(int arrayNew[])
    {
        for (int i : arrayNew)
        {
            System.out.print(i);
            System.out.print(' ');
        }
        System.out.println();
    }
}
```

The output of the CopyDemo program looks like this:

```
42 55 21
```

Line 9 is where the array is actually copied. Here, the number 3 specifies how many array elements to copy.

If you want, you can copy less than the full array. For example:

```
int arrayNew[] = Arrays.copyOf(arrayOriginal, 2);
```

Then, arrayNew has just two elements:

```
42 55
```

You can also copy more than the full array, in which case, additional elements are added with default values. For example:

```
int arrayNew[] = Arrays.copyOf(arrayOriginal, 8);
```

Then arrayNew has eight elements:

```
42 55 21 0 0 0 0 0
```

The copyOfRange method is even more versatile. If you execute the instructions

```
int arrayOriginal[] = {42, 55, 21, 16, 100, 88};
int arrayNew[] = Arrays.copyOfRange(arrayOriginal, 2, 5);
```

the values in arrayNew are

```
21 16 100
```

Sorting an array

The sort method is a quick way to sort an array in sequence. These statements create an array with 100 random numbers and then sort the array in sequence so that the random numbers are in order:

```
int[] lotto = new int[6];
for (int i = 0; i < 6; i++)
    lotto[i] = (int)(Math.random() * 100) + 1;
Arrays.sort(lotto);
```

Searching an array

The `binarySearch` method is an efficient way to locate an item in an array by its value. Suppose you want to find out whether your lucky number is in the `lotto` array created in the preceding example. You could just use a `for` loop, like this:

```
int lucky = 13x;
int foundAt = -1;
for (int i = 0; i < lotto.length; i++)
    if (lotto[i] == lucky)
        foundAt = i;
if (foundAt > -1)
    System.out.println("My number came up!");
else
    System.out.println("I'm not lucky today.");
```

Here the `for` loop compares each element in the array with your lucky number. This code works fine for small arrays, but what if the array had 1,000,000 elements instead of 6? In that case, it would take a while to look at each element. If the array is sorted in sequence, the `binarySearch` method can find your lucky number more efficiently and with less code:

```
int lucky = 13;
int foundAt = Arrays.binarySearch(lotto, lucky);
if (foundAt > -1)
    System.out.println("My number came up!");
else
    System.out.println("I'm not lucky today.");
```

TECHNICAL STUFF

The `binarySearch` method uses a technique similar to the strategy for guessing a number. If I say that I'm thinking of a number between 1 and 100, you don't start guessing the numbers in sequence starting with 1. Instead, you guess 50. If I tell you that 50 is low, you guess 75. Then if I tell you 75 is high, you guess halfway between 50 and 75, and so on until you find the number. The `binarySearch` method uses a similar technique, but it works only if the array is sorted first.

Comparing arrays

If you use the equality operator (==) to compare array variables, the array variables are considered to be equal only if both variables point to exactly the same array instance. To compare two arrays element by element, you should use the `Arrays.equals` method instead. For example:

```
if (Arrays.equals(array1, array2))
    System.out.println("The arrays are equal!");
```

Here the arrays array1 and array2 are compared element by element. If both arrays have the same number of elements, and all elements have the same value, the equals method returns true. If the elements are not equal, or if one array has more elements than the other, the equals method returns false.

If the array has more than one dimension, you can use the deepEquals method instead. It compares any two subarrays, element by element, to determine whether they're identical.

Converting arrays to strings

The toString method of the Arrays class is handy if you want to quickly dump the contents of an array to the console to see what it contains. This method returns a string that shows the array's elements enclosed in brackets, with the elements separated by commas.

Here's a routine that creates an array, fills it with random numbers, and then uses the toString method to print the array elements:

```
int[] lotto = new int[6];
for (int i = 0; i < 6; i++)
    lotto[i] = (int)(Math.random() * 100) + 1;
System.out.println(Arrays.toString(lotto));
```

Here's a sample of the console output created by this code:

```
[4, 90, 65, 84, 99, 81]
```

Note that the toString method works only for one-dimensional arrays. To print the contents of a two-dimensional array with the toString method, use a for loop to call the toString method for each subarray.

IN THIS CHAPTER

» **Working with the ArrayList class**

» **Creating an array list**

» **Introducing generics**

» **Adding elements to an array list**

» **Deleting elements from or modifying elements in an array list**

Chapter **3**

Using the ArrayList Class

S ome people love to collect things: nick-knacks, baseball cards, postage stamps, dolls — you name it, someone collects it.

If I were a collector of some random thing — say, old tin advertising signs — an array would be a poor choice for storing the data. That's because on any given day, I may find another tin sign at a yard sale. So if I had 87 tin signs before, and I had created an array big enough to hold all 87 signs, I'd have to change the array declaration to hold 88 signs.

Java's *collection classes* are designed to simplify the programming for applications that have to keep track of groups of objects. These classes are very powerful and surprisingly easy to use — at least the basics, anyway. The more advanced features of collection classes take some serious programming to get right, but for most applications, a few simple methods are all you need to use collection classes.

Unfortunately, Java's collection classes are organized according to a pretty complicated inheritance hierarchy that can be very confusing for beginners. Most of the Java books I have on my shelf start by explaining this inheritance scheme and showing how each of the various collection classes fits into this scheme, and why.

I'm not going to do that. I think it's very confusing for newcomers to collections to have to wade through a class hierarchy that doesn't make sense until they know some of the details of how the basic classes work. Instead, I just show you how to use two of the best of these classes.

In this chapter, you find out how to use the `ArrayList` class. Then, in the next chapter, you find out how to use its first cousin, the `LinkedList`. When you know how to use these two classes, you shouldn't have any trouble figuring out how to use the other collection classes from the API documentation.

TECHNICAL STUFF

This chapter also introduces you to an important feature in Java called *generics*. Simply put, the generics feature provides a way to make collection classes such as `ArrayList` and `LinkedList` type safe. In other words, with generics you can specify the type of data that can be stored in a collection and ensure that data of the wrong type can't accidentally be put into the collection.

Because generics are an integral part of how collections work, I incorporate the generics feature into this chapter from the very start. I point out the differences for using `ArrayList` without generics along the way, just in case you're using an older version of Java or are working with programs that were written before Java 1.5 became available. (For a complete explanation of how the generics feature works, you can move ahead to Book 4, Chapter 5.)

In addition to generics, Java *lambda expressions* can also be used with collections, especially when dealing with collections that have a large number of items. For more information about working with these collection features, see Chapter 6 of this minibook. And for more information about lambda expressions, see Book 3, Chapter 7.

Understanding the ArrayList Class

An *array list* is the most basic type of Java collection. You can think of an array list as being an array on steroids. It's similar to an array but averts many of the most common problems of working with arrays, specifically the following:

» **An array list automatically resizes itself whenever necessary.** If you create an array with 100 elements, and then fill it up and need to add a 101st element, you're out of luck. The best you can do is create a new array with 101 elements, copy the 100 elements from the old array to the new one, and then put the new data in the 101st element. With an array list, there's never a limit to how many elements you can create. You can keep adding elements as long as you want.

» **An array list lets you insert elements into the middle of the collection.** With an array, inserting elements is pretty hard to do. Suppose that you have an array that can hold 100 elements, but only the first 50 have data. If you need to insert a new element after the 25th item, you first must make a copy of elements 26 through 50 to make room for the new element. With an array list, you just tell the array list you want to insert the new element after the 25th item; the array list takes care of shuffling things around.

>> **An array list lets you delete items.** If you delete an item from an array, the deleted element becomes null, but the empty slot that was occupied by the item stays in the array. When you delete an item from an array list, any subsequent items in the array are automatically moved forward one position to fill the spot that was occupied by the deleted item.

TECHNICAL STUFF

>> **The** `ArrayList` **class actually uses an array internally to store the data you add to the array list.** The `ArrayList` class takes care of managing the size of this array. When you add an item to the array list, and the underlying array is full, the `ArrayList` class automatically creates a new array with a larger capacity and copies the existing items to the new array before it adds the new item.

TECHNICAL STUFF

>> **The** `ArrayList` **class, like other collection classes, cannot be used to store primitive data types.** Collections can store objects, not primitives. If you want to store primitive values in a collection, you must first encapsulate them in a wrapper class. For example, to store `int` values, you could use the `Integer` class. For more information about wrapper classes, refer to Book 2, Chapter 2.

The `ArrayList` class has several constructors and a ton of methods. For your reference, Table 3-1 lists the constructors and methods of the `ArrayList` class.

The rest of this chapter shows you how to use these constructors and methods to work with `ArrayList` objects.

TABLE 3-1 **The ArrayList Class**

Constructor	Explanation
`ArrayList()`	Creates an array list with an initial capacity of ten elements.
`ArrayList(int capacity)`	Creates an array list with the specified initial capacity.
`ArrayList(Collection c)`	Creates an array list and copies all the elements from the specified collection into the new array list.
Method	*Explanation*
`add(Object element)`	Adds the specified object to the array list. If you specified a type when you created the array list, the object must be of the correct type.
`add(int index, Object element)`	Adds the specified object to the array list at the specified index position. If you specified a type when you created the array list, the object must be of the correct type.
`addAll(Collection c)`	Adds all the elements of the specified collection to this array list.

(continued)

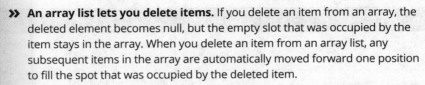

Using the ArrayList Class

TABLE 3-1 *(continued)*

Constructor	Explanation
addAll(int *index*, Collection *c*)	Adds all the elements of the specified collection to this array list at the specified index position.
clear()	Deletes all elements from the array list.
clone()	Returns a shallow copy of the array list. The elements contained in the copy are the same object instances as the elements in the original.
contains(Object *elem*)	Returns a boolean that indicates whether the specified object is in the array list.
containsAll(Collection *c*)	Returns a boolean that indicates whether this array list contains all the objects that are in the specified collection.
ensureCapacity(int *minCapacity*)	Increases the array list's capacity to the specified value. (If the capacity is already greater than the specified value, this method does nothing.)
get(int *index*)	Returns the object at the specified position in the list.
indexOf(Object *elem*)	Returns the index position of the first occurrence of the specified object in the array list. If the object isn't in the list, it returns –1.
isEmpty()	Returns a boolean value that indicates whether the array list is empty.
iterator()	Returns an iterator for the array list.
lastIndexOf(Object *elem*)	Returns the index position of the last occurrence of the specified object in the array list. If the object isn't in the list, it returns –1.
remove(int *index*)	Removes the object at the specified index and returns the element that was removed.
remove(Object *elem*)	Removes an object from the list. Note that more than one element refers to the object; this method removes only one of them. It returns a boolean that indicates whether the object was in the list.
remove(int *fromIndex*, int *toIndex*)	Removes all objects whose index values are between the values specified. Note that the elements at the *fromIndex* and *toIndex* positions are not themselves removed.
removeAll(Collection *c*)	Removes all the objects in the specified collection from this array list.
retainAll(Collection *c*)	Removes all the objects that are not in the specified collection from this array list.
set(int *index*, Object *elem*)	Sets the specified element to the specified object. The element that was previously at that position is returned as the method's return value.
size()	Returns the number of elements in the list.
toArray()	Returns the elements of the array list as an array of objects (Object[]).
toArray(type[] array)	Returns the elements of the array list as an array whose type is the same as the array passed via the parameter.

Creating an ArrayList Object

To create an array list, you first declare an `ArrayList` variable and then call the `ArrayList` constructor to instantiate an `ArrayList` object and assign it to the variable. You can do this on separate lines:

```
ArrayList signs;
signs = new ArrayList();
```

Alternatively, you can do it on a single line:

```
ArrayList signs = new ArrayList();
```

Here are a few things to note about creating array lists:

TIP

>> The `ArrayList` class is in the `java.util` package, so your program must import either `java.util.ArrayList` or `java.util.*`.

>> Unlike an array, an array list doesn't make you specify a capacity — though you can if you want. Here's a statement that creates an array list with an initial capacity of 100:

```
ArrayList signs = new ArrayList(100);
```

If you don't specify a capacity for the array list, the initial capacity is set to 10. Providing at least a rough estimate of how many elements each array list can hold when you create it is a good idea.

REMEMBER

>> The capacity of an array list is not a fixed limit. The `ArrayList` class automatically increases the list's capacity whenever necessary.

TIP

>> You can (and should) also specify the type of elements the array list is allowed to contain. This statement creates an array list that holds `String` objects:

```
ArrayList<String> signs = new ArrayList<String>();
```

The advantage of specifying a type when you declare an array list is that the compiler complains if you then try to add an object of the wrong type to the list. (This feature is called *generics* because it lets the Java API designers create generic collection classes that can be used to store any type of object. For more information, see Book 4, Chapter 5.)

TIP

>> As a shortcut, you can omit the explicit type specification on the right side of the equals sign if the types are identical. Instead, you just use the ‹ and › signs together (this is called the *diamond operator*), like this:

```
ArrayList<String> signs = new ArrayList<>();
```

> » The ArrayList class also has a constructor that lets you specify another collection object (typically, another array list) whose items are copied into the new array list. This provides an easy way to make a copy of an array list, but you can also use it to convert any other type of collection to an array list.

Adding Elements

After you create an array list, you can use the add method to add objects to the array list. Here's code that adds strings to an array list:

```
signs.add("Drink Pepsi");
signs.add("No minors allowed");
signs.add("Say Pepsi, Please");
signs.add("7-Up: You Like It, It Likes You");
signs.add("Dr. Pepper 10, 2, 4");
```

If you specified a type when you created the array list, the objects you add via the add method must be of the correct type.

You can insert an object at a specific position in the list by listing the position in the add method. Consider these statements:

```
ArrayList<String> nums = new ArrayList<String>();
nums.add("One");
nums.add("Two");
nums.add("Three");
nums.add("Four");
nums.add(2, "Two and a half");
```

After these statements execute, the nums array list contains the following strings:

```
One
Two
Two and a half
Three
Four
```

Here are some important points to keep in mind when you add elements to array lists:

TIP

> » If an array list is already at its capacity when you add an element, the array list automatically expands its capacity. Although this capacity is flexible, it's also inefficient. Whenever possible, you should anticipate how many elements you're adding to an array list and set the list's initial capacity accordingly.

(You can also change the capacity at any time by calling the *ensureCapacity method*.)

» Like arrays, array lists are indexed starting with zero. Keep this in mind when you use the version of the add method that accepts an index number.

WARNING

» The add method that inserts elements at a specific index position throws the unchecked exception IndexOutOfBoundsException if an object isn't already at the index position you specify.

Accessing Elements

To access a specific element in an array list, you can use the get method, which specifies the index value of the element you want to retrieve. Here's a for loop that prints all the strings in an array list:

```
for (int i = 0; i < nums.size(); i++)
    System.out.println(nums.get(i));
```

Here the size method is used to set the limit of the for loop's index variable.

The easiest way to access all the elements in an array list is to use an enhanced for statement, which lets you retrieve the elements without bothering with indexes or the get method. For example:

```
for (String s : nums)
    System.out.println(s);
```

Here each String element in the nums array list is printed to the console.

If you need to know the index number of a particular object in an array list, and you have a reference to the object, you can use the indexOf method. Here's an enhanced for loop that prints the index number of each string along with the string:

```
for (String s : nums)
{
    int i = nums.indexOf(s);
    System.out.println("Item " + i + ": " + s);
}
```

Depending on the contents of the array list, the output from this loop may look something like this:

```
Item 0: One
Item 1: Two
Item 2: Three
Item 3: Four
```

Printing an ArrayList

The toString method of the ArrayList class (as well as other collection classes) is designed to make it easy to quickly print out the contents of the list. It returns the contents of the array list enclosed in a set of brackets, with each element value separated by commas. The toString method of each element is called to obtain the element value.

Consider these statements:

```
ArrayList<String> nums = new ArrayList<String>();
nums.add("One");
nums.add("Two");
nums.add("Three");
nums.add("Four");
System.out.println(nums);
```

When you run these statements, the following is displayed on the console:

```
[One, Two, Three, Four]
```

Although this output isn't very useful for actual applications, it's convenient for testing purposes or for debugging problems in programs that use array lists.

Using an Iterator

Another way to access all the elements in an array list (or any other collection type) is to use an iterator. An *iterator* is a special type of object whose sole purpose in life is to let you step through the elements of a collection.

TIP

The enhanced `for` statement can simplify programs that use iterators. However, keep in mind that you can't modify the contents of an `ArrayList` when you use a `for-each` statement. So you still need to know how to use iterators.

An iterator object implements the `Iterator` interface, which is defined as part of the `java.util` package. As a result, to use an iterator, you must import either `java.util.Iterator` or `java.util.*`. The `Iterator` interface defines just three methods, as listed in Table 3-2. These methods are all you need to access each element of a collection. (Actually, you usually need just the `hasNext` and `next` methods. The `remove` method is gravy.)

TABLE 3-2   **The Iterator Interface**

Method	Explanation
`hasNext()`	Returns `true` if the collection has at least one element that hasn't yet been retrieved
`next()`	Returns the next element in the collection
`remove()`	Removes the most recently retrieved element

THE ITERATOR PATTERN

Java's iterators follow a common design pattern called the *Iterator pattern,* which is useful whenever you need to provide sequential access to a collection of objects. The Iterator pattern relies on interfaces so that the code that's using the iterator doesn't have to know what actual class is being iterated. As long as the class implements the iterator interface, it can be iterated.

The `Iterator` interface itself defines the methods used for sequential access. The common pattern is for this interface to provide at least two methods:

- `hasNext`: Returns a `boolean` value that indicates whether another item is available.
- `next`: Returns the next item.

Java also defines a third method for its `Iterator` interface: `remove`, which removes the most recently retrieved object.

In addition to the `Iterator` interface, the collection class itself needs a way to get an `iterator` object. It does so via the `iterator` method, which simply returns an `iterator` object for the collection. The `iterator` method is defined by the `Iterable` interface. Thus, any object that implements `Iterable` has an `iterator` method that provides an iterator for the object.

To use an iterator, you first call the array list's iterator method to get the iterator. Then you use the iterator's hasNext and next methods to retrieve each item in the collection. The normal way to do that is with a while loop. Here's an example:

```
ArrayList<String> nums = new ArrayList<String>();
nums.add("One");
nums.add("Two");
nums.add("Three");
nums.add("Four");

String s;
Iterator<String> e = nums.iterator();
while (e.hasNext())
{
    s = e.next();
    System.out.println(s);
}
```

Here the first five statements create an array list and add four strings to it. Next, the iterator method is called to get an iterator for the nums array list. The hasNext method is called in the while statement, and the next method is called to get the element to be printed.

REMEMBER

Note that the Iterator interface is generic, so you can supply a type when you declare it. In the preceding example, I declare the type as String so that the Iterator will work with String values returned from the ArrayList. If I had omitted the type when I declared the Iterator, I'd have to cast the result of the next method to a String.

Updating Elements

You can use the set method to replace an existing object with another object. Consider this example:

```
ArrayList<String> nums = new ArrayList<String>();
nums.clear();
nums.add("One");
nums.add("Two");
nums.add("Three");
System.out.println(nums);
nums.set(0, "Uno");
nums.set(1, "Dos");
nums.set(2, "Tres");
System.out.println(nums);
```

Here an array list is created with three strings, and the contents of the array list are printed to the console. Then each of the three strings is replaced by another string, and the contents print to the console again. When you run this code, the following is what you see printed on the console:

```
[One, Two, Three]
[Uno, Dos, Tres]
```

TIP

Because array lists contain references to objects, not the objects themselves, any changes you make to an object in an array list are automatically reflected in the list. As a result, you don't often have to use the set method.

For example:

```
ArrayList<Employee> emps = new ArrayList<Employee>();

// add employees to array list
emps.add(new Employee("Addams", "Gomez"));
emps.add(new Employee("Taylor", "Andy"));
emps.add(new Employee("Kirk", "James"));

// print array list
System.out.println(emps);

// change one of the employee's names
Employee e = emps.get(1);
e.changeName("Petrie", "Robert");

// print the array list again
System.out.println(emps);
```

This example uses the Employee class, whose constructor accepts an employee's last name and first name to create a new employee object, as well as a changeName method that also accepts a last name and a first name. In addition, the Employee class overrides the toString method to return the employee's first name and last name.

The main method begins by creating an ArrayList object and adding three employees. Then it prints out the contents of the array list. Next, it retrieves the employee with index number 1 and changes that employee's name. Finally, it prints the contents of the array list again.

Here's what this code produces on the console:

```
[Gomez Addams, Andy Taylor, James Kirk]
[Gomez Addams, Robert Petrie, James Kirk]
```

Notice that the second employee's name was changed, even though the program doesn't use the `set` method to replace the changed `Employee` object in the collection. That's because the array list merely stores references to the `Employee` objects.

Deleting Elements

The `ArrayList` class provides several methods that let you remove elements from the collection. To remove all the elements, use the `clear` method, like this:

```
emps.clear();
```

To remove a specific element, use the `remove` method. It lets you remove an element based on the index number, like this:

```
emps.remove(0);
```

Here the first element in the array list is removed.

Alternatively, you can pass the actual object you want removed. This is useful if you don't know the index of the object you want to remove, but you happen to have a reference to the actual object, as in this example:

```
ArrayList<Employee> emps = new ArrayList<Employee>();

// create employee objects
Employee emp1 = new Employee("Addams", "Gomez");
Employee emp2 = new Employee("Taylor", "Andy");
Employee emp3 = new Employee("Kirk", "James");

// add employee objects to array list
emps.add(emp1);
emps.add(emp2);
emps.add(emp3);

// print the array list
System.out.println(emps);
```

```
// remove one of the employees
emps.remove(emp2);

// print the array list again
System.out.println(emps);
```

Here's what this code produces on the console:

```
[Gomez Addams, Andy Taylor, James Kirk]
[Gomez Addams, James Kirk]
```

As you can see, the program was able to remove Andy Taylor from the list without knowing his index position.

Here are a few important details to keep in mind:

TIP

>> The clear and remove methods don't actually delete objects; they simply remove the references to the objects from the array list. Like any other objects, the objects in a collection are deleted automatically by the garbage collector — and then only if the objects are no longer being referenced by the program.

>> You can remove more than one element at the same time by using the remove Range method. On it, you specify the starting and ending index numbers. (Note that this method removes all elements between the elements you specify, but the elements you specify aren't themselves removed. removeRange(5, 8), for example, removes elements 6 and 7, but elements 5 and 8 aren't removed.)

>> You can also use the removeAll method to remove all the objects in one collection from another collection. A similar method, retainAll, removes all the objects that are *not* in another collection.

Chapter **4**

Using the LinkedList Class

The `ArrayList` class, which I cover in the preceding chapter, is a collection class that's based on an array. Arrays have their strengths and their weaknesses. The strength of an array is that it's very efficient — at least until you fill it up or try to reorganize it by inserting or deleting elements. Then it suddenly becomes very inefficient.

Over the years, computer scientists have developed various alternatives to arrays that are more efficient for certain types of access. One of the oldest of these alternatives is the linked list. A linked list is less efficient than an array for tasks such as directly accessing an element based on its index number, but linked lists run circles around arrays when you need to insert or delete items in the middle of the list.

In this chapter, you find out how to use Java's `LinkedList` class, which provides a collection that's based on a linked list rather than an array. You'll find that although the `LinkedList` class provides many of the same features as the `Array List` class, it also has some tricks of its own.

Understanding the LinkedList Class

A *linked list* is a collection in which every object in the list maintains with it a pointer to the following object in the list and another pointer to the preceding object in the list. No array is involved at all in a linked list. Instead, the list is managed entirely by these pointers.

TIP

Don't worry — you don't have to do any of this pointer management yourself. It's all taken care of for you by the LinkedList class.

This arrangement has some compelling advantages over arrays:

>> Because the ArrayList class uses an array to store list data, the ArrayList class frequently has to reallocate its array when you add items to the list. Not so with the LinkedList class. Linked lists don't have any size issues. You can keep adding items to a linked list until your computer runs out of memory.

>> Like the ArrayList class, the LinkedList class lets you insert items into the middle of the list. With the ArrayList class, however, this operation is pretty inefficient. It has to copy all the items past the insertion point one slot over to free a slot for the item you're inserting. Not so with the LinkedList class. To insert an item into the middle of a linked list, all you have to do is change the pointers in the preceding and the following objects.

>> With an array list, removing items from the list is pretty inefficient. The ArrayList class has to copy every item after the deleted item one slot closer to the front of the array to fill the gap left by the deleted item. Not so with the LinkedList class. To remove an item from a linked list, all that's necessary is to update the pointers in the items that were before and after the item to be removed.

If you want to remove the third item from a list that has 10,000 items in it, for example, the ArrayList class has to copy 9,997 items. By contrast, the LinkedList class does it by updating just two of the items. By the time the ArrayList class is done, the LinkedList class has had time to mow the lawn, read a book, and go to Disneyland.

TECHNICAL STUFF

>> The ArrayList class actually uses an array internally to store the data you add to the array list. The ArrayList class takes care of managing the size of this array. When you add an item to the array list, and the underlying array is full, the ArrayList class automatically creates a new array with a larger capacity and copies the existing items to the new array before it adds the new item.

There's no such thing as a free lunch, however. The flexibility of a linked list comes at a cost: Linked lists require more memory than arrays and are slower than arrays when it comes to simple sequential access.

Like the ArrayList class, the LinkedList class has several constructors and a ton of methods. For your reference, Table 4-1 lists the constructors and methods of the LinkedList class.

TABLE 4-1   ## The LinkedList Class

Constructor	Explanation
LinkedList()	Creates an empty linked list.
LinkedList(Collection c)	Creates a linked list and copies all the elements from the specified collection into the new linked list.
Method	*Explanation*
add(Object *element*)	Adds the specified object to the end of the linked list. If you specified a type when you created the linked list, the object must be of the correct type.
add(int *index*, Object *element*)	Adds the specified object to the linked list at the specified index position. If you specified a type when you created the linked list, the object must be of the correct type.
addAll(Collection *c*)	Adds all the elements of the specified collection to this linked list.
addAll(int *index*, Collection *c*)	Adds all the elements of the specified collection to this linked list at the specified index position.
addFirst(Object *element*)	Inserts the specified object at the beginning of the list. If you specified a type when you created the linked list, the object must be of the correct type.
addLast(Object *element*)	Adds the specified object to the end of the list. This method performs the same function as the add method. If you specified a type when you created the linked list, the object must be of the correct type.
clear()	Deletes all elements from the linked list.
clone()	Returns a copy of the linked list. The elements contained in the copy are the same object instances as the elements in the original.
contains(Object *elem*)	Returns a boolean that indicates whether the specified object is in the linked list.

(continued)

TABLE 4-1 *(continued)*

Constructor	Explanation
containsAll(Collection c)	Returns a boolean that indicates whether this linked list contains all the objects that are in the specified collection.
descendingIterator()	Returns an iterator that steps backward from the end to the beginning of the linked list.
element()	Retrieves the first element from the list. (The element is not removed.)
get(int index)	Returns the object at the specified position in the list.
getFirst()	Returns the first element in the list. If the list is empty, it throws NoSuchElementException.
getLast()	Returns the last element in the list. If the list is empty, it throws NoSuchElementException.
indexOf(Object elem)	Returns the index position of the first occurrence of the specified object in the list. If the object isn't in the list, it returns –1.
isEmpty()	Returns a boolean value that indicates whether the linked list is empty.
iterator()	Returns an iterator for the linked list.
lastIndexOf(Object elem)	Returns the index position of the last occurrence of the specified object in the linked list. If the object isn't in the list, it returns –1.
offer(Object elem)	Adds the specified object to the end of the list. This method returns a boolean value, which is always true.
offerFirst(Object elem)	Adds the specified object to the front of the list. This method returns a boolean value, which is always true.
offerLast(Object elem)	Adds the specified object to the end of the list. This method returns a boolean value, which is always true.
peek()	Returns (but does not remove) the first element in the list. If the list is empty, it returns null.
peekFirst()	Returns (but does not remove) the first element in the list. If the list is empty, it returns null.
peekLast()	Returns (but does not remove) the last element in the list. If the list is empty, it returns null.
poll()	Retrieves the first element and removes it from the list. It returns the element that was retrieved or, if the list is empty, null.
pollFirst()	Retrieves the first element and removes it from the list. It returns the element that was retrieved or, if the list is empty, null.

Constructor	Explanation
pollLast()	Retrieves the last element and removes it from the list. It returns the element that was retrieved or, if the list is empty, null.
pop()	Pops an element from the stack represented by this list.
push(Object elem)	Pushes an element onto the stack represented by this list.
remove()	Retrieves the first element and removes it from the list. It returns the element that was retrieved. If the list is empty, it throws NoSuchElementException.
remove(int index)	Removes the object at the specified index and returns the element that was removed.
remove(Object elem)	Removes an object from the list. Note that if more than one element refers to the object, this method removes only one of them. It returns a boolean that indicates whether the object was in the list.
removeAll(Collection c)	Removes all the objects in the specified collection from this linked list.
removeFirst()	Retrieves the first element and removes it from the list. It returns the element that was retrieved. If the list is empty, it throws NoSuchElementException.
removeFirstOccurrence (Object elem)	Finds the first occurrence of elem in the list and removes this occurrence from the list. It returns false if the list has no such occurrence.
removeLast()	Retrieves the last element and removes it from the list. It returns the element that was retrieved. If the list is empty, it throws NoSuchElementException.
removeLastOccurrence (Object elem)	Finds the last occurrence of elem in the list and removes this occurrence from the list. It returns false if the list has no such occurrence.
retainAll(Collection c)	Removes all the objects that are not in the specified collection from this linked list.
set(int index, Object elem)	Sets the specified element to the specified object. The element that was previously at that position is returned as the method's return value.
size()	Returns the number of elements in the list.
toArray()	Returns the elements of the linked list as an array of objects (Object[]).
toArray(type[] array)	Returns the elements of the linked list as an array whose type is the same as the array passed via the parameter.

As you look over these methods, you'll find several that seem to do the same thing. These similar methods usually have subtle differences. The getFirst and peek methods, for example, both return the first element from the list without removing the element. The only difference is what happens if the list is empty. In that case, getFirst throws an exception, but peek returns null.

In some cases, however, the methods are identical, such as the remove and removeFirst methods. In fact, if you're crazy enough to look at the source code for the LinkedList class, you'll find that the remove method consists of a single line: a call to the removeFirst method.

Creating a LinkedList

As with any other kind of object, creating a linked list is a two-step affair. First, you declare a LinkedList variable; then you call one of the LinkedList constructors to create the object, as in this example:

```
LinkedList officers = new LinkedList();
```

Here a linked list is created and assigned to the variable officers.

TIP

You can (and should) use the generics feature to specify a type when you declare the linked list. Here's a statement that creates a linked list that holds strings:

```
LinkedList<String> officers = new LinkedList<String>();
```

Then you can add only String objects to this list. If you try to add any other type of object, the compiler balks. (Base runners advance.)

Adding Items to a LinkedList

The LinkedList class gives you many ways to add items to the list. The most basic is the add method, which works pretty much the same way that it does for the ArrayList class. Here's an example:

```
LinkedList<String> officers = new LinkedList<String>();
officers.add("Blake");
officers.add("Burns");
officers.add("Houlihan");
officers.add("Pierce");
```

```
officers.add("McIntyre");
for (String s : officers)
    System.out.println(s);
```

The add method adds these items to the end of the list. So the resulting output is this:

```
Blake
Burns
Houlihan
Pierce
McIntyre
```

The addLast method works the same way, but the addFirst method adds items to the front of the list. Consider these statements:

```
LinkedList<String> officers = new LinkedList<String>();
officers.addFirst("Blake");
officers.addFirst("Burns");
officers.addFirst("Houlihan");
officers.addFirst("Pierce");
officers.addFirst("McIntyre");
for (String s : officers)
    System.out.println(s);
```

Here the resulting output shows the officers in reverse order:

```
McIntyre
Pierce
Houlihan
Burns
Blake
```

To insert an object into a specific position into the list, specify the index in the add method, as in this example:

```
LinkedList<String> officers = new LinkedList<String>();
officers.add("Blake");
officers.add("Burns");
officers.add("Houlihan");
officers.add("Pierce");
officers.add("McIntyre");
officers.add(2, "Tuttle");
for (String s : officers)
    System.out.println(s);
```

The console output from these statements is this:

```
Blake
Burns
Tuttle
Houlihan
Pierce
McIntyre
```

(In case you're not a *M*A*S*H* fan, Tuttle was a fictitious officer that Hawkeye and Trapper made up in one episode so that they could collect his paychecks and donate the money to the local orphanage. Unfortunately, the ruse got out of hand. When Tuttle won a medal, and a general wanted to present it in person, they arranged for Tuttle to die in an unfortunate helicopter accident.)

Here are some other thoughts to consider when you ponder how to add elements to linked lists:

REMEMBER

>> If you specified a type for the list when you created it, the items you add must be of the correct type. The compiler kvetches if they aren't.

>> Like arrays and everything else in Java, linked lists are indexed starting with zero.

>> If you specify an index that doesn't exist, the add method throws IndexOutOf BoundsException. This is an unchecked exception, so you don't have to handle it.

TECHNICAL STUFF

>> LinkedList also has weird methods named offer, offerFirst, and offerLast. The offer method adds an item to the end of the list and has a return type of boolean, but it always returns true. The offer method is defined by the Queue interface, which LinkedList implements. Some classes that implement Queue can refuse to accept an object added to the list via offer. In that case, the offer method returns false. But because a linked list never runs out of room, the offer method always returns true to indicate that the object offered to the list was accepted.

Retrieving Items from a LinkedList

As with the ArrayList class, you can use the get method to retrieve an item based on its index. If you pass it an invalid index number, the get method throws the unchecked IndexOutOfBoundsException.

You can also use an enhanced `for` loop to retrieve all the items in the linked list. The examples in the preceding section use this enhanced `for` loop to print the contents of the `officers` linked list:

```
for (String s : officers)
    System.out.println(s);
```

If you want, you can also use the `iterator` method to get an iterator that can access the list. (For more information about iterators, refer to Book 4, Chapter 3.)

The `LinkedList` class also has a variety of other methods that retrieve items from the list. Some of these methods remove the items as they are retrieved; some throw exceptions if the list is empty; others return `null`.

Nine methods retrieve the first item in the list:

- » `getFirst`: Retrieves the first item from the list. This method doesn't delete the item. If the list is empty, `NoSuchElement-Exception` is thrown.

- » `element`: Identical to the `getFirst` method. This strangely named method exists because it's defined by the `Queue` interface, and the `LinkedList` class implements `Queue`.

- » `peek`: Similar to `getFirst` but doesn't throw an exception if the list is empty. Instead, it just returns `null`. (The `Queue` interface also defines this method.)

- » `peekFirst`: Identical to `peek`. Only the name of the method is changed to protect the innocent.

- » `remove`: Similar to `getFirst` but also removes the item from the list. If the list is empty, it throws `NoSuchElementException`.

- » `removeFirst`: Identical to `remove`. If the list is empty, it throws `NoSuchElementException`.

- » `poll`: Similar to `removeFirst` but returns `null` if the list is empty. (This method is yet another method that the `Queue` interface defines.)

- » `pollFirst`: Identical to `poll` (well, identical except for the name of the method).

- » `pop`: Identical to `removeFirst` (but with a catchier name).

Four methods also retrieve the last item in the list:

- » `getLast`: Retrieves the last item from the list. This method doesn't delete the item. If the list is empty, `NoSuchElement-Exception` is thrown.

>> peekLast: Similar to getLast but doesn't throw an exception if the list is empty. Instead, it just returns null.

>> removeLast: Similar to getLast but also removes the item. If the list is empty, it throws NoSuchElementException.

>> pollLast: Similar to removeLast but returns null if the list is empty.

Updating LinkedList Items

As with the ArrayList class, you can use the set method to replace an object in a linked list with another object. In that *M*A*S*H* episode in which Hawkeye and Trapper made up Captain Tuttle, they quickly found a replacement for him when he died in that unfortunate helicopter accident. Here's how Java implements that episode:

```
LinkedList<String> officers = new LinkedList<String>();
// add the original officers
officers.add("Blake");
officers.add("Burns");
officers.add("Tuttle");
officers.add("Houlihan");
officers.add("Pierce");
officers.add("McIntyre");
System.out.println(officers);
// replace Tuttle with Murdock
officers.set(2, "Murdock");
System.out.println("\nTuttle is replaced:");
System.out.println(officers);
```

The output from this code looks like this:

```
[Blake, Burns, Tuttle, Houlihan, Pierce, McIntyre]
Tuttle is replaced:
[Blake, Burns, Murdock, Houlihan, Pierce, McIntyre]
```

TIP

As with an ArrayList, any changes you make to an object retrieved from a linked list are automatically reflected in the list. That's because the list contains references to objects, not the objects themselves. (For more information about this issue, refer to Book 4, Chapter 3.)

Removing LinkedList Items

You've already seen that several of the methods that retrieve items from a linked list also remove the items. In particular, the `remove`, `removeFirst`, and `poll` methods remove the first item from the list, and the `removeLast` method removes the last item.

You can also remove any arbitrary item by specifying either its index number or a reference to the object you want to remove on the `remove` method. To remove item 3, for example, use a statement like this:

```
officers.remove(3);
```

If you have a reference to the item that you want to remove, use the `remove` method, like this:

```
officers.remove(tuttle);
```

To remove all the items from the list, use the `clear` method:

```
officers.clear();        // Goodbye, Farewell, and Amen.
```

» Using generics in your own classes

» Working with wildcards in a generic class

» Examining a pair of classes that demonstrate generics

Chapter **5**

Creating Generic Collection Classes

I n the previous two chapters, you've seen how you can specify the type for an ArrayList or a LinkedList so the compiler can prevent you from accidentally adding the wrong type of data to the collection. The ArrayList and LinkedList classes can do this because they take advantage of a feature called *generics*. Generics first became available in Java 1.5.

In this chapter, I show you how the generics feature works and how to put it to use in your own classes. Specifically, you see examples of two classes that use the LinkedList class to implement a specific kind of collection. The first is a *stack*, a collection in which items are always added to the front of the list and retrieved from the front of the list. The second is a *queue*, a collection in which items are added to the end of the list and retrieved from the front.

**TECHNICAL
STUFF**

This is one of those chapters where the entire chapter could get a Technical Stuff icon. Frankly, generics is on the leading edge of object-oriented programming. You can get by without knowing any of the information in this chapter, so feel free to skip it if you're on your way to something more interesting. However, this chapter is worth looking at, even if you just want to get an idea of how the ArrayList and LinkedList classes use the new generics feature. And you might find that someday you want to create your own generic classes. Your friends will surely think you're a genius.

To be sure, I won't be covering all the intricacies of programming with generics. If your next job happens to be writing Java class libraries for Oracle, you'll need to know a lot more about generics than this chapter covers. I focus just on the basics of writing simple generic classes.

Why Generics?

If you don't specify otherwise, collection classes can hold any type of object. For example, the add method for the ArrayList class had this declaration:

```
public boolean add(Object o)
{
    // code to implement the add method
}
```

Thus, you can pass any type of object to the add method — and the array list gladly accepts it.

When you retrieve an item from a collection, you must cast it to the correct object type before you can do anything with it. For example, if you have an array list named empList with Employee objects, you'd use a statement like this one to get the first Employee from the list:

```
Employee e = (Employee)empList.get(0);
```

The trouble is, what if the first item in the list isn't an Employee? Because the add method accepts any type of object, there's no way to guarantee that only certain types of objects could be added to the collection.

That's where generics come into play. With generics, you can declare the ArrayList like this:

```
ArrayList<Employee> empList = new ArrayList<Employee>();
```

Here empList is declared as an ArrayList that can hold only Employee types. Now the add method has a declaration that is the equivalent of this:

```
public boolean add(Employee o)
{
    // code to implement the add method
}
```

Thus you can only add `Employee` objects to the list. And the `get` method has a declaration that's equivalent to this:

```
public Employee get(int index)
{
    // code to implement the get method
}
```

Thus the `get` method returns `Employee` objects. You don't have to cast the result to an `Employee` because the compiler already knows the object is an `Employee`.

Creating a Generic Class

Generics let you create classes that can be used for any type specified by the programmer at compile time. To accomplish that, the Java designers introduced a new feature to the language, called *formal type parameters*. To create a class that uses a formal type parameter, you list the type parameter after the class name in angle brackets. The type parameter has a name — Oracle recommends you use single uppercase letters for type parameter names — that you can then use throughout the class anywhere you'd otherwise use a type.

For example, here's a simplified version of the class declaration for the `ArrayList` class:

```
public class ArrayList<E>
```

I left out the `extends` and `implements` clauses to focus on the formal type parameter: `<E>`. The E parameter specifies the type of the elements that are stored in the list. Oracle recommends the type parameter name E (for Element) for any parameter that specifies element types in a collection.

So consider this statement:

```
ArrayList<Employee> empList;
```

Here the E parameter is `Employee`, which simply means that the element type for this instance of the `ArrayList` class is `Employee`.

Now take a look at the declaration for the add method for the `ArrayList` class:

```
public boolean add(E o)
{
    // body of method omitted (thank you)
}
```

Where you normally expect to see a parameter type, you see the letter E. Thus, this method declaration specifies that the type for the o parameter is the type specified for the formal type parameter E. If E is Employee, that means the add method only accepts Employee objects.

So far, so good. Now take a look at how you can use a formal type parameter as a return type. Here's the declaration for the get method:

```
public E get(int index)
{
    // body of method omitted (you're welcome)
}
```

Here E is specified as the return type. That means that if E is Employee, this method returns Employee objects.

One final technique you need to know before moving on: You can use the formal type parameter within your class to create objects of any other class that accepts formal type parameters. For example, the clone method of the `ArrayList` class is written like this:

```
public Object clone()
{
    try
    {
        ArrayList<E> v = (ArrayList<E>) super.clone();
        v.elementData = (E[])new Object[size];
        System.arraycopy(elementData, 0,
            v.elementData, 0, size);
        v.modCount = 0;
        return v;
    }
    catch (CloneNotSupportedException e)
    {
        // this shouldn't happen since we're Cloneable
        throw new InternalError();
    }
}
```

You don't need to look much at the details in this method; just notice that the first statement in the try block declares an ArrayList of type <E>. In other words, the ArrayList class uses its own formal type parameter to create another array list object of the same type. If you think about it, that makes perfect sense. After all, that's what the clone method does: It creates another array list just like this one.

The key benefit of generics is that this typing happens at compile time. Thus, after you specify the value of a formal type parameter, the compiler knows how to do the type checking implied by the parameter. That's how it knows not to let you add String objects to an Employee collection.

A Generic Stack Class

Now that you've seen the basics of creating generic classes, in this section you look at a simple generic class that implements a stack. A *stack* is a simple type of collection that lets you add objects to the top of the collection and remove them from the top. I name this Stack class in this section GenStack, and it has five methods:

>> push: This method adds an object to the top of the stack.

>> pop: This method retrieves the top item from the stack. The item is removed from the stack in the process. If the stack is empty, this method returns null.

>> peek: This method lets you peek at the top item on the stack. In other words, it returns the top item without removing it. If the stack is empty, it returns null.

>> hasItems: This method returns a boolean value of true if the stack has at least one item in it.

>> size: This method returns an int value that indicates how many items are in the stack.

The GenStack class uses a LinkedList to implement the stack. For the most part, this class simply exposes the various methods of the LinkedList class using names that are more appropriate for a stack. The complete code for the GenStack class is shown in Listing 5-1.

LISTING 5-1: **The GenStack Class**

```java
import java.util.*;

public class GenStack<E>                                        →3
{
    private LinkedList<E> list = new LinkedList<E>();           →5

    public void push(E item)                                    →7
    {
        list.addFirst(item);
    }

    public E pop()                                              →12
    {
        return list.poll();
    }

    public E peek()                                             →17
    {
        return list.peek();
    }

    public boolean hasItems()                                   →22
    {
        return !list.isEmpty();
    }

    public int size()                                           →27
    {
        return list.size();
    }
}
```

The following paragraphs highlight the important details in this class:

» →3 The class declaration specifies the formal type parameter <E>. Thus users of this class can specify the type for the stack's elements.

» →5 This class uses a private LinkedList object list to keep the items stored in the stack. The LinkedList is declared with the same type as the GenStack class itself. Thus, if the E type parameter is Employee, the type for this LinkedList is Employee.

» →7 The push method accepts a parameter of type E. It uses the linked list's addFirst method to add the item to the beginning of the list.

>> →12 The pop method returns a value of type E. It uses the linked list's poll method, which removes and returns the first element in the linked list. If the list is empty, the poll method — and therefore the pop method — returns null.

>> →17 The peek method also returns a value of type E. It simply returns the result of the linked list's peek method.

>> →22 The hasItems method returns the opposite of the linked list's isEmpty method.

>> →27 The size method simply returns the result of the linked list's size method.

That's all there is to it. The following program gives the GenStack class a little workout to make sure it functions properly:

```
public class GenStackTest
{
    public static void main(String[] args)
    {
        GenStack<String> gs = new GenStack<String>();

        System.out.println(
            "Pushing four items onto the stack.");
        gs.push("One");
        gs.push("Two");
        gs.push("Three");
        gs.push("Four");

        System.out.println("There are "
            + gs.size() + " items in the stack.\n");

        System.out.println("The top item is: "
            +gs.peek() + "\n");

        System.out.println("There are still "
            + gs.size() + " items in the stack.\n");

        System.out.println("Popping everything:");
        while (gs.hasItems())
            System.out.println(gs.pop());

        System.out.println("There are now "
            + gs.size() + " items in the stack.\n");

        System.out.println("The top item is: "
            +gs.peek() + "\n");

    }
}
```

This program creates a GenStack object that can hold String objects. It then pushes four strings onto the stack and prints the number of items in the stack. Next, it uses the peek method to print the top item and again prints the number of items in the stack, just to make sure the peek method doesn't accidentally remove the item. Next, it uses a while loop to pop each item off the stack and print it. Then, once again, it prints the number of items (which should now be zero), and it peeks at the top item (which should be null).

Here's the output that results when you run this program:

```
Pushing four items onto the stack.
There are 4 items in the stack.

The top item is: Four

There are still 4 items in the stack.

Popping everything:
Four
Three
Two
One
There are now 0 items in the stack.

The top item is: null
```

TIP

Notice that when the program pops the items off the stack, they come out in reverse order in which they were pushed onto the stack. That's normal behavior for stacks. In fact, stacks are sometimes called *Last-In, First-Out* (LIFO) collections for this very reason.

Using Wildcard-Type Parameters

Suppose you have a method that's declared like this:

```
public void addItems(ArrayList<Object> list)
{
    // body of method not shown
}
```

Thought question: Does the following statement compile?

```
addItems(new ArrayList<String>());
```

Answer: Nope.

That's surprising because String is a subtype of Object. So you'd think that a parameter that says it accepts an ArrayList of objects accepts an ArrayList of strings.

Unfortunately, inheritance doesn't work quite that way when it comes to formal type parameters. Instead, you have to use another feature of generics, called *wildcards*.

In short, if you want to create a method that accepts any type of ArrayList, you have to code the method like this:

```
public void addItems(ArrayList<?> list)
```

In this case, the question mark indicates that you can code any kind of type here.

That's almost as good as inheritance, but what if you want to actually limit the parameter to collections of a specific superclass? For example, suppose you're working on a payroll system that has an Employee superclass with two subclasses named HourlyEmployee and SalariedEmployee, and you want this method to accept an ArrayList of Employee objects, HourlyEmployee objects, or Salaried Employee objects?

In that case, you can add an extends clause to the wildcard, like this:

```
public void addItems(ArrayList<? extends Employee> list)
```

Then you can call the addItems method with an ArrayList of type Employee, HourlyEmployee, or SalariedEmployee.

Alternatively, suppose you want to allow the parameter to accept HourlyEmployee or its superclass, Employee. You could code it like this:

```
public void addItems(ArrayList<? super HourlyEmployee> list)
```

Here, the parameter can be an HourlyEmployee or an Employee. But it can't be a SalariedEmployee, because SalariedEmployee is not a superclass of HourlyEmployee.

Now, before you call it a day, take this example one step further: Suppose this addItems method appears in a generic class that uses a formal type parameter <E> to specify the type of elements the class accepts, and you want the addItems method to accept an ArrayList of type E or any of its subclasses. To do that, you'd declare the addItems method like this:

```
public void addItems(ArrayList<? extends E> list)
```

Here the wildcard type parameter <? extends E> simply means that the Array List can be of type E or any type that extends E.

A Generic Queue Class

Now that you've seen how to use wildcards in a generic class, this section presents a generic class that implements a queue. A *queue* is another type of collection that lets you add objects to the end of the collection and remove them from the top. Queues are commonly used in all sorts of applications, from data processing applications to sophisticated networking systems.

This queue class is named GenQueue and has the following methods:

- » enqueue: This method adds an object to the end of the queue.

- » dequeue: This method retrieves the first item from the queue. The item is removed from the queue in the process. If the queue is empty, this method returns null.

- » hasItems: This method returns a boolean value of true if the queue has at least one item in it.

- » size: This method returns an int value that indicates how many items are in the stack.

- » addItems: This method accepts another GenQueue object as a parameter. All the items in that queue are added to this queue. In the process, all the items from the queue passed to the method are removed. The GenQueue parameter must be of the same type as this queue or a subtype of this queue's type.

The GenQueue class uses a LinkedList to implement its queue. The complete code for the GenQueue class is shown in Listing 5-2.

LISTING 5-2: **The GenQueue Class**

```
import java.util.*;

public class GenQueue<E>                                    →3
{
    private LinkedList<E> list = new LinkedList<E>();        →5

    public void enqueue(E item)                             →7
    {
        list.addLast(item);
    }

    public E dequeue()                                      →11
    {
        return list.poll();
    }

    public boolean hasItems()                               →16
    {
        return !list.isEmpty();
    }

    public int size()                                       →21
    {
        return list.size();
    }

    public void addItems(GenQueue<? extends E> q)           →26
    {
        while (q.hasItems())
            list.addLast(q.dequeue());
    }
}
```

The following paragraphs point out the highlights of this class:

>> →3 The class declaration specifies the formal type parameter <E>. Thus, users of this class can specify the type for the elements of the queue.

>> →5 Like the GenStack class, this class uses a private LinkedList object list to keep its items.

>> →7 The enqueue method accepts a parameter of type E. It uses the linked list's addLast method to add the item to the end of the queue.

>> →11 The dequeue method returns a value of type E. Like the pop method of the GenStack class, this method uses the linked list's poll method to return the first item in the list.

>> →16 The `hasItems` method returns the opposite of the linked list's `isEmpty` method.

>> →21 The `size` method returns the result of the linked list's `size` method.

>> →26 The `addItems` method accepts a parameter that must be another `GenQueue` object whose element type is either the same type as this `GenQueue` object's elements or a subtype of this `GenQueue` object's element type. This method uses a `while` loop to remove all the items from the `q` parameter and add them to this queue.

The following program exercises the `GenQueue` class:

```java
public class GenQueueTest
{
    public static void main(String[] args)
    {
        GenQueue<Employee> empList;
        empList = new GenQueue<Employee>();

        GenQueue<HourlyEmployee> hList;
        hList = new GenQueue<HourlyEmployee>();
        hList.enqueue(new HourlyEmployee(
            "Trump", "Donald"));
        hList.enqueue(new HourlyEmployee(
            "Gates", "Bill"));
        hList.enqueue(new HourlyEmployee(
            "Forbes", "Steve"));

        empList.addItems(hList);

        while (empList.hasItems())
        {
            Employee emp = empList.dequeue();
            System.out.println(emp.firstName
                + " " + emp.lastName);
        }
    }
}
class Employee
{
    public String lastName;
    public String firstName;

    public Employee() {}

    public Employee(String last, String first)
    {
```

```
        this.lastName = last;
        this.firstName = first;
    }

    public String toString()
    {
        return firstName + " " + lastName;
    }
}
class HourlyEmployee extends Employee
{
    public double hourlyRate;

    public HourlyEmployee(String last, String first)
    {
        super(last, first);
    }
}
```

This program begins by creating a GenQueue object that can hold Employee objects. This queue is assigned to a variable named empList.

Next, the program creates another GenQueue object. This one can hold Hourly Employee objects (HourlyEmployee is a subclass of Employee) and is assigned to a variable named hList.

Then three rookie employees are created and added to the hList queue. The addItems method of the empList queue is then called to transfer these employees from the hList queue to the empList queue. Because HourlyEmployee is a subclass of Employee, the addItems method of the empList queue accepts hList as a parameter.

Finally, a while loop is used to print the employees that are now in the empList queue.

When this program is run, the following is printed on the console:

```
Donald Trump
Bill Gates
Steve Forbes
```

Thus the addItems method successfully transferred the employees from the hlist queue, which held HourlyEmployee objects, to the empList queue, which holds Employee objects.

Chapter **6**

Using Bulk Data Operations with Collections

One of the most common things to do with a collection is to iterate over it, performing some type of operation on all of its elements. For example, you might use a for each loop to print all of the elements. The body of the foreach loop might contain an if statement to select which elements to print. Or it might perform a calculation such as accumulating a grand total or counting the number of elements that meet a given condition.

In Java, for each loops are easy to create and can be very powerful. However, they have one significant drawback: They iterate over the collection's elements one at a time, beginning with the first element and proceeding sequentially to the last element. As a result, a for each loop must be executed sequentially within a single tread.

That's a shame, since modern computers have multicore processors that are capable of doing several things at once. Wouldn't it be great if you could divide a for each loop into several parts, each of which can be run independently on one of the processor cores? For a small collection, the user probably wouldn't notice the difference. But if the collection is extremely large (say, a few million elements), unleashing the power of a multicore processor could make the program run much faster.

While you can do that with earlier versions of Java, the programming is tricky. You have to master one of the more difficult aspects of programming in Java: working with threads, which are like separate sections of your program that can be executed simultaneously. You'll learn the basics of working with threads in Book 5, Chapter 1. For now, take my word for it: Writing programs that work with large collections of data and take advantage of multiple threads was a difficult undertaking. At least until Java 8.

With Java 8 or later, you can use a feature called *bulk data operations* that's designed specifically to attack this very problem. When you use bulk data operations, you do not directly iterate over the collection data using a for each loop. Instead, you simply provide the operations that will be done on each of the collection's elements and let Java take care of the messy details required to spread the work over multiple threads.

At the heart of the bulk data operations feature is a new type of object called a *stream*, defined by the Stream interface. A stream is simply a sequence of elements of any data type which can be processed sequentially or in parallel. The Stream interface provides methods that let you perform various operations such as filtering the elements or performing an operation on each of the elements.

TIP

File streams are used to read and write data to disk files. The streams described in this chapter are used to process data extracted from collection classes.

Streams rely on the use of lambda expressions to pass the operations that are performed on stream elements. In fact, the primary reason Java's developers introduced lambda expressions into the Java language was to facilitate streams. If you haven't yet read Book 3, Chapter 7, I suggest you do so now, before reading further into this chapter. Otherwise you'll find yourself hopelessly confused by the peculiar syntax of the lambda expressions used throughout this chapter.

In this chapter, you learn the basics of using streams to perform simple bulk data operations.

Looking At a Basic Bulk Data Operation

Suppose you have a list of spells used by a certain wizard who, for copyright purposes, we'll refer to simply as HP. The spells are represented by a class named Spell, which is defined as follows:

```
public class Spell
{

    public String name;
    public SpellType type;
    public String description;

    public enum SpellType {SPELL, CHARM, CURSE}

    public Spell(String spellName, SpellType spellType,
        String spellDescription)
    {
        name = spellName;
        type = spellType;
        description = spellDescription;
    }

    public String toString()
    {
        return name;
    }
}
```

As you can see, the Spell class has three public fields that represent the spell's name, type (SPELL, CHARM, or CURSE), and description, as well as a constructor that lets you specify the name, type, and description for the spell. Also, the toString() method is overridden to return simply the spell name.

Let's load a few of HP's spells into an ArrayList:

```
ArrayList<Spell> spells = new ArrayList<>();
spells.add(new Spell("Aparecium", Spell.SpellType.SPELL,
    "Makes invisible ink appear."));
spells.add(new Spell("Avis", Spell.SpellType.SPELL,
    "Launches birds from your wand."));
spells.add(new Spell("Engorgio", Spell.SpellType.CHARM,
    "Enlarges something."));
spells.add(new Spell("Fidelius", Spell.SpellType.CHARM,
    "Hides a secret within someone."));
```

```
spells.add(new Spell("Finite Incatatum", Spell.SpellType.SPELL,
    "Stops all current spells."));
spells.add(new Spell("Locomotor Mortis", Spell.SpellType.CURSE,
    "Locks an opponent's legs."));
```

Now, suppose you want to list the name of each spell on the console. You could do that using a for each loop like this:

```
for (Spell spell : spells)
    System.out.println(spell.name);
```

Written with streams, the code would look like this:

```
spells.stream().forEach(s -> System.out.println(s));
```

Here, I first use the stream method of the ArrayList class to convert the Array List to a stream. All of the classes that inherit from java.Collection implement a stream method that returns a Stream object. That includes not only ArrayList, but also LinkedList and Stack.

Next, I use the stream's forEach method to iterate the stream, passing a lambda expression that calls System.out.println for each item in the stream. The forEach method processes the entire stream, writing each element to the console.

Suppose you want to list just the spells, not the charms or curses. Using a traditional for each loop, you'd do it like this:

```
for (Spell spell : spells)
{
    if (spell.type == Spell.SpellType.SPELL)
        System.out.println(spell.name);
}
```

Here an if statement selects just the spells so that the charms and curses aren't listed.

Here's the same thing using streams:

```
spells.stream()
    .filter(s -> s.type == Spell.SpellType.SPELL)
    .forEach(s -> System.out.println(s));
```

In this example, the stream method converts the ArrayList to a stream. Then the stream's filter method is used to select just the SPELL items. Finally, the forEach method sends the selected items to the console. Notice that lambda expressions are used in both the forEach method and the filter method.

The `filter` method of the `Stream` class returns a `Stream` object. Thus, it is possible to apply a second filter to the result of the first filter, like this:

```
spells.parallelStream()
    .filter(s -> s.type == Spell.SpellType.SPELL)
    .filter(s -> s.name.toLowerCase().startsWith("a"))
    .forEach(s -> System.out.println(s));
```

In this example, just the spells that start with the letter A are listed.

TECHNICAL STUFF

The term *pipeline* is often used to describe a sequence of method calls that start by creating a stream, then manipulate the stream in various ways by calling methods such as `filter`, and finally end by calling a method that does not return another stream object, such as `forEach`.

Looking Closer at the Stream Interface

The `Stream` interface defines about 40 methods. In addition, three related interfaces — `DoubleStream`, `IntStream`, and `LongStream` — extend the `Stream` interface to define operations that are specific to a single data type: `double`, `int`, and `long`. Table 6-1 lists the most commonly used methods of these interfaces.

TABLE 6-1   **The Stream and Related Interfaces**

Methods that Return Streams	Explanation
`Stream distinct()`	Returns a stream consisting of distinct elements of the input stream. In other words, duplicates are removed.
`Stream limit(long maxSize)`	Returns a stream having no more than `maxSize` elements derived from the input stream.
`Stream filter(Predicate<? super T> predicate)`	Returns a stream consisting of those elements in the input stream that match the conditions of the predicate.
`Stream sorted()`	Returns the stream elements in sorted order using the natural sorting method for the stream's data type.
`Stream sorted(Comparator<? super T> comparator)`	Returns the stream elements in sorted order using the specified `Comparator` function. The `Comparator` interface accepts two parameters and returns a negative value if the first is less than the second, zero if they are equal, and a positive value if the first is greater than the second.

(continued)

TABLE 6-1 *(continued)*

Mapping Methods	Explanation
`<R> Stream<R> map(Function<? super T,? extends R> mapper`	Returns a stream created by applying the mapper function to each element of the input stream.
`DoubleStream mapToDouble(ToDoubleFunction<? super T> mapper)`	Returns a `DoubleStream` created by applying the mapper function to each element of the input stream.
`IntStream mapToInt(ToIntFunction<? super T> mapper)`	Returns an `IntStream` created by applying the mapper function to each element of the input stream.
`LongStream mapToLong(ToLongFunction<? super T> mapper)`	Returns a `LongStream` created by applying the mapper function to each element of the input stream.

Terminal and Aggregate Methods	Explanation
`void forEach(Consumer<? super T> action)`	Executes the action against each element of the input stream.
`void forEachOrdered (Consumer<? super T> action)`	Executes the action against each element of the input stream, ensuring that the elements of the input stream are processed in order.
`long count()`	Returns the number of elements in the stream.
`Optional<T> max(Comparator<? super T> comparator)`	Returns the largest element in the stream.
`Optional<T> min(Comparator<? super T> comparator)`	Returns the smallest element in the stream.
`OptionalDouble average()`	Returns the average value of the elements in the stream. Valid only for `DoubleStream`, `IntStream`, and `Longstream`.
`resultType sum()`	Returns the sum of the elements in the stream. Result type is `double` for `DoubleStream`, `int` for `IntStream`, and `long` for `LongStream`.
`resultType summaryStatistics()`	Returns a summary statistics object that includes property methods named `getCount`, `getSum`, `getAverage`, `getMax`, and `getMmin` of the elements in the stream. The result type is `IntSummaryStatistics` for an `IntStream`, `DoubleSummaryStatistics` for a `DoubleStream`, and `LongSummaryStatistics` for a `LongStream`.

The first group of methods in Table 6-1 define methods that return other Stream objects. Each of these methods manipulates the stream in some way, then passes the altered stream down the pipeline to be processed by another operation.

The filter method is one of the most commonly used stream methods. It's argument, called a *predicate*, is a function that returns a boolean value. The function is called once for every element in the stream and is passed a single argument that contains the element under question. If the method returns true, the element is passed on to the result stream. If it returns false, the element is *not* passed on.

The easiest way to implement a filter predicate is to use a lambda expression that specifies a conditional expression. For example, the following lambda expression inspects the name field of the stream element and returns true if it begins with the letter *a* (upper- or lowercase):

```
s -> s.name.toLowerCase().startsWith("a")
```

The other methods in the first group let you limit the number of elements in a stream or sort the elements of the stream. To sort a stream, you can use either the element's natural sorting order, or you can supply your own comparator, either as a function or as an object that implements the Comparator interface.

The second group of methods in Table 6-1 are called *mapping methods* because they convert a stream whose elements are of one type to a stream whose elements are of another type. The mapping function, which you must pass as a parameter, is responsible for converting the data from the first type to the second.

One common use for mapping methods is to convert a stream of complex types to a stream of simple numeric values of type double, int, or long, which you can then use to perform an aggregate calculation such as sum or average. For example, suppose HP's spells were for sale and the Spell class included a public field named price. To calculate the average price of all the spells, you would first have to convert the stream of Spell objects to a stream of doubles. To do that, you use the mapToDouble method. The mapping function would simply return the price field:

```
.mapToDouble(s -> s.price)
```

Methods in the last group in Table 6-1 are called *terminal methods* because they do not return another stream. As a result, they are always the last methods called in stream pipelines. Note that if you don't call a terminal method, no data from the stream will be processed — the terminal method is what gets the ball rolling.

You have already seen the forEach method in action; it provides a function that is called once for each element in the stream. Note that in the examples so far, the function to be executed on each element has consisted of just a single method call, so I've included it directly in the lambda expression. If the function is more complicated, you can isolate it in its own method. Then the lambda expression should call the method that defines the function.

Aggregate methods perform a calculation on all of the elements in the stream, then return the result. Of the aggregate methods, count is straightforward: It simply returns the number of elements in the stream. The other aggregate methods need a little explanation because they return an optional data type. An *optional data type* is an object that might contain a value, or it might not.

For example, the average method calculates the average value of a stream of integers, longs, or doubles and returns the result as an OptionalDouble. If the stream was empty, the average is undefined, so the OptionalDouble contains no value. You can determine if the OptionalDouble contains a value by calling its isPresent method, which returns true if there is a value present. If there is a value, you can get it by calling the getAsDouble method.

WARNING

Note that getAsDouble will throw an exception if no value is present, so you should always call isPresent before you call getAsDouble.

Here's an example that calculates the average price of spells:

```
OptionalDouble avg = spells.stream()
        .mapToDouble(s -> s.price)
        .average();
```

Here is how you would write the average price to the console:

```
if (avg.isPresent())
{
    System.out.println("Average = "
        + avg.getAsDouble());
}
```

Using Parallel Streams

Streams come in two basic flavors: *sequential* and *parallel*. Elements in a sequential stream are produced by the stream method and create streams that are processed one element after the next. Parallel streams, in contrast, can take full advantage of multicore processors by breaking its elements into two or more smaller streams, performing operations on them, and then recombining the separate streams to create the final result stream. Each of the intermediate streams can be processed by a separate thread, which can improve performance for large streams.

By default, streams are sequential. But creating a parallel stream is easy: Just use the `parallelStream` method instead of the `stream` method at the beginning of the pipeline.

For example, to print all of HP's spells using a parallel stream, use this code:

```
spells.parallelStream()
    .forEach(s -> System.out.println(s));
```

Note that when you use a parallel stream, you can't predict the order in which each element of the stream is processed. That's because when the stream is split and run on two or more threads, the order in which the processor executes the threads is not predictable.

To demonstrate this point, consider this simple example:

```
System.out.println("First Parallel stream: ");
spells.parallelStream()
    .forEach(s -> System.out.println(s));
System.out.println("\nSecond Parallel stream: ");
spells.parallelStream()
    .forEach(s -> System.out.println(s));
```

When you execute this code, the results will look something like this:

```
First parallel stream:
Fidelius
Finite Incatatum
Engorgio
Locomotor Mortis
Aparecium
Avis

Second parallel stream:
Fidelius
Engorgio
Finite Incatatum
Locomotor Mortis
Avis
Aparecium
```

Notice that although the same spells are printed for each of the streams, they are printed in a different order.

5 Programming Techniques

Contents at a Glance

Chapter **1**

Programming Threads

Have you ever seen a plate-spinning act, in which a performer spins plates or bowls on top of poles, keeping multiple plates spinning at the same time, running from pole to pole to give each plate a little nudge — just enough to keep it going? The world record is 108 plates kept simultaneously spinning.

In Java, *threads* are the equivalent of plate spinning. Threads let you divide the work of an application into separate pieces, all of which then run simultaneously. The result is a faster and more efficient program, but along with the increased speed come more difficult programming and debugging.

Truthfully, the subtleties of threaded programming are a topic for computer science majors, but the basics of working with threads aren't all that difficult to understand. In this chapter, I focus on those basics and leave the advanced techniques for the grad students.

WARNING

The main application I use to illustrate threading in this chapter simulates the countdown clock for the spacecraft. Working with threads isn't really rocket science, but threading is used to solve difficult programming problems. You invariably find yourself trying to get two or more separate pieces of code to coordinate their activities, and that's not as easy as you might think at first guess. As a result,

I can't possibly talk about threading without getting into some challenging mental exercises, so be prepared to spend some mental energy figuring out how it works.

TIP

The listings in this chapter, as well as throughout the book, are available at www. dummies.com/go/javaaiofd6e.

Understanding Threads

A *thread* is a single sequence of executable code within a larger program. All the programs shown so far in this book have used just one thread — the *main thread* that starts automatically when you run the program — but Java lets you create programs that start additional threads to perform specific tasks.

You're probably familiar with programs that use threads to perform several tasks at the same time. Here are some common examples:

>> Web browsers can download files while letting you view web pages. When you download a file in a web browser, the browser starts a separate thread to handle the download.

>> Email programs don't make you wait for all your messages to download before you can read the first message. Instead, these programs use separate threads to display and download messages.

>> Word processors can print long documents in the background while you continue to work. These programs start a separate thread to handle print jobs.

>> Word processors can also check your spelling as you type. Depending on how the word processor is written, it may run the spell check in a separate thread.

>> Game programs commonly use several threads to handle different parts of the game to improve the overall responsiveness of the game.

TIP

>> All GUI-based programs use at least two threads — one thread to run the application's main logic and another thread to monitor mouse and keyboard events. You find out about creating GUI programs in Java in Book 6.

>> Indeed, the Java Virtual Machine itself uses threading for some of its house-keeping chores. The garbage collector, for example, runs as a separate thread so it can constantly monitor the state of the VM's memory and decide when it needs to create some free memory by removing objects that are no longer being used.

Creating a Thread

Suppose you're developing software for NASA, and you're in charge of the program that controls the final 20 seconds of the countdown for a manned spacecraft. Your software has to coordinate several key events that occur when the clock reaches certain points:

>> **T minus 16 seconds:** Flood launch pad. This event releases 350,000 gallons of water onto the launch pad, which helps protect the spacecraft systems during launch.

>> **T minus 6 seconds:** Start the main engines. Huge clamps hold the spacecraft in place while the engines build up thrust.

>> **T minus 0:** Lift off! The clamps are released, and the spacecraft flies into space.

For this program, I don't actually start any rocket engines or release huge amounts of water. Instead, I just display messages on the console to simulate these events. But I do create four separate threads to make everything work. One thread manages the countdown clock. The other three threads fire off their respective events at T minus 16 seconds (flood the pad), T minus 6 seconds (fire the engines), and T minus 0 (launch).

For the first attempt at this program, I just get the countdown clock up and running. The countdown clock is represented by a class named CountDownClock. All this class does is count down from 20 to 0 at 1-second intervals, displaying messages such as T minus 20 on the console as it counts. This version of the program doesn't do much of anything, but it does demonstrate how to get a thread going. We'll start by looking at the Thread class.

Understanding the Thread class

The Thread class lets you create an object that can be run as a thread in a multithreaded Java application. The Thread class has quite a few constructors and methods, but for most applications, you need to use only the ones listed in Table 1-1. (Note that this table is here to give you an overview of the Thread class and to serve as a reference. Don't worry about the details of each constructor and method just yet. By the end of this chapter, I explain each of the constructors and methods.)

TABLE 1-1   Constructors and Methods of the Thread Class

Constructor	Explanation
`Thread()`	Creates an instance of the `Thread` class. This constructor is the basic `Thread` constructor without parameters.
`Thread(String name)`	Creates a `Thread` object and assigns the specified name to the thread.
`Thread(Runnable target)`	Turns any object that implements an API interface called `Runnable` into a thread. You see how this more-advanced constructor is used later in this chapter.
`Thread(Runnable target, String name)`	Creates a thread from any object that implements `Runnable` and assigns the specified name to the thread.
`static int activeCount()`	Returns the number of active threads.
`static int enumerate(Thread[] t)`	Fills the specified array with a copy of each active thread. The return value is the number of threads added to the array.
`String getName()`	Returns the name of the thread.
`int getPriority()`	Returns the thread's priority.
`void interrupt()`	Interrupts this thread.
`boolean isInterrupted()`	Checks whether the thread has been interrupted.
`void setPriority(int priority)`	Sets the thread's priority.
`void setName(String name)`	Sets the thread's name.
`static void Sleep`	Causes the currently executing thread `(int milliseconds)` to sleep for the specified number of milliseconds.
`void run()`	Is called when the thread is started. Place the code that you want the thread to execute inside this method.
`void start()`	Starts the thread.
`static void yield()`	Causes the currently executing thread to yield to other threads that are waiting to execute.

Extending the Thread class

The easiest way to create a thread is to write a class that extends the `Thread` class. Then all you have to do to start a thread is create an instance of your thread class and call its `start` method.

Listing 1-1 is a version of the `CountDownClock` class that extends the `Thread` class.

LISTING 1-1: **The CountDownClock Class (Version 1)**

```
public class CountDownClock extends Thread                      →1
{
    public void run()                                           →3
    {
        for (int t = 20; t >= 0; t--)                           →5
        {
            System.out.println("T minus " + t);
            try
            {
                Thread.sleep(1000);                             →10
            }
            catch (InterruptedException e)
            {}
        }
    }
}
```

Here are a few key points to notice in this class:

» →1 The CountDownClock class extends the Thread class. Thread is defined in the java.lang package, so you don't have to provide an import statement to use it.

» →3 The CountDownClock class has a single method, named run. This method is called by Java when the clock thread has been started. All the processing done by the thread must either be in the run method or in some other method called by the run method.

» →5 The run method includes a for loop that counts down from 20 to 0.

» →10 The CountDownClock class uses the sleep method to pause for 1 second. Because the sleep method throws Interrupted Exception, a try/catch statement handles this exception. If the exception is caught, it is simply ignored.

REMEMBER

At some point in its execution, the run method should either call sleep or yield to give other threads a chance to execute.

Programming Threads

Creating and starting a thread

After you define a class that defines a `Thread` object, you can create and start the thread. Here's the main class for the first version of the countdown application:

```java
public class CountDownApp
{
    public static void main(String[] args)
    {
        Thread clock = new CountDownClock();
        clock.start();
    }
}
```

Here a variable of type `Thread` is declared, and an instance of the `Count Down-Clock` is created and assigned to it. This creates a `Thread` object, but the thread doesn't begin executing until you call its `start` method.

When you run this program, the thread starts counting down in 1-second increments, displaying messages such as the following on the console:

```
T minus 20
T minus 19
T minus 18
```

And so on, all the way to zero. So far, so good.

Implementing the Runnable Interface

For the threads that trigger specific countdown events such as flooding the launch pad, starting the events, and lifting off, I create another class called `LaunchEvent`. This class uses another technique for creating and starting threads — one that requires a few more lines of code but is more flexible.

The problem with creating a class that extends the `Thread` class is that a class can have one superclass. What if you'd rather have your thread object extend some other class? In that case, you can create a class that implements the `Runnable` interface rather than extends the `Thread` class. The `Runnable` interface marks an object that can be run as a thread. It has only one method, `run`, that contains the code that's executed in the thread. (The `Thread` class itself implements `Runnable`, which is why the `Thread` class has a `run` method.)

Using the Runnable interface

To use the Runnable interface to create and start a thread, you have to do the following:

1. **Create a class that implements Runnable.**

2. **Provide a run method in the class you created in Step 1.**

3. **Create an instance of the Thread class and pass your Runnable object to its constructor as a parameter.**

A Thread object is created that can run your Runnable class.

4. **Call the Thread object's start method.**

The run method of your Runnable object is called and executes in a separate thread.

The first two of these steps are easy. The trick is in the third and fourth steps, because you can complete them in several ways. Here's one way, assuming that your Runnable class is named RunnableClass:

```
RunnableClass rc = new RunnableClass();
Thread t = new Thread(rc);
t.start();
```

Java programmers like to be as concise as possible, so you often see this code compressed to something more like

```
Thread t = new Thread(new RunnableClass());
t.start();
```

or even just this:

```
new Thread(new RunnableClass()).start();
```

This single-line version works — provided that you don't need to access the thread object later in the program.

Creating a class that implements Runnable

To sequence the launch events for the NASA application, I create a Runnable object named LaunchEvent. The constructor for this class accepts two parameters: the countdown time at which the event fires and the message that is displayed when the time arrives. The run method for this class uses Thread.sleep to wait until the desired time arrives. Then it displays the message.

Listing 1-2 shows the code for this class.

LISTING 1-2:	The LaunchEvent Class (Version 1)

```
public class LaunchEvent implements Runnable                          →1
{
    private int start;
    private String message;

    public LaunchEvent(int start, String message)                    →6
    {
        this.start = start;
        this.message = message;
    }

    public void run()
    {
        try
        {
            Thread.sleep(20000 - (start * 1000));                     →16
        }
        catch (InterruptedException e)
        {}
        System.out.println(message);                                 →20
    }
}
```

The following paragraphs draw your attention to the listing's key lines:

» →1 This class implements the Runnable interface.

» →6 The constructor accepts two parameters: an integer representing the start time (in seconds) and a string message that's displayed when the time arrives. The constructor simply stores these parameter values in private fields.

» →16 In the run method, the Thread.sleep method is called to put the thread to sleep until the desired countdown time arrives. The length of time that the thread should sleep is calculated by the expression 20000 - (start * 1000). The countdown clock starts at 20 seconds, which is 20,000 milliseconds. This expression simply subtracts the number of milliseconds that corresponds to the desired start time from 20,000. Thus, if the desired start time is 6 seconds, the sleep method sleeps for 14,000 milliseconds — that is, 14 seconds.

» →20 When the thread wakes up, it displays the message passed via its constructor on the console.

Using the CountDownApp class

Now that you've seen the code for the LaunchEvent and CountDownClock classes, Listing 1-3 shows the code for a CountDownApp class that uses these classes to launch a spacecraft.

LISTING 1-3: **The CountDownApp Class (Version 2)**

```
public class CountDownApp
{
    public static void main(String[] args)
    {
        Thread clock = new CountDownClock();               →5

        Runnable flood, ignition, liftoff;                 →7
        flood = new LaunchEvent(16, "Flood the pad!");
        ignition = new LaunchEvent(6, "Start engines!");
        liftoff = new LaunchEvent(0, "Liftoff!");

        clock.start();                                     →12

        new Thread(flood).start();                         →14
        new Thread(ignition).start();
        new Thread(liftoff).start();
    }
}
```

The following paragraphs summarize how this program works:

- » →5 The main method starts by creating an instance of the Count DownClock class and saving it in the clock variable.

- » →7 Next, it creates three LaunchEvent objects to flood the pad at 16 seconds, start the engines at 6 seconds, and lift off at 0 seconds. These objects are assigned to variables of type Runnable named flood, ignition, and liftoff.

- » →12 The clock thread is started. The countdown starts ticking.

- » →14 Finally, the program starts the three LaunchEvent objects as threads. It does this by creating a new instance of the Thread class, passing the LaunchEvent objects as parameters to the Thread constructor, and then calling the start method to start the thread. Note that because this program doesn't need to do anything with these threads after they're started, it doesn't bother creating variables for them.

When you run this program, output similar to the following is displayed on the console:

```
T minus 20
T minus 19
T minus 18
T minus 17
T minus 16
Flood the pad!
T minus 15
T minus 14
T minus 13
T minus 12
T minus 11
T minus 10
T minus 9
T minus 8
T minus 7
T minus 6
Start engines!
T minus 5
T minus 4
T minus 3
T minus 2
T minus 1
Liftoff!
T minus 0
```

As you can see, the LaunchEvent messages are interspersed with the Count DownClock messages. Thus, the launch events are triggered at the correct times.

Note that the exact order in which some of the messages appear may vary slightly. For example, "Flood the pad!" might sometimes come before "T minus 16" because of slight variations in the precise timing of these independently operating threads. Later in this chapter, the section "Creating Threads That Work Together" shows you how to avoid such inconsistencies.

TIP

You can improve the main method for this class by using an ArrayList to store the Runnable objects. Then you can start all the LaunchEvent threads by using an enhanced for loop. Here's what the improved code looks like:

```
public static void main(String[] args)
{
    Thread clock = new CountDownClock();
    ArrayList<Runnable> events
        = new ArrayList<Runnable>();
    events.add(new LaunchEvent(16, "Flood the pad!"));
```

```
        events.add(new LaunchEvent(6, "Start engines!"));
        events.add(new LaunchEvent(0, "Liftoff!"));
        clock.start();
        for (Runnable e : events)
            new Thread(e).start();
    }
```

The advantage of this technique is that you don't need to create a separate variable for each LaunchEvent. (Don't forget to add an import statement for the java.util.* to gain access to the ArrayList class.)

Creating Threads That Work Together

Unfortunately, the countdown application presented in the preceding section has a major deficiency: The CountDownClock and LaunchEvent threads depend strictly on timing to coordinate their activities. After these threads start, they run independently of one another. As a result, random variations in their timings can cause the thread behaviors to change. If you run the program several times in a row, you'll discover that sometimes the Start engines! message appears after the T minus 6 message, and sometimes it appears *before* the T minus 6 message. That might not seem like a big deal to you, but it probably would be disastrous for the astronauts on the spacecraft. What these classes really need is a way to communicate.

Listing 1-4 shows an improved version of the countdown application that incorporates several enhancements. The CountDownClock class in this version adds a new method named getTime that gets the current time in the countdown. Then the LaunchEvent class checks the countdown time every 10 milliseconds and triggers the events only when the countdown clock actually says that it's time. This version of the application runs consistently.

In addition, you want to enable the LaunchEvent class to monitor the status of the CountDownClock, but you don't want to couple the LaunchEvent and CountDownClock classes too closely. Suppose that later, you develop a better countdown clock. If the LaunchEvent class knows what class is doing the counting, you have to recompile it if you use a different countdown class.

The solution is to use an interface as a buffer between the classes. This interface defines a method that gets the current status of the clock. Then the Count DownClock class can implement this interface, and the LaunchEvent class can use any object that implements this interface to get the time.

LISTING 1-4: **The Coordinated CountDown Application**

```java
import java.util.ArrayList;
// version 2.0 of the Countdown application

public class CountDownApp
{
    public static void main(String[] args)
    {
        CountDownClock clock = new CountDownClock(20);          →8
        ArrayList<Runnable> events =
            new ArrayList<Runnable>();                          →10

        events.add(new LaunchEvent(16,                          →12
            "Flood the pad!", clock));
        events.add(new LaunchEvent(6,
            "Start engines!", clock));
        events.add(new LaunchEvent(0,
            "Liftoff!", clock));

        clock.start();                                         →19

        for (Runnable e : events)                              →21
            new Thread(e).start();
    }
}

interface TimeMonitor                                          →26
{
    int getTime();
}

class CountDownClock extends Thread
    implements TimeMonitor                                     →32
{
    private int t;                                             →34

    public CountDownClock(int start)                           →36
    {
        this.t = start;
    }

    public void run()
    {
        for (; t >= 0; t--)                                   →43
        {
            System.out.println("T minus " + t);
            try
```

```
                {
                    Thread.sleep(1000);
                }
                catch (InterruptedException e)
                {}
            }
        }

    public int getTime()                                                    →55
    {
        return t;
    }
}

class LaunchEvent implements Runnable                                        →61
{
    private int start;
    private String message;
    TimeMonitor tm;                                                         →65

    public LaunchEvent(int start, String message,
        TimeMonitor monitor)
    {
        this.start = start;
        this.message = message;
        this.tm = monitor;
    }

    public void run()
    {
        boolean eventDone = false;
        while (!eventDone)
        {
            try
            {
                Thread.sleep(10);                                           →82
            }
            catch (InterruptedException e)
            {}
            if (tm.getTime() <= start)                                      →86
            {
                System.out.println(this.message);
                eventDone = true;
            }
        }
    }
}
```

The following paragraphs describe the high points of this version:

» →8 As you see in line 35, the constructor for the CountDownClock class now accepts a parameter to specify the starting time for the countdown. As a result, this line specifies 20 as the starting time for the CountDownClock object.

» →10 An ArrayList of LaunchEvent objects is used to store each launch event.

» →12 The lines that create the LaunchEvent objects pass the CountDownClock object as a parameter to the LaunchEvent constructor. That way the LaunchEvent objects can call the clock's abort method if necessary.

» →19 The clock is started!

» →21 An enhanced for loop starts threads to run the LaunchEvent objects.

» →26 The TimeMonitor interface defines just one method, named getTime. This method returns an integer that represents the number of seconds left on the countdown timer.

» →32 The CountDownClock class implements the TimeMonitor interface.

» →34 A private field named t is used to store the current value of the count-down clock. That way, the current clock value can be accessed by the con-structor, the run method, and the getTime method.

» →36 The constructor for the CountDownClock class accepts the starting time for the countdown as a parameter. Thus, this countdown clock doesn't have to start at 20 seconds. The value passed via this parameter is saved in the t field.

» →43 The for loop in the run method tests and decrements the t variable. But because this variable is already initialized, it doesn't have an initialization expression.

» →55 The getTime() method simply returns the value of the t variable.

» →61 This line is the start of the LaunchEvent class.

» →65 A private field of type TimeMonitor is used to access the countdown clock. A reference to this object is passed to the LaunchEvent class via its constructor. The constructor simply stores that reference in this field.

» →82 The while loop includes a call to Thread.sleep that sleeps for just 10 milliseconds. Thus, this loop checks the countdown clock every 10 milliseconds to see whether its time has arrived.

» →86 This statement calls the getTime method of the countdown clock to see whether it's time to start the event. If so, a message is displayed, and eventDone is set to true to terminate the thread.

Using an Executor

The countdown application in Listings 1-1 through 1-4 uses Java's original threading mechanisms — tools that were available in the mid-1990s, when Java was in diapers. Since then, Java programmers have longed for newer, more sophisticated threading techniques. The big breakthrough came in 2004, with the release of Java 1.5. The Java API gained a large assortment of classes for fine-grained control of the running of threads. Since then, subsequent releases of Java have added even more classes to give you even better control over how threads execute.

A full discussion of Java threading would require another 850 pages. (How about *Java Threading All-in-One For Masochists*?) This chapter presents only a small sampling of these Java threading features.

Listings 1-5 through 1-7 repeat the work done by Listings 1-1 through 1-4, but Listings 1-5 through 1-7 use Java 1.5 threading classes.

LISTING 1-5: **A New CountDownClock**

```java
public class CountDownClockNew implements Runnable
{
    int t;
    public CountDownClockNew(int t)
    {
        this.t = t;
    }
    public void run()
    {
        System.out.println("T minus " + t);
    }
}
```

LISTING 1-6: **A New Event Launcher**

```java
public class LaunchEventNew implements Runnable
{
    private String message;
    public LaunchEventNew(String message)
    {
        this.message = message;
    }
    public void run()
```

(continued)

LISTING 1-6: *(continued)*

```
        {
            System.out.println(message);
        }
    }
}
```

LISTING 1-7: **A New CountDown Application**

```
import java.util.concurrent.ScheduledThreadPoolExecutor;
import java.util.concurrent.TimeUnit;
class CountDownAppNew
{
    public static void main(String[] args)
    {
        ScheduledThreadPoolExecutor pool =
            new ScheduledThreadPoolExecutor(25);
        Runnable flood, ignition, liftoff;
        flood = new LaunchEventNew("Flood the pad!");
        ignition = new LaunchEventNew("Start engines!");
        liftoff = new LaunchEventNew("Liftoff!");
        for (int t = 20; t >= 0; t--)
            pool.schedule(new CountDownClockNew(t),
                (long) (20 - t), TimeUnit.SECONDS);
        pool.schedule(flood, 3L, TimeUnit.SECONDS);
        pool.schedule(ignition, 13L, TimeUnit.SECONDS);
        pool.schedule(liftoff, 19L, TimeUnit.SECONDS);
        pool.shutdown();
    }
}
```

In the new version of the countdown application, Listing 1-7 does all the busy-work. The listing uses the ScheduledThreadPoolExecutor class. The class's long name tells much of the story:

>> Scheduled: Using this class, you can schedule a run of code for some future time.

>> ThreadPool: This class typically creates several threads (a pool of threads) at the same time. When you want to run some code, you grab an available thread from the pool and use that thread to run your code.

>> Executor: An Executor executes something. No big surprise here!

The loop in Listing 1-7 spawns 20 threads, each with its own initial delay. The fifth loop iteration, for example, calls

```
pool.schedule(new CountDownClockNew(16),
    (long) (20 - 16), TimeUnit.SECONDS);
```

In the `pool.schedule` method call, the number `(long) (20 - 16)` tells Java to wait 4 seconds before scheduling the `T minus 16` thread. Each of the `T minus` threads has a different delay, so each thread runs at the appropriate time. The same is true of the `flood`, `ignition`, and `liftoff` events.

Synchronizing Methods

Whenever you work on a program that uses threads, you have to consider the nasty issue of concurrency. In particular, what if two threads try to access a method of an object at precisely the same time? Unless you program carefully, the result can be disastrous. A method that performs a simple calculation returns inaccurate results. In an online banking application, you might discover that some deposits are credited twice and some withdrawals aren't credited at all. In an online ordering system, one customer's order might get recorded in a different customer's account.

The key to handling concurrency issues is recognizing methods that update data and that might be called by more than one thread. After you identify those methods, the solution is simple. You just add the `synchronized` keyword to the method declaration, like this:

```
public synchronized void someMethod()...
```

This code tells Java to place a *lock* on the object so that no other methods can call any other synchronized methods for the object until this method finishes. In other words, it temporarily disables multithreading for the object. (I discuss locking in the section "Creating a Lock," later in this chapter.)

The next several listings present some concrete examples. Listing 1-8 creates an instance of the `CountDownClock` class (the class in Listing 1-1).

Programming Threads

LISTING 1-8: **Creating Two CountDownClock Threads**

```
import java.util.concurrent.ScheduledThreadPoolExecutor;
public class DoTwoThings
{
    ScheduledThreadPoolExecutor pool =
        new ScheduledThreadPoolExecutor(2);
    CountDownClock clock = new CountDownClock(20);
    public static void main(String[] args)
```

(continued)

LISTING 1-8: *(continued)*

```
    {
        new DoTwoThings();
    }
    DoTwoThings()
    {
        pool.execute(clock);
        pool.execute(clock);
        pool.shutdown();
    }
}
```

The resulting output is an unpredictable mishmash of two threads' outputs, with some of the counts duplicated and others skipped altogether, like this:

```
T minus 20
T minus 20
T minus 19
T minus 19
T minus 18
T minus 17
T minus 16
T minus 15
T minus 13
T minus 13
T minus 12
T minus 12
T minus 11
T minus 11
T minus 10
T minus 9
T minus 7
T minus 7
T minus 6
T minus 5
T minus 4
T minus 3
T minus 2
T minus 2
T minus 1
T minus 0
```

The two threads execute their loops simultaneously, so after one thread displays its T minus 20, the other thread displays its own T minus 20. The same thing happens for T minus 19, T minus 18, and so on.

Then Listing 1-9 spawns two threads, each of which runs a copy of the CountDownClock instance's code.

LISTING 1-9: **Creating Two More CountDownClock Threads**

```
import java.util.concurrent.ScheduledThreadPoolExecutor;
public class DoTwoThingsSync
{
    ScheduledThreadPoolExecutor pool =
        new ScheduledThreadPoolExecutor(2);
    CountDownClockSync clock =
        new CountDownClockSync(20);
    public static void main(String[] args)
    {
        new DoTwoThingsSync();
    }
    DoTwoThingsSync()
    {
        pool.execute(clock);
        pool.execute(clock);
        pool.shutdown();
    }
}
```

In Listing 1-10, Java's synchronized keyword ensures that only one thread at a time calls the run method. The resulting output shows one complete execution of the run method followed by another.

LISTING 1-10: **Using the synchronized Keyword**

```
class CountDownClockSync extends Thread
{
    private int start;
    public CountDownClockSync(int start)
    {
        this.start = start;
    }
    synchronized public void run()
    {
        for (int t = start; t >= 0; t--)
```

(continued)

Programming Threads

LISTING 1-10: *(continued)*

```
        {
            System.out.println("T minus " + t);
            try
            {
                Thread.sleep(1000);
            }
            catch (InterruptedException e)
            {}
        }
    }
}
```

The two threads' calls to the run method are not interleaved, so the output counts down from 20 to 0 and then counts down a second time from 20 to 0:

```
T minus 20
T minus 19
T minus 18
```

And so on, down to

```
T minus 2
T minus 1
T minus 0
T minus 20
T minus 19
T minus 18
```

And so on, down to

```
T minus 2
T minus 1
T minus 0
```

WARNING

The tough part is knowing which methods to synchronize. When I said that any method that updates data can be synchronized, I didn't mean just any method that updates a database. Any method that updates instance variables is at risk — and needs to be synchronized. That's because when two or more threads run a method at the same time, the threads have a common copy of the method's instance variables.

TECHNICAL STUFF

Even methods that consist of just one line of code are at risk. Consider this method:

```
int sequenceNumber = 0;
public int getNextSequenceNumber()
```

```
{
    return sequenceNumber++;
}
```

You'd think that because this method has just one statement, some other thread could not interrupt it in the middle. Alas, that's not the case. This method must get the value of the sequenceNumber field, add 1 to it, save the updated value back to the sequenceNumber field, and return the value. In fact, this single Java statement compiles to 11 bytecode instructions. If the thread is preempted between any of those bytecodes by another thread calling the same method, the serial numbers get munged.

For safety's sake, why not just make all the methods synchronized? You have two reasons not to do so:

» Synchronizing methods takes time. Java has to acquire a lock (see the next section) on the object being synchronized, run the method, and then release the lock. But before it can do that, it has to check to make sure that some other thread doesn't already have a lock on the object. All this work takes time.

» More important, synchronizing all your methods defeats the purpose of multithreading, so you should synchronize only those methods that require it.

REMEMBER

The synchronized keyword doesn't block all access to an object. Other threads can still run unsynchronized methods of the object while the object is locked.

TECHNICAL STUFF

The Object class provides three methods that can let synchronized objects coordinate their activities. The wait method puts a thread in the waiting state until some other thread calls either the object's notify or (more commonly) notifyAll method. These methods are useful when one thread has to wait for another thread to do something before it can proceed. The classic example is a banking system in which one thread makes withdrawals and the other makes deposits. If a customer's account balance drops to zero, the thread that makes withdrawals can call wait; then the thread that makes deposits can call notifyAll. That way, each time a deposit is made, the withdrawal thread can recheck the customer's account balance to see whether it now has enough money for the customer to make a withdrawal.

Creating a Lock

A few years back, Java version 1.5 introduced many new threading features. One such feature was the introduction of locks. A lock can take the place of Java's synchronized keyword, but a lock is much more versatile. Listings 1-11 and 1-12 illustrate the use of a lock.

LISTING 1-11: **Creating CountDownClock Threads (Again)**

```java
import java.util.concurrent.ScheduledThreadPoolExecutor;
public class DoTwoThingsLocked {
    ScheduledThreadPoolExecutor pool =
        new ScheduledThreadPoolExecutor(2);
    CountDownClockLocked clock =
        new CountDownClockLocked();
    public static void main(String[] args)
    {
        new DoTwoThingsLocked();
    }
    DoTwoThingsLocked()
    {
        pool.execute(clock);
        pool.execute(clock);
        pool.shutdown();
    }
}
```

LISTING 1-12: **Using a Lock**

```java
import java.util.concurrent.locks.ReentrantLock;
public class CountDownClockLocked extends Thread
{
    ReentrantLock lock = new ReentrantLock();
    public void run()
    {
        lock.lock();
        for (int t = 20; t >= 0; t--)
        {
            System.out.println("T minus " + t);
            try
            {
                Thread.sleep(1000);
            }
            catch (InterruptedException e)
            {}
        }
        lock.unlock();
    }
}
```

Listing 1-12 is remarkably similar to Listing 1-10. The only significant difference is the replacement of the synchronized keyword by calls to ReentrantLock methods.

At the start of Listing 1-12, the code declares the variable `lock` — an instance of the `ReentrantLock` class. This `lock` object is like a gas station's restroom key: Only one thread at a time can have the `lock` object. When one thread gets the `lock` object — by calling `lock.lock()` at the start of the `run` method — no other thread can get past the `lock.lock()` call. A second thread must wait at the `lock.lock()` call until the "restroom key" becomes available. In Listing 1-12, the key becomes available only when the first thread reaches the `lock.unlock()` statement. After the first thread calls `lock.unlock()`, the second thread proceeds into the method's `for` loop.

The overall result is the same as the output of Listings 1-9 and 1-10. In this example, using a lock is no better than using Java's `synchronized` keyword. But Java 1.5 has several kinds of locks, and each kind of lock has its own useful features.

Coping with Threadus Interruptus

You can interrupt another thread by calling its `interrupt` method, provided that you have a reference to the thread, as in this example:

```
t.interrupt();
```

Here the thread referenced by the `t` variable is interrupted. Now all the interrupted thread has to do is find out that it has been interrupted and respond accordingly. That's the topic of the following sections.

Finding out whether you've been interrupted

As you've already seen, several methods of the `Thread` class, including `sleep` and `yield`, throw `InterruptedException`. Up until now, I've told you to simply ignore this exception — and in many cases, that's appropriate. Many (if not most) threads, however, should respond to `InterruptedException` in one way or another. In most cases, the thread should terminate when it's interrupted.

Unfortunately, finding out whether a thread has been interrupted isn't as easy as it sounds. `InterruptedException` is thrown when another thread calls the `interrupt` method on this thread while the thread is not executing. That's why the methods that can cause the thread to give up control to another thread throw this exception. That way, when the thread resumes execution, you know that it was interrupted.

The yield and sleep methods aren't the only way for control to be wrested away from a thread, however. Sometimes the thread scheduler just steps in and says, "You've had enough time; now it's someone else's turn to play." If that happens and then some other thread calls your thread's interrupt method, InterruptedException isn't thrown. Instead, a special flag called the *interrupted flag* is set to indicate that the thread was interrupted. You can test the status of this flag by calling the static interrupted method.

Unfortunately, that means your threads have to check twice to see whether they have been interrupted. The usual way to do that is to follow this form:

```
public void run()
{
    boolean done = false
    boolean abort = false;
    while(!done)
    {
        // do the thread_s work here
        // set done to true when finished
        try
        {
            sleep(100);  // sleep a bit
        }
        catch(InterruptedException e)
        {
            abort = true;
        }
        if (Thread.interrupted())
            abort = true;
        if (abort)
            break;
    }
}
```

Here the boolean variable abort is set to true if InterruptedException is thrown or if the interrupted flag is set. Then, if abort has been set to true, a break statement is executed to leave the while loop. This scheme has a million variations, of course, but this one works in most situations.

Aborting the countdown

To illustrate how you can interrupt threads, Listing 1-13 shows yet another version of the countdown application. This version aborts the countdown if something goes wrong with any of the launch events.

To simplify the code a bit, I assume that things aren't going well at NASA, so every launch event results in a failure that indicates a need to abort the countdown.

Thus, whenever the start time for a LaunchEvent arrives, the LaunchEvent class attempts to abort the countdown. It goes without saying that in a real launch-control program, you wouldn't want to abort the launch unless something actually *does* go wrong.

LISTING 1-13: **The Countdown Application with Aborts**

```
import java.util.ArrayList;

public class CountDownApp                                          →3
{
    public static void main(String[] args)
    {
        CountDownClock clock = new CountDownClock(20);
        ArrayList<Runnable> events =
            new ArrayList<Runnable>();
        events.add(new LaunchEvent(16,
            "Flood the pad!", clock));
        events.add(new LaunchEvent(6,
            "Start engines!", clock));
        events.add(new LaunchEvent(0,
            "Liftoff!", clock));
        clock.start();
        for (Runnable e : events)
            new Thread(e).start();
    }
}

interface TimeMonitor
{
    int getTime();
    void abortCountDown();                                         →25
}

class CountDownClock extends Thread
    implements TimeMonitor
{
    private int t;

    public CountDownClock(int start)
    {
        this.t = start;
    }

    public void run()
    {
        boolean aborted = false;                                   →40
        for (; t >= 0; t--)
```

(continued)

LISTING 1-13: *(continued)*

```java
        {
            System.out.println("T minus " + t);
            try
            {
                Thread.sleep(1000);
            }
            catch (InterruptedException e)
            {
                aborted = true;                              →50
            }
            if (Thread.interrupted())
                aborted = true;                              →53
            if (aborted)                                     →54
            {
                System.out.println(
                    "Stopping the clock!");
                break;
            }
        }
    }

    public int getTime()
    {
        return t;
    }

    public synchronized void abortCountDown()               →68
    {
        Thread[] threads =
            new Thread[Thread.activeCount()];                →71
        Thread.enumerate(threads);                           →72
        for(Thread t : threads)                              →73
            t.interrupt();
    }
}

class LaunchEvent implements Runnable
{
    private int start;
    private String message;
    TimeMonitor tm;

    public LaunchEvent(int start, String message,
        TimeMonitor monitor)
    {
        this.start = start;
        this.message = message;
        this.tm = monitor;
    }
```

```
public void run()
{
    boolean eventDone = false;
    boolean aborted = false;                                    →95
    while (!eventDone)
    {
        try
        {
            Thread.sleep(10);
        }
        catch (InterruptedException e)
        {
            aborted = true;                                     →104
        }
        if (tm.getTime() <= start)
        {
            System.out.println(this.message);
            eventDone = true;
            System.out.println("ABORT!!!!");
            tm.abortCountDown();
        }
        if (Thread.interrupted())
            aborted = true;                                     →114
        if (aborted)                                            →115
        {
            System.out.println(
                "Aborting " + message);
            break;
        }
    }
}
```

The following paragraphs point out the highlights of this program:

» →3 The CountDownApp class itself hasn't changed. That's the beauty of object-oriented programming. Although I changed the implementations of the CountDownClock and LaunchEvent classes, I didn't change the public interfaces for these classes. As a result, no changes are needed in the CountDownApp class.

» →25 The LaunchEvent class needs a way to notify the CountDown Timer class that the countdown should be aborted. To do that, I added an abortCountDown method to the TimeMonitor interface.

» →40 The run method of the CountDownClass uses a boolean variable named aborted to indicate whether the thread has been interrupted. This variable is set to true in line 50 if InterruptedException is caught. It's also set to true in line 53 if Thread.interrupted() returns true.

» →54 If the aborted field has been set to true, it means that the thread has been interrupted, so the message Stopping the clock! is displayed, and a break statement exits the loop. Thus the thread is terminated.

» →68 The abortCountDown method is synchronized. That happens because any of the LaunchEvent objects can call it, and there's no guarantee that they won't all try to call it at the same time.

» →71 The abortCountDown method starts by creating an array of Thread objects that's large enough to hold all the active threads. The number of active threads is provided by the activeCount method of the Thread class.

» →72 The abortCountDown method calls the enumerate method of the Thread class to copy all the active threads into this array. Note that this method is static, so you don't need a reference to any particular thread to use it. (The activeCount method used in line 69 is static too.)

» →73 An enhanced for loop is used to call the interrupt method on all the active threads. That method shuts down everything.

» →95 Like the CountDownClock class, the LaunchEvent class uses a boolean variable to indicate whether the thread has been interrupted. This variable is set if InterruptedException is caught in line 104 or if Thread.interrupted() returns true in line 114; then it's tested in line 115. If the aborted variable has been set to true, the thread prints a message indicating that the launch event has been aborted, and a break statement is used to exit the loop and (therefore) terminate the thread.

When you run this version of the countdown application, the console output will appear something like this (minor variations might occur because of the synchronization of the threads):

```
T minus 20
T minus 19
T minus 18
T minus 17
T minus 16
Flood the pad!
ABORT!!!!
Stopping the clock!
Aborting Flood the pad!
Aborting Start engines!
Aborting Liftoff!
```

Chapter **2**

Using Regular Expressions

Regular expressions are not expressions that have a lot of fiber in their diet. Instead, a *regular expression* is a special type of pattern-matching string that can be very useful for programs that do string manipulation. Regular expression strings contain special pattern-matching characters that can be matched against another string to see whether the other string fits the pattern. You'll find that regular expressions are very handy for doing complex data validation — for making sure that users enter properly formatted phone numbers, email addresses, or Social Security numbers, for example.

Regular expressions are also useful for many other purposes, including searching text files to see whether they contain certain patterns (can you say, Google?), filtering email based on its contents, or performing complicated search-and-replace functions.

In this chapter, you find out the basics of using regular expressions. I emphasize validation and focus on comparing strings entered by users against patterns specified by regular expressions to see whether they match up. For more complex uses of regular expressions, you have to turn to a more extensive regular expression reference. You can find several in-depth tutorials using a search engine such as Google; search for **regular expression tutorial**.

WARNING

Regular expressions are constructed in a simple but powerful mini-language, so they're like little programs unto themselves. Unfortunately, this mini-language is terse — very terse — to the point of sometimes being downright arcane. Much of it depends on single characters packed with meaning that's often obscure. So be warned — the syntax for regular expressions takes a little getting used to. After you get your mind around the basics, however, you'll find that simple regular expressions aren't that tough to create and can be very useful.

Also be aware that this chapter covers only a portion of all you can do with regular expressions. If you find that you need to use more complicated patterns, you can find plenty of helpful information on the Internet. Just search any search engine for *regular expression*.

TIP

A regular expression is often called a *regex*. Most people pronounce that with a soft *g*, as though it were spelled *rejex*, and some people pronounce it as though it were spelled *rejects*.

Creating a Program for Experimenting with Regular Expressions

Before I get into the details of putting together regular expressions, let me direct your attention to Listing 2-1, which presents a short program that can be very useful while you're learning how to create regular expressions. First, this program lets you enter a regular expression. Next, you can enter a string, and the program tests it against the regular expression and lets you know whether the string matches the regex. Then the program prompts you for another string to compare. You can keep entering strings to compare with the regex you've already entered. When you're done, just press the Enter key without entering a string. The program asks whether you want to enter another regular expression. If you answer yes (y), the whole process repeats. If you answer no (n), the program ends.

LISTING 2-1: **The Regular Expression Test Program**

```java
import java.util.regex.*;
import java.util.Scanner;
public final class Reg {
    static String r, s;
    static Pattern pattern;
    static Matcher matcher;
    static boolean match, validRegex, doneMatching;
    private static Scanner sc =
        new Scanner(System.in);
    public static void main(String[] args)
    {
        System.out.println("Welcome to the "
          + "Regex Tester\n");
        do
        {
            do
            {
                System.out.print("\nEnter regex:  ");
                r = sc.nextLine();
                validRegex = true;
                try
                {
                    pattern = Pattern.compile(r);
                }
                catch (Exception e)
                {
                    System.out.println(e.getMessage());
                    validRegex = false;
                }
            } while (!validRegex);
            doneMatching = false;
            while (!doneMatching)
            {
                System.out.print("Enter string: ");
                s = sc.nextLine();
                if (s.length() == 0)
                    doneMatching = true;
                else
                {
                    matcher = pattern.matcher(s);
                    if (matcher.matches())
                        System.out.println("Match.");
                    else
                        System.out.println(
                            "Does not match.");
                }
            }
```

(continued)

LISTING 2-1:   *(continued)*

```
        } while (askAgain());
    }
    private static boolean askAgain()
    {
        System.out.print("Another? (Y or N) ");
        String reply = sc.nextLine();
        if (reply.equalsIgnoreCase("Y"))
            return true;
        return false;
    }
}
```

Here's a sample run of this program. For now, don't worry about the details of the regular expression string. Just note that it should match any three-letter word that begins with *f*; ends with *r*; and has *a*, *i*, or *o* in the middle.

```
Welcome to the Regex Tester
Enter regex:  f[aio]r
Enter string: for
Match.
Enter string: fir
Match.
Enter string: fur
Does not match.
Enter string: fod
Does not match.
Enter string:
Another? (Y or N) n
```

In this test, I entered the regular expression **f[aio]r**. Then I entered the string **for**. The program indicated that this string matched the expression and asked for another string. So I entered **fir**, which also matched. Then I entered **fur** and **fod**, which didn't match. Next, I entered a blank string, so the program asked whether I wanted to test another regex. I entered **n**, so the program ended.

This program uses the Pattern and Matcher classes, which I don't explain until the end of the chapter. I suggest that you use this program alongside this chapter, however. Regular expressions make a lot more sense if you actually try them out to see them in action. Also, you can learn a lot by trying simple variations as you go. (You can always download the source code for this program from this book's website at www.dummies.com/go/javaaiofd6e if you don't want to enter it yourself.)

In fact, I use portions of console output from this program throughout the rest of this chapter to illustrate regular expressions. There's no better way to see how regular expressions work than to see an expression and some samples of strings that match and don't match the expression.

Performing Basic Character Matching

Most regular expressions simply match characters to see whether a string complies with a simple pattern. You can check a string to see whether it matches the format for Social Security numbers (xxx-xx-xxxx), phone numbers [(xxx) xxx-xxxx], or more complicated patterns such as email addresses. (Well, actually, Social Security and phone numbers are more complicated than you may think — more on that in the section "Using predefined character classes," later in this chapter.) In the following sections, you find out how to create regex patterns for basic character matching.

Matching single characters

The simplest regex patterns match a string literal exactly, as in this example:

```
Enter regex:    abc
Enter string:   abc
Match.
Enter string:   abcd
Does not match.
```

Here the pattern abc matches the string abc but not abcd.

Using predefined character classes

A *character class* represents a particular type of character rather than a specific character. A regex pattern lets you use two types of character classes: predefined classes and custom classes. The predefined character classes are shown in Table 2-1.

TABLE 2-1

Character Classes

Regex	Matches
.	Any character
\d	Any digit (0–9)
\D	Any nondigit (anything other than 0–9)
\s	Any white-space character (space, tab, new line, return, or backspace)
\S	Any character other than a white-space character
\w	Any word character (a–z, A–Z, 0–9, or an underscore)
\W	Any character other than a word character

The period is like a wildcard that matches any character, as in this example:

```
Enter regex:  c.t
Enter string: cat
Match.
Enter string: cot
Match.
Enter string: cart
Does not match.
```

Here c.t matches any three-letter string that starts with c and ends with t. In this example, the first two strings (cat and cot) match, but the third string (cart) doesn't because it's more than three characters.

The \d class represents a digit and is often used in regex patterns to validate input data. Here's a simple regex pattern that validates a U.S. Social Security number, which must be entered in the form xxx-xx-xxxx:

```
Enter regex:  \d\d\d-\d\d-\d\d\d\d
Enter string: 779-54-3994
Match.
Enter string: 550-403-004
Does not match.
```

Here the regex pattern specifies that the string must contain three digits, a hyphen, two digits, another hyphen, and four digits.

TECHNICAL
STUFF

Note that this regex pattern isn't enough to validate real Social Security numbers because the government places more restrictions on these numbers than just the pattern xxx-xx-xxxx. No Social Security number can begin with 779, for example. Thus the number 779-54-3994 entered in the preceding example isn't a valid Social Security number.

Note that the \d class has a counterpart: \D. The \D class matches any character that is *not* a digit. Here's a first attempt at a regex for validating droid names:

```
Enter regex:  \D\d-\D\d
Enter string: R2-D2
Match.
Enter string: C2-D0
Match.
Enter string: C-3PO
Does not match.
```

Here the pattern matches strings that begin with a character that isn't a digit, followed by a character that is a digit, followed by a hyphen, followed by another nondigit character, and ending with a digit. Thus, R2-D2 and C3-P0 match.

Unfortunately, this regex is far from perfect, as any *Star Wars* fan can tell you, because the proper spelling of the shiny gold protocol droid's name is C-3PO, not C3-P0. Typical.

The \s class matches white-space characters including spaces, tabs, newlines, returns, and backspaces. This class is useful when you want to allow the user to separate parts of a string in various ways, as in this example. (Note that in the fourth line, I use the Tab key to separate abc from def.)

```
Enter regex: ...\s...
Enter string: abc def
Match.
Enter string: abc     def
Match.
```

Here the pattern specifies that the string can be two groups of any three characters separated by one white-space character. In the first string that's entered, the groups are separated by a space; in the second group, they're separated by a tab. The \s class also has a counterpart: \S. It matches any character that isn't a white-space character.

TIP

If you want to limit white-space characters to actual spaces, use a space in the regex, like this:

```
Enter regex: ... ...
Enter string: abc def
Match.
Enter string: abc     def
Does not match.
```

Here the regex specifies two groups of any character separated by a space. The first input string matches this pattern, but the second does not because the groups are separated by a tab.

The last set of predefined classes is \w and \W. The \w class identifies any character that's normally used in words, including uppercase and lowercase letters, digits, and underscores. An example shows how all that looks:

```
Enter regex:  \w\w\w\W\w\w\w
Enter string: abc def
Match.
Enter string: 123 456
Match.
Enter string: 123_456
Does not match.
```

Here the pattern calls for two groups of word characters separated by a nonword character.

Isn't it strange that underscores are considered to be word characters? I don't know of too many words in the English language (or any other language, for that matter) that have underscores in them. I guess that's the computer-nerd origins of regular expressions showing through.

Using custom character classes

To create a custom character class, you simply list all the characters that you want to include in the class within a set of brackets. Here's an example:

```
Enter regex:   b[aeiou]t
Enter string: bat
Match.
Enter string: bet
Match.
Enter string: bit
Match.
Enter string: bot
Match.
Enter string: but
Match.
Enter string: bmt
Does not match.
```

Here the pattern specifies that the string must start with the letter b, followed by a class that can include a, e, i, o, or u, followed by t. In other words, it accepts three-letter words that begin with *b*, end with *t*, and have a vowel in the middle.

If you want to let the pattern include uppercase letters as well as lowercase letters, you have to list them both:

```
Enter regex:   b[aAeEiIoOuU]t
Enter string: bat
Match.
Enter string: BAT
Does not match.
Enter string: bAt
Match.
```

You can use as many custom groups on a line as you want. Here's an example that defines classes for the first and last characters so that they too can be uppercase or lowercase:

```
Enter regex:   [bB][aAeEiIoOuU][tT]
Enter string: bat
Match.
Enter string: BAT
Match.
```

This pattern specifies three character classes. The first can be b or B, the second can be any uppercase or lowercase vowel, and the third can be t or T.

Using ranges

Custom character classes can also specify *ranges* of letters and numbers, like this:

```
Enter regex:   [a-z][0-5]
Enter string: r2
Match.
Enter string: b9
Does not match.
```

Here the string can be two characters long. The first must be a character from a–z, and the second must be 0–5.

You can also use more than one range in a class, like this:

```
Enter regex:   [a-zA-Z][0-5]
Enter string: r2
Match.
Enter string: R2
Match.
```

Here the first character can be lowercase a–z or uppercase A–Z.

TIP

You can use ranges to build a class that accepts only characters that appear in real words (as opposed to the \w class, which allows underscores):

```
Enter regex:   [a-zA-Z0-9]
Enter string: a
Match.
Enter string: N
Match.
Enter string: 9
Match.
```

Using negation

Regular expressions can include classes that match any character *but* the ones listed for the class. To do that, you start the class with a caret, like this:

```
Enter regex:    [^cf]at
Enter string: bat
Match.
Enter string: cat
Does not match.
Enter string: fat
Does not match.
```

Here the string must be a three-letter word that ends in at but isn't fat or cat.

Matching multiple characters

The regex patterns described so far in this chapter require that each position in the input string match a specific character class. The pattern \d\W[a-z], for example, requires a digit in the first position, a white-space character in the second position, and one of the letters a–z in the third position. These requirements are pretty rigid.

To create more flexible patterns, you can use any of the quantifiers listed in Table 2-2. *Quantifiers* let you create patterns that match a variable number of characters at a certain position in the string.

TABLE 2-2

Quantifiers

Regex	Matches the Preceding Element
?	Zero times or one time
*	Zero or more times
+	One or more times
{n}	Exactly *n* times
{n,}	At least *n* times
{n,m}	At least *n* times but no more than *m* times

To use a quantifier, you code it immediately after the element you want it to apply to. Here's a version of the Social Security number pattern that uses quantifiers:

```
Enter regex:  \d{3}-\d{2}-\d{4}
Enter string: 779-48-9955
Match.
Enter string: 483-488-9944
Does not match.
```

The pattern matches three digits, followed by a hyphen, followed by two digits, followed by another hyphen, followed by four digits.

TIP

Simply duplicating elements rather than using a quantifier is just as easy, if not easier. \d\d is just as easy as \d{2}.

The ? quantifier lets you create an optional element that may or may not be present in the string. Suppose you want to allow the user to enter Social Security numbers without the hyphens. You could use this pattern:

```
Enter regex:  \d{3}-?\d{2}-?\d{4}
Enter string: 779-48-9955
Match.
Enter string: 779489955
Match.
Enter string: 779-489955
Match.
Enter string: 77948995
Does not match.
```

The question marks indicate that the hyphens are optional. Notice that this pattern lets you include or omit either hyphen. The last string entered doesn't match because it has only eight digits, and the pattern requires nine.

Using escapes

In regular expressions, certain characters have special meaning. What if you want to search for one of those special characters? In that case, you *escape* the character by preceding it with a backslash. Here's an example:

```
Enter regex:  \(\d{3}\) \d{3}-\d{4}
Enter string: (559) 555-1234
Match.
Enter string: 559 555-1234
Does not match.
```

Here \(represents a left parenthesis, and \) represents a right parenthesis. Without the backslashes, the regular expression treats the parenthesis as a grouping element.

Here are a few additional points to ponder about escapes:

» Strictly speaking, you need to use the backslash escape only for characters that have special meanings in regular expressions. I recommend, however, that you escape any punctuation character or symbol, just to be sure.

» You can't escape alphabetic characters (letters) because a backslash followed by certain alphabetic characters represents a character, a class, or some other regex element.

» To escape a backslash, code two slashes in a row. The regex \d\d\\\d\d, for example, accepts strings made up of two digits followed by a backslash and two more digits, such as 23\88 and 95\55.

Using parentheses to group characters

You can use parentheses to create groups of characters to apply other regex elements to, as in this example:

```
Enter regex:  (bla)+
Enter string: bla
Match.
Enter string: blabla
Match.
Enter string: blablabla
Match.
Enter string: bla bla bla
Does not match.
```

Here the parentheses treat bla as a group, so the + quantifier applies to the entire sequence. Thus, this pattern looks for one or more occurrences of the sequence bla.

Here's an example that finds U.S. phone numbers that can have an optional area code:

```
Enter regex:  (\(\d{3}\)\s?)?\d{3}-\d{4}
Enter string: 555-1234
Match.
Enter string: (559) 555-1234
Match.
Enter string: (559)555-1239
Match.
```

This regex pattern is a little complicated, but if you examine it element by element, you should be able to figure it out. It starts with a group that indicates the optional area code: `(\(\d{3}\)\s?)?`. This group begins with the left parenthesis, which marks the start of the group. The characters in the group consist of an escaped left parenthesis, three digits, an escaped right parenthesis, and an optional white-space character. Then a right parenthesis closes the group, and the question mark indicates that the entire group is optional. The rest of the regex pattern looks for three digits followed by a hyphen and four more digits.

When you mark a group of characters with parentheses, the text that matches that group is *captured* so that you can use it later in the pattern. The groups that are captured are called *capture groups* and are numbered beginning with 1. Then you can use a backslash followed by the capture-group number to indicate that the text must match the text that was captured for the specified capture group.

Suppose that droids named following the pattern `\w\d-\w\d` must have the same digit in the second and fifth characters. In other words, r2-d2 and b9-k9 are valid droid names, but r2-d4 and d3-r4 are not.

Here's an example that can validate that type of name:

```
Enter regex:  \w(\d)-\w\1
Enter string: r2-d2
Match.
Enter string: d3-r4
Does not match.
Enter string: b9-k9
Match.
```

Here `\1` refers to the first capture group. Thus the last character in the string must be the same as the second character, which must be a digit.

Using the pipe symbol

The vertical bar (|) symbol defines an or operation, which lets you create patterns that accept any of two or more variations. Here's an improvement of the pattern for validating droid names:

```
Enter regex:  (\w\d-\w\d)|(\w-\d\w\w)
Enter string: r2-d2
Match.
Enter string: c-3po
Match.
```

The | character indicates that either the group on the left or the group on the right can be used to match the string. The group on the left matches a word character, a digit, a hyphen, a word character, and another digit. The group on the right matches a word character, a hyphen, a digit, and two word characters.

You may want to use an additional set of parentheses around the entire part of the pattern that the | applies to. Then you can add pattern elements before or after the | groups. What if you want to let a user enter the area code for a phone number with or without parentheses? Here's a regex pattern that does the trick:

```
Enter regex:  ((\d{3} )|(\(\d{3}\) ))?\d{3}-\d{4}
Enter string: (559) 555-1234
Match.
Enter string: 559 555-1234
Match.
Enter string: 555-1234
Match.
```

The first part of this pattern is a group that consists of two smaller groups separated by a | character. The first of these groups matches an area code without parentheses followed by a space, and the second matches an area code with parentheses followed by a space. So the outer group matches an area code with or without parentheses. This entire group is marked with a question mark as optional; then the pattern continues with three digits, a hyphen, and four digits.

Using Regular Expressions in Java Programs

So far, this chapter has shown you the basics of creating regular expressions. The following sections show you how to put them to use in Java programs.

Understanding the String problem

Before getting into the classes for working with regular expressions, I want to clue you in about a problem that Java has in dealing with strings that contain regular expressions. As you've seen throughout this chapter, regex patterns rely on the backslash character to mark different elements of a pattern. The bad news is that Java treats the backslash character in a string literal as an escape character. Thus, you can't just quote regular expressions in string literals, because Java steals the backslash characters before they get to the regular expression classes.

In most cases, the compiler simply complains that the string literal is not correct. The following line won't compile:

```
String regex = "\w\d-\w\d";    // error: won't compile
```

The compiler sees the backslashes in the string and expects to find a valid Java escape sequence, not a regular expression.

Unfortunately, the solution to this problem is ugly: You have to double the backslashes wherever they occur. Java treats two backslashes in a row as an escaped backslash and places a single backslash in the string. Thus you have to code the statement shown in the preceding example like this:

```
String regex = "\\w\\d-\\w\\d";    // now it will
                                   // compile
```

Here each backslash I want in the regular expression is coded as a pair of backslashes in the string literal.

TIP

If you're in doubt about whether you're coding your string literals right, just use `System.out.println` to print the resulting string. Then you can check the console output to make sure that you wrote the string literal right. If I followed the preceding statement with `System.out.println(regex)`, the following output would appear on the console:

```
\w\d-\w\d
```

Thus I know that I coded the string literal for the regular expression correctly.

Using regular expressions with the String class

If all you want to do with a regular expression is check whether a string matches a pattern, you can use the `matches` method of the `String` class. This method accepts a regular expression as a parameter and returns a boolean that indicates whether the string matches the pattern.

Here's a static method that validates droid names:

```
private static boolean validDroidName(String droid)
{
    String regex = "(\\w\\d-\\w\\d)|(\\w-\\d\\w\\w)";
    return droid.matches(regex);
}
```

Here the name of the droid is passed via a parameter, and the method returns a boolean that indicates whether the droid's name is valid. The method simply creates a regular expression from a string literal and then uses the matches method of the droid string to match the pattern.

You can also use the split method to split a string into an array of String objects based on delimiters that match a regular expression. One common way to do that is to simply create a custom class of characters that can be used for delimiters, as in this example:

```
String s = "One:Two;Three|Four\tFive";
String regex = "[:;|\\t]";
String strings[] = s.split(regex);
for (String word : strings)
    System.out.println(word);
```

Here a string is split into words marked by colons, semicolons, vertical bars, or tab characters. When you run this program, the following text is displayed on the console:

```
One
Two
Three
Four
Five
```

Using the Pattern and Matcher classes

The matches method is fine for occasional use of regular expressions, but if you want your program to do a lot of pattern matching, you should use the Pattern and Matcher classes instead. The Pattern class represents a regular expression that has been compiled into executable form. (Remember that regular expressions are like little programs.) Then you can use the compiled Pattern object to create a Matcher object, which you can use to match strings.

The Pattern class itself is pretty simple. Although it has about ten methods, you usually use just these two:

>> static Pattern compile (String *pattern*): Compiles the specified pattern. This static method returns a Pattern object. It throws PatternSyntaxException if the pattern contains an error.

>> Matcher matcher (String *input*): Creates a Matcher object to match this pattern against the specified string.

First, you use the compile method to create a Pattern object. (Pattern is one of those weird classes that doesn't have constructors. Instead, it relies on the static compile method to create instances.) Because the compile method throws PatternSyntaxException, you must use a try/catch statement to catch this exception when you compile a pattern.

After you have a Pattern instance, you use the matcher method to create an instance of the Matcher class. This class has more than 30 methods that let you do all sorts of things with regular expressions that aren't covered in this chapter, such as finding multiple occurrences of a pattern in an input string or replacing text that matches a pattern with a replacement string. For purposes of this book, I'm concerned only with the matches method: static boolean matches() returns a boolean that indicates whether the entire string matches the pattern.

To illustrate how to use these methods, here's an enhanced version of the valid-DroidName method that creates a pattern for the droid-validation regex and saves it in a static class field:

```
private static Pattern droidPattern;
private static boolean validDroidName(String droid)
{
    if (droidPattern == null)
    {
        String regex = "(\\w\\d-\\w\\d)|"
            + "(\\w-\\d\\w\\w)";
        droidPattern = Pattern.compile(regex);
    }
    Matcher m = droidPattern.matcher(droid);
    return m.matches();
}
```

Here the private class field droidPattern saves the compiled pattern for validating droids. The if statement in the validDroidName method checks whether the pattern has already been created. If not, the pattern is created by calling the static compile method of the Pattern class. Then the matcher method is used to create a Matcher object for the string passed as a parameter, and the string is validated by calling the matches method of the Matcher object.

Chapter **3**

Using Recursion

Recursion is a basic programming technique in which a method calls itself to solve some problem. A method that uses this technique is called *recursive.* Many programming problems can be solved only by recursion, and some problems that can be solved by other techniques are better solved by recursion.

I'm not sure, but I think that the term *recursion* comes from the Latin *recurse, recurset, recursum,* which means to curse repeatedly. I do know that that's exactly what many programmers feel like doing when they're struggling with complex recursive programming problems.

True, the concept of recursion can get a little tricky. Many programmers steer clear of it, looking for other techniques to solve the problem at hand, and in many cases, a nonrecursive solution is best. Many problems just cry out for recursion, however.

Calculating the Classic Factorial Example

One of the classic problems for introducing recursion is calculating the factorial of an integer. The *factorial* of any given integer — I'll call it *n* so that I sound mathematical — is the product of all the integers from 1 to *n*. Thus the factorial of 5 is 120: $5 \times 4 \times 3 \times 2 \times 1$.

The nonrecursive solution

You don't have to use recursion to calculate factorials. Instead, you can use a simple for loop. Here's a method that accepts an int number and returns the number's factorial as a long:

```
private static long factorial(int n)
{
    long f = 1;
    for (int i = 1; i <=n; i++)
        f = f * i;
    return f;
}
```

This method uses a for loop to count from 1 to the number, keeping track of the product as it goes. Here's a snippet of code that calls this method and displays the result:

```
int n = 5;
long fact;
fact = factorial(n);
System.out.println("The factorial of "+ n + " is "
    + fact + ".");
```

If you run this code, the following line is displayed on the console:

```
The factorial of 5 is 120.
```

Factorials get big fast. You should use a long rather than an int to calculate the result. Also, you should use the NumberFormat class to format the result. If int is 20 instead of 5, the preceding code prints this on the console:

```
The factorial of 20 is 2432902008176640000.
```

If you use the NumberFormat class to format the result, the console output is more readable:

```
The factorial of 20 is 2,432,902,008,176,640,000.
```

The recursive solution

The nonrecursive solution to the factorial problem works, but it isn't much fun. The recursive solution is based on the notion that the factorial for any number n is equal to n times the factorial of $n - 1$, provided that n is greater than 1. If n is 1, the factorial of n is 1.

This definition of factorial is recursive because the definition includes the factorial method itself. It also includes the most important part of any recursive method: an end condition. The *end condition* indicates when the recursive method should stop calling itself. In this case, when n is 1, I just return 1. Without an end condition, the recursive method keeps calling itself forever.

Here's the recursive version of the factorial method:

```
private static long factorial(int n)
{
    if (n == 1)
        return 1;
    else
        return n * factorial(n–1);
}
```

This method returns exactly the same result as the version in the preceding section, but it uses recursion to calculate the factorial.

One way to visualize how recursion works is to imagine that you have five friends: Jordan, Jeremy, Jacob, Justin, and Bob. Your friends aren't very smart, but they're very much alike. In fact, they're clones of one another. Cloning isn't a perfect process yet, so these clones have limitations. Each can do only one multiplication problem and can ask one of its clones one question.

Now suppose that you walk up to Jordan and ask, "Jordan, what's the factorial of 5?"

Jordan says, "I don't know, but I do know it's 5 × the factorial of 4. Jeremy, what's the factorial of 4?"

Jeremy says, "I don't know, but I do know it's 4 × the factorial of 3. Jacob, what's the factorial of 3?"

Jacob says, "I don't know, but I do know it's 3 × the factorial of 2. Justin, what's the factorial of 2?"

Justin says, "I don't know, but I do know it's 2 × the factorial of 1. Hey, Bob! What's the factorial of 1?"

Bob, being the most intelligent of the bunch on account of not having a J-name, replies, "Why, 1, of course." He tells Justin his answer.

Justin says, "Ah — 2 × 1 is 2." He tells Jacob his answer.

Jacob says, "Thanks — 3 × 2 is 6." Jacob tells Jeremy his answer.

Jeremy says, "Dude — 4×6 is 24." Jeremy tells Jordan his answer.

Jordan says, "Very good — 5×24 is 120." He tells you the answer.

That's pretty much how recursion works.

Displaying Directories

Recursion lends itself well to applications that have to navigate directory structures, such as a Windows or Unix file system. In a file system, a directory is a list of files and other directories. Each of those directories is itself a list of files and other directories, and so on. Directories can be snugly nestled inside other directories and have no limit in number.

Listing 3-1, at the end of this section, shows a program that uses a recursive method to list all the directories that are found starting from a given path. I use indentation to show the directory structure.

Here's the console output for the directories I used to organize the documents for this book:

```
Welcome to the Directory Lister
Enter a path: C:\Java AIO

Listing directory tree of:
C:\Java AIO
   Apps
      Book 1
      Book 2
      Book 3
      Book 4
      Book 5
   Manuscript
      Book 1
      Book 2
      Book 3
      Book 4
      Book 5
      Front
   Plans
Another? (Y or N) n
```

As you can see, I haven't done Books 6–8 yet. By the time you read this chapter, there will be even more directories to list!

Don't enter **c:** unless you're prepared to wait a long time for the program to finish listing *all* the directories on your hard drive. (Of course, you can always press Ctrl+C to stop the program, or just close the console window.)

The Directory Listing application is remarkably simple. Before I explain its details, though, I want to point out that this program uses the File class, which is part of the java.io package. The File class represents a single file or directory. You can find out much more about this class in the bonus content online. For now, you just need to know these five details:

>> The constructor for this class accepts a directory path as a parameter and creates an object that represents the specified directory.

>> You can use the exists method to find out whether the directory specified by the path parameter exists.

>> The listFiles method returns an array of File objects that represent every file and directory in the current File object.

>> The isDirectory method returns a boolean that indicates whether the current File object is a directory. If this method returns false, you can assume that the File object is a file.

>> The getName method returns the name of the file.

Note that Chapter 1 of the bonus content shows how to use the Path class along with an interface called FileVisitor to automatically traverse all the files in a directory tree, including files in any subdirectories. When you use the Path class and the FileVisitor interface, you don't have to write the recursive code yourself.

LISTING 3-1: **The Directory Listing Application**

```
import java.io.File;                                              →1
import java.util.Scanner;

public class DirList
{
    static Scanner sc = new Scanner(System.in);

    public static void main(String[] args)
    {
        System.out.print(
            "Welcome to the Directory Lister");
        do
        {
            System.out.print("\nEnter a path: ");
            String path = sc.nextLine();                         →15
```

(continued)

Using Recursion

LISTING 3-1: *(continued)*

```
                        File dir = new File(path);                          →17
                        if (!dir.exists() || !dir.isDirectory())            →18
                            System.out.println(
                                "\nThat directory doesn't exist.");
                        else
                        {
                            System.out.println(
                                "\nListing directory tree of:");
                            System.out.println(dir.getPath());              →25
                            listDirectories(dir, "  ");                     →26
                        }
                    } while(askAgain());                                    →28
            }

            private static void listDirectories(                           →31
                File dir, String indent)
            {
                File[] dirs = dir.listFiles();                             →34
                for (File f : dirs)                                        →35
                {
                    if (f.isDirectory())                                   →37
                    {
                        System.out.println(
                            indent + f.getName());                         →40
                        listDirectories(f, indent + "  ");                 →41
                    }
                }
            }

            private static boolean askAgain()
            {
                System.out.print("Another? (Y or N) ");
                String reply = sc.nextLine();
                if (reply.equalsIgnoreCase("Y"))
                    return true;
                return false;
            }
        }
```

The following paragraphs point out the highlights of how this program works:

» →1 This import statement is required to use the File class.

» →15 A Scanner object is used to get the pathname from the user.

» →17 The pathname is passed to the File class constructor to create a new File object for the directory entered by the user.

» →18 The exists and isDirectory methods are called to make sure that the path entered by the user exists and points to a directory rather than a file.

» →25 If the user entered a good path, the getPath method is called to display the name of the path represented by the File object. (I could just as easily have displayed the path variable here.)

» →26 The listDirectories method is called to list all the subdirectories in the directory specified by the user.

» →28 The user is asked whether he wants to list another directory, and the loop repeats if the user answers Y.

» →31 This line is the start of the listDirectories method. This method takes two parameters: a File object representing the directory to be listed and a String object that provides the spaces used to indent each line of the listing. When this method is first called from the main method, the indentation is set to two spaces by a string literal.

» →34 The listFiles method is called to get an array of all the File objects in this directory.

» →35 An enhanced for loop is used to process all the File objects in the array.

» →37 This if statement checks to see whether a file is a directory rather than a file.

» →40 If the File object is a directory, the indentation string is printed, followed by the name of the directory as returned by the getName method.

» →41 Next, the listDirectories method is called recursively to list the contents of the f directory. Two spaces are added to the indentation string, however, so that any directories in the f directory are indented two spaces to the right of the current directory.

If you're having trouble understanding how the recursion in this program works, think of it this way: The listDirectory method lists all the subdirectories in a single directory. For each directory, this method does two things: (1) prints the directory's name and (2) calls itself to print any subdirectories of that directory.

Earlier in this chapter, I mention that all recursive methods must have some type of condition test that causes the method to stop calling itself. In this program, the condition test may not be obvious. Eventually, however, the listDirectories method is passed a directory that doesn't have any subdirectories. When that happens, the recursion ends — at least for that branch of the directory tree.

Writing Your Own Sorting Routine

The world is full of computer science majors who don't know anything more about computers than you do, but they once attended a class in which the instructor explained how sorting algorithms worked. They may have received a C in that class, but it was good enough to graduate.

Now you have a chance to find out what you missed by not majoring in computer science. I'm going to show you how one of the most commonly used sorting techniques actually works. This technique is called *Quicksort*, and it's a very ingenious use of recursion. I even show you a simple Java implementation of it.

Quicksort is easily the most technical part of this entire book. If you never wanted to major in computer science (and if you don't even want to talk to people who did), you may want to skip the rest of this chapter now.

For most of us, figuring out how sorting algorithms such as Quicksort work is merely an intellectual exercise. The Java API has sorting already built in. (Check out the `Arrays.sort` method, for example.) Those sort routines are way better than any that you or I will ever write.

Understanding how Quicksort works

The Quicksort technique sorts an array of values by using recursion. Its basic steps are thus:

1. **Pick an arbitrary value that lies within the range of values in the array.**

   This value is the *pivot point*. The most common way to choose the pivot point is to simply pick the first value in the array. Folks have written doctoral degrees on more-sophisticated ways to pick a pivot point that results in faster sorting. I like to stick with using the first element in the array.

2. **Rearrange the values in the array so that all the values that are less than the pivot point are on the left side of the array and all the values that are greater than or equal to the pivot point are on the right side of the array.**

   The *pivot value* indicates the boundary between the left side and the right side of the array. It probably won't be dead center, but that doesn't matter. This step is called *partitioning,* and the left and right sides of the arrays are called *partitions*.

3. **Now treat each of the two sections of the array as a separate array, and start over with Step 1 for that section.**

   That's the recursive part of the algorithm.

The hardest part of the Quicksort algorithm is the partitioning step, which must rearrange the partition so that all values that are smaller than the pivot point are on the left and all elements that are larger than the pivot point are on the right. Suppose that the array has these ten values:

```
38 17 58 22 69 31 88 28 86 12
```

Here the pivot point is 38, and the task of the partitioning step is to rearrange the array to something like this:

```
17 12 22 28 31 38 88 69 86 58
```

Notice that the values are still out of order. The array, however, has been divided around the value 38: All values that are less than 38 are to the left of 38, and all values that are greater than 38 are to the right of 38.

Now you can divide the array into two partitions at the value 38 and repeat the process for each side. The pivot value itself goes with the left partition, so the left partition is this:

```
17 12 22 28 31 38
```

This time, the partitioning step picks 17 as the pivot point and rearranges the elements as follows:

```
12 17 22 28 31 38
```

As you can see, this portion of the array is sorted now. Unfortunately, Quicksort doesn't realize that at this point, so it takes a few more recursions to be sure. But that's the basic process.

Using the sort method

In the remainder of this chapter, I present a class that implements the Quicksort algorithm to sort an array of 100 random int values. The array is stored as a static class variable named simply a, so it's visible throughout the class.

The actual code that drives a Quicksort routine is surprisingly simple:

```
public static void sort(int startIndex, int endIndex)
{
    if (startIndex >= endIndex)
        return;
    int pivotIndex = partition(startIndex, endIndex);
```

```
        sort (startIndex, pivotIndex);
        sort (pivotIndex+1, endIndex);
    }
```

This method sorts the portion of an array indicated by the starting and ending index values passed to it (startIndex and endIndex). Ignoring the if statement for now, the sort method works by calling a partition method. This method rearranges the array into two partitions so that all the values in the left partition are smaller than all the values in the right partition. The partition method returns the index of the end of the left partition. Then the sort method calls itself twice: once to sort the left partition and again to sort the right partition.

To get the sort method started, you call it with 0 as the starting index value and the array length minus 1 as the ending index value — in other words, the indexes of the first and last element in the array. Thus, the sort method begins by sorting the entire array.

Each time the sort method executes, it calls itself twice to sort smaller partitions of the array. This is the recursive portion of the algorithm.

The if statement at the beginning of the sort method provides the exit condition that indicates when the recursion should stop. It compares the starting index (startIndex) with the ending index (endIndex). If startIndex is equal to or greater than endIndex, the partition has only one element (or perhaps no elements) and is, therefore, already sorted. In that case, the sort method simply returns without calling itself again. Thus ends the recursion.

Using the partition method

The sort method itself is the simple part of the Quicksort technique. The hard part is the partition method. This method accepts two parameters: the starting index and ending index that designate the portion of the array that should be sorted. The basic outline of the partition method goes something like this:

1. Pick a pivot value; for simplicity, use the value of the first element in the partition.

2. Move all elements that are less than the pivot value to the left side of the partition.

3. Move all elements that are greater than the pivot value to the right side of the partition.

4. Return the index of the pivot point.

The most common technique for partitioning the array is to maintain two index variables, named i and j, that work from both ends of the array toward the center. First, i starts at the beginning of the array and moves forward until it encounters a value that's greater than the pivot value. Then j starts at the opposite end of the array and moves backward until it finds a value that's less than the pivot value. At that point, the partition method has a value that's greater than the pivot point on the left side of the array and a value that's less than the pivot point on the right side of the array. So it swaps those two array elements.

Note that these two for statements are not nested. The first for statement runs to completion, finding an element that should be swapped on the left side of the array. Then, the second for statement runs to completion, finding an element that should be swapped on the right side of the array. Then the two elements are swapped.

Next, the cycle repeats: i is incremented until it finds another value that's greater than the pivot value, j is decremented until it finds another value that's less than the pivot value, and the elements are swapped. This process repeats until j is less than i, which means that the indexes have crossed and the partitioning is done.

Here's some code that puts everything together:

```
public static int partition(int startIndex, int endIndex)
{
    int pivotValue = a[startIndex];

    int i = startIndex - 1;
    int j = endIndex + 1;

    while (i < j)
    {
        for (i++; a[i] < pivotValue; i++);
        for (j--; a[j] > pivotValue; j--);
        if (i < j)
            swap(i, j);
    }
    return j;
}
```

Remember that the array being sorted is a static int array named a that's visible throughout the class, so it isn't passed into this method as an argument.

The starting and ending index of the partition to be sub-partitioned are passed in as arguments, and the method starts by choosing the first element in the partition to use as the pivot value. For example, if the value of the first element is 42, the partition method will shuffle up the array such that all values that are 42 or less appear on the left side of the array and all values that are greater than 42 are on the right side of the array.

After picking the pivot value, we initialize the index variables i and j from the arguments. Notice that 1 is subtracted from the starting index and 1 is added to the ending index. The index variables take one step back from the array before the looping starts so they can get a good start.

The while loop is used to indicate when the partitioning is finished. It repeats as long as i is less than j. After these index variables cross, the partitioning is done, and the value of j is returned to indicate the index point that divides the left partition from the right partition.

In the body of the while loop are two strange bodyless for loops. These for loops don't have bodies because their only purpose is to move their index values until they find a value that's either less than or greater than the pivot value.

The first for loop increments the i index variable until it finds a value that's greater than the pivot point. This for loop finds the first value that might need to be moved to the other side of the array.

Next, the second for loop decrements the j index variable until it finds a value that's less than the pivot point. So this loop finds a value that may need to be swapped with the value found by the first for loop.

Finally, the if statement checks whether the indexes have crossed. Assuming that they haven't, a swap method is called to swap the elements. The swap method is mercifully simple:

```
public static void swap(int i, int j)
{
    int temp = a[i];
    a[i] = a[j];
    a[j] = temp;
}
```

This method moves the i element to a temporary variable, moves the j element to the i element, and then moves the temporary variable to the j element.

Putting it all together

Now that you've seen the basic steps necessary to create a Quicksort program, Listing 3-2 shows a program that gives these methods a workout. This program creates an array of 100 randomly selected numbers with values from 1–100. It prints the array, uses the sorting methods shown in the previous sections to sort the array, and then prints the sorted array. Here's a sample run:

```
Unsorted array:

76  35  89  96  33  22  72  18  79  14   6   7  31  91   3  58  56  63  77   9
82  31  52  65  14  84  33  47  50   6   1  73  94  63  96  29  12  84  48  85
69  55   4   4  72  80  65  44  82  97  39  42  28  71  96   7  31   3  32  77
78  72  66   4  18  35  27   7  31  67  96  96  30  86   8  98  73  40  55  72
21  98  31  13  87  55  82  42  95  97  15  68  43  70  71  26  32  35  78  99


Sorted array:

 1   3   3   4   4   4   6   6   7   7   7   8   9  12  13  14  14  15  18  18
21  22  26  27  28  29  30  31  31  31  31  31  32  32  33  33  35  35  35  39
40  42  42  43  44  47  48  50  52  55  55  55  56  58  63  63  65  65  66  67
68  69  70  71  71  72  72  72  72  73  73  76  77  77  78  78  79  80  82  82
82  84  84  85  86  87  89  91  94  95  96  96  96  96  96  97  97  98  98  99
```

As you can see, the first array is in random order, but the second array is nicely sorted. (Your results will vary, of course, because the unsorted array will be in a different order every time you run the program as the numbers are generated randomly.)

LISTING 3-2: **A Sorting Program**

```java
public class QuickSortApp
{
    private static int[] a = new int[100];                      →4

    public static void main(String[] args)
    {
        for (int i = 0; i<a.length; i++)                        →8
            a[i] = (int)(Math.random() * 100) + 1;

        System.out.println("Unsorted array:");
        printArray();                                           →12

        sort(0, a.length - 1);                                  →14

        System.out.println("\n\nSorted array:");
        printArray();                                           →17
    }

    private static void printArray()                            →20
    {
        System.out.println();
        for (int i = 0; i < a.length; i++)
```

(continued)

LISTING 3-2: *(continued)*

```
        {
            if (a[i] < 100)
                System.out.print(" ");

            if (a[i] < 10)
                System.out.print(" ");

            System.out.print(a[i] + " ");

            if ((i+1) % 20 == 0)
                System.out.println();
        }
    }


    public static void sort(int startIndex, int endIndex)        →38
    {
        if (startIndex >= endIndex)
            return;

        int pivotIndex = partition(startIndex, endIndex);

        sort (startIndex, pivotIndex);
        sort (pivotIndex+1, endIndex);
    }

    public static int partition(int startIndex, int endIndex)    →49
    {
        int pivotValue = a[startIndex];

        int i = startIndex - 1;
        int j = endIndex + 1;

        while (i < j)
        {
            for (i++; a[i] < pivotValue; i++);
            for (j--; a[j] > pivotValue; j--);
            if (i < j)
                swap(i, j);
        }
        return j;
    }
```

```
public static void swap(int i, int j)                                    →66
{
    int temp = a[i];
    a[i] = a[j];
    a[j] = temp;
}
}
```

Most of the code in this program has already been explained, so I just point out a few of the highlights here:

>> →4 The array is created as a static class variable named a.

>> →8 This for loop assigns 100 random values to the array.

>> →12 The printArray method is called to print the unsorted array.

>> →14 he sort method is called to sort the array.

>> →17 The printArray method is called again to print the sorted array.

>> →20 The printArray method uses a for loop to print array elements. Each element is separated by one space. An additional space, however, is printed before each element if the element's value is less than 100 and yet another space is printed if the value is less than 10. That way, the values line up in columns whether the value is one digit, two digits, or three digits in length. Also, the remainder operator (%) is used to call the println method every 20 elements. Thus this method prints 5 lines with 20 values on each line.

>> →38 This line declares the sort method, which sorts the partition indicated by the parameters. (The operation of this method is explained in detail in the section "Using the sort method," earlier in this chapter.)

>> →49 The partition method is explained in detail in the preceding section.

>> →66 The swap method simply exchanges the two indicated values.

Remember the cool XOR technique for exchanging two integer values without the need for a temporary variable? You can improve the performance of your sort ever so slightly by replacing the swap method with this code:

```
public static void swap(int i, int j)
{
    a[i] ^= a[j];
    a[j] ^= a[i];
    a[i] ^= a[j];
}
```

Chapter **4**

Working with Dates and Times

D*oes anybody really know what time it is? Does anybody really care about time?*

So mused Robert Lamm of *The Chicago Transit Authority* (later known as simply *Chicago*) in 1969.

I'm not sure who cared much about time in 1969, but I do know that the people who designed the original version of Java in 1995 didn't care much about it, at least as evidenced by the weak classes they provided for working with times and dates in the `Java.util` package. Java programmers have long struggled with simple calculations involving dates and times, such as determining what the date will be 45 days from today or calculating the number of days between two given dates.

Java finally got with the times with the release of Java 8, which introduced an entirely new framework for working with dates and times, usually referred to as the Date-Time API. This new API is pretty complicated, involving about

50 new classes and interfaces and hundreds of new methods. In this chapter, I'll introduce you to just a few of the most important and useful classes of the new Date-Time API. Then you can explore the rest online via Oracle's documentation at `https://docs.oracle.com/en/java/javase/14/docs/api/index.html`.

Pondering How Time is Represented

Before I launch into the details of the new Date-Time API's classes, let's review a few basic concepts about time. Probably the most important basic concept to understand about time (at least from a programming point of view) is that computers and humans use two entirely different methods of keeping track of time. Humans measure time using a system of progressively longer units, starting with seconds and increasing to minutes, hours, days, weeks, months, years, decades, and centuries.

Our human time units are intuitively familiar to us, but their precise definitions are more complicated than you might guess. All kinds of factors muck up the way we represent time: leap days, time zones, and daylight-saving time. And did you know that about once every 18 months, scientists pick a day (they usually choose June 30 or December 31) that they add one second to? This is necessary because the speed of the earth's rotation varies ever so slightly, throwing our clocks off.

In contrast, the way computers keep track of time is much simpler: Computers simply count the number of units (typically milliseconds or nanoseconds) that have elapsed since a given start time. Thus, to a computer, a time is just a number.

In Java, machine time is set as the number of nanoseconds that have elapsed since midnight, January 1, 1970. Why January 1, 1970? There's no particular reason other than historical: Java inherited that date from the Unix operating system, which was developed in the 1970s. (For more information, see the sidebar "And you thought Y2K was bad!")

TIP

The designers of Microsoft Windows decided to use January 1, 1601 as the start day for their clock. I guess that was a big year for William Shakespeare; in 1601 he was able to use the first version of Microsoft Word to write his plays.

The difference between how humans and computers keep track of time makes any computer program that deals with dates and times a bit tricky. For example, suppose you want to schedule a phone call between two people, one in Los Angeles, the other in New York. Obviously, the time that you agree upon for the call must take into account the time zone of each participant. Thus you might agree to make the call at 1 p.m. local time for the West Coast participant and 4 p.m. local time for the East Coast participant. So, how would you represent that appointment time in a database?

AND YOU THOUGHT Y2K WAS BAD!

Java counts time in nanoseconds starting with January 1, 1970. Why that particular date? It turns out that January 1, 1970 was the date the original designers of Unix chose as their origin time for the Unix operating system. That tradition was carried on into Linux, and then into Java.

The original Unix used a signed 32-bit number to count time, and counted time in full seconds rather than nanoseconds. Thus, in Unix time was represented as the number of seconds that have elapsed since January 1, 1970. Because the largest number that can be represented by a signed 32-bit number is relatively small, Unix time will come to an end on January 19, 2038.

Today, most Unix- and Linux-based systems have been upgraded to use 64-bit numbers to count time, so the world will not end in 2038 for those systems. But systems that still use old-style 32-bit time values will break in 2038.

I am already looking forward to the end-of-the-world party.

Or suppose you want to calculate the due date for an invoice that is dated January 27, 2015, when the payment due date is 45 days after the invoice date. The algorithm that calculates the due date must be aware that January has 31 days and that February has 28 days in 2015.

Fortunately, the new Date–Time API is designed to handle all those nuances for you. The Date–Time API includes all the classes you need to represent dates and times in just about any imaginable context, for performing calculations and comparisons between date and time objects, and for converting dates and times to string representations in just about any imaginable format.

Picking the Right Date and Time Class for Your Application

The first order of business when developing an application that must work with dates or times (or both) is picking the Date–Time class to represent your date and time values. The `java.time` package defines ten distinct classes used to represent different types of times and dates, as described in Table 4-1.

Each of these classes has many different methods that let you create date and time objects, perform calculations on them, compare them, and convert them to strings that can be displayed and read by humans. You can find complete documentation of the methods for each of these classes online at https://docs.oracle.com/en/java/javase/14/docs/api/index.html.

TABLE 4-1   Ten Date-Time Classes in java.time

Class	What It Represents
LocalTime	A time (hours, minutes, and seconds to nanosecond precision) without an associated time zone.
LocalDate	A date (year, month, and day) without an associated time zone.
LocalDateTime	A date and time without an associated time zone.
OffsetTime	A time and an offset from UTC (Coordinated Universal Time, also known as Greenwich Mean Time), such as 12:30:00-8.00, which means the time is 12:30 with an offset of -8 hours from UTC.
OffsetDateTime	A date and time with an offset value from UTC.
ZonedDateTime	A date and time with an associated time zone, such as America/Los_Angeles.
MonthDay	A month and day without an associated year. You can use a MonthDay object to represent a date such as a birthday, anniversary, or holiday.
YearMonth	A year and month, such as December, 2015. No day, time, or time zone values are associated with the year and month.
Year	A year, such as 2038. No month, day, time, or time zone values are associated with the year.
Instant	A single point of time, represented internally as the number of nanoseconds that have elapsed since midnight, January 1, 1970. The value assumes a UTC/GMT time offset of 0.

Using the now Method to Create a Date-Time Object

All Date-Time classes have a static now method, which creates an object representing the current date and/or time. For example, to get the current date, you would use code similar to this:

```
LocalDate date = LocalDate.now();
```

To get the current date and time with time zone, use this code:

```
ZonedDateTime datetime = ZonedDateTime.now();
```

The following program displays the current time using all ten classes, creating an object of each class using now() and printing it with toString():

```
import java.util.*;
import java.time.*;

public class TimeTester
{
    public static void main(String[] args)
    {
        System.out.println("\nLocalTime: "
            + LocalTime.now().toString());
        System.out.println("\nLocalDateTime: "
            + LocalDateTime.now().toString());
        System.out.println("\nZonedDateTime: "
            + ZonedDateTime.now().toString());
        System.out.println("\nOffsetTime: "
            + OffsetTime.now().toString());
        System.out.println("\nOffsetDateTime: "
            + OffsetDateTime.now().toString());
        System.out.println("\nMonthDay: "
            + MonthDay.now().toString());
        System.out.println("\nYearMonth: "
            + YearMonth.now().toString());
        System.out.println("\nInstant: "
            + Instant.now().toString());
    }
}
```

If you compile and run this program, the output will appear something like this:

```
LocalTime: 20:56:26.325

LocalDateTime: 2013-10-07T20:56:26.388

ZonedDateTime: 2013-10-07T20:56:26.388-07:00[America/Los_Angeles]

OffsetTime: 20:56:26.388-07:00

OffsetDateTime: 2013-10-07T20:56:26.388-07:00
```

```
MonthDay: --10-07

YearMonth: 2013-10

Instant: 2013-10-08T03:56:26.388Z
```

From this output, you can get an idea of the information represented by the various Date-Time classes.

Using the parse Method to Create a Date-Time Object

Another way to create a Date-Time object is to use the static parse method, which creates a Date-Time object from a string that represents a specific date or time. For example, the following code creates a LocalDate object representing December 15, 2014:

```
LocalDate d = LocalDate.parse("2014-12-15");
```

To create a LocalDateTime object that represents a specific time on a specific date, use the parse method. Here's an example that sets the time to 3:45 p.m. on December 15, 2014:

```
LocalDateTime dt;
dt = LocalDateTime.parse("2014-12-15T15:45");
```

Note that the letter T separates the date from the time, and the time is expressed in 24-hour clock format. If you need to be more precise, you can also specify seconds, as in this example:

```
dt = LocalDateTime.parse("2014-12-15T15:45:13.5");
```

Here the time is set to 13.5 seconds after 2:45 p.m.

If the string is not in the correct format, the parse method throws a DateTime ParseException. Whenever you use the parse method, you should enclose it in a try block and catch this exception, as in this example:

```
LocalDateTime dt;
try
{
```

```
    dt = LocalDateTime.parse("2014-12-15T03:45PM");
}
catch (DateTimeParseException ex)
{
    System.out.println(ex.toString());
}
```

The parse method is especially useful for converting user input to a Date-Time object. For example, you might use it along with the Scanner class to read a date from the console, or you can use parse in a Swing application to read a date from a text box. When you do, you should prompt the user with the expected date format and catch DateTimeParseException in case the user enters the date in the wrong format.

Using the of Method to Create a Date-Time Object

A third way to create Date-Time objects is to use the static of method to create a Date-Time object from its constituent parts. For example, you can create a LocalDate object by supplying integers that represent the year, month, and day like this:

```
LocalDate date = LocalDate.of(2014,12,15);
```

Each of the Date-Time classes has one or more variations of the of method, as spelled out in Table 4-2.

TABLE 4-2 **Date-Time of Methods**

Class	Method
LocalTime	of(int hour, int minute)
	of(int hour, int minute, int second)
	of(int hour, int minute, int second, int nanoOfSecond)
LocalDate	of(int year, int month, int dayOfMonth)
	of(int year, Month month, int dayOfMonth)

(continued)

TABLE 4-2 *(continued)*

Class	Method
LocalDateTime	of(int year, int month, int dayOfMonth, int hour, int minute)
	of(int year, int month, int dayOfMonth, int hour, int minute, int second)
	of(int year, int month, int dayOfMonth, int hour, int minute, int second, int nanoOfSecond)
	of(int year, Month month, int dayOfMonth, int hour, int minute)
	of(int year, Month month, int dayOfMonth, int hour, int minute, int second)
	of(int year, Month month, int dayOfMonth, int hour, int minute, int second, int nanoOfSecond)
	of(LocalDate date, LocalTime time)
OffsetTime	of(int hour, int minute, int second, int nanoOfSecond, ZoneOffset offset)
	of(LocalTime time, ZoneOffset offset)
OffsetDateTime	of(int year, int month, int dayOfMonth, int hour, int minute, int second, int nanoOfSecond, ZoneOffset offset)
	of(LocalDate date, LocalTime time, ZoneOffset offset)
	of(LocalDateTime dateTime, ZoneOffset offset)
MonthDay	of(int month, int dayOfMonth)
	of(Month month, int dayOfMonth)
YearMonth	of(int year, int month)
	of(int year, Month month)
Year	of(int year)

Note that several of the methods in Table 4-2 use the additional types Month, ZoneOffset, and ZoneId. These types are described in the following sections.

Using the Month enumeration

Several of the methods listed in Table 4-2 let you specify the month as a Month object. Month is an enumeration that represents the twelve months of the year, as follows:

```
Month.JANUARY      Month.MAY       Month.SEPTEMBER
Month.FEBRUARY     Month.JUNE      Month.OCTOBER
Month.MARCH        Month.JULY      Month.NOVEMBER
Month.APRIL        Month.AUGUST    Month.DECEMBER
```

Thus you can create a date like this:

```
LocalDate date = LocalDate.of(2014,Month.DECEMBER,15);
```

Interestingly, the `Month` enumeration has some interesting methods which you might find occasionally useful. For example, you can print the number of days in December like this:

```
System.out.println("December hath "
    + Month.DECEMBER.length(false) + " days.");
```

The `boolean` argument in the `length` method indicates whether the calculation should be for a leap year. Consult the online documentation for other useful methods of the `Month` enumeration.

Using the ZoneId class

To create a `ZonedDateTime`, you must first create a time zone object by using the `ZoneId` class. To create a time zone, you must know the standard name of the time zone you want to create. Unfortunately, there are more than 500 distinct zone IDs, and they periodically change. So listing them here would be impractical, but you can easily list them all by using this handy bit of code:

```
for (String id : ZoneId.getAvailableZoneIds())
    System.out.println(id);
```

This `for` loop will write the names of each `ZoneId` to the console.

Once you know the name of the ZoneId you want to use, you can create it using `ZoneId.of`, then use it to create a `ZonedDateTime` as in this example:

```
ZoneId z = ZoneId.of("America/Los_Angeles");
ZonedDateTime zdate;
zdate = ZonedDateTime.of(2014, 12, 15, 0, 0, 0, 0, z);
```

Or, if you prefer, you can create the `ZoneId` directly when you create the ZonedDateTime:

```
zdate = ZonedDateTime.of(2014, 12, 15, 0, 0, 0, 0,
    ZoneId.of("America/Los_Angeles"));
```

Using the ZoneOffset class

The of method OffsetTime and OffsetDateTime classes use an additional class named ZoneOffset to indicate the offset from UTC. You can create a ZoneOffset by using any of the following methods of the ZoneOffset class:

» of(String offsetId)

» ofHours(int hours)

» ofHoursMinutes(int hours, int minutes)

» ofHoursMinutesSeconds(int hours, int minutes, int seconds)

» ofTotalSeconds(int totalSeconds)

For example, you can create a ZoneOffset of –8 hours like this:

```
ZoneOffset z = ZoneOffset.ofHours(-8);
```

Alternatively, you could specify the offset as a string, as in this example:

```
ZoneOffset z = ZoneOffset.of("-08:00");
```

Note that when you use a string offset, you must provide two digits for the hours, minutes, and (optionally) seconds.

Once you have a ZoneOffset object, you can use it to create an OffsetTime, as in this example:

```
OffsetTime time = OffsetTime.of(10, 30, 0, 0, z);
```

Or if you prefer, you can create the ZoneOffset directly in the call to the Offset-Time's of method:

```
OffsetTime time = OffsetTime.of(10, 30, 0, 0,
    ZoneOffset.ofHours(-8));
```

Looking Closer at the LocalDate Class

The basic java.time classes are similar enough that once you learn how to use one of them, you'll find it easy to learn how to use the rest. Thus, for the rest of this chapter, I'll focus on just one: the LocalDate class. This class represents a date (year, month, and day) without an associated time. In this section and in the sections that follow, you'll learn how to use many of the methods of this class to extract information about a date, to compare two dates, and to perform calculations on a date.

Table 4-3 shows the most commonly used methods of the LocalDate class. For your convenience, this table includes the methods used to create LocalDate objects, even though those methods have already been covered earlier in this chapter.

TABLE 4-3 ## Methods of the LocalDate Class

Method	Explanation
Methods that create a LocalDate object	
LocalDate now()	Creates a LocalDate object that represents the current date.
LocalDate of(int year, int month, int dayOfMonth)	Creates a LocalDate object with a given year, month, and day.
LocalDate of(int year, Month month, int dayOfMonth)	Creates a LocalDate object with a given year, month, and day.
LocalDate parse(String text)	Creates a LocalDate object by parsing the text string.
Methods that extract information about a date	
int getYear()	Returns the year.
Month getMonth()	Returns the month as a Month object.
int getMonthValue()	Returns the month as an int from 1 through 12.
int getDayOfMonth()	Returns the day of the month.
DayOfWeek getDayOfWeek()	Returns the day of the week as a DayOfWeek object.
int getDayOfYear()	Returns the day of the year.
int lengthOfMonth()	Returns the number of days in this month.
int lengthOfYear()	Returns the number of days in this year.
Methods that compare dates	
boolean isAfter(LocalDate other)	Returns true if this date is after the other date.
boolean isBefore(LocalDate other)	Returns true if this date is before the other date.
boolean isEqual(LocalDate other)	Returns true if this date and other represent the same date.
Methods that perform date calculations	

(continued)

TABLE 4-3 *(continued)*

Method	Explanation
`LocalDate plusDays(long days)`	Returns a copy of the LocalDate with the specified number of days added.
`LocalDate plusNMonths(long days)`	Returns a copy of the LocalDate with the specified number of months added.
`LocalDate plusWeeks(long months)`	Returns a copy of the LocalDate with the specified number of weeks added.
`LocalDate plusYears(long days)`	Returns a copy of the LocalDate with the specified number of years added.
`LocalDate minusDays(long days)`	Returns a copy of the LocalDate with the specified number of days subtracted.
`LocalDate minusMonths(long months)`	Returns a copy of the LocalDate with the specified number of months subtracted.
`LocalDate minusWeeks(long months)`	Returns a copy of the LocalDate with the specified number of weeks subtracted.
`LocalDate minusYears(long years)`	Returns a copy of the LocalDate with the specified number of years subtracted.
`long until(LocalDate endDate, ChronoUnit unit)`	Returns the difference between this date and the specified date measured in the specified units.

Extracting Information About a Date

Several methods of the LocalDate class let you extract useful information about a given date. For instance, the following example shows how you can extract the current year, month, and day:

```
LocalDate date = LocalDate.now();
int year = date.getYear();
int month = date.getMonthValue();
int day = date.getDayOfMonth();
```

If you need to know how many days into the year a particular date is, you can use this code:

```
LocalDate date = LocalDate.parse("2016-04-09");
System.out.println(date.getDayOfYear());
```

This example will print the number 100, as April 9 is the 100th day of 2016.

The getDayOfWeek method returns a value of type DayOfWeek, which is an enumeration with the following values:

SUNDAY	THURSDAY
MONDAY	FRIDAY
TUESDAY	SATURDAY
WEDNESDAY	

Here's an example of how you might use this method:

```
LocalDate date = LocalDate.parse("2016-04-09");
System.out.println(date.getDayOfWeek());
```

In this example, the string SATURDAY will be printed because in 2016, April 9 falls on a Saturday.

The lengthOfMonth and lengthOfYear are useful if you want to know the number of days in the month or year represented by a LocalDate. Both methods take into account leap years.

Comparing Dates

You can't compare Date-Time objects using Java's standard comparison operators. Consider the following example:

```
if (LocalDate.now() == LocalDate.now())
    System.out.println("All is right in the universe.");
else
    System.out.println("There must be a disturbance " +
        "in the space-time continuum!");
```

If you run this code, There must be a disturbance in the space-time continuum! will be printed. That's because when used on objects, the equality operator tests whether two expressions refer to the same object, not to objects with the same value.

To test the equality of two dates, you must use the isEqual method, as in this example:

```
if (LocalDate.now().isEqual(LocalDate.now()))
    System.out.println("All is right in the universe.");
```

Similarly, you must use either the isBefore or the isAfter method to determine whether one date falls before or after another date.

Note that you *can* use built-in operators with methods that return integer results. Thus, the following code will work just as you would expect:

```
if (LocalDate.now().getDayOfMonth() < 15)
    System.out.println("It is not yet the 15th.");
```

Because the getDayOfMonth method returns an integer, you can use the < operator to determine if the 15th of the month has yet arrived.

Calculating with Dates

Just as you cannot use Java's built-in comparison operators with dates, you also may not use built-in mathematical operators. Instead, you can perform addition and subtraction on dates using the various plus and minus methods, and you can determine the difference between two dates by using the until method.

An important fact to consider when doing date and time calculations is that Date-Time objects are *immutable*. That means that once you create a Date-Time object, you cannot change its value. When you perform a calculation on a Date-Time object, the result is a new Date-Time object with a new value.

The plus and minus methods let you add various date and time units to a Date-Time object. Table 4-3 lists four variants of each for the LocalDate class, allowing you to add or subtract years, months, weeks, and days to a LocalDate object. The following code prints the current date, tomorrow's date, and the date one week, one month, and one year from now:

```
System.out.println("Today: " + LocalDate.now());
System.out.println("Tomorrow: " + LocalDate.now().plusDays(1));
System.out.println("Next week: " + LocalDate.now().plusWeeks(1));
System.out.println("Next month: " + LocalDate.now().plusMonths(1));
System.out.println("Next year: " + LocalDate.now().plusYears(1));
```

TIP

To determine the difference between two dates, use the until method. It calculates the difference between a date and the date passed as the first parameter, measured in the units indicated by the second parameter. For example, the following code determines the number of days between May 16, 2014 and December 15, 2014:

```
LocalDate date1 = LocalDate.parse("2014-05-16");
LocalDate date2 = LocalDate.parse("2014-12-15");
System.out.println(date1.until(date2, ChronoUnit.DAYS));
```

Some date calculations can be a bit more complex. For example, consider a business that prepares invoices on the 15th of each month. The following snippet of code displays the number of days from the current date until the next invoicing date:

```
LocalDate today = LocalDate.now();
LocalDate invDate = LocalDate.of(today.getYear(),
    today.getMonthValue(), 15);
if (today.getDayOfMonth() > 15)
    invDate = invDate.plusMonths(1);
long daysToInvoice = today.until(invDate,
    ChronoUnit.DAYS);
System.out.println(daysToInvoice
    + " until next invoice date.");
```

This example works by first getting the current date, then creating a new LocalDate object that represents the 15th of the current month. Then, if the current day of the month is greater than 15, it adds one month to the invoicing date. In other words, if it is the 16th or later, invoicing occurs on the 15th of the *following* month, not of this month. Then it uses the until method to determine the number of days between the current date and the next invoicing date.

TECHNICAL
STUFF

ChronoUnit is an enumeration that defines the various units of time that can be used in date and time calculations. The possible values are:

```
CENTURIES
DAYS
DECADES
ERAS
FOREVER
HALF-DAYS
HOURS
MICROS
MILLENNIA
MILLIS
MINUTES
MONTHS
NANOS
SECONDS
WEEKS
YEARS
```

Most of these are self-explanatory, but two of them are a bit peculiar:

>> ERA indicates whether the date refers to the Common Era (CE, also known as AD) or Before Era (BCE, also known as BC).

>> FOREVER represents the largest value that can be represented as a duration. Sadly, Java won't let you live forever. The following code throws an exception:

```
LocalDate birthday = LocalDate.parse("1959-05-16);
birthday = birthday.plus(1,ChronoUnit.FOREVER);
```

Note that ChronoUnit is in the java.time.temporal package, so be sure to include the following statement at the top of any program that uses ChronoUnit:

```
import java.time.temporal.*;
```

Formatting Dates

If you use the toString() method to convert a LocalDate to a string, you get a string such as 2014-10-31. What if you want to display the date in a different format, such as 10-31-2014 or October 31, 2014? To accomplish that, you can use the format method of the LocalDate class along with a custom formatter you create using the DateTimeFormatter class. To specify the format you want to use, you pass the DateTimeFormatter class a pattern string, using the formatting symbols listed in Table 4-4.

TABLE 4-4

Formatting Characters for the DateTimeFormatter Class

Format Pattern	Explanation
y	Year (two or four digits)
M	Month (one or two digits or three or more letters)
d	Day of month (such as 1, 28)
H	Hour
m	Minute
s	Second (0 to 59)
h	Clock hour (1 to 12)
a	AM or PM
V	Time zone ID (such as America/Los_Angeles)
z	Time zone name (such as Pacific Daylight Time)

The easiest way to create a `DateTimeFormatter` object is to use the static `ofPattern` method along with a pattern string. For example:

```
DateTimeFormatter formatter;
formatter = DateTimeFormatter.ofPattern("dd MMM YYYY");
```

This formatter produces dates formatted like 04 SEP 2014. You can then use the formatter to produce a formatted date string like this:

```
LocalDate date = LocalDate.now();
String formattedDate = date.format(formatter);
```

Here's a simple program that prints the current date in several different formats:

```
import java.util.*;
import java.time.*;
import java.time.format.*;

public class FormatDateTime
{
    public static void main(String[] args)
    {
        LocalDateTime  now = LocalDateTime.now();
        printDate(now, "YYYY-MM-dd");
        printDate(now, "MM-dd-YYYY");
        printDate(now, "dd MMM YYYY");
        printDate(now, "MMMM d, YYYY");
        printDate(now, "HH:mm");
        printDate(now, "h:mm a");
    }

    public static void printDate(LocalDateTime date, String pattern)
    {
        DateTimeFormatter f;
        f = DateTimeFormatter.ofPattern(pattern);
        pattern = (pattern + "              ").substring(0, 14);
        System.out.println(pattern + " " + date.format(f));
    }
}
```

When you run this program, you'll get console output that resembles this:

```
YYYY-MM-dd     2013-10-09
MM-dd-YYYY     10-09-2013
dd MMM YYYY    09 Oct 2013
MMMM d, YYYY   October 9, 2013
```

```
HH:mm          20:29
h:mm a         8:29 PM
```

Did you notice the cool formatting trick I used? I forced the `System.out.println()` patterns to print 14-character-long strings so all the dates would line up. The padding is accomplished by this slick line of code:

```
pattern = (pattern + "              ").substring(0, 14);
```

Here a string of 14 spaces is added to the `pattern` string, then a 14-character-long substring is taken starting at the first character. I figured the nice spacing in the output would make it easier for you to see the effect of each of the pattern strings.

Looking at a Fun Birthday Calculator

Now that you've seen the techniques for working with Date-Time objects, it's time to look at a complete programming example. Listing 4-1 presents a program that prompts the user to enter his or her birthday and then prints a variety of interesting information deduced from the date, including:

>> The day of the week on which the user was born

>> The user's age in years

>> The date of the user's next birthday

>> The number of days until the user's next birthday

>> The user's half-birthday (six months from his or her birthday)

Here's an example of the `BirthdayFun` application in action:

```
Today is October 9, 2013.

Please enter your birthdate (yyyy-mm-dd): 1959-12-15

December 15, 1959 was a very good day!
You were born on a TUESDAY.
You are 53 years young.
Your next birthday is December 15, 2013.
That's just 67 days from now!
Your half-birthday is June 15.

Another? (Y or N) N
```

LISTING 4-1: **The BirthdayFun Application**

```java
import java.util.*;
import java.time.*;                                                    →2
import java.time.format.*;
import java.time.temporal.*;

public class BirthdayFun
{
    static Scanner sc = new Scanner(System.in);                       →8

    public static void main(String[] args)
    {
        do
        {
            LocalDate birthDate;                                       →14

            DateTimeFormatter fullFormat =                            →16
                DateTimeFormatter.ofPattern("MMMM d, YYYY");
            DateTimeFormatter monthDayFormat =
                DateTimeFormatter.ofPattern("MMMM d");

            System.out.println("Today is "                           →21
                + LocalDate.now().format(fullFormat) + ".");

            System.out.println();
            System.out.print("Please enter your birthdate "
                + "(yyyy-mm-dd): ");
            String input = sc.nextLine();
            try
            {
                birthDate = LocalDate.parse(input);                   →30

                if (birthDate.isAfter(LocalDate.now()))               →32
                {
                    System.out.println("You haven't been born yet!");
                    continue;
                }

                System.out.println();

                System.out.println(birthDate.format(fullFormat)       →40
                    + " was a very good day!");

                DayOfWeek birthDayOfWeek = birthDate.getDayOfWeek();  →43
                System.out.println("You were born on a "
                    + birthDayOfWeek + ".");
```

(continued)

LISTING 4-1: *(continued)*

```
        long years = birthDate.until(LocalDate.now(),          →47
            ChronoUnit.YEARS);
        System.out.println("You are " + years + " years young.");

        LocalDate nextBDay = birthDate.plusYears(years + 1);    →51
        System.out.println("Your next birthday is "
            + nextBDay.format(fullFormat) + ".");

        long wait = LocalDate.now().until(nextBDay,             →55
            ChronoUnit.DAYS);
        System.out.println("That's just " + wait
            + " days from now!");

        LocalDate halfBirthday = birthDate.plusMonths(6);       →60
        System.out.println("Your half-birthday is "
            + halfBirthday.format(monthDayFormat) + ".");
    }
    catch (DateTimeParseException ex)
    {
        System.out.println("Sorry, that is not a valid date.");
    }
} while(askAgain());
}

private static boolean askAgain()
{
    System.out.println();
    System.out.print("Another? (Y or N) ");
    String reply = sc.nextLine();
    if (reply.equalsIgnoreCase("Y"))
    {
        return true;
    }
    return false;
}
}
```

The following paragraphs explain the most important lines in this program:

» →2 The program uses classes from three packages: java.time, java.time.format, and java.time.temporal.

» →8 A Scanner is used to get the user input. The Scanner is defined as a class variable so that it can be accessed by both the main and the askAgain methods.

>> →14 The `birthdate` variable is used to store the birthdate entered by the user.

>> →16 The program uses two formatters: `fullFormat` formats the date in full-text format (such as December 15, 1959) and `monthDay Format` formats the date as just a month and day.

>> →21 This line displays the current date.

>> →30 The string entered by the user is parsed.

>> →32 The `if` statement ensures that the user has not entered a date in the future.

>> →40 The date entered by the user is displayed.

>> →43 The day of the week is calculated and displayed.

>> →47 The person's age is calculated by determining the difference between the current date and the birthdate in years.

>> →51 The date of the person's next birthday is calculated by adding the person's age plus 1 year to the original birthdate.

>> →55 The number of days until the person's next birthday is calculated.

>> →60 The person's half birthday is calculated by adding six months to the original birthdate.

Chapter **5**

IoT Programming with Raspberry Pi

Welcome to the Internet of Things (IoT)!

The world has exploded with small devices powered by microprocessors and connected to the Internet, including smart TVs, air conditioning and heating controllers, smart-home hubs and controllers, smart speakers, refrigerators, home security systems, toys and games, and more.

The programs that run on IoT devices are often referred to as *embedded programs* because the device is typically locked down to prevent other types of programs from being loaded and run. In other words, these devices are single-purpose, with one and only one program that runs on the device. For example, an IoT thermostat has just one program, designed to turn your air conditioner and/or heater on and off to keep your home comfortable. Similarly, a backyard sprinkler timer has just one program, designed to turn your sprinkler valves on and off according to a schedule.

Many IoT devices run the Linux operating system, configured in a way that completely hides the presence of the operating system from the end user. These devices automatically launch Linux when they're turned on; Linux, in turn, automatically launches the program that controls the device. The user interacts only with this program, using whatever sorts of input/output (I/O) components the device may possess — usually, a small display, as well as some buttons the user can press to interact with the program.

This chapter shows you how to write Java programs that run under Linux on a popular platform for IoT devices: the Raspberry Pi, a complete computer-on-a-board that sells for as little as $45.

This chapter presents a brief introduction to Raspberry Pi and shows you the basic programming techniques for working with the most interesting feature of the Pi: the I/O pins that let you wire up your own circuits to incorporate external devices such as LEDs, push buttons, and sensors like temperature or proximity sensors.

TECHNICAL STUFF

To run the programs in this chapter, you need to do a bit of electronic tinkering. Don't worry — I provide detailed instructions on how to build the simple and inexpensive circuits required by these programs. But you do need to purchase a handful of resistors, LEDs, and a few other bits and pieces. And, of course, you need a Raspberry Pi. See the sidebar "Where to get the parts" for information about sources.

For more complete information about building electronic circuits and working with Raspberry Pi, pick up a copy of *Electronics All-in-One For Dummies*, 2nd Edition, by yours truly (published by Wiley).

Introducing the Raspberry Pi

A Raspberry Pi is a small computer about the size of a deck of cards. It contains most of the components found in a traditional desktop computer. The latest model of the Raspberry Pi, called the Raspberry Pi 4, is shown in Figure 5-1. You can buy one of these online for $40 to $65, depending on the amount of random access memory (RAM).

The Raspberry Pi 4 includes all the following packed onto the board:

>> **Central processing unit (CPU):** A quad-core 64-bit ARM Cortex-A72 microprocessor running at 1.5GHz.

>> **RAM:** 2GB, 4GB, or 8GB.

FIGURE 5-1:
A Raspberry Pi 4.

>> **USB ports:** Two standard USB 2.0 ports and two standard USB 3.0 ports mounted on the board. These ports can be used to connect any USB device, such as a keyboard, a mouse, or a flash drive.

>> **Video:** A built-in graphics processor that can support 4K resolution on two monitors.

>> **HDMI:** Two micro-HDMI ports for dual-monitor support.

>> **Display serial interface (DSI):** A display interface designed to connect to small LCDs via a 15-pin ribbon cable.

>> **MicroSDHC card:** The MicroSDHC card acts as the computer's disk drive. The operating system (Linux) is installed on the MicroSD card, along with any other software you want to use.

>> **Ethernet networking:** A built-in 1 Gbps RJ-45 connector for networking.

>> **802.11ac wireless network:** A built-in wireless network connection. The antenna is actually built into the board itself, so no external antenna is needed.

>> **Bluetooth:** Built-in Bluetooth networking for wireless devices such as a keyboard, a mouse, and headphones.

>> **Camera serial interface (CSI):** A special interface designed to connect to a camera device via a 15-pin ribbon cable.

>> **Audio:** A 3.5mm audio jack for sound applications.

>> **Power:** The Raspberry Pi is powered by a 5V supply connected to the board via a USB-C connection.

>> **General-purpose input/output (GPIO) header:** The most interesting thing about the Raspberry Pi from a hobbyist's perspective is the 40-pin GPIO header, which provides access to a variety of features including 26 GPIO pins. You can use these GPIO pins as output pins to connect to devices such as LEDs or servo or stepper motors or as input pins to read input from external switches, potentiometers, temperature sensors, proximity sensors, and many other types of devices.

TECHNICAL
STUFF

Each of the 26 GPIO pins is analogous to a Boolean value, which can be true or false. In the world of electronics, true and false are usually called HIGH and LOW, respectively, and are represented by the presence or absence of positive voltage. In the case of the Raspberry Pi GPIO pins, HIGH (true) is indicated by the presence of +3.3V; LOW (false) is 0 volts.

TIP

Older versions of the Raspberry Pi are more than adequate for the programs presented in this chapter. So, if you already have one lying around, you don't need to buy a newer one. (All the examples in this chapter were tested on both a Raspberry Pi 4 and a Raspberry Pi 3B.)

Setting Up a Raspberry Pi

Before you can fire up your Raspberry Pi and start building projects, you need to do some basic setup work. Start by setting up the hardware. You need the following to set up your Pi so that you can program it for the projects presented in this chapter:

>> **A Raspberry Pi:** I recommend either the latest version (Pi 4) or its predecessor (Pi 3B).

>> **A suitable power supply:** The Raspberry Pi requires a 5V power supply connected via a micro-USB (Pi 3) or USB-C (Pi 4) connection on the card. For a Pi 3B, use at least a 2.5A power supply; for a Pi 4, 3.0A or more is recommended.

>> **A monitor:** You don't need a large monitor, but I suggest at least 17 inches. The monitor should have an HDMI input.

>> **An HDMI cable:** You'll need an HDMI cable of the correct format for your Raspberry Pi model — for a Pi 3B, a standard HDMI; for a Pi 4, a micro-HDMI.

>> **A USB keyboard:** Any keyboard with a USB connector will do.

>> **A USB mouse:** Any mouse with a USB connector will do.

>> **A microSD card with the Raspbian Linux distribution:** The Raspberry Pi uses a microSD card instead of a disk drive. Buy a microSD card with at least

8GB capacity. The maximum size for a Pi 3B is 32GB; for a Pi 4, the limit is 64GB.

After you have the card, download and run the Raspberry Pi Installer to install Raspbian. You can find the installer at www.raspberrypi.org/downloads.

» **A network connection:** A network connection is essential to download several of the support packages you need for the projects in this book. You can connect your Pi to a network using either the built-in Wi-Fi or the wired Ethernet port.

That's all you need to get started. Plug the monitor, mouse, and keyboard into your Pi's HDMI and USB ports, insert the microSD card into the microSD slot, and then plug in the power connector. Your Pi will start right up.

Installing Java on a Raspberry Pi

Before you can write Java programs on a Raspberry Pi, you need to install the JDK and configure a development environment. For this chapter, we'll use Geany, a fast and lightweight IDE that is included in Raspbian.

 To install Java, open a terminal window by clicking the icon shown in the margin. Then, enter the following commands:

```
sudo apt update
sudo apt install openjdk-8-jdk
```

The first of these two lines updates the list of available packages; the second downloads and installs the current version of the JDK. (If you're prompted to continue, just enter **Y**.)

To verify that Java was installed, enter this command:

```
java -version
```

If Java installed correctly, you'll see something like this:

```
openjdk version "11.0.7" 2020-04-14
OpenJDK Runtime Environment (build 11.0.7+10-post-Raspbian-3deb10u1)
OpenJDK Server VM (build 11.0.7+10-post-Raspbian-3deb10u1, mixed mode)
```

If you encounter an error that your Pi doesn't support this version of Java, you can uninstall jdk-11.

Installing the Pi4J Library

After you've successfully installed Java, the next step is to install a library named Pi4J, which will allow your programs to access the GPIO pins on the Pi. You'll learn more about this library later in this chapter.

You can install Pi4J by entering this command:

```
curl -sSL https://Pi4J.com/install | sudo bash
```

This command downloads an installation script that handles all the details of downloading and installing Pi4J.

Configuring the Geany Integrated Development Environment for Java Development

The final step in preparing your Raspberry Pi for Java is to configure the built-in Geany integrated development environment (IDE) to work with Java and the Pi4J Library. Follow these steps:

1. **Open Geany by clicking the Pi icon at the top of the desktop (shown in the margin) and then choosing Programming⇨Geany.**

   Geany, shown in Figure 5-2, appears.

2. **Choose Build⇨Set Build Commands.**

   The Set Build Commands window, shown in Figure 5-3, appears.

3. **Confirm the Compile and Execute settings.**

   The Compile Command should be

   ```
   sudo javac "%f" -classpath .:/opt/P\pi4j/lib/'*'
   ```

   The Execute Command should be

   ```
   sudo java  -classpath .:/opt/Pi4J/lib/'*' "%e"
   ```

   If these commands aren't already set this way, reset them now.

4. **Click OK to save the changes.**

Congratulations! You're now ready to start writing Java programs on your Raspberry Pi.

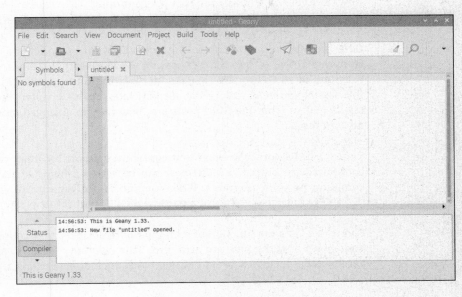

FIGURE 5-2:
Geany, the IDE for Raspberry Pi programming.

FIGURE 5-3:
The Set Build Commands window.

Examining GPIO Ports

Now that your Pi is ready for Java programming, let's take a deeper look at the GPIO ports that make Raspberry Pi different from other computers that run Java.

GPIO ports allow you to control your own electronic circuits and devices directly from the Pi. The 40-pin GPIO header on the Raspberry Pi 3 and 4 provides access to a total of 26 GPIO ports that can be controlled in Java using Pi4J. I explain all the details you need to know about Pi4J later in this chapter, but for now let's take a look at what the GPIO ports are, how they're arranged, and what you can do with them.

To use a GPIO port, you must first configure the port for input or output. When configured for output, a GPIO port can be set to a high (+3.3 V) or low (0 V) condition by your programs. When configured for input, a program can easily determine the current state of a particular GPIO port (high or low).

Figure 5-4 lists the function of each pin on the Raspberry Pi 40-pin header block and shows how the pins are arranged. This diagram applies to all Raspberry Pi versions that include a 40-pin header block, including versions 2, 3, and 4.

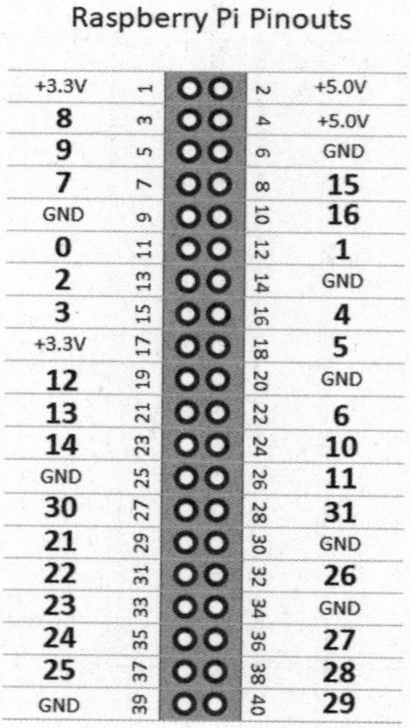

Raspberry Pi Pinouts

FIGURE 5-4: Raspberry Pi header pins (Pi versions 2, 3, and 4).

Unfortunately, the pins on the header block are not labeled in any way on the Raspberry Pi board. If you orient the board so that the header block is on the right edge of the board, just above the USB ports, pins 1 and 2 are at the top of the header block. The odd-numbered pins are on the left side of the header block; even-numbered pins are on the right side.

In addition to GPIO ports, the 40-pin header also provides pins for power and ground. Pins 1 and 2 provide +3.3V and +5.0V, respectively. You won't use these pins in any of the circuits in this chapter, because the GPIO ports provide the positive voltage the circuits require. But you *will* use the ground pins to complete the circuits. The 40-pin header provides a total of eight ground pins, but the projects in this chapter will use just the one on pin 6. (All the ground pins are the same; there are eight of them on the header just for convenience.)

You've probably already noticed that the pins on the header block don't appear to be laid out in any particular logical order. The GPIO ports are scattered about the header haphazardly, interspersed with 3.3V, 5V, and ground pins at seemingly random locations.

Table 5-1 lists all the GPIO ports and their pin locations, sorted by GPIO number.

TABLE 5-1

GPIO Ports and Pin Locations

GPIO Port	Header Pin	GPIO Port	Header Pin
0	11	13	21
1	12	14	23
2	13	15	8
3	15	16	10
4	16	21	29
5	18	22	31
6	22	23	33
7	7	24	35
8	3	25	37
9	5	26	32
10	24	27	36
11	26	28	38
12	19	29	40

You may notice in Table 5-1 that GPIO ports 17, 18, 19, and 20 are missing. Those ports exist internally but aren't connected to any of the pins on the 40-pin header.

Be aware that there are three distinct ways of numbering the GPIO ports on a Raspberry Pi:

>> **Board:** When you use board numbering, you refer to pins using the actual pin number on the 40-pin connector, starting with 1 at the upper left and ending with 40 at the lower right. In this chapter, I use the board numbers in the instructions for constructing circuits that connect to the Pi.

>> **BCM:** BCM, which stands for *Broadcom SOC Channel,* is based on the GPIO port numbering used by the CPU, which is made by Broadcom. Most Raspberry Pi programming languages use the BCM numbering scheme, but pi4j doesn't because the GPIO numbers sometimes change when new versions of the Raspberry Pi, based on different Broadcom CPUs, are released. (The only reason I mention it here is that when you search for information about Raspberry Pi pin numbers, most of the hits will list the BCM numbering. BCM port numbers are not used in this book.)

>> **WiringPi:** This is the GPIO numbering scheme used by pi4j. It abstracts the GPIO port numbering from both the physical pin numbers and the CPU port numbers. With each release of pi4j, the GPIO numbering scheme remains the same.

TIP

All the programs and examples in this book use WiringPi port numbering. I mention the corresponding board pin number whenever it's needed to help you assemble the projects.

The pin diagram shown in Figure 5-4 is helpful to see which pin corresponds to which function.

Connecting an LED to a GPIO Port

If you configure a GPIO port as an output port, you can use the port to drive a *light-emitting diode* (LED). In fact, one of the most common uses of GPIO ports is to control one or more LEDs to act as status indicators for your project or perhaps flash in interesting patterns purely for the fun of it.

An LED has two wire leads protruding from its bottom. It's important to realize that these two leads are not interchangeable. Current can flow in only one direction through the LED. The longer lead is the positive voltage side of the LED. It's called the *anode.* The shorter lead, called the *cathode,* is the negative side of the LED.

When connecting an LED in a circuit, the anode goes on the positive voltage side of the circuit and the cathode goes on the negative side. Current flows from positive to negative, so when an LED is connected in this way and voltage is present on the anode, the LED will light up.

However, LEDs allow current to flow in only one direction. If you reverse the LED in the circuit — in other words, if you apply positive voltage on the cathode instead of the anode — current will not flow through the LED and it will not light up.

WARNING

When you use an LED in a circuit, you must provide a resistor in the circuit to limit the amount of current that flows through the LED. When used in this way, the resistor is referred to as a *current-limiting resistor*. Without a current-limiting resistor, the LED will flash brightly for an instant, and then join the nether regions where dead LEDs gather for an eternity of cursing their former owners who didn't protect them from the perils of excessive current.

The current-limiting resistor can go on either side of the LED — that is, on the anode side or the cathode side. The resistor will limit the amount of current flowing through the LED in either position.

Figure 5-5 is an electronic schematic diagram that shows how an LED can be connected to a Raspberry Pi. In this case, the LED's anode (the long lead) is connected directly to GPIO port 1 (pin 12) on the Pi. A 220-ohm resistor is used to limit the current flowing through the LED, and the LED's cathode (the short side) is connected to pin 6 on the Raspberry Pi, which is a ground pin.

WARNING

The current-limiting resistor is essential in this circuit. Without it, the LED will fry in an instant. In Figure 5-5, a 220-ohm resistor is placed between the cathode and the ground. However, the resistor could just as effectively be placed between the GPIO port and the anode.

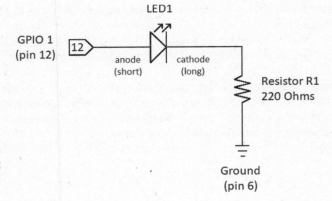

FIGURE 5-5: Connecting an LED to a GPIO port.

Speaking of resistors, the resistance value of a resistor is measured in units called *ohms*, often represented by the Greek letter omega (Ω). The resistance value of a resistor is painted on the resistor as a series of colored stripes. Only the first three stripes (starting from the left) are important for our purposes. I won't get into the complete color coding now, but the following table shows the color codes for the three resistor values needed for this chapter.

Resistor Value	Color Stripes
220Ω	red-red-black
1KΩ	brown-black-red
10KΩ	brown-black-orange

You'll also need a small solderless breadboard to build this circuit, as well as the other circuits in this chapter. If you're not familiar with how solderless breadboards work, check out the nearby sidebar, "Using a solderless breadboard."

TIP

One more detail you should be aware of is that the header block on the Raspberry Pi uses male pins rather than female sockets as are found on Arduino and BASIC Stamp microcontrollers. As a result, to connect a breadboard circuit to pins on the Raspberry Pi, you need jumper wires that have male pins on one end (to plug into a breadboard hole) and female sockets on the other end (to plug into the male pins on the Raspberry Pi header).

If you're wondering where you can get all these parts, please refer to the helpful and cleverly named sidebar "Where to get parts."

USING A SOLDERLESS BREADBOARD

You'll need to use a small solderless breadboard to build the projects in this chapter. The board consists of several hundred little holes called *contact holes* that are spaced 0.1 inch apart. You can insert electronic components such as resistors or LEDs into the holes. You can also insert *jumper wires*, which let you make cross-connection between holes on the board or connect the board to the 40-pin header on the Raspberry Pi.

Beneath the plastic surface of the solderless breadboard, the contact holes are connected to one another inside the breadboard. These connections are made according to a specific pattern that's designed to make it easy to construct even complicated circuits. The following diagram shows how this pattern works.

The holes in the middle portion of a solderless breadboard are connected in groups of five that are called *terminal strips.* These terminal strips are arranged in two groups, with a long open slot between the two groups, like a little ditch. Note that the rows on the left and the rows on the right are not connected to each other through the ditch. So, there are a total of 60 terminal strips — 30 on the left of the ditch and 30 on the right.

The holes on the outside edges of the breadboard are called *bus strips.* There are two bus strips on either side of the breadboard. On each side of the breadboard, the bus strips are labelled with + and – signs to indicate their intended usage: + is for positive voltage, and – is for negative voltage or ground.

Most solderless breadboards use numbers and letters to designate the individual connection holes in the terminal strips. The rows are labeled with numbers (such as from 1 to 30), and the columns are identified with the letters A through J. So, the connection hole in the upper-left corner of the terminal strip area is A1, and the hole in the lower-right corner is J30. The holes in the bus strips are not typically numbered.

WHERE TO GET PARTS

One question you may be asking yourself as you read this chapter is, "Where do I get all this stuff?"

Back in the day, when there was a Radio Shack on every corner, you could get it all at your local Radio Shack store. But since the demise of Radio Shack's retail stores, you'll probably have to go online to buy what you need.

Fortunately, there are many sources for electronic parts on the Internet. You can find everything you need on Amazon, as well as from many other sources.

One of my favorite sources for Raspberry Pi and related paraphernalia is Adafruit (www.adafruit.com), which specializes in providing the information and resources for learning electronics. For your convenience, I've put together a parts list to build all the projects in this chapter, along with the Adafruit Part ID and price as of May 2020. The complete bill of materials should range from $60 to $125, depending on where you get the parts and which Raspberry Pi model you choose.

Description	Adafruit Part ID	Price
Half-size breadboard with 78-piece jumper wire bundle	3314	9.95
Female–male jumper cable (40)	826	$3.95
Resistor, 220-ohm (25)	2780	$0.75
Resistor, 1K-ohm (25)	4294	$0.75
Resistor, 10K-ohm (25)	2784	$0.75
LED, 5mm red (25)	299	$4.00
Tactile button 6mm switch (20)	367	$2.50
Raspberry Pi 4 Model B with 2GB RAM	4292	$35.00
Raspberry Pi 4 Model B with 4GB RAM	4296	$55.00
Raspberry Pi 4 Power USB-C	4298	$7.95

Building a Raspberry Pi LED Circuit

In this section, you build a simple breadboard circuit that connects an LED to a Raspberry Pi via GPIO Port 1, which is located on pin 8 of the header. Figure 5-6 shows a diagram of the completed project.

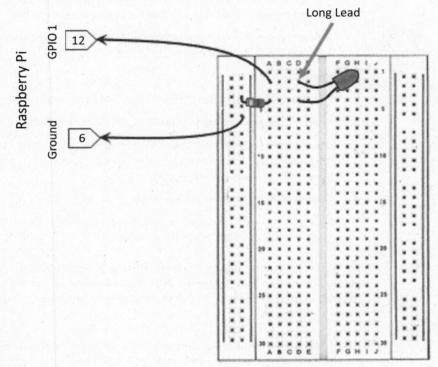

IoT Programming with Raspberry Pi

FIGURE 5-6:
The Raspberry Pi LED flasher.

Parts

You need the following parts:

>> One Raspberry Pi 3 or 4 with Raspbian, Java, and Pi4J installed, connected to a monitor, keyboard, and power

>> One small solderless breadboard

>> One 5mm red LED

>> One 220Ω resistor (red-red-black)

>> Two jumper wires (male–female)

Steps

Refer to Figure 5-6 throughout these steps for the proper location of the resistor, LED, and jumper wires.

TIP

1. **Insert the resistor.**

   Connect one or the resistor's wire lead to the breadboard hole at position A4; connect the other lead to any nearby hole along the ground bus.

   Resistors have no polarity requirements, so you don't have to worry about installing the resistor backward.

2. **Insert the LED.**

   The anode (long lead) goes in position D2. The cathode (short lead) goes in D4.

3. **Connect the ground bus to ground at pin 6 on the Raspberry Pi header.**

   Use the male end of a jumper wire to connect any hole in the ground bus on the breadboard. Then slide the female end of the jumper onto pin 6 of the header. (Pin 6 on the header is a ground pin.)

4. **Connect GPIO 1 (header pin 12) to A2.**

   Slide the female end of a jumper wire onto pin 12 of the header (GPIO 1). Then insert the male end into A5 on the breadboard.

You're done! You're almost ready to write a program to light up the LED. But before you can do that, you need to look at the some of the important classes in the Pi4J Library to see how Pi4J enables your Java program to connect and control the Pi's GPIO pins.

Note that when you complete the circuit, the LED may turn on initially. The reason for this is simply that the initial state of a GPIO port is unpredictable. There are ways to mitigate the fact that the initial state is unpredictable, but the techniques are beyond the scope of this chapter.

Examining the Pi4J Library

The Pi4J Library is a set of packages that allow you to access the Pi's GPIO ports in a Java program. You can find detailed information about the library at http://pi4j.com, including a short tutorial on many of the functions, sample programs, and a complete Javadoc specification for the library's packages, classes, and interfaces.

You can find instructions for installing the library earlier in this chapter, in the "Installing Pi4J" section.

In this chapter, I focus on the most important package in Pi4J: `pi4j.io.gpio`. Table 5-2 lists types from this package that are used in the programs shown in this chapter.

TABLE 5-2   ## Commonly Used GPIO Members

Event Class or Interface	Type	Description
`GpioController`	Interface	This interface describes all the operations you can perform with GPIO ports.
`GpioFactory`	Class	A factory class that is used to create an instance of a `GpioController`.
`GpioPinDigitalOutput`	Interface	Describes the capabilities of a GPIO pin that is used for output.
`GpioPinDigitalInput`	Interface	Describes the capabilities of a GPIO pin that is used for input.
`PinState`	Enum	Lists the states of a pin (HIGH and LOW).
`Pin`	Interface	Describes the general features a GPIO pin.
`RaspiPin`	Class	Defines a total of 32 fields that represent the GPIO ports on a Raspberry Pi according to the WiringPi numbering scheme. The fields are named GPIO_00 through GPIO_31. Each field is of type `Pin`. Note that on a Pi with a 40-pin header, GPIO_17 through 20 and GPIO_30 and 31 are not available for use.
`PinPullResistance`	Enum	Defines the options for pin pull resistance for input pins. (For more information, see the section "Working with Input Pins," later in this chapter.)

Importing GPIO Types

As with any Java library, you must import the Pi4J types that your program will require. Most of the programs so far in this book have been liberal about importing Java types, using an asterisk wildcard to import entire packages. You could do that with the `pi4j.io.gpio` package, as follows:

```
import com.pi4j.io.gpio.*;
```

Using the asterisk wildcard is a bit lazy. If you prefer to be explicit, you'll need the minimum the following `import` statements:

```
import com.pi4j.io.gpio.GpioController;
import com.pi4j.io.gpio.GpioFactory;
import com.pi4j.io.gpio.GpioPinDigitalOutput;
import com.pi4j.io.gpio.PinState;
import com.pi4j.io.gpio.RaspiPin;
```

If you're working with input pins, you'll also need the following:

```
import com.pi4j.io.gpio.GpioPinDigitalInput;
import com.pi4j.io.gpio.PinPullResistance;
```

As you learn more about Pi4J, you'll need additional `import` statements to reference other Pi4J types.

Instantiating a GpioController

Before you can do any operations on GPIO pins, you must create an instance of `GpioController`. You can do that easily with the following code in your `main` method:

```
final GpioController gpio = GpioFactory.getInstance();
```

In this example, the `GpioController` is assigned to a variable name `gpio`.

Table 5-3 shows the `getInstance` method of the `GpioFactory` class.

TABLE 5-3 **The GpioFactory Class**

Method	Description
`GpioController getInstance()`	Returns a `GpioController` instance.

Provisioning GPIO Pins

Let me start here by acknowledging the confusion between the terms *ports* and *pins*. Whenever you're working in Java, the term *pin* refers to a GPIO port, not to an actual pin on the 40-pin header. The Pi4J library regularly uses the terms *GPIO pin* or just *pin* when referring to GPIO ports. Just remember that *pin* in a Java context refers to ports, not header pins.

To work with a GPIO pin, you must first *provision* the pin using one of the various provisioning methods provided by GpioController. When you provision a pin, you establish several important characteristics of the pin:

>> **Pin number:** Remember, this is actually the port number, not the physical pin number on the header. Also, remember that we're using the WiringPi port numbering scheme. (Refer to Figure 5-4 for help on this if you need it.)

>> **Pin name:** This is an optional setting that lets you provide a name for the pin. Note that the pin name is not the same as the name of the variable you use to refer to the pin; instead, the pin name is a value that's held internally as part of the pin object. The pin name is useful in some programming situations, but I don't use it in any of the programs I present in this chapter.

>> **Mode:** Any GPIO pin (port) can be provisioned in one of several modes. The modes specify whether the pin is a digital pin or an analog pin and whether it's an input pin or an output pin. Because analog pins require additional electronics to work with and this isn't an electronics book, I'll stick with digital pins.

>> **Default state:** Specifies whether the pin's initial state is HIGH or LOW.

>> **Pull-down resistance:** This one is for input pins only. For more information, refer to the section "Working with Input Pins," later in this chapter.

GpioController provides several methods available for provisioning GPIO pins; the most common ones are shown in Table 5-4. I don't use all these methods in this chapter, but I want you to see the options available.

Table 5-5 lists all the possible values for the PinMode enum. In this chapter, I only use digital input and output pins, but again I want to show you the other types of pin modes that are possible on Raspberry Pi GPIO pins. If you're interested in working with nondigital pin modes, you'll find plenty of information on the Internet.

With all that information in hand, let's turn to the code required to provision GPIO pins. In the following examples, I assume you've instantiated a GpioController referenced by a variable named gpio, as follows:

```
final GpioController gpio = GpioFactory.getInstance();
```

TABLE 5-4   **The GpioController Interface**

Method	Description
`GpioPin provisionPin(Pin, PinMode)`	Creates a pin on the specified GPIO pin with the specified mode. Refer to Table 5-5 for allowable pin modes.
`GpioPin provisionPin(Pin, String, PinMode)`	Creates a pin on the specified GPIO pin with the specified name and mode. Refer to Table 5-5 for allowable pin modes.
`GpioPinDigitalOutput provisionDigitalOutputPin(Pin)`	Creates an output pin on the specified GPIO pin.
`GpioPinDigitalOutput provisionDigitalOutputPin(Pin, PinState)`	Creates an output pin on the specified GPIO pin with the specified default state.
`GpioPinDigitalOutput provisionDigitalOutputPin(Pin, String)`	Creates an output pin on the specified GPIO pin with the specified name.
`GpioPinDigitalOutput provisionDigitalOutputPin(Pin, String, PinState)`	Creates an output pin on the specified GPIO pin with the specified name and default state.
`GpioPinDigitalInput provisionDigitalInputPin(Pin)`	Creates an output pin on the specified GPIO pin.
`GpioPinDigitalInput provisionDigitalInputPin(Pin, PinPullResistance)`	Creates an input pin on the specified GPIO pin with the specified pin pull resistance. (Refer to Table 5-6 for allowable pin pull resistance options, and to the section "Working with Input Pins," later in this chapter, for information about pin pull resistance.)
`GpioPinDigitalInput provisionDigitalInputPin(Pin, String)`	Creates an input pin on the specified GPIO pin with the specified name.
`GpioPinDigitalInput provisionDigitalInputPin(Pin, String, PinPullResistance)`	Creates an input pin on the specified GPIO pin with the specified name and pin pull resistance.

The simplest way to provision a digital output pin is to use the `provision DigitalOutputPin` method, like this:

```
final GpioPinDigitalOutput pin =
    gpio.provisionDigitalOutputPin(RaspiPin.GPIO_01);
```

Here, a GPIO pin 1 is provisioned as a digital output pin and assigned to the variable `pin`.

TABLE 5-5　　**The PinMode enum**

Constant	Description
ANALOG_INPUT	An analog input pin that can read the voltage level on the pin.
ANALOG_OUTPUT	An analog output pin that can place a variable voltage level on the pin.
DIGITAL_INPUT	A digital input pin whose value can be HIGH or LOW.
DIGITAL_OUTPUT	A digital output pin whose value can be HIGH or LOW.
GPIO_CLOCK	A pin that has regular pulses at an interval you can specify.
PWM_OUTPUT	Pulse-width modulation, often used to control servos, for audio effects or telecommunications applications.
PWM_INPUT	Pulse-width modulation input.
SOFT_PWM_OUTPUT	Pulse-width modification generated by software rather than by the Pi's hardware.
SOFT_TONE_OUTPUT	Creates tone pulses on an output pin.

It's always a good idea to assign an initial state to an output pin. Otherwise, the initial state of the pin will be out of your control. To do so, use this form of the provisionDigitalOutputPin method:

```
final GpioPinDigitalOutput pin =
    gpio.provisionDigitalOutputPin(RaspiPin.GPIO_02,
        PinState.LOW);
```

And to assign a name to the pin, use something like this:

```
final GpioPinDigitalOutput pin =
    gpio.provisionDigitalOutputPin(RaspiPin.GPIO_02, "GPIO 2",
        PinState.LOW);
```

Controlling the Pin State

After you've provisioned a digital output pin, setting its state is easy. Table 5-6 lists the most commonly used methods of the GpioPinDigitalOutput interface, which are used to change the state of a pin.

TABLE 5-6   **The GpioPinDigitalOutput Interface**

Method	Description
`void high()`	Sets the pin state to HIGH.
`void low()`	Sets the pin state to LOW.
`void toggle()`	Inverts the pin state: If it is HIGH, changes it to LOW and vice versa.
`void setState(boolean state)`	Sets the state to true or false.
`void setState(PinState state)`	Sets the state to `PinState.HIGH` or `PinState.LOW`.
`Future<?> pulse(long duration)`	Sets the pin state to HIGH for the specified duration and then sets it to LOW.
`Future<?> blink(long delay)`	Blinks the pin state from HIGH to LOW at the rate specified by `delay`.
`Future<?> blink(long delay, long duration)`	Blinks the pin state at the rate specified by `delay` for the specified `duration`.

Note: The Future type is not covered in this book; in most cases the return value of the pulse method is ignored.

The following examples all assume that you've provisioned a digital output pin and assigned it to a variable named `pin`, and that an LED is connected to the pin as previously shown in Figure 5-5.

Turning a pin on and off is simple using the `high` and `low` methods. For example:

```
pin.high();
```

Here's a simple `while` loop to alternately turn the LED on and off at intervals of a quarter of a second:

```
while (true)
{
    pin.high();
    Thread.sleep(250);
    pin.low();
    Thread.sleep(250);
}
```

Here's an alternative that does the same thing in less code using the `toggle` method:

```
while (true)
{
    pin.toggle();
    Thread.sleep(250);
}
```

Here's a version that accomplishes the same thing using the `pulse` method:

```
while (true)
{
    pin.pulse(250);
    Thread.sleep(250);
}
```

And here's a version that doesn't need a `while` loop at all; instead, it just sets the LED blinking by calling the `blink` method:

```
pin.blink(250);
```

TECHNICAL STUFF

Both the pulse and blink methods return a value of type `Future<V>`. The `Future` type, which isn't covered in this book, provides a means to get a value from a long-running computation. For the `pulse` and `blink` methods, it can be used to pulse or blink a pin for a long duration, and then cancel the blink or pulse later by calling the `Future` object's `cancel` method (for example, when a user pushes a button because he or she has become bored with the blinking LED). In most cases, the return value of the `pulse` and `blink` methods are simply ignored.

The Morse Code Program

If you've read this chapter from the beginning, you've now seen all the basic techniques you need to create a simple Java program that will play with an LED on a Raspberry Pi, using the circuit that you build earlier in the chapter, in the "Connecting an LED to a GPIO Port" section and illustrated in Figure 5-5.

Listing 5-1 shows a program that flashes the familiar Morse code distress signal SOS on the LED. The Morse code for SOS is three dots, three dashes, and three dots. This program simply blinks an LED in the familiar pattern: dot-dot-dot dash-dash-dash dot-dot-dot.

TIP

The proper way to pronounce Morse code is *dit* for dots and *dah* for dashes, so the correct way to say SOS in Morse code is *dit dit dit dah dah dah dit dit dit.*

According to the rules of Morse code, a dash is three times as long as a dot, and the gap between words is seven times as long as a dash. Morse code standards also call for a three-dot gap between the individual letters of a word. However, for the special code SOS, the interletter gaps are skipped.

LISTING 5-1: **The Morse Code Program**

```
import com.pi4j.io.gpio.GpioController;                                    →1
import com.pi4j.io.gpio.GpioFactory;
import com.pi4j.io.gpio.GpioPinDigitalOutput;
import com.pi4j.io.gpio.PinState;
import com.pi4j.io.gpio.RaspiPin;

// Flashes the Morse Code for SOS on an LED connected to GPIO Pin 1.

public class MorseCode                                                     →9
{
    public static void main(String args[]) throws InterruptedException
    {
        final GpioController gpio = GpioFactory.getInstance();             →13

        final GpioPinDigitalOutput pin =                                  →15
            gpio.provisionDigitalOutputPin(RaspiPin.GPIO_01,
                PinState.LOW);

        int speed = 100;                                                  →19
        // Dot length is one unit
        // Dash length is three units
        // Element gap is one unit
        // Letter gap is three units
        // Word gap is seven units

        while(true)                                                      →26
        {
            // Signal SOS
            // Note that the SOS distress signal is transmitted
            // without the usual gaps between letters.

            showDots(3, speed, pin);        // "S"                       →32
            showDashes(3, speed, pin);      // "O"                       →33
            showDots(3, speed, pin);        // "S"                       →34
            showWordGaps(speed);                                        →35
        }
    }

    public static void showDots(int count, int speed,                    →39
            GpioPinDigitalOutput pin)
        throws InterruptedException
    {
```

```
        for (int i = 0; i < count; i++)                          →43
        {
            pin.high();
            Thread.sleep(speed);
            pin.low();
            Thread.sleep(speed);
        }


    }

    public static void showDashes(int count, int speed,         →53
                GpioPinDigitalOutput pin)
                throws InterruptedException
    {
        for (int i = 0; i < count; i++)
        {
            pin.high();
            Thread.sleep(speed * 3);
            pin.low();
            Thread.sleep(speed);
        }
    }

    public static void showLetterGaps(int speed)                →66
                throws InterruptedException
    {
        Thread.sleep(speed * 3);
    }

    public static void showWordGaps(int speed)                  →72
                throws InterruptedException
    {
        Thread.sleep(speed * 7);
    }
}
```

Here are the key aspects of this program:

>> →1 The import statements import the five com.pi4j.io.gpio types that will be used by the program.

>> →9 There's nothing unusual about the main method except that it throws InterruptedException. This is necessary because several of the programs methods use Thread.sleep, which can throw InterruptedException.

>> →13 This line creates a GpioController instance and assigns it to the variable gpio.

» →15 This line provisions a digital output pin on GPIO_01 with an initial state of LOW, and assigns the pin to the variable pin. This pin is connected to an LED that will light when the pin is HIGH.

» →19 This line defines a variable named speed, which will be used to govern the tempo of the Morse code flashed on the LED. The speed indicates the length of a single dot. The other elements of Morse code are governed by the following standards:

- A dash is three times as long as a dot.
- The gap between individual dots and dashes of a single letters is the same duration as a dot.
- The gap between individual letters is three times as long as a dot.
- The gap between individual words is seven times as long as a dot.

» →26 This while loop runs forever so that the program flashes SOS on the LED repeatedly until the program is terminated.

» →32 This line calls a method named showDots, which shows one or more dots on the LED. The number of dots to show, the speed, and the pin are passed as arguments. Three dots — Morse code for the letter *S* — are shown.

» →33 This line calls a method named showDashes to display three dashes — Morse code for the letter *O*.

» →34 The showDots method is called again to display the letter *S* — three dots.

» →35 A method named showWordGaps is called to dim the LED for the duration of a word gap (equivalent to seven dots). Note that no pin is passed to this method because the LED remains dark.

» →39 The showDots method accepts three parameters: an int representing the number of dots to display, an int that represents the speed of the display, and a GpioPinDigitalOutput that specifies the pin used to flash the LED. This method also throws InterruptedException, because it calls the Thread.Sleep method, which can throw that exception.

» →43 The showDots method features a for loop that turns LED on, sleeps for the duration of the sleep argument, turns the LED off, and sleeps again for the duration of the sleep argument. The first Thread.sleep call sets the duration of the dot, while the second represents the length of the gap between dots. The for loop iterates once for each dot to be displayed, as indicated by the count argument.

» →53 The showDashes method is almost identical to the showDots method, with just one difference: The first Thread.Sleep call keeps the LED on for three times the duration of a single dot as indicated by the sleep argument.

>> →66 The showLetterGaps method is not called anywhere in the program, but I included it out of a compulsive need for completeness. This method implements the three-dot-length gap between individual letters in a normal word in Morse code. For the special code SOS, the interletter gap is not used, but I thought it would be nice to implement it anyway.

>> →72 Finally, the showWordGap method implements the standard seven-dot-length gap that appears between words in Morse code. It does this simply by sleeping for seven times the length of a dot as passed via the speed argument.

That's it for the simple Morse code program. If you'd like a little programming challenge, extend this program so that it translates a text string passed as a command-line argument into Morse code displayed on the LED. To do that, you'll have to devise a data structure to represent the Morse code values for each letter and numeral; then parse the input string, character by character, and display the appropriate sequence of dots and dashes on the LED. Have fun with that one!

The Cylon Eyes Program

The Morse code program in the previous section is interesting, but it demonstrates how to use only one GPIO pin at a time. In many cases, using more than one pin is simply a matter of provisioning additional digital output pins and assigning them unique variable names. However, there are cases where you may want to hold a collection of pins in an array or a list and iterate over them in a for loop.

The program in this section shows you how to do that by using eight LEDs to simulate the creepy flashing electronic eyes of the Cylons from the old, campy science-fiction show *Battlestar Galactica*. (If you didn't watch *Battlestar Galactica*, maybe you watched *Knight Rider* — the famous KITT car had similar blinking lights on the front end. If you didn't watch either show, well, you now know how much older I am than you.)

Assembling the Cylon Eyes circuit

In this section, you build a breadboard circuit that connects eight LEDs to a Raspberry Pi on GPIO ports 1 through 8 (using the Wiring Pi numbering convention). Figure 5-7 shows a diagram of how the LEDs and current-limiting resistors are assembled on the breadboard and how the breadboard should be connected to the Pi. The LEDs are connected as shown in Table 5-7.

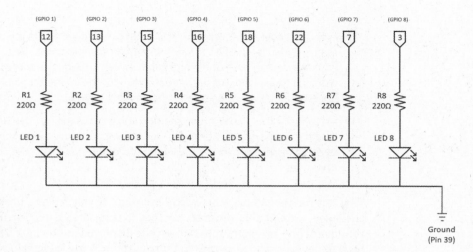

FIGURE 5-7:
A schematic
diagram for the
Cylon Eyes circuit.

TABLE 5-7

How the LEDs Are Connected

LED	WiringPi Pin	Physical pin
LED1	GPIO_01	12
LED2	GPIO_02	13
LED3	GPIO_03	15
LED4	GPIO_04	16
LED5	GPIO_05	18
LED6	GPIO_06	22
LED7	GPIO_07	7
LED8	GPIO_08	3

Figure 5-8 shows the breadboard layout for the circuit. You can use this diagram as a guide when you assemble the parts on the breadboard.

You need the following parts to build the circuit, all of which are readily available on the Internet:

>> One Raspberry Pi 3 or 4

>> One small solderless breadboard

>> Eight 5mm red LEDs

» Eight 220Ω resistors

» 9 male–female jumper wires

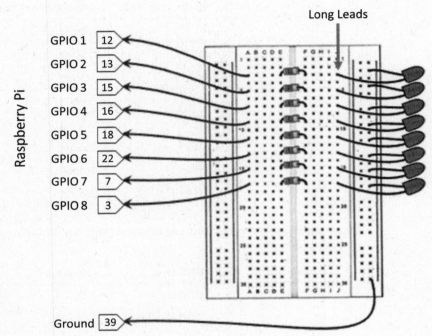

FIGURE 5-8:
The breadboard
layout for the
Cylon Eyes circuit.

Follow these steps to assemble the circuit:

1. **Insert the resistors into the correct pins on the breadboard:**

Resistor	From	To
R1	E3	F3
R2	E5	F5
R3	E7	F7
R4	E9	F9
R5	E11	F11
R6	E13	F13
R7	E15	F15
R8	E17	F17

Note that the resistors span the ditch that runs down the middle of the breadboard.

2. **Insert the LEDs into the correct pins on the breadboard:**

LED	Anode (Long lead)	Cathode (Short lead)
LED1	J3	Ground
LED2	J5	Ground
LED3	J7	Ground
LED4	J9	Ground
LED5	J11	Ground
LED6	J13	Ground
LED7	J15	Ground
LED8	J17	Ground

3. **Connect jumper wires from the breadboard to the GPIO pins on the Raspberry Pi.**

Breadboard	GPIO Port	Physical Pin
A3	GPIO 1	12
A5	GPIO 2	13
A7	GPIO 3	15
A9	GPIO 4	16
A11	GPIO 5	18
A13	GPIO 6	22
A15	GPIO 7	7
A17	GPIO 8	3

4. **Connect the ground bus to pin 39 on the Raspberry Pi.**

   You can now use the Cylon Eyes program (see the next section).

Running the Cylon Eyes program

Listing 5-2 presents the complete code for the Cylon Eyes program.

LISTING 5-2: **The Cylon Eyes Program**

```java
import com.pi4j.io.gpio.GpioController;                                      →1
import com.pi4j.io.gpio.GpioFactory;
import com.pi4j.io.gpio.GpioPinDigitalOutput;
import com.pi4j.io.gpio.PinState;
import com.pi4j.io.gpio.RaspiPin;
import java.util.ArrayList;

// Creates a bouncing light effect on 8 LEDs connected to GPIO 1-8.

public class CylonPi
{
    public static void main(String args[]) throws InterruptedException
    {
        final GpioController gpio = GpioFactory.getInstance();              →14

        ArrayList<GpioPinDigitalOutput> pins = new ArrayList<GpioPinDigitalOut
put>();                                                                     →17

        pins.add(gpio.provisionDigitalOutputPin(RaspiPin.GPIO_01,           →19
                PinState.LOW));
        pins.add(gpio.provisionDigitalOutputPin(RaspiPin.GPIO_02,
                PinState.LOW));
        pins.add(gpio.provisionDigitalOutputPin(RaspiPin.GPIO_03,
                PinState.LOW));
        pins.add(gpio.provisionDigitalOutputPin(RaspiPin.GPIO_04,
                PinState.LOW));
        pins.add(gpio.provisionDigitalOutputPin(RaspiPin.GPIO_05,
                PinState.LOW));
        pins.add(gpio.provisionDigitalOutputPin(RaspiPin.GPIO_06,
                PinState.LOW));
        pins.add(gpio.provisionDigitalOutputPin(RaspiPin.GPIO_07,
                PinState.LOW));
        pins.add(gpio.provisionDigitalOutputPin(RaspiPin.GPIO_08,
                PinState.LOW));

        while(true)                                                         →36
        {
            for(int i = 0; i < 8; i++)                                      →38
            {
                pins.get(i).high();
                Thread.sleep(50);
                pins.get(i).low();
            }
            for(int i = 6; i > 0; i--)                                      →44
            {
```

(continued)

IoT Programming with
Raspberry Pi

LISTING 5-2: **(continued)**

```
            pins.get(i).high();
            Thread.sleep(50);
            pins.get(i).low();
        }
    }
  }
}
```

The following paragraphs highlight the pertinent aspects of this program:

>> →1 These are the standard import statements for a GPIO program. In addition, I've imported the `java.util.ArrayList` so that I can use an `ArrayList` to easily manage the eight pins.

>> →14 This line creates an instance of `GpioController` and saves it in the variable named `gpio`.

>> →17 Creates a new `ArrayList` of type `GpioPinDigitalOutput`. This `ArrayList` will hold references to the eight pins used by the Cylon Eyes program.

>> →19 This statement and the seven similar statements that follow define the GPIO pins and add them to the `ArrayList`.

>> →36 The `while` loop makes the Cylon Eyes bounce back and forth forever.

>> →38 Each back-and-forth cycle of the Cylon Eyes is done with two `for` loops. The first loop iterates over the eight pins in the `ArrayList` in the forward direction, from item 0 to 7. The `ArrayList` `get` method is used to retrieve the pin by index number to call the pin's `high` method to turn the LED on for 50 milliseconds, and then the `low` method turns the LED off.

>> →44 The second `for` loop iterates backward over the `ArrayList`, but omitting the last pin and the first pin. Thus, the index counts backward from 6 to 1, not from 7 to 0. The reason for this is to avoid the appearance that bouncing light gets "stuck" at both ends of the string of LEDs.

Working with Input Pins

Up to this point in this chapter, I explain how to use the Raspberry Pi GPIO pins for output. But what if you want to use a GPIO port for input rather than for output? In other words, what if you want the Raspberry Pi to react to the status of an external device instead of the other way around? The easiest way to do that is to connect a push button to a GPIO port, and then use Pi4J to program the port as an input port rather than as an output port. You can then read the status of the port and respond accordingly.

Understanding active-high and active-low inputs

Before we dig into the details of provisioning and using digital input pins, let's look at two basic types of input circuits that can be connected to a digital input pin:

- » **Active-high:** This type of connection places +3.3V on the input port when the push button is pressed. When the button is released, the input port sees 0V.

- » **Active-low:** This type of connection sees +3.3V at the input port when the push button is not pressed. When you press the push button, the +3.3V is removed, and the input pin sees 0V.

In other words, an active-high circuit is HIGH when the button is pressed. An active-low circuit is LOW when the button is pressed. Neither option is better than the other, but a program that works with an input port needs to know whether the input is active-high or active-low.

Figure 5-9 shows sample circuits that include active-high and active-low push buttons. The two circuits are very similar: They both have connections to the Pi's +3.3V source, ground, and a GPIO pin, and they both include a switch and two resistors. But there's a subtle but crucial difference: The switch and resistor R2 are reversed in the second circuit.

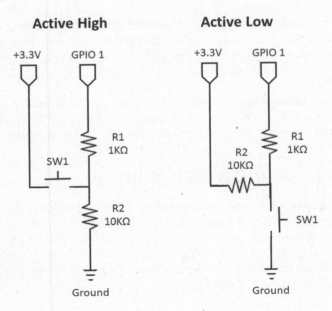

FIGURE 5-9: Active-high and active-low input circuits.

This resistor R2 is called a *pin-pull resistor*. It's designed to prevent a situation in which nothing at all is connected to a GPIO port. If the circuit allows nothing at all to be connected to a GPIO port, the Pi can't reliably determine whether a port is HIGH or LOW; this is called a *floating port*.

There are two types of pin-pull resistors, corresponding to the two types of input circuits:

>> **Pull-down resistor:** Used in active-high circuits. It's called a pull-down because it pulls current from the GPIO port down to ground when the switch is not pressed.

>> **Pull-up resistor:** Used in active-low circuits. It's called a pull-up resistor because it pulls current from the +3.3V source up to the GPIO pin when the switch is not pressed.

In the active-high circuit, the input port is connected to ground through resistors R1 and R2 when the push button is not pressed. Thus, the voltage at the input port is 0, and the combination of the two resistors prevents a short circuit between the +3.3V source and ground. When the push button is pressed, the input port is connected to +3.3V through R1, causing the input port to see +3.3V. As a result, the input port is HIGH when the button is pressed and LOW when the button is not pressed.

In the active-low circuit, the input port is connected to +3.3V through resistors R1 and R2, causing the input port to be HIGH. When the button is pressed, the current from the +3.3V source is passed through R1 to ground, causing the voltage at the input port to drop to near zero. Thus, the input pin is LOW when the button is pressed and HIGH when the button is not pressed.

WARNING

Note that in both circuits, resistor R2 is connected between the +3.3V and ground to prevent a short circuit when the button is pressed. Without this resistor, the Raspberry Pi would likely be damaged when the button is pressed.

Provisioning a digital input

The distinction between active-high and active-low input comes into play when you provision a digital input port. If you look back to Table 5-5, you'll see four variants of the provisionDigitalInputPin method. Two of these methods require a PinPullResistance argument. PinPullResistance is an enum that specifies whether the circuit will have a pull-down or a pull-up resistor or no pull resistor at all, as described in Table 5-8.

TABLE 5-8 **The PinPullResistance enum**

Constant	Description
PULL_DOWN	The input circuit is active-high with a pull-down resistor.
PULL_UP	The input circuit is active-low with a pull-up resistor.
OFF	The input circuit type is neither active-high or active-low, so no pull resistor is used.

Here's how to provision a digital input pin with an active-high circuit:

```
final GpioPinDigitalInput button =
    gpio.provisionDigitalInputPin(RaspiPin.GPIO_01,
                        PinPullResistance.PULL_DOWN);
```

And here's how to provision for an active-low circuit:

```
final GpioPinDigitalInput button =
    gpio.provisionDigitalInputPin(RaspiPin.GPIO_01,
                        PinPullResistance.PULL_UP);
```

Note that in order to use the `PinPullResistance` enum, you need to import it at the top of your program, like this:

```
import com.pi4j.io.gpio.PinPullResistance;
```

Reading the state of a digital input pin

To read the state of a digital input pin, you use the `GpioPinDigitalInput` interface. It's methods are described in Table 5-9.

The methods for reading the state of an input pin are pretty straightforward, but using them in a program to determine, for example, when a user has pressed a button on an input circuit, can be a little tricky.

TABLE 5-9 **The GpioPinDigitalInput Interface**

Method	Description
PinState getState()	Gets the current state of the pin.
boolean isHigh()	True if the input state is HIGH.
boolean isLow()	True if the input state is LOW.
boolean isState(PinState state)	True if the input state matches the specified PinState.

Building a circuit with a digital input pin

Let's now build a simple circuit that will allow you to test a program that provisions and uses a digital input pin. For this circuit, you'll use two LEDs and one push button. The program you'll run with this circuit will display one of the two LEDs; the user can press the push button to toggle the LEDs. In other words, each time the user pushes the button, the LED that's on will go off, and the LED that's off will go on.

The schematic diagram for the circuit is shown in Figure 5-10, and a diagram to help you build the circuit on a solderless breadboard is shown in Figure 5-11.

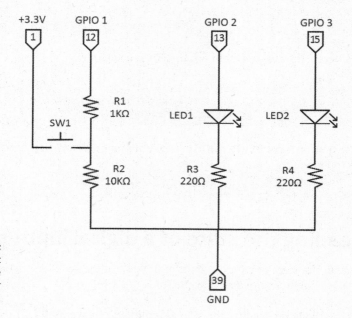

FIGURE 5-10: A schematic diagram for the Button Switcher circuit.

To build this circuit, you'll need the following parts:

>> One Raspberry Pi with software installed, connected to a monitor, keyboard, and mouse

>> One small solderless breadboard

>> Two 5mm red LEDs

>> One normally open DIP breadboard push button

>> Two 220Ω resistors (yellow-violet-brown)

>> One 10kΩ resistor (brown-black-orange)

» One 1kΩ resistor (brown-black-red)

» Six male–female jumper wires

» One short (½-inch) jumper wire

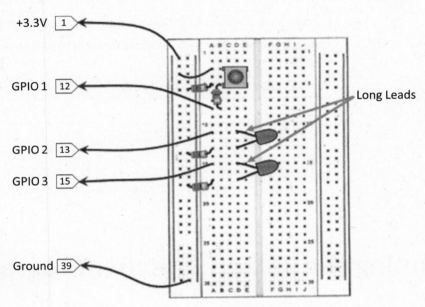

+3.3V 1

GPIO 1 12

Long Leads

GPIO 2 13

GPIO 3 15

FIGURE 5-11:
The completed
Button Switcher
circuit.

Ground 39

Follow these steps to assemble the board:

1. **Insert the resistors into the breadboard.**

Resistor	From	To
R1 (1KΩ)	B5	B8
R2 (10KΩ)	A5	Ground
R3 (220Ω)	A13	Ground
R4 (220kΩ)	A17	Ground

2. **Insert the LEDs.**

LED	Anode (Long Lead)	Cathode (Short Lead)
LED1	D11	D13
LED2	D15	D17

3. **Insert the push button.**

   The pins should be inserted in C3, E3, C5, and E5 such that the switch opens and closes across rows 3 and 5.

   TIP

   You may need to get creative with a pair of pliers to get the pins to fit in the breadboard: Just squeeze the pins together if they're spaced too wide to fit.

4. **Connect a jumper from A3 to the positive voltage bus on the breadboard.**

5. **Connect jumpers from the breadboard to the GPIO pins on the Raspberry Pi.**

Breadboard	GPIO Port/Power	Physical Pin
A8	GPIO_01	12
A11	GPIO_02	13
A15	GPIO_03	15
Positive bus	+3.3V	1
Ground bus	Ground	39

Running the Button Switcher Program

At last, we've arrived at the point where we can create a program that will respond when the user has pushed a button by toggling the status of the LEDs. Each time the user presses the push button, the LED that's on goes off and the LED that's off goes on. The program is shown in Listing 5-3.

LISTING 5-3: **The Button Switcher Program**

```
import com.pi4j.io.gpio.GpioController;
import com.pi4j.io.gpio.GpioFactory;
import com.pi4j.io.gpio.GpioPinDigitalOutput;
import com.pi4j.io.gpio.GpioPinDigitalInput;
import com.pi4j.io.gpio.PinState;
import com.pi4j.io.gpio.RaspiPin;
import com.pi4j.io.gpio.PinPullResistance;


//  Toggles LEDs on GPIO2 and GPIO3 when the user
//  presses a push button on GPIO1.

public class ButtonSwitcher
{
```

```
public static void main(String args[]) throws InterruptedException
{
    final GpioController gpio = GpioFactory.getInstance();

    final GpioPinDigitalInput button =                              →19
        gpio.provisionDigitalInputPin(RaspiPin.GPIO_01,
                                  PinPullResistance.PULL_DOWN);

    final GpioPinDigitalOutput led1 =                               →23
        gpio.provisionDigitalOutputPin(RaspiPin.GPIO_02, PinState.LOW);

    final GpioPinDigitalOutput led2 =
        gpio.provisionDigitalOutputPin(RaspiPin.GPIO_03, PinState.LOW);

    // Flash both LEDs briefly to confirm that the circuit is working
    led1.high();                                                    →30
    led2.high();
    Thread.sleep(250);
    led1.low();
    led2.low();
    Thread.sleep(250);

    while(true)
    {
        if (button.isHigh())                                        →39
        {
            led1.toggle();                                          →41
            led2.toggle();
            Thread.sleep(250);                                      →43
        }
    }
}
}
```

Here are the highlights of this program:

» →19 This statement provisions GPIO pin 1 as a digital input pin, which uses an active-high circuit and assigns the pin to a variable named button.

» →23 This line and the next line provision two digital output pins on GPIO pins 2 and 3, with a default state of LOW, and assigns them to variables named led1 and led2.

» →30 These lines briefly flash both LEDs — a simple startup test to ensure that the circuit is working.

» →39 This `if` statement tests to see if the button has been pressed. Placing this test inside of a `while` loop is a technique known as *polling,* in which the program periodically checks the state of the input pin. When the user presses the button, the statements in the `if` block are executed.

» →41 This statement and the next one toggle the state of the two LEDs. Thus, when the user presses the push button, the LEDs reverse themselves.

» →43 This statement pauses the loop for a quarter of a second to ensure that the user doesn't push the button too fast.

Finding a Better Way to Handle Input Events

The first version of the Button Switcher program (refer to Listing 5-3) uses a polling loop to constantly check the status of GPIO pin 1 to see if the user has pressed the button. That technique works for a simple program that has nothing better to do with its time than wait for the user to press a button. But for more significant programs, there are two problems with the polling loop technique:

» **What if the program has a more important task to perform while the user isn't pressing the button?** For example, suppose the real purpose of the program is to find the answer to the Ultimate Question of Life, the Universe, and Everything.

» **The call to** `Thread.sleep` **introduces a quarter of a second pause.** This is pause is a waste of time: A program can't do anything while it's sleeping.

The reason for the time-wasting pause in the Button Switcher program is to prevent a single press of the button from actually registering as two or more separate button presses. This can happen because of an annoying characteristic of mechanical switches called *bounce.* Bounce happens when the two pieces of metal within the switch initially make contact — they literally bounce off of each other.

Even though this bouncing happens extremely fast and is extremely small, it can result in a less-than-clean transition from LOW to HIGH on the pin (or vice versa for an active-low circuit). Without the call to `Thead.Sleep()`, the LEDs may toggle several times for each press of the button.

Compensating for bounce is called *debouncing.* Debouncing ensures that every time the user presses a button, the program sees one and only one button press. Without debouncing, your programs will behave in erratic and random ways.

The `Thread.Sleep` call is very crude way to debounce an input switch. In the next section, I show you a much better way.

Crafting a state change event listener

In this section, I show you an alternative to polling a GPIO pin to determine if a button has been pressed. This technique allows the program to get on with other work while waiting for the user to press a button. And it solves the debounce problem by eliminating the call to `Thread.sleep`.

The basics of this technique are presented in Book 3, Chapter 5. Essentially, what you do is create a listener that is automatically invoked whenever the state of the button's GPIO pin changes.

The listener must implement s the interface `GpioPinListenerDigital`, which is contained in the `com.pi4j.io.gpio.event` package. This interface defines just one method:

```
void handleGpioPinDigitalStateChangeEvent(
    GpioPinDigitalStateChangeEvent event);
```

This method is referred to as an *event handler* because it's automatically called whenever the state of the input pin changes. You must provide an implementation of the event handler method in your listener to handle the state change event.

The argument passed to the event handler, `GpioPinDigitalStateChangeEvent`, represents the event itself. The most important bit of information included in the event object is the state of the input pin, which you can obtain by calling the `getState` method. The `getState` method returns a value of type `PinState` value; you can use this value to determine whether the event was raised because the user pressed the button or because the user released it.

REMEMBER

This is an important point to remember: Every press of a push button will raise *two* state change events. The first event is raised when the user presses the button; the state will be HIGH for this event. The second event is raised when the user releases the button; the state will be LOW for this event.

Here's how you can toggle the LEDs when the user presses the button but not when the user releases the button:

```
PinState ps = event.getState();
if (ps == PinState.isHigh())
{
```

```
        led1.toggle();
        led2.toggle();
    }
```

In this example, the pin state is obtained from the event and stored in a variable named ps. Then, the value of ps is tested. If the pin state is HIGH, the LEDs are toggled.

Here's a more precise way to do the same thing:

```
if (event.getState().isHigh())
{
    led1.toggle();
    led2.toggle();
}
```

Putting it together, a complete implementation of the event handler method might look like this:

```
public void handleGpioPinDigitalStateChangeEvent(
    GpioPinDigitalStateChangeEvent event)
{
    if (event.getState().isHigh())
    {
        led1.toggle();
        led2.toggle();
    }
}
```

Adding an event handler to a pin

The final detail you need to know about setting up a state change event listener is how to add the listener to the pin. GpioPinDigitalInput has a simple method to do this: addListener. All you need to do is create an event listener instance that implements GpioPinListenerDigital; then pass it as an argument to addListener.

If you wanted to, you could create a separate class for the event listener. Then, you would create an instance of your event handler class, and pass the instance to addListener.

That's pretty cumbersome, though. The more common technique is to use an anonymous inner class. Here's an example that adds a listener to a digital input pin named button:

```
button.addListener(new GpioPinListenerDigital()
    {
        public void handleGpioPinDigitalStateChangeEvent(
            GpioPinDigitalStateChangeEvent event)
        {
            if (event.getState().isHigh())
            {
                led1.toggle();
                led2.toggle();
            }
        }
    });
```

Here, the event listener instance is created directly in the call to addListener. The new keyword creates an instance that implements GpioPinListenerDigital. This instance includes an implementation of handleGpioPinDigitalState ChangeEvent, which takes the event as an argument and toggles the LEDs if the event state is HIGH.

Better yet, you can use a lambda expression to create the event listener, like this:

```
button.addListener((GpioPinListenerDigital) event ->
                {
                    if (event.getState().isHigh())
                    {
                        led1.toggle();
                        led2.toggle();
                    }
                });
```

As you can see, the lambda is much briefer.

TIP

If you're not up to speed on the use of anonymous inner classes, refer to Book 3, Chapter 7.

Using automatic debounce

One of the pluses of using a state change event listener is that you can enable automatic debouncing for the push-button circuit. That way, you don't have to unnecessarily pause the program to ensure that each button press is processed just one time.

To enable automatic debouncing, just call the pin's setDebounce method and specify a debounce interval in milliseconds. Here's an example:

```
button.setDebounce(250);
```

In this example, the pin is protected from debounce for a quarter of a second.

Automatic debouncing works by suppressing all subsequent state change events for the designated interval after the first state change event occurs. So, when the user presses the button, the pin state changes to HIGH, a state change event is raised, and the state change handler method is called. Then, all subsequent state change events in which the state changes to HIGH are ignored until the debounce interval has expired.

Note that a state change to HIGH does not suppress LOW state change events. The first time the state changes to LOW, subsequent LOW state changes will also be suppressed. In this way, the automatic debounce feature ensures that each time the button is pressed and released, one and only one of each type of state change — HIGH and LOW — will be raised.

Working with the EventSwitcher program

Listing 5-4 presents a version of the Button Switcher program, called EventSwitcher, which uses the event listener and automatic debounce features described in the previous sections. This program uses the same circuit as the first version, so you don't have to build a new circuit to run this program.

LISTING 5-4: **The EventSwitcher Program**

```
import com.pi4j.io.gpio.GpioController;
import com.pi4j.io.gpio.GpioFactory;
import com.pi4j.io.gpio.GpioPinDigitalOutput;
import com.pi4j.io.gpio.GpioPinDigitalInput;
import com.pi4j.io.gpio.PinState;
import com.pi4j.io.gpio.RaspiPin;
import com.pi4j.io.gpio.PinPullResistance;
import com.pi4j.io.gpio.event.GpioPinDigitalStateChangeEvent;        →8
import com.pi4j.io.gpio.event.GpioPinListenerDigital;

// Toggles LEDs on GPIO2 and GPIO3 when the user
// presses a pushbutton on GPIO1.
// Uses an event listener and an automatic debounce.

public class EventSwitcher
{
    public static void main(String args[]) throws InterruptedException
    {
        final GpioController gpio = GpioFactory.getInstance();
```

```
final GpioPinDigitalInput button =                                    →21
    gpio.provisionDigitalInputPin(RaspiPin.GPIO_01,
        PinPullResistance.PULL_DOWN);

button.setDebounce(250);                                              →25

final GpioPinDigitalOutput led1 =
    gpio.provisionDigitalOutputPin(RaspiPin.GPIO_02, PinState.LOW);
final GpioPinDigitalOutput led2 =
    gpio.provisionDigitalOutputPin(RaspiPin.GPIO_03, PinState.LOW);

button.addListener((GpioPinListenerDigital) event ->                 →32
        {
                if (event.getState().isHigh())                        →34
                {
                    led1.toggle();
                    led2.toggle();
                }
        });

// Flash both LEDs briefly to confirm that the circuit is working
led1.high();
led2.high();
Thread.sleep(1000);
led1.low();
led2.low();
Thread.sleep(1000);

led1.high();

while(true)                                                           →51
{
    // Loop forever and perhaps answer the Ultimate Question
    // of Life, the Universe, and Everything.

    boolean isQuestionKnown = true;                                   →56
    if (!isQuestionKnown)
        {
            int answer = 42;
        }


}
    }
}
```

Here are the key points of this program:

» →8 In addition to the standard GPIO pin classes, this program also imports `GpioPinDigitalStateChangeEvent` and `GpioPinListenerDigital`, which are part of the `com.pi4j.io.gpio.event` package.

» →21 The digital input pin is provisioned as usual.

» →25 The `setDebounce` method is called to set the automatic debounce interval to 250 milliseconds.

» →32 Here's where the state change event listener is defined and added to the input pin. In this line, the `addListener` method is called and a lambda expression is used to create the listener.

» →34 In these lines, `eventgetState().isHigh()` is called to ensure that the button has been pressed rather than released. If so, both of the LEDs are toggled.

» →51 Although this program uses an event listener rather than polling to watch for button presses, a `while` loop is still used to allow the program to run continuously.

» →56 *Spoiler alert:* If you haven't read the book or seen the movie, the answer to *The Ultimate Question of Life, The Universe, and Everything* is 42. The `if` statement is there as a precaution because it is not possible to know both the question and the answer in the same universe without destroying said universe.

JavaFX

6

Contents at a Glance

Chapter **1**

Hello, JavaFX!

U p to this point in this book, all the programs are console-based, like something right out of the 1980s. Console-based Java programs have their place, especially when you're just starting with Java. But eventually you'll want to create programs that work with a graphical user interface (GUI).

This chapter gets you started in that direction using a Java GUI package called JavaFX to create simple GUI programs that display simple buttons and text labels. Along the way, you learn about several of the key JavaFX classes that let you create the layout of a GUI.

TECHNICAL STUFF

Prior to JavaFX, the main way to create GUIs in Java was through the Swing API. JavaFX is similar to Swing in many ways, so if you've ever used Swing to create a user interface (UI) for a Java program, you have a good head start at learning JavaFX.

JavaFX has become the de facto replacement for Swing. Although Swing is still supported in Java 8 and will be supported for the foreseeable future, Oracle is concentrating new features on JavaFX. Eventually, Swing will become obsolete.

Beginning with Java version 11, JavaFX was modularized and decoupled from the rest of Java. It's now a separate feature that you must download and install before you can create your own JavaFX programs. This complicates the task of compiling and testing even simple JavaFX programs. However, when you get it figured out, you'll have no trouble learning this powerful tool for creating GUI applications. Don't worry — I show you everything you need to know to set up JavaFX on your computer and how you need to structure your JavaFX applications so they compile and run.

Perusing the Possibilities of JavaFX

One of the basic strengths of JavaFX is its ability to let you easily create complicated GUIs with all the classic UI gizmos everyone knows and loves. Thus, JavaFX provides a full range of controls — dozens of them in fact, including the classics such as buttons, labels, text boxes, check boxes, drop-down lists, and menus, as well as more exotic controls such as tabbed panes and accordion panes. Figure 1-1 shows a typical JavaFX UI that uses several of these control types to create a form for data entry.

FIGURE 1-1: A typical JavaFX program.

Besides basic data-entry forms such as the one shown in Figure 1-1, JavaFX has many other powerful features for creating advanced UIs. In particular:

» **Cascading Style Sheets (CSS):** Style sheets allow you to control the appearance of your UI. When you use CSS, all the formatting details of your application are placed in in a separate file dubbed a *style sheet.* A style sheet is a simple text file that provides a set of rules for formatting the various elements of the UI. You can use CSS to control literally hundreds of formatting properties. For example, you can easily change the text properties such as font, size, color, and weight, and you can add a background image, gradient fills, borders, and special effects such as shadows, blurs, and light sources.

» **Visual effects:** You can add a wide variety of visual effects to your UI elements, including shadows, reflections, blurs, lighting, and perspective effects.

» **Animation:** You can specify animation effects that apply transitions gradually over time.

» **Charts:** You can create bar charts, pie charts, and many other chart types using the many classes of the `javafx.scene.chart` package.

» **3-D objects:** You can draw three-dimensional objects such as cubes, cylinders, spheres, and more complex shapes.

» **Touch interface:** JavaFX can handle touchscreen devices, such as smartphones and tablet computers with ease.

» **Property bindings:** JavaFX lets you create *properties,* which are special data types that can be bound to UI controls. For example, you can create a property that represents the price of an item being purchased and then bind a label to it. Then, whenever the value of the price changes, the value displayed by the label is updated automatically.

TECHNICAL STUFF

When you first encounter properties in JavaFX, you may be confused because JavaFX uses the term *property* differently from how you may be accustomed to using the term. In colloquial usage, the term *property* refers to an attribute of a class. The closest thing the Java language has to properties in this sense are getter and setter methods that expose fields that are internal to a class; for example, a `Customer` class may have a "property" called `Address`, which is accessed via methods named `getAddress` and `setAddress`.

In JavaFX, properties are much more powerful than that. As you find out in Chapter 6 of this minibook, a JavaFX property not only manages an attribute of a class (such as whether a check box has been checked) but also allows you to bind an *observer* to the property such that the observer is called into action whenever the value of the property changes.

Chapter 6 of this minibook covers properties, but the rest of this chapter and the four chapters that follow focus on the basics of creating simple UIs with JavaFX.

TIP

If you want to go deeper into JavaFX, pick up a copy of my book *JavaFX For Dummies* (Wiley).

Getting Ready to Run JavaFX

With version 11 of Java, the JavaFX system was fully modularized and removed from the standard Java distribution. That means that if you simply try to compile and run the JavaFX programs featured in this minibook without first setting up JavaFX, you'll encounter a plethora of error messages.

To properly compile and test a JavaFX program, first you need to install JavaFX. To do that, go to `http://openjfx.io`, click the Download link, and then download the JavaFX Windows software development kit (SDK). Unzip the download in any location on your hard disk (I suggest `c:\javafx14.0.1`, or whatever version number variant is current when you download the file) so that the Java modules will be readily available. If you look inside this folder, the key subfolder is `lib`, which contains the `.jar` files that make up the various JavaFX modules.

Having downloaded JavaFX and placed it in a readily accessible location, there are a few complications you'll have to deal with when you create and run your JavaFX programs. This is because JavaFX program must use the Java Module System to properly package the program's classes with the required JavaFX modules. This may be a good time to review Book 3, Chapter 8 if you aren't familiar with the Java Module System.

Here's a brief summary of what you need to do to set up a JavaFX project:

>> Every program must be part of a named package. For the programs in this minibook, the package names all begin with `com.lowewriter`. For example, the first program you see (later in this chapter) is in a package named `com.lowewriter.clickme`.

TIP

Remember that package names should consist of lowercase letters; this helps distinguish them from class names.

>> The folder structure for a program should include a root folder with subfolders that correspond to the parts of the package name. For example, the ClickMe program is in a root folder named `com.lowerwiter.clickme`.

This folder the contains a folder named `com`, which in turn contains a folder named `lowewriter`, which in turn contains a folder named `clickme`.

>> The class files for a JavaFX program should be in the bottommost folder of this structure. For example, the `ClickMe.java` file is found in `clickme\com\lowewriter\clickme`.

>> The root folder should contain a `module-info.java` file with the following content:

```
module com.lowewriter.clickme
{
        requires javafx.controls;
        exports com.lowewriter.clickme;
}
```

>> (The name of the module and the exported package will be different for each program.)

>> The root folder should also contain a manifest file that names the main class. For example, the ClickMe application has a manifest file named `clickme.mf` that contains the following line:

```
Main-Class: com.lowewriter.clickme.ClickMe
```

Many integrated development environments (IDEs), such as Eclipse or Netbeans, will set up this project structure (or a similar structure). If you're not using a development environment, you'll need to set up these folders and files manually.

When you get your project set up according to these guidelines, you can compile the java files, create a jar file, and then run the jar file to test the program. Again, an IDE such as Eclipse or Netbeans will handle the details of the correct command options for you. But if you're working from the command line, you'll need to know the following points:

>> To compile a JavaFX program, you must add two switches to the `javac` command: `--module-path` and `--add-modules`. The `--module-path` switch provides the path to the JavaFX lib folder, and the `--add-modules` switch names the JavaFX modules that must be added (usually just `javafx.controls`.) Here's what a `javac` command would look like to compile a package named `com.lowewriter.ClickMe`:

```
javac module-info.java com\lowewriter\ClickMe\*.java
    --module-path "c:\javafx14.0.1\lib"
    --add-modules javafx.controls
```

» To create the jar file, you need to use the switches `cmvf` and name the manifest file, the jar file to create, and the classes to add. Here's an example:

```
jar cmvf ClickMe.mf com.lowewriter.ClickMe.jar *.class com\lowewriter\
    ClickMe\*.class
```

» Finally, to run the jar file, you must once again provide the `--module-path` and `--add-modules` switches. Here's an example:

```
java --module-path c:\javafx14.0.1\lib
    --add-modules javafx.controls --jar com.lowewriter.clickme.jar
```

Whew! That seems like a lot of work just to get JavaFX programs to run. But when you get your mind around these details, you can start to focus on the task of learning how to write beautiful GUIs with JavaFX. Read on to get started with the fun stuff.

Looking at a Simple JavaFX Program

Figure 1-2 shows the UI for a very simple JavaFX program that includes just a single button. Initially, the text of this button says `Click me please!` When you click it, the text of the button changes to `You clicked me!` If you click the button again, the text changes back to `Click me please!` Thereafter, each time you click the button, the text cycles between `Click me please!` and `You clicked me!` To quit the program, simply click the Close button (the X at the upper-right corner).

FIGURE 1-2:
The Click Me
program.

Listing 1-1 shows the complete listing for this program.

The Click Me Program

```java
package com.lowewriter.clickme;

import javafx.application.*;
import javafx.stage.*;
import javafx.scene.*;
import javafx.scene.layout.*;
import javafx.scene.control.*;

public class ClickMe extends Application
{
    public static void main(String[] args)
    {
        launch(args);
    }

    Button btn;

    @Override public void start(Stage primaryStage)
    {
        // Create the button
        btn = new Button();
        btn.setText("Click me please!");
        btn.setOnAction(e -> buttonClick());

        // Add the button to a layout pane
        BorderPane pane = new BorderPane();
        pane.setCenter(btn);
        // Add the layout pane to a scene
        Scene scene = new Scene(pane, 300, 250);

        // Finalize and show the stage
        primaryStage.setScene(scene);
        primaryStage.setTitle("The Click Me App");
        primaryStage.show();
    }

    public void buttonClick()
    {
        if (btn.getText() == "Click me please!")
        {
            btn.setText("You clicked me!");
        }
        else
        {
```

(continued)

LISTING 1-1:   *(continued)*

```
                btn.setText("Click me please!");
        }
    }
}
```

The sections that follow describe the remaining key aspects of this basic JavaFX program.

Importing JavaFX Packages

After the required `package` statement, JavaFX programs begin with a series of `import` statements that reference the various JavaFX packages that the program will use. The Click Me program includes the following five `import` statements:

```
import javafx.application.*;
import javafx.stage.*;
import javafx.scene.*;
import javafx.scene.layout.*;
import javafx.scene.control.*;
```

As you can see, all the JavaFX packages begin with `javafx`. The Click Me program uses classes from five distinct JavaFX packages:

» `javafx.application`: This package defines the core class on which all JavaFX applications depend: `Application`. You read more about the `Application` class in the section "Extending the Application Class" later in this chapter.

» `javafx.stage`: The most important class in this package is the `Stage` class, which defines the top-level container for all UI objects. `Stage` is a JavaFX application's highest-level window, within which all the application's user-interface elements are displayed.

» `javafx.scene`: The most important class in this package is the `Scene` class, which is a container that holds all the UI elements displayed by the program.

» `javafx.scene.layout`: This package defines a special type of user-interface element called a *layout manager.* The job of a layout manager is to determine the position of each control displayed in the UI.

» `java.scene.control`: This package contains the classes that define individual UI controls such as buttons, text boxes, and labels. The Click Me program uses just one class from this package, `Button`, which represents a button that the user can click.

Extending the Application Class

A JavaFX application is a Java class that extends the `javafx.application.Application` class. Thus, the declaration for the Click Me application's main class is this:

```
public class ClickMe extends Application
```

Here, the Click Me application is defined by a class named `ClickMe`, which extends the `Application` class.

Because the entire `javafx.application` package is imported in line 1 of the Click Me program, the `Application` class does not have to be fully qualified. If you omit the `import` statement for the `javafx.application` package, the `ClickMe` class declaration would have to look like this:

```
public class ClickMe
    extends javafx.application.Application
```

The `Application` class is responsible for managing what is called the *life cycle* of a JavaFX application. The life cycle consists of the following steps:

1. **Create an instance of the `Application` class.**

2. **Call the `init` method.**

   The default implementation of the `init` method does nothing, but you can override the `init` method to provide any processing you want to be performed before the application's UI displays.

3. **Call the `start` method.**

   The `start` method is an abstract method, which means that there is no default implementation provided as a part of the `Application` class. Therefore, you must provide your own version of the `start` method. The `start` method is responsible for building and displaying the UI. (For more information, see the section "Overriding the start Method" later in this chapter.)

4. **Wait for the application to end, which typically happens when the user signals the end of the program by closing the main application window or choosing the program's exit command.**

   During this time, the application isn't really idle. Instead, it's busy performing actions in response to user events, such as clicking a button or choosing an item from a drop-down list.

5. **Call the** stop **method.**

Like the init method, the default implementation of the stop method doesn't do anything, but you can override it to perform any processing necessary as the program terminates, such as closing database resources or saving files.

Launching the Application

As you know, the standard entry-point for Java programs is the main method. Here is the main method for the Click Me program:

```
public static void main(String[] args)
{
    launch(args);
}
```

As you can see, the main method consists of just one statement, a call to the Application class' launch method.

The launch method is what actually starts a JavaFX application. The launch method is a static method, so it can be called in the static context of the main method. It creates an instance of the Application class and then starts the JavaFX lifecycle, calling the init and start methods, waiting for the application to finish, and then calling the stop method.

The launch method doesn't return until the JavaFX application ends. Suppose you wrote the main method for the Click Me program like this:

```
public static void main(String[] args)
{
    System.out.println("Launching JavaFX");
    launch(args);
    System.out.println("Finished");
}
```

Then, you would see Launching JavaFX displayed in the console window while the JavaFX application window opens. When you close the JavaFX application window, you would then see Finished in the console window.

Overriding the start Method

Every JavaFX application must include a start method. You write the code that creates the UI elements your program's user will interact with in the start method. For example, the start method in Listing 1-1 contains code that displays a button with the text Click Me!

When a JavaFX application is launched, the JavaFX framework calls the start method after the Application class has been initialized.

The start method for the Click Me program looks like this:

```java
@Override public void start(Stage primaryStage)
{
    // Create the button
    btn = new Button();
    btn.setText("Click me please!");
    btn.setOnAction(e -> buttonClick());

    // Add the button to a layout pane
    BorderPane pane = new BorderPane();
    pane.setCenter(btn);

    // Add the layout pane to a scene
    Scene scene = new Scene(pane, 300, 250);

    // Finalize and show the stage
    primaryStage.setScene(scene);
    primaryStage.setTitle("The Click Me App");
    primaryStage.show();
}
```

To create the UI for the Click Me program, the start method performs the following four basic steps:

1. **Create a button control named btn. The button's text is set to Click me please!, and a method named buttonClick will be called when the user clicks the button.**

    For a more detailed explanation of this code, see the sections "Creating a Button" and "Handling an Action Event" later in this chapter.

2. **Create a layout pane named pane and add the button to it.**

    For more details, see the section "Creating a Layout Pane" later in this chapter.

3. **Create a scene named** scene **and add the layout pane to it.**

For more details, see the "Making a Scene" section later in this chapter.

4. **Finalize the stage by setting the scene, setting the stage title, and showing the stage.**

See the "Setting the Stage" section later in this chapter for more details.

You find pertinent details of each of these blocks of code later in this chapter. But before I proceed, I want to point out a few additional salient details about the start method:

TIP

>> The start method is defined as an abstract method in the Application class, so when you include a start method in a JavaFX program, you're actually overriding the abstract start method.

Although it isn't required, it's always a good idea to include the @override annotation to explicitly state that you're overriding the start method. If you omit this annotation and then make a mistake in spelling the method named (for example, Start instead of start) or if you list the parameters incorrectly, Java thinks you're defining a new method instead of overriding the start method.

>> Unlike the main method, the start method is not a static method. When you call the launch method from the static main method, the launch method creates an instance of your Application class and then calls the start method.

>> The start method accepts one parameter: the Stage object on which the application's UI will display. When the application calls your start method, the application passes the main stage — known as the *primary stage* — via the primaryStage parameter. Thus, you can use the primaryStage parameter later in the start method to refer to the application's stage.

Creating a Button

The button displayed by the Click Me program is created using a class named Button. This class is one of many classes that you can use to create UI controls. The Button class and most of the other control classes are found in the package javafx.scene.control.

To create a button, simply define a variable of type Button and then call the Button constructor like this:

```
Button btn;
btn = new Button();
```

In the code in Listing 1-1, the btn variable is declared as a class variable outside of the start method but the Button object is actually created within the start method. Controls are often declared as class variables so that you can access them from any method defined within the class. As you discover in the following section ("Handling an Action Event"), a separate method named buttonClicked is called when the user clicks the button. By defining the btn variable as a class variable, both the start method and the buttonClicked method have access to the button.

To set the text value displayed by the button, call the setText method, passing the text to be displayed as a string:

```
btn.setText("Click me please!");
```

Here are a few additional tidbits about buttons:

>> The Button constructor allows you to pass the text to be displayed on the button as a parameter, as in this example:

```
Btn = new Button("Click me please!");
```

If you set the button's text in this way, you don't need to call the setTitle method.

TECHNICAL STUFF

>> The Button class is one of many classes that are derived from a parent class known as javafx.scene.control.Control. Many other classes derive from this class, including Label, TextField, ComboBox, CheckBox, and RadioButton.

TECHNICAL STUFF

>> The Control class is one of several different classes that are derived from higher-level parent class called javafx.scene.Node. Node is the base class of all user-interface elements that can be displayed in a scene. A control is a specific type of node, but there are other types of nodes. In other words, all controls are nodes, but not all nodes are controls. You can read more about several other types of nodes later in this minibook.

Handling an Action Event

When the user clicks a button, an *action event* is triggered. Your program can respond to the event by providing an *event handler*, which is simply a bit of code that will be executed whenever the event occurs. The Click Me program works by

Hello, JavaFX!

setting up an event handler for the button; the code for the event handler changes the text displayed on the button.

There are several ways to handle events in JavaFX. The most straightforward is to simply specify that a method be called whenever the event occurs and then provide the code to implement that method.

To specify the method to be called, you call the setOnAction method of the button class. Here's how it's done in Listing 1-1:

```
btn.setOnAction(e -> buttonClick());
```

If the syntax used here seems a little foreign, that's because it uses a *lambda expression*, which is a feature that was introduced into Java in version 1.8. As used in this example, there are three elements to this new syntax:

TIP

>> The argument e represents an object of type ActionEvent, which the program can use to get detailed information about the event.

The Click Me program ignores this argument, so you can ignore it too, at least for now.

>> The arrow operator (->) is a new operator introduced in Java 8 for use with Lambda expressions.

>> The method call buttonClick() simply calls the method named buttonClick.

You'll see more examples of Lambda expressions used to handle events in Chapter 2 of this minibook, and you can find a more complete discussion of Lambda expressions in Book 3, Chapter 7.

After buttonClick has been established as the method to call when the user clicks the button, the next step is to code the buttonClick method. You find it near the bottom of Listing 1-1:

```
public void buttonClick()
{
    if (btn.getText() == "Click me please!")
    {
        btn.setText("You clicked me!");
    }
    else
    {
        btn.setText("Click me please!");
    }
}
```

This method uses an `if` statement to alternately change the text displayed by the button to either `You clicked me!` or `Click me please!`. In other words, if the button's text is `Click me please!` when the user clicks the button, the `buttonClicked` method changes the text to `You clicked me!`. Otherwise, the `if` statement changes the button's text back to `Click me please!`.

The `buttonClicked` method uses two methods of the `Button` class to perform its work:

>> `getText`: Returns the text displayed by the button as a string

>> `setText`: Sets the text displayed by the button

For more information about handling events, see Chapter 2 of this minibook.

Creating a Layout Pane

By itself, a button is not very useful. You must actually display it on the screen for the user to be able to click it. And any realistic JavaFX program will have more than one control. The moment you add a second control to your UI, you need a way to specify how the controls are positioned relative to one another. For example, if your application has two buttons, do you want them to be stacked vertically, one above the other, or side by side?

That's where layout panes come in. A *layout pane* is a container class to which you can add one or more user-interface elements. The layout pane then determines exactly how to display those elements relative to each other.

To use a layout pane, you first create an instance of the pane. Then, you add one or more controls to the pane. When you do so, you can specify the details of how the controls will be arranged when the pane is displayed. After you add all the controls to the pane and arrange them just so, you add the pane to the scene.

JavaFX provides a total of eight distinct types of layout panes, all defined by classes in the package `javafx.scene.layout`. The Click Me program uses a type of layout called a *border pane*, which arranges the contents of the pane into five general regions: top, left, right, bottom, and center. The `BorderPane` class is ideal for layouts in which you have elements such as a menu and toolbar at the top, a status bar at the bottom, optional task panes or toolbars on the left or right, and a main working area in the center of the screen.

The lines that create the border pane in the Click Me program are

```
BorderPane pane = new BorderPane();
pane.setCenter(btn);
```

Here, a variable of type BorderPane is declared with the name pane, and the Border Pane constructor is called to create a new BorderPane object. Then, the setCenter method is used to display the button (btn) in the center region of the pane.

Here are a few other interesting details about layout panes:

TECHNICAL STUFF

>> Layout panes automatically adjust the exact position of the elements they contain based on the size of the elements contained in the layout as well as on the size of the space in which the layout pane is displayed.

>> I said earlier that controls are a type of node, and that you would read about other types of nodes later in this book. Well, you just read about one: A layout pane is also a type of node.

>> Each region of a border pane can contain a node. Because a layout pane itself is a type of node, each region of a border pane can contain another layout pane. For example, suppose you want to display three controls in the center region of a border pane. To do that, you'd create a second layout pane and add the three controls to it. Then, you'd set the second layout pane as the node to be displayed in the center region of the first layout pane.

>> You read more about the BorderPane class and a few other commonly used layout panes in Chapter 4 of this minibook.

Making a Scene

After you create a layout pane that contains the controls you want to display, the next step is to create a scene that will display the layout pane. You can do that in a single line of code that declares a variable of type Scene and calls the Scene class constructor. Here's how I did it in the Click Me program:

```
Scene scene = new Scene(pane, 300, 250);
```

The Scene constructor accepts three arguments:

>> A node object that represents the *root node* to be displayed by the scene.

A scene can have only one root node, so the root node is usually a layout pane, which in turn contains other controls to be displayed. In the

Click Me program, the root note is the border layout pane that contains the button.

>> The width of the scene in pixels.

>> The height of the scene in pixels.

Note: If you omit the width and height, the scene will be sized automatically based on the size of the elements contained within the root node.

You can find out about some additional capabilities of the Scene class in Chapter 3 of this minibook.

Setting the Stage

If the *scene* represents the nodes (controls and layout panes) that are displayed by the application, the *stage* represents the window in which the scene is displayed. When the JavaFX framework calls your application's start method, it passes you an instance of the Stage class that represents the application's primary stage — that is, the stage that represents the application's main window. This reference is passed via the primaryStage argument.

Having created your scene, you're now ready to finalize the primary stage so that the scene can be displayed. To do that, you must do at least two things:

>> Call the setScene method of the primary stage to set the scene to be displayed.

>> Call the show method of the primary stage to display the scene.

After you call the show method, your application's window becomes visible to the user and the user can then begin to interact with its controls.

It's also customary to set the title displayed in the application's title bar. You do that by calling the setTitle method of the primary stage. The last three lines of the start method for the Click Me application perform these functions:

```
primaryStage.setScene(scene);
primaryStage.setTitle("The Click Me App");
primaryStage.show();
```

When the last line calls the show method, the Stage displays.

You can read about additional capabilities of the Stage class in Chapter 3 of this minibook.

Examining the Click Counter Program

Now that I've explained the details of every line of the Click Me program, I look at a slightly enhanced version of the Click Me program called the Click Counter program. In the original Click Me program that was shown earlier in this chapter (Listing 1-1), the text displayed on the button changes when the user clicks the button. In the Click Counter program, an additional type of control called a *label* displays the number of times the user has clicked the button.

Figure 1-3 shows the Click Counter program in operation. The window at the top of this figure shows how the Click Counter program appears when you first start it. As you can see, the text label at the top of the window displays the text You have not clicked the button. The second window shows what the program looks like after you click the button the first time. Here, the label reads You have clicked the button once. When the button is clicked a second time, the label changes again, as shown in the third window. Here, the label reads You have clicked the button 2 times. After that, the number displayed by the label updates each time you click the button to indicate how many times the button has been clicked.

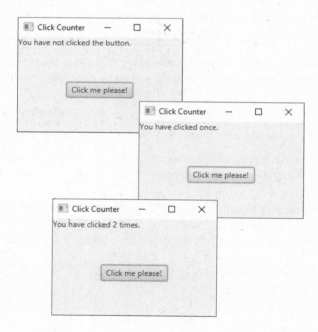

FIGURE 1-3:
The Click Counter program in action.

Listing 1-2 shows the source code for the Click Counter program, and the following paragraphs describe the key points of how it works:

LISTING 1-2:	**The Click Counter Program**

```
package com.lowewriter.clickcounter;                            →1

import javafx.application.*;                                    →3
import javafx.stage.*;
import javafx.scene.*;
import javafx.scene.layout.*;
import javafx.scene.control.*;

public class ClickCounter extends Application                  →9
{
    public static void main(String[] args)                     →11
    {
        launch(args);                                          →13
    }

    Button btn;                                                →16
    Label lbl;                                                 →17
    int iClickCount = 0;                                       →18

    @Override public void start(Stage primaryStage)            →20
    {
        // Create the button
        btn = new Button();                                    →23
        btn.setText("Click me please!");                       →24
        btn.setOnAction(e -> buttonClick());                   →25

        // Create the Label
        lbl = new Label();                                     →28
        lbl.setText("You have not clicked the button.");       →29

        // Add the label and the button to a layout pane
        BorderPane pane = new BorderPane();                    →32
        pane.setTop(lbl);                                      →33
        pane.setCenter(btn);                                   →34

        // Add the layout pane to a scene
        Scene scene = new Scene(pane, 250, 150);               →37

        // Add the scene to the stage, set the title
        // and show the stage
        primaryStage.setScene(scene);                          →41
        primaryStage.setTitle("Click Counter");                →42
        primaryStage.show();                                   →43
```

(continued)

LISTING 1-2: *(continued)*

```
    }

    public void buttonClick()                                →46
    {
        iClickCount++;                                       →48
        if (iClickCount == 1)                                →49
        {
            lbl.setText("You have clicked once.");           →51
        }
        else
        {
            lbl.setText("You have clicked "                  →55
                + iClickCount + " times." );
        }
    }

}
```

The following paragraphs explain the key points of the Click Me program:

➤ →1 The package statement names the package: com.lowewriter.clickcounter.

➤ →3 The import statements reference the javafx packages that will be used by the Click me program.

➤ →9 The ClickMe class extends javafx.application.Application, thus specifying that the ClickMe class is a JavaFX application.

➤ →11 As with any Java program, the main method is the main entry point for all JavaFX programs.

➤ →13 The main method calls the launch method, which is defined by the Application class. The launch method, in turn, creates an instance of the ClickMe class and then calls the start method.

➤ →16 A variable named btn of type javafx.scene.control.Button is declared as a class variable. Variables representing JavaFX controls are commonly defined as class variables so that they can be accessed by any method in the class.

➤ →17 A class variable named lbl of type javafx.scene.control.Label represents the Label control so that it can be accessed from any method in the class.

➤ →18 A class variable named iClickCount will be used to keep track of the number of times the user clicks the button.

» →20 The declaration of the start method uses the @override annotation, indicating that this method overrides the default start method provided by the Application class. The start method accepts a parameter named primaryStage, which represents the window in which the Click Me application will display its UI.

» →23 The start method begins by creating a Button object and assigning it to a variable named btn.

» →24 The button's setText method is called to set the text displayed by the button to Click Me Please!.

» →25 The setOnAction is called to create an event handler for the button. Here, a Lambda expression is used to simply call the buttonClick method whenever the user clicks the button.

» →28 The constructor of the Label class is called to create a new label.

» →29 The label's setText method is called to set the initial text value of the label to You have not clicked the button.

» →32 A border pane object is created by calling the constructor of the BorderPane class, referencing the border pane via a variable named pane. The border pane will be used to control the layout of the controls displayed on the screen.

» →33 The border pane's setTop method is called to add the label to the top region of the border pane.

» →34 The border pane's setCenter method is called to add the button to the center region of the border pane.

» →37 A scene object is created by calling the constructor of the Scene class, passing the border pane created in line 32 to the constructor to establish the border pane as the root node of the scene. In addition, the dimensions of the scene are set to 300 pixels in width and 250 pixels in height.

» →41 The setScene method of the primaryStage is used to add the scene to the primary stage.

» →42 The setTitle method is used to set the text displayed in the primary stage's title bar.

» →43 The show method is called to display the primary stage. When this line is executed, the user can begin to interact with the program.

» →46 The buttonClick method is called whenever the user clicks the button.

» →48 The iClickCount variable is incremented to indicate that the user has clicked the button.

>> →49 An `if` statement is used to determine whether the button has been clicked one or more times.

>> →51 If the button has been clicked once, the label text is set to `You have clicked once`.

>> →55 Otherwise, the label text is set to a string that indicates how many times the button has been clicked.

That's all there is to it. If you understand the details of how the Click Counter program works, you're ready to move on to Chapter 2. If you're still struggling with a few points, I suggest you spend some time reviewing this chapter and experimenting with the Click Counter program in TextPad, Eclipse, or NetBeans.

The following paragraphs help clarify some of the key sticking points that might be tripping you up about the Click Counter program and JavaFX in general:

>> **When does the program switch from static to non-static?** Like every Java program, the main entry point of a JavaFX program is the static `main` method.

In most JavaFX programs, the static `main` method does just one thing: It calls the `launch` method to start the JavaFX portion of the program. The `launch` method creates an instance of the `ClickCounter` class and then calls the `start` method. At that point, the program is no longer running in a static context because an instance of the `ClickCounter` class has been created.

>> **Where does the `primaryStage` variable come from?** The `primaryStage` variable is passed to the `start` method when the `launch` method calls the `start` method. Thus, the `start` method receives the `primaryStage` variable as a parameter.

That's why you won't find a separate variable declaration for the `primaryStage` variable.

>> **What does the** `module-info.java` **file look like for this program?**

```
module com.lowewriter.clickcounter
{
    requires javafx.controls;
    exports com.lowewriter.clickcounter;
}
```

>> **What does the manifest file look like for this program?**

```
Main-Class: com.lowewriter.clickcounter.ClickCounter
```

Chapter **2**

Handling Events

n Chapter 1 of this minibook, you examined two programs that display simple scenes that include a button and that respond when the user clicks the button. These programs respond to the event triggered when the user clicks the button by providing an *event handler* that's executed when the event occurs.

In this chapter, you read more details about how event handling works in JavaFX. I discuss how events are generated and how they're dispatched by JavaFX so that your programs can respond to them. You discover the many varieties of events that can be processed by a JavaFX program. And you figure out several programming techniques for handling JavaFX events.

Finally, in this chapter you're introduced to the idea of *property bindings,* which let you write code that responds to changes in the value of certain types of class fields, dubbed *property fields.*

TIP

Although event handling is used mostly to respond to button clicks, it can also be used to respond to other types of user interactions. You can use event handling, for example, to write code that's executed when the user makes a selection from a combo box, moves the mouse over a label, or presses a key on the keyboard. The event-handling techniques in this chapter work for those events as well.

Examining Events

An *event* is an object that's generated when the user does something noteworthy with one of your user-interface components. Then this event object is passed to a special method you create, called an event handler. The *event handler* can examine the event object, determine exactly what type of event occurred, and respond accordingly. If the user clicks a button, the event handler might write any data entered by the user via text fields to a file. If the user passes the mouse cursor over a label, the event handler might change the text displayed by the label. And if the user selects an item from a combo box, the event handler might use the value that was selected to look up information in a database. The possibilities are endless!

An event is represented by an instance of the class `javafx.event.Event` or one of its many subclasses. Table 2-1 lists the most commonly used event classes.

TABLE 2-1   ## Commonly Used Event Classes

Event Class	Package	Description
ActionEvent	javafx.event	Created when the user performs an action with a button or other component. Usually this means that the user clicked the button, but the user can also invoke a button action by tabbing to the button and pressing the Enter key. This is the most commonly used event class, as it represents the most common types of user-interface events.
InputEvent	javafx.scene.input	Created when an event that results from user input, such as a mouse or key click, occurs.
KeyEvent	javafx.scene.input	Created when the user presses a key on the keyboard. This event can be used to watch for specific keystrokes entered by the user. (KeyEvent is a subclass of InputEvent.)
MouseEvent	javafx.scene.input	Created when the user does something interesting with the mouse, such as clicking one of the buttons, dragging the mouse, or simply moving the mouse cursor over another object. (MouseEvent is a subclass of InputEvent.)
TouchEvent	javafx.scene.input	Created when a user initiates a touch event on a device that allows touch input.
WindowEvent	javafx.stage	Created when the status of the window (stage) changes.

TIP

Here are four important terms you need to know:

>> **Event:** An object that's created when the user does something noteworthy with a component, such as clicking it.

>> **Event source:** The object on which the event initially occurred.

>> **Event target:** The node that the event is directed at.

 This is usually the button or other control that the user clicked or otherwise manipulated. (In most cases, the event source and the event target are the same.)

>> **Event handler:** The object that listens for events and handles them when they occur.

 The event-listener object must implement the EventHandler functional interface, which defines a single method named handle (see Table 2-2). The EventHandler interface is defined in the package javafx.event.

TABLE 2-2

The EventHandler Interface

Method	Description
void handle<T event>	Called when an event occurs

Handling Events

Now that you know the basic classes and interfaces that are used for event handling, you're ready to figure out how to wire them to create a program that responds to events.

In this section, I discuss how to implement the event handler by coding the program's Application so that in addition to extending the Application class, it also implements the EventHandler interface. In subsequent sections of this chapter, I discuss alternative techniques to implement event handlers that are more concise and, in many cases, easier to work with.

Note that the programs that were shown in Chapter 1 of this minibook use the concise lambda expressions technique, and most of the programs featured throughout the rest of this minibook also use lambda expressions. But it's important that you know the other techniques so that you have a complete understanding of how event handling actually works.

Here are three steps you must take to handle a JavaFX event:

1. **Create an event source.**

An *event source* is simply a control, such as a button, that can generate events. Usually, you declare the variable that refers to the event source as a private class field, outside the start method or any other class methods:

```
private Button btn;
```

Then, in the start method, you can create the button like this:

```
btn = new Button();
btn.setText("Click me please!");
```

2. **Create an event handler.**

To create an event handler, you must create an object that implements an interface appropriately named EventHandler. This object must provide an implementation of the handle method.

Here are four ways to create an event handler:

- Add implements EventHandler to the program's Application class and provide an implementation of the handle method. You find out how to use this technique in the section "Implementing the EventHandler Interface."

- Create an inner class that implements EventHandler within the Application class. You figure out how to use this technique in the section "Handling Events with Inner Classes."

- Create an anonymous class that implements EventHandler. I show you how to use this technique in the section "Handling Events with Anonymous Inner Classes."

- Use a lambda expression to implement the handle method. You read about how to use this technique in the section "Using lambda Expressions to Handle Events."

3. **Register the event handler with the event source.**

The final step is to register the event handler with the event source so that the handle method is called whenever the event occurs.

Every component that serves as an event source provides a method that lets you register event handlers to listen for the event. For example, a Button control provides a setOnAction method that lets you register an event handler for the action event. In the setOnAction method, you specify the event handler object as a parameter. The exact way you do that depends on which of the various techniques you used to create the event handler.

Implementing the EventHandler Interface

To see how all these elements work together in a complete program, Figure 2-1 shows the output from a simple program called AddSubtract1. This program displays a label and two buttons, one titled Add and the other titled Subtract. The label initially displays the number 0. Each time the user clicks the Add button, the value displayed by the label is increased by one; each time the user clicks the Subtract button, the value is decreased by one.

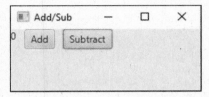

FIGURE 2-1:
The AddSubtract1 program.

Listing 2-1 shows the complete code for this program.

LISTING 2-1: **The AddSubtract1 Program**

```
package com.lowewriter.addsubtract1;

import javafx.application.*;
import javafx.stage.*;
import javafx.scene.*;
import javafx.scene.layout.*;
import javafx.scene.control.*;
import javafx.event.*;                                          →8

public class AddSubtract extends Application
    implements EventHandler <ActionEvent>                       →11
{
    public static void main(String[] args)                      →13
    {
        launch(args);
    }

    Button btnAdd;                                              →18
    Button btnSubtract;
    Label lbl;
    int iCounter = 0;                                           →21

    @Override public void start(Stage primaryStage)             →23
```

(continued)

LISTING 2-1:   *(continued)*

```
    {
        // Create the Add button
        btnAdd = new Button();                                       →26
        btnAdd.setText("Add");
        btnAdd.setOnAction(this);                                    →28

        // Create the Subtract button
        btnSubtract = new Button();                                  →31
        btnSubtract.setText("Subtract");
        btnSubtract.setOnAction(this);

        // Create the Label                                          →35
        lbl = new Label();
        lbl.setText(Integer.toString(iCounter));

        // Add the buttons and label to an HBox pane
        HBox pane = new HBox(10);                                    →40
        pane.getChildren().addAll(lbl, btnAdd, btnSubtract);         →41

        // Add the layout pane to a scene
        Scene scene = new Scene(pane, 200, 75);                      →44

        // Add the scene to the stage, set the title
        // and show the stage
        primaryStage.setScene(scene);                               →48
        primaryStage.setTitle("Add/Sub");
        primaryStage.show();
    }

    @Override public void handle(ActionEvent e)                      →53
    {
        if (e.getSource()==btnAdd)                                   →55
        {
            iCounter++;
        }
        else
        {
            if (e.getSource()==btnSubtract)                          →61
            {
                iCounter--;
            }
        }
        lbl.setText(Integer.toString(iCounter));                     →66
    }

}
```

The following paragraphs point out some key lines of the program:

» →8 The program must import the javafx.event package, which defines the ActionEvent class and the EventHandler interfaces.

» →11 As in any JavaFX program, the AddSubtract1 class extends the Application class. However, the AddSubtract1 class also implements the EventHandler interface so that it can define a handle method that will handle ActionEvent events that are generated by the buttons. The EventHandler interface is a generic interface, which means that you must specify the specific event type that the interface will implement. In this case, the class will handle ActionEvent events.

» →13 The main method is required as usual. This method simply calls the launch method to create an instance of the AddSubtract class, which in turn calls the start method.

» →18 Two buttons (btnAdd and btnSubtract) and a label (lbl) are defined as class fields so that they can be accessed throughout the class.

» →21 The iCounter variable keeps track of the value displayed by the label. The value will be incremented when the user clicks the btnAdd button and decremented when the user clicks the btnSubtract button.

» →23 The start method is called when the application is started.

» →26 This line and the next line create the Add button and set its text to display the word *Add*.

» →28 This line sets the current object as the event handler for the btnAdd button. The this keyword is used here because the AddSubtract class implements the EventHandler. In effect, the AddSubtract class itself handles any events that are created by its own controls.

» →31 These lines create the Subtract button, set its text to the word *Subtract*, and set the current object (this) as the event handler for the button.

» →35 These two lines create the label and set its initial text value to a string equivalent of the iCounter variable.

» →40 For this program, a border pane is not the appropriate type of layout pane. Instead, for this program, use a new type of layout pane called an HBox. An HBox pane arranges any controls you add to it in a horizontal row. The parameter 10 indicates that the controls should be separated from one another by a space ten pixels wide.

» →41 This line adds the label and the two buttons to the horizontal box. The code required to do this is admittedly a bit convoluted. First, you must call the getChildren method to get a list of all the child nodes that are in the HBox. Then, you call the addAll method to add one or more controls. In this case, three controls are added: the label (lbl), the Add button (btnAdd), and the Subtract button (btnSubtract).

» →44 This line creates a new scene, using the HBox pane as its root node.

» →48 This line sets the scene created in line 42 as the primary scene for the stage, sets the stage title, and then shows the stage.

» →53 The handle method must be coded because the AddSubtract class implements the EventHandler interface. This method is called by either of the button objects whenever the user clicks one of the buttons. The ActionEvent parameter is the event generated by the button click and passed to the handle method.

» →55 The getSource method of the ActionEvent parameter is called to determine the event source. If the event source is btnAdd, the iCounter variable is incremented.

» →61 If, on the other hand, the event source is btnSubtract, the iCounter variable is decremented.

» →66 The label's text value is set to the string equivalent of the iCounter variable.

Handling Events with Inner Classes

An *inner class* is a class that's nested within another class. Inner classes are commonly used for event handlers. That way, the class that defines the application doesn't also have to implement the event handler. Instead, it includes an inner class that handles the events.

Listing 2-2 shows the AddSubtract2 program, which uses an inner class to handle the action event for the buttons.

LISTING 2-2: **The AddSubtract2 Program with an Inner Class**

```
package com.lowewriter.addsubtract2;

import javafx.application.*;
import javafx.stage.*;
import javafx.scene.*;
import javafx.scene.layout.*;
import javafx.scene.control.*;
import javafx.event.*;
```

```java
public class AddSubtract2 extends Application                                    →10
{
    public static void main(String[] args)
    {
        launch(args);
    }

    Button btnAdd;
    Button btnSubtract;
    Label lbl;
    int iCounter = 0;

    @Override public void start(Stage primaryStage)
    {
        // Create a ClickHandler instance
        ClickHandler ch = new ClickHandler();                                   →25

        // Create the Add button
        btnAdd = new Button();
        btnAdd.setText("Add");
        btnAdd.setOnAction(ch);                                                 →30

        // Create the Subtract button
        btnSubtract = new Button();
        btnSubtract.setText("Subtract");
        btnSubtract.setOnAction(ch);                                           →35

        // Create the Label
        lbl = new Label();
        lbl.setText(Integer.toString(iCounter));

        // Add the buttons and label to an HBox pane
        HBox pane = new HBox(10);
        pane.getChildren().addAll(lbl, btnAdd, btnSubtract);

        // Add the layout pane to a scene
        Scene scene = new Scene(pane, 200, 75);

        // Add the scene to the stage, set the title
        // and show the stage
        primaryStage.setScene(scene);
        primaryStage.setTitle("Add/Sub");
        primaryStage.show();
    }

    private class ClickHandler                                                 →55
        implements EventHandler <ActionEvent>
```

(continued)

Handling Events

LISTING 2-2: *(continued)*

```
{
    @Override public void handle(ActionEvent e)                    →58
    {
        if (e.getSource()==btnAdd)
        {
            iCounter++;
        }
        else
        {
            if (e.getSource()==btnSubtract)
            {
                iCounter--;
            }
        }
        lbl.setText(Integer.toString(iCounter));
    }
}
```

This program works essentially the same way as the program shown in Listing 2-1, so I don't review every detail. Instead, I just highlight the differences:

» →10 The AddSubtract2 class still extends Application but doesn't implement EventHandler.

» →25 This statement creates an instance of the ClickHandler class (the inner class) and assigns it to the variable ch.

» →30 This statement sets ch as the action listener for the Add button.

» →35 This statement sets ch as the action listener for the Subtract button.

» →55 The ClickHandler class is declared as an inner class by placing its declaration completely within the AddSubtract2 class. The ClickHandler class implements the EventHandler interface so that it can handle events.

» →58 The handle method here is identical to the handle method in the AddSubtract1 program (see Listing 2-1) but resides in the inner ClickHandler class instead of in the outer class.

Handling Events with Anonymous Inner Classes

An *anonymous inner class*, usually just called an *anonymous class*, is a class that's defined on the spot, right at the point where you need it. Because you code the body of the class right where you need it, you don't have to give it a name; that's why it's called an *anonymous* class.

Anonymous classes are often used for event handlers to avoid the need to create a separate class that explicitly implements the EventHandler interface.

One advantage of using anonymous classes for event handlers (and lambdas as well) is that you can easily create a separate event handler for each control that generates events. Then, in the handle method for those event handlers, you can dispense with the if statements that check the event source.

Consider the event handler for the AddSubtract2 program shown earlier in Listing 2-2: It must check the event source to determine whether to increment or decrement the iCounter variable. By using anonymous classes, you can create separate event handlers for the Add and Subtract buttons. The event handler for the Add button increments iCounter, and the event handler for the Subtract button decrements it. Neither event handler needs to check the event source because the event handler's handle method will be called only when an event is raised on the button with which the handler is associated.

Listing 2-3 shows the AddSubtract3 program, which uses anonymous inner classes in this way.

LISTING 2-3: **The AddSubtract3 Program with Anonymous Inner Classes**

```
package com.lowewriter.addsubtract3;

import javafx.application.*;
import javafx.stage.*;
import javafx.scene.*;
import javafx.scene.layout.*;
import javafx.scene.control.*;
import javafx.event.*;

public class AddSubtract3 extends Application
```

(continued)

Handling Events

LISTING 2-3: *(continued)*

```java
{
    public static void main(String[] args)
    {
        launch(args);
    }

    Button btnAdd;
    Button btnSubtract;
    Label lbl;
    int iCounter = 0;

    @Override public void start(Stage primaryStage)
    {
        // Create the Add button
        btnAdd = new Button();
        btnAdd.setText("Add");
        btnAdd.setOnAction(
            new EventHandler<ActionEvent>()                          →28
            {
                public void handle(ActionEvent e)                    →30
                {
                    iCounter++;                                      →32
                    lbl.setText(Integer.toString(iCounter));
                }
            } );

        // Create the Subtract button
        btnSubtract = new Button();
        btnSubtract.setText("Subtract");
        btnSubtract.setOnAction(
            new EventHandler<ActionEvent>()                          →41
            {
                public void handle(ActionEvent e)                    →43
                {
                    iCounter--;
                    lbl.setText(Integer.toString(iCounter));
                }
            } );

        // Create the Label
        lbl = new Label();
        lbl.setText(Integer.toString(iCounter));

        // Add the buttons and label to an HBox pane
        HBox pane = new HBox(10);
        pane.getChildren().addAll(lbl, btnAdd, btnSubtract);
```

```
    // Add the layout pane to a scene
    Scene scene = new Scene(pane, 200, 75);

    // Add the scene to the stage, set the title
    // and show the stage
    primaryStage.setScene(scene);
    primaryStage.setTitle("Add/Sub");
    primaryStage.show();
    }
}
```

The following paragraphs highlight the key points of how this program uses anonymous inner classes to handle the button events:

» →28 This line calls the setOnAction method of the Add button and creates an anonymous instance of the EventHandler class, specifying ActionEvent as the type.

» →30 The handle method must be defined within the body of the anonymous class.

» →32 Because this handle method will be called only when the Add button is clicked (not when the Subtract button is clicked), it does not need to determine the event source. Instead, the method simply increments the counter variable and sets the label text to display the new value of the counter.

» →41 This line calls the setOnAction method of the Subtract button and creates another anonymous instance of the EventHandler class.

» →43 This time, the handle method decrements the counter variable and updates the label text to display the new counter value.

Using Lambda Expressions to Handle Events

Java 8 introduces a new feature that in some ways is similar to anonymous classes, but with more concise syntax. More specifically, a lambda expression lets you create an anonymous class that implements a specific type of interface — a *functional interface* — which has one and only one abstract method.

The EventHandler interface used to handle JavaFX events meets that definition: It has just one abstract method, handle. Thus, EventHandler is a functional interface and can be used with lambda expressions.

A lambda expression is a concise way to create an anonymous class that implements a functional interface. Instead of providing a formal method declaration that includes the return type, method name, parameter types, and method body, you simply define the parameter types and the method body. The Java compiler infers the rest based on the context in which you use the lambda expression.

The parameter types are separated from the method body by a new operator — the *arrow operator* — which consists of a hyphen followed by a greater-than symbol. Here's an example of a lambda expression that implements the EventHandler interface:

```
e ->
    {
        iCounter++;
        lbl.setText(Integer.toString(iCounter);
    }
```

In this case the lambda expression implements a functional interface whose single method accepts a single parameter, identified as e. When the method is called, the iCounter variable is incremented and the label text is updated to display the new counter value.

Here's how you'd register this lambda expression as the event handler for a button:

```
btnAdd.setOnAction( e ->
    {
        iCounter++;
        lbl.setText(Integer.toString(iCounter));
    } );
```

One of the interesting things about lambda expressions is that you don't need to know the name of the method being called. This is possible because a functional interface used with a lambda expression can have only one abstract method. In the case of the EventHandler interface, the method is named handle.

You also do not need to know the name of the interface being implemented. This is possible because the interface is determined by the context. The setOnAction method takes a single parameter of type EventHandler. Thus, when you use a lambda expression in a call to setOnAction, the Java compiler can deduce that the lambda expression will implement the EventHandler interface. And because the only abstract method of EventHandler is the handle method, the compiler can deduce that the method body you supply is an implementation of the handle method.

In a way, lambda expressions take the concept of anonymous classes two steps further. When you use an anonymous class to set an event handler, you must know and specify the name of the class (EventHandler) and the name of the method to be called (handle), so the only sense in which the class is anonymous is that you don't need to provide a name for a variable that will reference the class. But when you use a lambda expression, you don't have to know or specify the name of the class, the method, or a variable used to reference it. All you have to do, essentially, is provide the body of the handle method.

Listing 2-4 shows the AddSubtract4 program, which uses lambda expressions to handle the button clicks.

LISTING 2-4: **The AddSubtract4 Program with Lambda Expressions**

```
package com.lowewriter.AddSubtract4;

import javafx.application.*;
import javafx.stage.*;
import javafx.scene.*;
import javafx.scene.layout.*;
import javafx.scene.control.*;
import javafx.event.*;

public class AddSubtract4 extends Application
{
    public static void main(String[] args)
    {
        launch(args);
    }

    Button btnAdd;
    Button btnSubtract;
    Label lbl;
    int iCounter = 0;

    @Override public void start(Stage primaryStage)
    {
        // Create the Add button
        btnAdd = new Button();
        btnAdd.setText("Add");
        btnAdd.setOnAction( e ->                               →27
            {
                iCounter++;
                lbl.setText(Integer.toString(iCounter));
            } );
```

(continued)

LISTING 2-4: *(continued)*

```
                // Create the Subtract button
                btnSubtract = new Button();
                btnSubtract.setText("Subtract");
                btnSubtract.setOnAction( e ->                              →36
                    {
                        iCounter--;
                        lbl.setText(Integer.toString(iCounter));
                    } );

                // Create the Label
                lbl = new Label();
                lbl.setText(Integer.toString(iCounter));

                // Add the buttons and label to an HBox pane
                HBox pane = new HBox(10);
                pane.getChildren().addAll(lbl, btnAdd, btnSubtract);

                // Add the layout pane to a scene
                Scene scene = new Scene(pane, 200, 75);

                // Add the scene to the stage, set the title
                // and show the stage
                primaryStage.setScene(scene);
                primaryStage.setTitle("Add/Sub");
                primaryStage.show();
            }

    }
```

This program works essentially the same way as the program shown in Listing 2-3, so I just point out the features directly related to the use of the lambda expression:

>> →27 This statement uses a lambda expression to add an event handler to the Add button. The method body of this lambda expression increments the counter variable and then sets the label text to reflect the updated value.

>> →36 This statement uses a similar lambda expression to create the event handler for the Subtract button. The only difference between this lambda expression and the one for the Add button is that here the counter variable is decremented instead of incremented.

Note that in this example, the lambda expressions for the two event handlers are simple because very little processing needs to be done when either of the buttons in this program are clicked. What would the program look like, however, if the processing required for one or more of the button clicks required hundreds of lines of Java code to implement? The lambda expression would become unwieldy. For this reason, I often prefer to isolate the actual processing to be done by an event handler in a separate method. Then, the lambda expression itself includes just one line of code that simply calls the method.

Listing 2-5 shows another variation of the AddSubtract5 program implemented using that technique. Note that the technique used in Listing 2-5 is the technique that most of the remaining programs in this book use.

LISTING 2-5: **The AddSubtract5 Program with Lambda Expressions**

```java
package com.lowewriter.addsubtract5;

import javafx.application.*;
import javafx.stage.*;
import javafx.scene.*;
import javafx.scene.layout.*;
import javafx.scene.control.*;
import javafx.event.*;

public class AddSubtract5 extends Application
{
    public static void main(String[] args)
    {
        launch(args);
    }

    Button btnAdd;
    Button btnSubtract;
    Label lbl;
    int iCounter = 0;

    @Override public void start(Stage primaryStage)
    {
        // Create the Add button
        btnAdd = new Button();
        btnAdd.setText("Add");
        btnAdd.setOnAction( e -> btnAdd_Click() );                    →27

        // Create the Subtract button
        btnSubtract = new Button();
        btnSubtract.setText("Subtract");
        btnSubtract.setOnAction( e -> btnSubtract_Click() );          →32

        // Create the Label
        lbl = new Label();
        lbl.setText(Integer.toString(iCounter));
```

(continued)

LISTING 2-5:   *(continued)*

```
        // Add the buttons and label to an HBox pane
        HBox pane = new HBox(10);
        pane.getChildren().addAll(lbl, btnAdd, btnSubtract);

        // Add the layout pane to a scene
        Scene scene = new Scene(pane, 200, 75);

        // Add the scene to the stage, set the title
        // and show the stage
        primaryStage.setScene(scene);
        primaryStage.setTitle("Add/Sub");
        primaryStage.show();
    }

    private void btnAdd_Click()                                        →52
    {
        iCounter++;
        lbl.setText(Integer.toString(iCounter));
    }

    private void btnSubtract_Click()                                   →58
    {
        iCounter--;
        lbl.setText(Integer.toString(iCounter));
    }

}
```

The following paragraphs highlight the important points of this version of the program:

» →27 The setOnAction method for the Add button uses a lambda expression to specify that the method named btnAdd_Click should be called when the user clicks the button.

» →32 The setOnAction method for the Subtract button uses a lambda expression to specify that the method named btnSubtract_Click should be called when the user clicks the button.

» →52 The btnAdd_Click method increments the counter and updates the label's text to reflect the updated counter value.

» →58 Likewise, the btnSubtract_Click method decrements the counter and updates the label's text accordingly.

Chapter **3**

Setting the Stage and Scene Layout

O for a Muse of fire, that would ascend

The brightest heaven of invention,

A kingdom for a stage, princes to act,

And monarchs to behold the swelling scene!

So begins William Shakespeare's play *Henry V*, and so also begins this chapter, in which I explore the various ways to manipulate the appearance of a JavaFX application by manipulating its stage and its swelling scenes.

Specifically, this chapter introduces you to important details about the Stage class and the Scene class so that you can control such things as whether the window is resizable and if so, whether it has a maximum or a minimum size. You also learn how to coerce your programs into displaying additional stages beyond the primary stage, such as an alert or confirmation dialog box. And finally, you learn the proper way to end a JavaFX program by handling the events generated when the user closes the stage.

Examining the Stage Class

A stage, which is represented by the Stage class, is the topmost container in which a JavaFX user interface appears. In Windows, on a Mac, or in Linux, a stage is usually a window. On mobile devices, the stage may be the full screen or a tiled region of the screen.

When a JavaFX application is launched, a stage known as the *primary stage* is automatically created. A reference to this stage is passed to the application's start method via the primaryStage parameter:

```
@Override public void start(Stage primaryStage)
{
    // primaryStage refers to the
    // application's primary stage.
}
```

You can then use the primary stage to create the application's user interface by adding a scene, which contains one or more controls or other user-interface nodes.

TIP

In many cases, you will need to access the primary stage outside of the scope of the start method. You can easily make this possible by defining an instance variable (visible throughout the entire class) and using it to reference the primary stage. You see an example of how to do that later in this chapter, in the section "Switching Scenes."

The primary stage initially takes on the default characteristics of a normal windowed application, which depends on the operating system within which the program will run. You can, if you choose, change these defaults to suit the needs of your application. At the minimum, you should always set the window title. You may also want to change such details as whether the stage is resizable and various aspects of the stage's appearance.

The Stage class comes equipped with many methods that let you manipulate the appearance and behavior of a stage. Table 3-1 lists the ones you're most likely to use.

TABLE 3-1

Commonly Used Methods of the Stage Class

Method	Description
`void close()`	Closes the stage.

`void initModality(Modality modality)`	Sets the modality of the stage. This method must be called before the show method is called. The modality can be one of the following:
	
	`Modality.NONE`
	
	`Modality.APPLICATION_MODAL`
	
	`Modality.WINDOW_MODAL`

`void initStyle(StageStyle style)`	Sets the style for the stage. This method must be called before the show method is called. The style can be one of the following:
	
	`StageStyle.DECORATED`
	
	`StageStyle.UNDECORATED`
	
	`StageStyle.TRANSPARENT`
	
	`StageStyle.UTILITY`

`void getMaxHeight(double maxheight)`	Gets the maximum height for the stage.
`void getMaxWidth(double maxwidth)`	Gets the maximum width for the stage.
`void getMinHeight(double maxheight)`	Gets the minimum height for the stage.
`void getMinWidth(double maxwidth)`	Gets the minimum width for the stage.
`void setFullScreen(boolean fullscreen)`	Sets the fullscreen status of the stage.
`void setIconified(boolean iconified)`	Sets the iconified status of the stage.
`void setMaximized(boolean maximized)`	Sets the maximized status of the stage.
`void setMaxHeight(double maxheight)`	Sets the maximum height for the stage.
`void setMaxWidth(double maxwidth)`	Sets the maximum width for the stage.
`void setMinHeight(double maxheight)`	Sets the minimum height for the stage.
`void setMinWidth(double maxwidth)`	Sets the minimum width for the stage.
`void setResizable(boolean resizable)`	Controls whether the user can resize the stage.
`void setScene(Scene scene)`	Sets the scene to be displayed on the stage.
`void setTitle(String title)`	Sets the title to be displayed in the stage's title bar, if a title bar is visible.

(continued)

TABLE 3-1 *(continued)*

Method	Description
void show()	Makes the stage visible.
void showAndWait()	Makes the stage visible and then waits until the stage is closed before continuing.
void toFront()	Forces the stage to the foreground.
void toBack()	Forces the stage to the background.

The following paragraphs point out some of the ins and outs of using the Stage class methods listed in Table 3-1:

» **For many (if not most) applications, the only three methods from Table 3-1 you need to use are** setScene, setTitle, **and** show.

- Every stage needs a scene. Otherwise, nothing will appear in the stage. And there's nothing sadder than an empty stage.

- Every stage should also have a title.

- And finally, there's not much point in creating a stage if you don't intend on showing it to the user.

The other methods in the table let you change the appearance or behavior of the stage, but the defaults are acceptable in most cases.

» **If you want to prevent the user from resizing the stage, use the** setResizable **method like this:**

```
primaryStage.setResizable(false);
```

Then, the user can't change the size of the window. (By default, the stage is resizable. Thus, you don't need to call the setResizable method unless you want to make the stage non-resizable.)

» **If the stage is resizable, you can set the minimum and maximum size for the window.** For example:

```
primaryStage.setResizable(true);
primaryStage.setMinWidth(200);
primaryStage.setMinHeight(200);
primaryStage.setMaxWidth(600);
primaryStage.setMaxHeight(600);
```

In this example, the user can resize the window, but the smallest allowable size is 200-x-200 pixels and the largest allowable size is 600-x-600 pixels.

>> **If you want to display the stage in a maximized window, call** `setMaximized`:

```
primaryStage.setMaximized(true);
```

A maximized window still has the usual decorations (a title bar, window borders, and Minimize, Restore, and Close buttons). If you want the stage to completely take over the screen with no such decorations, use the set-FullScreen method instead:

```
primaryStage.setFullScreen(true);
```

TIP

When your stage enters fullscreen mode, JavaFX displays a message advising the user on how to exit fullscreen mode.

>> **If, for some reason, you want to start your program minimized to an icon, use the** `setIconified` **method:**

```
primaryStage.setIconified(true);
```

>> **For more information about the** `close` **method, see the section "Exit, Stage Right" later in this chapter.**

>> **The** `initModality` **and** `initStyle` **methods are interesting because they can be called only** *before* **you call the** `show` **method.** The `initModality` method allows you to create a modal dialog box — that is, a window that must be closed before the user can continue using other functions within the program. And the `initStyle` method lets you create windows that do not have the usual decorations such as a title bar or Minimize, Restore, and Close buttons. You typically use these methods when you need to create additional stages for your application beyond the primary stage. You can read more about how that works later in this chapter, in the section "Creating a Dialog Box."

Examining the Scene Class

Like the Scene class, the Scene class is fundamental to JavaFX programs. In every JavaFX program, you use at least one instance of the Scene class to hold the user-interface controls that your users will interact with as they use your program.

Table 3-2 lists the more commonly used constructors and methods of the Scene class.

TABLE 3-2   **Commonly Used Constructors and Methods of the Scene class**

Constructor	Description
Scene(Parent root)	Creates a new scene with the specified root node
Scene(Parent root, double width, double height)	Creates a new scene with the specified root node, width, and height

Method	Description
double getHeight()	Gets the height of the scene
double getWidth()	Gets the width of the scene
double getX()	Gets the horizontal position of the scene
double getY()	Gets the vertical position of the screen
void setRoot(Parent root)	Sets the root node

The following paragraphs explain some of the more interesting details of the constructors and methods of the Scene class:

» **All the Scene class constructors require that you specify the root node.**
You can change the root node later by calling the setRoot method, but it's not possible to create a scene without a root node.

» **You might be wondering why the root node is an instance of the Parent class rather than an instance of the Node class.** The Parent class is actually a subclass of the Node class, which represents a node that can have child nodes. There are several other subclasses of Node, which represent nodes that can't have children; those nodes can't be used as the root node for a scene.

» **You can set the scene's initial size when you create it by specifying the Width and Height parameters.**
If you don't set the size, the scene will determine its own size based on its content.

» **You can retrieve the size of the scene via the getHeight and getWidth methods.**
There are no corresponding set methods that let you set the height or width.

>> **In general, the size of the scene determines the size of the stage,** provided that that scene is not smaller than the minimum size specified for the stage or larger than the maximum size.

>> **If the user resizes the stage, the size of the scene is resized accordingly.**

Switching Scenes

The primary stage of a JavaFX program (or any other stage, for that matter) can have only one scene displayed within it at any given time. However, that doesn't mean that your program can't create several scenes and then swap them as needed. For example, suppose you're developing a word-processing program and you want to let the user switch between an editing view and a page preview view. You could do that by creating two distinct scenes, one for each view. Then, to switch the user between views, you simply call the stage's setScene method to switch the scene.

In Chapter 1 of this minibook, you read about a ClickCounter program whose scene displays a label and a button and then updates the label to indicate how many times the user has clicked the button. Then, in Chapter 2 of this minibook, you saw several variations of an AddSubtract program whose scene displayed a label and two buttons: One button added one to a counter when clicked, the other subtracted one from the counter.

Listing 3-1 shows a program named SceneSwitcher that combines the scenes from the ClickCounter and AddSubtract programs into a single program. Figure 3-1 shows this program in action:

>> **When the SceneSwitcher program is first run,** it displays the ClickCounter scene as shown on the left side of the figure.

>> **When the user clicks the Switch Scene button,** the scene switches to the AddSubtract scene, as shown in the right side of the figure.

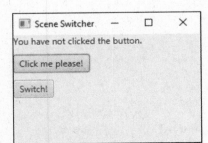
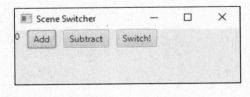

FIGURE 3-1:
The Scene Switcher program.

LISTING 3-1: **The SceneSwitcher Program**

```java
package com.lowewriter.sceneswitcher;

import javafx.application.*;
import javafx.stage.*;
import javafx.scene.*;
import javafx.scene.layout.*;
import javafx.scene.control.*;
import javafx.event.*;

public class SceneSwitcher extends Application
{
    public static void main(String[] args)
    {
        launch(args);
    }

    // class fields for Click-Counter scene            →17
    int iClickCount = 0;
    Label lblClicks;
    Button btnClickMe;
    Button btnSwitchToScene2;
    Scene scene1;

    // class fields for Add-Subtract scene             →24
    int iCounter = 0;
    Label lblCounter;
    Button btnAdd;
    Button btnSubtract;
    Button btnSwitchToScene1;
    Scene scene2;

    // class field for stage
    Stage stage;                                        →33

    @Override public void start(Stage primaryStage)
    {
        stage = primaryStage;                           →37

        // Build the Click-Counter scene                →39

        lblClicks = new Label();
        lblClicks.setText("You have not clicked the button.");

        btnClickMe = new Button();
        btnClickMe.setText("Click me please!");
        btnClickMe.setOnAction(
            e -> btnClickMe_Click() );
```

```
btnSwitchToScene2 = new Button();
btnSwitchToScene2.setText("Switch!");
btnSwitchToScene2.setOnAction(
    e -> btnSwitchToScene2_Click() );

VBox pane1 = new VBox(10);
pane1.getChildren().addAll(lblClicks, btnClickMe,
    btnSwitchToScene2);

scene1 = new Scene(pane1, 250, 150);


// Build the Add-Subtract scene                              →61

lblCounter = new Label();
lblCounter.setText(Integer.toString(iCounter));

btnAdd = new Button();
btnAdd.setText("Add");
btnAdd.setOnAction(
    e -> btnAdd_Click() );

btnSubtract = new Button();
btnSubtract.setText("Subtract");
btnSubtract.setOnAction(
    e -> btnSubtract_Click() );

btnSwitchToScene2 = new Button();
btnSwitchToScene2.setText("Switch!");
btnSwitchToScene2.setOnAction(
    e -> btnSwitchToScene1_Click() );

HBox pane2 = new HBox(10);
pane2.getChildren().addAll(lblCounter, btnAdd,
    btnSubtract, btnSwitchToScene2);

scene2 = new Scene(pane2, 300, 75);

// Set the stage with scene 1 and show the stage             →87
primaryStage.setScene(scene1);
primaryStage.setTitle("Scene Switcher");
primaryStage.show();
}


// Event handlers for scene 1                                →94
```

(continued)

LISTING 3-1:   *(continued)*

```
public void btnClickMe_Click()
{
    iClickCount++;
    if (iClickCount == 1)
    {
        lblClicks.setText("You have clicked once.");
    }
    else
    {
        lblClicks.setText("You have clicked "
            + iClickCount + " times." );
    }
}

private void btnSwitchToScene2_Click()
{
    stage.setScene(scene2);
}


// Event handlers for scene 2

private void btnAdd_Click()
{
    iCounter++;
    lblCounter.setText(Integer.toString(iCounter));
}

private void btnSubtract_Click()
{
    iCounter--;
    lblCounter.setText(Integer.toString(iCounter));
}

private void btnSwitchToScene1_Click()
{
    stage.setScene(scene1);
}

}
```

→116

The following paragraphs point out some key sections of the program:

» →17 The section of the programs defines class fields that will be used by the scene for the Click-Counter portion of the program. These fields include iClickCount, used to count the number of times the user has clicked the

Click Me! Button; the label used to display the count of how many times the Click Me! button has been clicked; the Click Me! button itself; and the button used to switch to the Add-Subtract scene. Also included is a Scene field named scene1 that will be used to reference the Click Counter scene.

» →24 These lines define class variables used by the Add-Subtract portion of the program, including the counter (iCounter), the label used to display the counter, the two buttons used to increment and decrement the counter, the button used to switch back to the Click-Counter scene, and a Scene field named scene2 that will be used to reference the Add-Subtract scene.

» →33 A class field named stage is used to hold a reference to the primary stage so that it can be accessed throughout the program.

» →37 This line sets stage class field to reference the primary stage.

» →39 This section of the program builds the Click-Counter scene. First, it creates the label and buttons displayed by the scene. Then it creates a VBox layout pane (which lays out its controls in a vertical stack) and adds the label and buttons to the pane. Finally, it creates the scene using the VBox pane as its root.

» →61 This section of the program builds the Add-Subtract scene by creating the label and the buttons displayed by the scene, arranging them in an HBox layout pane, and creating the scene using the HBox pane as its root.

» →87 These lines set the Click-Counter scene as the root scene for the primary stage, sets the stage title, and then shows the stage.

» →94 This section of the program provides the event handlers for the buttons in the Click-Counter scene. The event handler for the Click Me! button increments the click counter, then sets the label to display an appropriate message. The handler for btnSwitchToScene2 simply switches the scene of the primary stage to scene2, which instantly switches the display to the Add–Subtract scene as shown in the right side of Figure 3-1.

» →116 This section of the program provides the event handlers for the buttons in the Add-Subtract scene. The event handler for the Add and Subtract buttons increment or decrement the counter and update the text displayed by the label. The handler for btnSwitchToScene1 switches the scene back to scene1, which switches the display back to the Click-Counter scene shown in the left right side of Figure 3-1.

Creating an Alert Box

JavaFX provides a simple means of displaying a basic message box by using the `Alert` class, which is similar to the `JOptionPane` class you learned back in Book 2, Chapter 2. Table 3-3 shows the commonly used constructors and methods for this class.

TABLE 3-3 **Commonly Used Constructors and Methods of the Alert class**

Constructor	Description
`Alert(Alert.AlertType)`	Creates a new alert of the specified type
`Alert(Alert.AlertType, String text, ButtonType type...)`	Creates a new alert and optionally sets one or more buttons to be displayed
Method	**Description**
`void setTitle(String text)`	Sets the title
`Optional<ButtonType> showAndWait()`	Shows the alert and waits for the user's response, which is returned as a `ButtonType` object

The `AlertType` parameter lets you specify one of several types of Alert dialogs:

» `AlertType.CONFIRMATION`, which prompts the user to confirm an action.

» `AlertType.ERROR`, which display an error message.

» `AlertType.INFORMATION`, which displays an information dialog box.

» `AlertType.WARNING`, which displays a warning message.

» `AlertType.NONE`, which display a generic alert dialog.

Here's a snippet of code that displays a simple informational message using the `Alert` class:

```
Alert a = new Alert(Alert.AlertType.INFORMATION, "You have clicked once.");
a.showAndWait();
```

Figure 3-2 shows the resulting alert box.

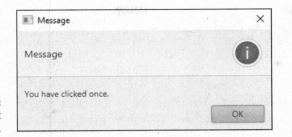

FIGURE 3-2:
An Alert
dialog box.

You can control what buttons appear on the Alert box by using the optional ButtonType parameter. You can choose from among the following types of buttons:

>> ButtonType.APPLY

>> ButtonType.CANCEL

>> ButtonType.CLOSE

>> ButtonType.FINISH

>> ButtonType.NEXT

>> ButtonType.NO

>> ButtonType.OK

>> ButtonType.PREVIOUS

>> ButtonType.YES

You can include more than one button on an Alert box by simply specifying more than one ButtonType parameters in the constructor. For example:

```
Alert a = new Alert(Alert.AlertType.INFORMATION, "Are you certain?",
          ButtonType.YES, ButtonType.NO);
```

In this example, the Alert box will include both a YES and a NO button.

To determine which button the user clicked, you must test the value returned by the showAndWait method. This value is an object of type Optional, since the user can close the dialog box without clicking any of the displayed buttons. You should first test whether the user clicked a button by calling the isPresent method. Then, you can call the get method to retrieve the actual result.

Here's an example that determines whether the user clicked the YES button:

```
Alert a = new Alert(Alert.AlertType.INFORMATION, "Are you certain?",
          ButtonType.YES, ButtonType.NO);
Optional<ButtonType> r = a.showAndWait();
```

```
if (r.isPresent() && r.get() == ButtonType.YES)
{
    // The user clicked OK!
}
```

To demonstrate how you might use the ALERT class, the program shown in Listing 3-2 is a variation of the ClickCounter program that was originally discussed in Chapter 2 of this minibook. The original version of this program displayed a label and a button, using the label to display a count of how many times the user has clicked the button. This version of the program dispenses with the label and instead uses the Alert class to display a message indicating how many times the user has clicked the button.

LISTING 3-2: **The ClickCounterAlert Program**

```
package com.lowewriter.clickcounteralert;

import javafx.application.*;
import javafx.stage.*;
import javafx.scene.*;
import javafx.scene.layout.*;
import javafx.scene.control.*;

public class ClickCounterAlert extends Application
{
    public static void main(String[] args)
    {
        launch(args);
    }

    Button btn;
    Label lbl;
    int iClickCount = 0;

    @Override public void start(Stage primaryStage)
    {
    // Create the button
        btn = new Button();
        btn.setText("Click me please!");
        btn.setOnAction(e -> buttonClick());

        // Add the button to a layout pane
        BorderPane pane = new BorderPane();
        pane.setCenter(btn);
```

```
        // Add the layout pane to a scene
        Scene scene = new Scene(pane, 250, 150);

        // Add the scene to the stage, set the title
        // and show the stage
        primaryStage.setScene(scene);
        primaryStage.setTitle("Click Counter");
        primaryStage.show();
    }

    public void buttonClick()
    {
        iClickCount++;
        if (iClickCount == 1)
        {
            Alert a = new Alert(Alert.AlertType.INFORMATION, "You have clicked
once." );
            a.showAndWait();
        }
        else
        {
            Alert a = new Alert(Alert.AlertType.INFORMATION, "You have clicked "
                + iClickCount + " times.");
            a.showAndWait();
        }
    }
}
```

This program is nearly identical to the version that was presented in Chapter 2 of this minibook (in Listing 2-2). In fact, here are the only two differences:

>> No label is defined in this program because a message box, not a label, is used to display the number of times the button has been clicked.

>> In the buttonClick method of the Chapter 2 version, the label's setText method was called to display the number of times the button has been clicked. In this version, an Alert box is used instead.

Figure 3-2 shows the new version of the ClickCounter program in action. Here, you can see the alert box displayed when the user clicks the button the first time.

Exit, Stage Right

Because I started this chapter by quoting Shakespeare, I thought it'd be nice to end it by quoting Snagglepuss, the famous pink mountain lion from the old Hanna-Barbera cartoons. He'd often leave the scene by saying, "Exit, stage left" or "Exit, stage right."

Heavens to Mergatroyd!

There's a right way and a wrong way to exit the stage, even. And so far, none of the programs presented in this book have done it the right way. The only mechanism the programs you've seen so far have provided to quit the program is for the user to click the standard window close button, typically represented by an X in the upper-right corner of the window's title bar. That is almost always the wrong way to exit a program.

In most cases, the correct way to exit a program involves the following details:

>> **Adding a button, menu command, or other way for the user to signal that she wishes to close the program.**

 Many programs include a button labeled Exit or Close, and programs that use a menu usually have an Exit command.

>> **Optionally displaying a confirmation box that verifies whether the user really wants to close the program.** You can do that by using the ConfirmationBox class shown in the preceding section or by using a similar class.

 Depending on the program, you might want to display this dialog box only if the user has made changes to a document, database, or other file that have not yet been saved.

>> **If the user really does want to close the program, the program should perform any necessary clean-up work,** such as

 ● Saving changes to documents, databases, or other files.

 ● Properly closing databases and other resources.

>> **After you've done any necessary clean-up work, you can close the application by calling the primary stage's** close **method.**

>> **The verification and clean-up steps should be taken whether the user attempts to close the program** by using a button or menu command you've provided in your user interface or by clicking the built-in window close button.

In the following sections, you read about how to add a Close button to your application, how to prevent the window close button from unceremoniously terminating your application, and how to put these two elements together in a complete program.

Creating a Close button

To add a button or other user-interface element that allows the user to close the button, all you have to do is provide an action event handler that calls the stage's close method.

For example, suppose you create a Close button using the following code:

```
Button btnClose = new Button();
btnClose.setText("Close");
btnClose.setOnAction( e -> primaryStage.close() );
```

In this case, the action event handler simply calls primaryStage.close() to close the application.

If you want to do more than simply call the close method in the action event handler, you may want to isolate the event handler in a separate method, as in this example:

```
btnClose.setOnAction( e -> btnClose_Clicked());
```

Because the btnClose_Clicked method will need to access the primary stage to close it, you need to define a class field of type Stage and use it to reference the primary stage. Then, your btnClose_Clicked method can easily perform additional tasks. For example:

```
private void btnClose_Click()
{
    boolean reallyQuit = false;
    reallyQuit = ConfirmationBox.show(
        "Are you sure you want to quit?",
        "Confirmation",
        "Yes", "No");
    if (reallyQuit)
    {
        // Perform cleanup tasks here
        // such as saving files or freeing resources
        stage.close();
    }
}
```

In this example, a confirmation box is displayed to make sure the user really wants to exit the program.

Handling the CloseRequest event

Providing a Close button is an excellent way to allow your users to cleanly exit from your program. However, the user can bypass your exit processing by simply closing the window — that is, by clicking the window close button, usually represented as an X in the upper-right corner of the window border. Unless you provide otherwise, clicking this button unceremoniously terminates the application, bypassing all your nice code that confirms whether the user wants to save his work, closes any open resources, and otherwise provides for a graceful exit.

Fortunately, you can easily avoid such ungraceful exits. Whenever the user attempts to close the window within which a JavaFX stage is displayed, JavaFX generates a CloseRequest event, which is sent to the stage. You can provide an event handler for this event by calling the setOnCloseRequest method of the Stage class. Then, the event handler is called whenever the user tries to close the window.

You might be tempted to create a single method that can serve as the event handler for both the Action event of your Close button and the CloseRequest event, like this:

```
btnClose.setText("Close");
btnClose.setOnAction( e -> btnClose_Click () );
primaryStage.setOnCloseRequest( e -> btnClose_Click () );
```

Here, the intent is to handle the CloseRequest event exactly as if the user had clicked the btnClose button.

That's a good idea, but it doesn't work if the btnClose_Click event displays a confirmation box and closes the stage only if the user confirms that she really wants to quit the program. That's because when the event handler for the CloseRequest event ends, JavaFX automatically closes the stage if the event handler doesn't explicitly close the stage.

To prevent that from happening, you call the consume method of the CloseRequest event object. Consuming the event causes it to be stopped in its tracks within the event handler, thus preventing JavaFX from automatically closing the stage when the event handler ends.

In the lambda expression passed to the setOnCloseRequest method, the CloseRequest event object is represented by the argument e. Thus, you can consume the CloseRequest event by calling e.consume().

An easy way to provide a method that handles both the Action event for a Close button and the CloseRequest event for a stage is to craft the lambda expression for the setOnCloseRequest method so that it consumes the event before calling the method that will handle the event:

```
btnClose.setText("Close");
btnClose.setOnAction( e -> btnClose_Click () );
primaryStage.setOnCloseRequest(
```

```
    e -> {
            e.consume();
            btnClose_Click ();
        } );
```

Here, the event handler for the CloseRequest event first consumes the event and then calls btnClose_Click. The btnClose_Click method, in turn, displays a confirmation box and closes the stage if the user confirms that this is indeed what he wishes to do.

Putting it all together

Now that you know how to add a Close button to a scene and how to handle the CloseRequest event, I look at a program that puts together these two elements to demonstrate the correct way to exit a JavaFX program.

This section presents a variation of the ClickCounter program that includes a Close button in addition to the Click Me! button. When the user clicks the Click Me! button, a message box displays to indicate how many times the button has been clicked. But when the user attempts to exit the program, whether by clicking the Close button or by simply closing the window, the ConfirmationBox class that was shown in Listing 3-3 is used to ask the user whether she really wants to exit the program. Then, the stage is closed only if the user clicks the Yes button in the confirmation box.

The source code for this program is shown in Listing 3-3.

LISTING 3-3: **The ClickCounterExit program**

```
package com.lowewriter.clickcounterexit;

import javafx.application.*;
import javafx.stage.*;
import javafx.scene.*;
import javafx.scene.layout.*;
import javafx.scene.control.*;
import javafx.geometry.*;
import java.util.*;

public class ClickCounterExit extends Application
{
    public static void main(String[] args)
    {
        launch(args);
```

```
    }

    Stage stage;
    int iClickCount = 0;


    @Override public void start(Stage primaryStage)
    {
        stage = primaryStage;

        // Create the Click Me button
        Button btnClickMe = new Button();
        btnClickMe.setText("Click me please!");
        btnClickMe.setOnAction(e -> btnClickMe_Click());

        // Create the Close button
        Button btnClose = new Button();
        btnClose.setText("Close");
        btnClose.setOnAction(e -> btnClose_Click());

        // Add the buttons to a layout pane
        VBox pane = new VBox(10);
        pane.getChildren().addAll(btnClickMe, btnClose);
        pane.setAlignment(Pos.CENTER);

        // Add the layout pane to a scene
        Scene scene = new Scene(pane, 250, 150);

        // Finish and show the stage
        primaryStage.setScene(scene);
        primaryStage.setTitle("Click Counter");
        primaryStage.setOnCloseRequest( e ->
            {
                e.consume();
                btnClose_Click();
            } );
        primaryStage.show();
    }

    public void btnClickMe_Click()
    {
        iClickCount++;
        if (iClickCount == 1)
        {
            Alert a = new Alert(Alert.AlertType.INFORMATION, "You have clicked
once." );
            a.showAndWait();
        }
```

(continued)

LISTING 3-3:   *(continued)*

```
        else
        {
            Alert a = new Alert(Alert.AlertType.INFORMATION, "You have clicked "
                + iClickCount + " times.");
            a.showAndWait();
        }
    }

    public void btnClose_Click()
    {
        Alert a = new Alert(Alert.AlertType.CONFIRMATION,
            "Are you sure you want to quit?",
            ButtonType.YES, ButtonType.NO);
        Optional<ButtonType> confirm = a.showAndWait();
        if (confirm.isPresent() && confirm.get() == ButtonType.YES)
        {
            stage.close();
        }
    }

}
```

Chapter **4**

Using Layout Panes to Arrange Your Scenes

Controlling the layout of components in a scene is often one of the most difficult aspects of working with JavaFX. In fact, at times it can be downright exasperating. Often the components almost seem to have minds of their own. They get stubborn and refuse to budge. They line up on top of one another when you want them to be side by side. You make a slight change to a label or text field, and the whole scene seems to rearrange itself. At times, you want to put your fist through the monitor.

WARNING

I recommend against putting your fist through your monitor. You'll make a mess, cut your hand, and have to spend money on a new monitor — and when you get your computer working again, the components *still* won't line up the way you want them to be.

The problem isn't with the components; it's with the *layout panes*, which determine where each component appears in its frame or panel. Layout panes are special classes whose sole purpose in life is to control the arrangement of the nodes that appear in a scene. JavaFX provides several distinct types of layout panes; each type uses a different approach to controlling the arrangement of nodes. The trick to successfully lay out a scene is to use the layout panes in the correct combination to achieve the arrangement you want.

Working with Layout Panes

Understanding layout panes is the key to creating JavaFX frames that are attractive and usable.

Introducing five JavaFX layout panes

JavaFX provides many different layout panes for you to work with. I explain the following five in this chapter:

>> HBox: This layout pane arranges nodes horizontally, one next to the other. You use it to create controls arranged neatly in rows.

>> VBox: This layout pane arranges nodes vertically, one above the other. You use it to create controls arranged neatly in columns.

>> FlowPane: This layout pane arranges nodes next to each other until it runs out of room; then, it wraps to continue layout nodes. You can configure a FlowPane to arrange nodes horizontally in rows or vertically in columns.

>> Border: This layout pane divides the pane into five regions: Top, Left, Center, Right, and Bottom. When you add a node, you can specify which region you want to place the node in.

>> GridPane: This layout pane divides the pane into a grid, affording you complete control of the arrangement of elements in rows and columns.

To give you a general idea of the results that can be achieved with the first four layout types, Figure 4-1 shows four sample windows that each use one of the layout panes. You'll see a detailed example of how the GridPane layout works later in this chapter.

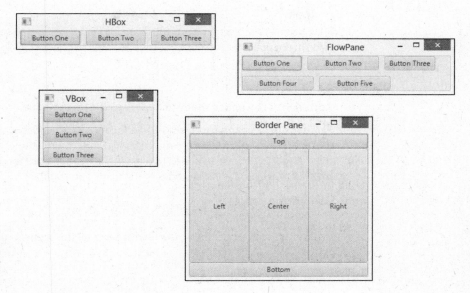

FIGURE 4-1:
Four commonly
used types of
layout panes.

Creating layout panes

The basic process of working with layout panes is simple. Here is the general procedure for creating a layout node:

1. Create the controls or other nodes you want to add to the pane.

For example, if the layout pane will contain two buttons, you should create the two buttons using code similar to this:

```
Button btnOK = new Button();
btnOK.setText("OK");
btnOK.setOnAction( e -> btnOK_Click() );
Button btnCancel = new Button();
btnCancel.setText("Cancel");
btnCancel.setOnAction( e -> btnCancel_Click() );
```

2. Create a layout pane by calling its constructor.

For example:

```
HBox pane = new HBox();
```

3. Fine-tune the optional settings used by the layout pane.

Each type of layout pane has a unique assortment of optional parameters that govern the details of how nodes are laid out within the pane. For example, the

HBox pane lets you set the number of pixels that will be used to separate each node in the pane. You can set this value as follows:

```
HBox.setSpacing(10);
```

4. **Add each of the nodes that will appear in the layout pane.**

Each type of layout pane provides a method for adding nodes to the pane. For the HBox pane, you must first call the getChildren method to get a list of all the nodes that have been added to the pane. Then, you call the addAll method to add one or more nodes to the pane. For example:

```
pane.getChildren().addAll(btnOK, btnCancel);
```

5. **Create the scene, specifying the layout pane as the scene's root node.**

For example:

```
Scene scene = new Scene(pane, 300, 400);
```

In this example, pane is added as the root node for the scene.

Combining layout panes

You can combine several layout panes to create layouts that are more complicated than a single layout pane can provide. For example, suppose you want to create a layout that has a horizontal row of buttons at the bottom and a vertical column of buttons at the right. To do that, you could create an HBox for the buttons at the bottom and a VBox for the buttons at the right. Then, you could create a Border-Pane and add the HBox to the bottom region and the VBox to the right region.

Combinations like this are possible because all the layout panes inherit the base class javafx.scene.layout.Pane, which in turn inherits the class javafx.scene.node. In other words, all panes are also nodes. Each node that you add to a layout pane can be another layout pane. You can nest layout panes within layout panes as deeply as you need to achieve the exact layout you need for your application.

Using the HBox Layout

The HBox class provides one of the simplest of all JavaFX's layout managers: It arranges one or more nodes into a horizontal row. Table 4-1 presents the most commonly used constructors and methods of the HBox class.

TABLE 4-1 **HBox Constructors and Methods**

Constructor	Description
HBox()	Creates an empty HBox.
HBox(double spacing)	Creates an empty HBox with the specified spacing.
HBox(Node... children)	Creates an HBox with the specified child nodes. This constructor lets you create an HBox and add child nodes to it at the same time.
HBox(double spacing, Node... children)	Creates an HBox with the specified spacing and child nodes.

Method	Description
ObservableList<Node> getChildren()	Returns the collection of all child nodes that have been added to the HBox. The collection is returned as an ObservableList type, which includes the method addAll, letting you add one or more nodes to the list.
static void setAlignment (Pos alignment)	Sets the alignment for child nodes within the HBox. See Table 4-5 for an explanation of the Pos enumeration. For more information, see the section "Aligning Nodes in a Layout Pane" later in this chapter.
static void setHgrow(Node child, Priority priority)	Sets the growth behavior of the given child node. See Table 4-3 for an explanation of the Priority enumeration. For more information, see the section "Adding Space by Growing Nodes" later in this chapter.
static void setMargin(Node child, Insets value)	Sets the margins for a given child node. See Table 4-2 for the constructors of the Insets class. For more information, see the section "Adding Space with Margins" later in this chapter.
void setPadding(Insets value)	Sets the padding around the inside edges of the Hbox. See Table 4-2 for the constructors of the Insets class. For more information, see the section "Spacing Things Out" later in this chapter.
void setSpacing(double value)	Sets the spacing between nodes displayed within the HBox. For more information, see the section "Spacing Things Out" later in this chapter.

The HBox class is defined in the `javafx.scene.layout` package, so you should include the following `import` statement in any program that uses an HBox:

```
import javafx.scene.layout.*;
```

The easiest way to create an HBox is to first create the nodes that you want to place in the HBox and then call the HBox constructor and pass the nodes as arguments. For example:

```
Button btn1 = new Button("Button One");
Button btn2 = new Button("Button Two");
Button btn3 = new Button("Button Three");
HBox hbox = new HBox(btn1, btn2, btn3);
```

If you prefer to create the HBox control in an initially empty state and later add the controls, you can do so like this:

```
HBox hbox = new HBox();
Hbox.getChildren().addAll(btn1, btn2, btn3);
```

Here, the `getChildren` method is called, which returns a collection of all the children added to the HBox pane. This collection is defined by the class `ObservableList`, which includes a method named `addAll` that you can use to add one or more nodes to the list.

Spacing Things Out

By default, child nodes in a layout pane are arranged immediately next to one another, with no empty space in between. If you want to provide space between the nodes in the pane, you can do so in four ways:

>> Adding spacing between elements within the pane

>> Adding padding around the inside edges of the pane

>> Adding margins to the individual nodes in the pane

>> Creating spacer nodes that can grow to fill available space

In this section, I show you how to add spacing and padding to a pane. Then, the next three sections show you how to use the other two techniques.

Note that although I illustrate the techniques in these sections using the HBox layout pane, the techniques apply to other types of panes as well.

To set the spacing for an HBox pane, you can use the spacing parameter on the HBox constructor or by calling the setSpacing method. For example, this statement creates an HBox pane with a default spacing of 10 pixels:

```
HBox hbox = new HBox(10);
```

This example creates an HBox pane with 10-pixel spacing and adds three buttons:

```
HBox hbox = new HBox(10, btn1, btn2, btn3);
```

And this example creates an HBox pane using the default constructor, and then calls the setSpacing method to set the spacing to 10 pixels:

```
HBox hbox = new HBox();
Hbox.setSpacing(10);
```

Although spacing adds space between nodes in an HBox pane, it doesn't provide any space between the nodes and the edges of the pane itself. For example, if you set the spacing to 10 pixels and add three buttons to the pane, the three buttons will be separated from one another by a gap of 10 pixels. However, there won't be any space at all between the left edge of the first button and the left edge of the pane itself. Nor will there be any space between the top of the buttons and the top of the pane. In other words, the three buttons will be crowded tightly into the pane.

To add space around the inside perimeter of the layout pane, use the setPadding method. This method takes as a parameter an object of type Insets, which represents the size of the padding (in pixels) for the top, right, bottom, and left edge of an object. You can create an Insets object using either of the two constructors listed in Table 4-2. The first provides an even padding for all four edges of an object; the second lets you set a different padding value for each edge.

TABLE 4-2   ## Insets Constructors

Constructor	Description
Insets(double value)	Creates an Insets object that uses the same value for the top, right, bottom, and left margins.
Insets(double top, double right, double bottom, double left)	Creates an Insets object that uses the specified top, right, bottom, and left margins.

To set the padding to a uniform 10 pixels, call the setPadding method like this:

```
hbox.setPadding(new Insets(10));
```

To set a different padding value for each edge, call it like this:

```
hbox.setPadding(new Insets(20, 10, 20, 10));
```

In this example, the top and bottom padding is set to 20 and the right and left padding is set to 10.

The Insets class is defined in the javafx.geometry package, so you should include the following import statement in any program that uses Insets:

```
import javafx.geometry.*;
```

Adding Space with Margins

Another way to add space around the nodes in a layout pane is to create margins around the individual nodes. This technique allows you to set a different margin size for each node in the layout pane, giving you complete control over the spacing of each node.

To create a margin, call the setMargin method for each node you want to add a margin to. You might think that because each node can have its own margin, the setMargin method would belong to the Node class. Instead, the setMargin method is defined by the HBox class. The setMargin method accepts two parameters:

>> The node you want to add the margin to

>> An Insets object that defines the margins you want to add

Here's an example that sets a margin of 10 pixels for all sides of a button named btn1:

```
Hbox.setMargin(btn1, new Insets(10));
```

Here's an example that sets a different margin for each side of the pane:

```
Hbox.setMargin(btn1, new Insets(10, 15, 20, 10));
```

In this example, the top margin is 10 pixels, the right margin is 15 pixels, the bottom margin is 20 pixels, and the left margin is 10 pixels.

Note that margins, spacing, and padding can work together. Thus, if you create a 5-pixel margin on all sides of two buttons, add those two buttons to a pane whose spacing is set to 10 pixels and whose padding is set to 10 pixels, the buttons will be separated from one another by a space of 20 pixels and from the inside edges of the pane by 15 pixels.

Adding Space by Growing Nodes

A third way to add space between nodes in an HBox is to create a node whose sole purpose is to add space between two HBox nodes. Then, you can configure the spacer node that will automatically grow to fill any extra space within the pane. By configuring only the spacer node and no other nodes in this way, only the spacer node will grow. This has the effect of pushing the nodes on either side of the spacer node apart from one another.

For example, suppose you want to create an HBox layout pane that contains three buttons. Instead of spacing all three buttons evenly within the pane, you want the first two buttons to appear on the left side of the pane and the third button to appear on the right side of the pane. The amount of space between the second and third buttons will depend entirely on the size of the pane. Thus, if the user drags the window to expand the stage, the amount of space between the second and third buttons should increase accordingly.

The easiest way to create a spacer node is by using the Region class. The Region class is the base class for both the Control class, from which controls such as Button and Label derive. It is also the base class for the Pane class, from which all the layout panes described in this chapter derive.

For my purposes here, I just use the simple default constructor of the Region class to create a node that serves as a simple spacer in a layout pane. I don't provide a specific size for the region. Instead, I configure it so that it will grow horizontally to fill any unused space within its container.

To do that, you use the static setHgrow method of the HBox class, specifying one of the three constant values defined by an enumeration named Priority enumeration. Table 4-3 lists these constants and explains what each one does.

The Priority enumeration is defined in the javafx.scene.layout package; the same package that defines the layout managers that require it.

TABLE 4-3 The Priority Enumeration

Constant	Description
Priority.NEVER	Indicates that the width of the node should never be adjusted to fill the available space in the pane. This is the default setting. Thus, by default, nodes are *not* resized based on the size of the layout pane that contains them.
Priority.ALWAYS	Indicates that the width of the node should always be adjusted if necessary to fill available space in the pane. If you set two or more nodes to ALWAYS, the adjustment will be split equally among each of the nodes.
Priority.SOMETIMES	Indicates that the node's width may be adjusted if necessary to fill out the pane. However, the adjustment will be made only if there are no other nodes that specify ALWAYS.

The following example creates three buttons and a spacer, set the margins for all three buttons to 10 pixels, and then add the three buttons and the spacer to an HBox such that the first two buttons appear on the left of the HBox and the third button appears on the right:

```
// Create the buttons
Button btn1 = new Button("One");
Button btn2 = new Button("Two");
Button btn3 = new Button("Three");

// Create the spacer
Region spacer = new Region();

// Set the margins
HBox.setMargin(btn1, new Insets(10));
HBox.setMargin(btn2, new Insets(10));
HBox.setMargin(btn3, new Insets(10));

// Set the Hgrow for the spacer
HBox.setHgrow(spacer, Priority.ALWAYS);

// Create the HBox layout pane
HBox hbox = new HBox(10, btn1, btn2, spacer, btn3);
```

Figure 4-2 shows how this pane appears when added to a stage. So that you can see how the spacer works, the first shows three incarnations of the pane, each with the window dragged to a different size. Notice how the spacing between the second and third buttons is adjusted automatically so that the first two buttons are on the left side of the pane and the third button is on the right.

Like the setMargin method, setHgrow is a static method, so it should be called from the class, not an instance of the class.

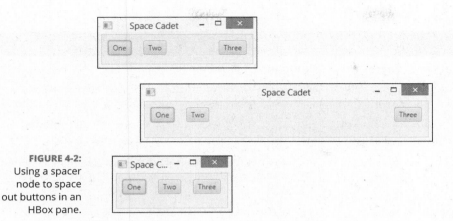

FIGURE 4-2:
Using a spacer
node to space
out buttons in an
HBox pane.

Using the VBox Layout

The VBox class is similar to the HBox class, but instead of arranging nodes horizontally in a row, it arranges them vertically in a column. Table 4-4 shows the most commonly used constructors and methods of the VBox class.

TABLE 4-4   **VBox Constructors and Methods**

Constructor	Description
VBox()	Creates an empty VBox.
VBox(double spacing)	Creates an empty VBox with the specified spacing.
VBox(Node... children)	Creates an VBox with the specified child nodes. This constructor lets you create a VBox and add child nodes to it at the same time.
VBox(double spacing, Node... children)	Creates a VBox with the specified spacing and child nodes.

Method	Description
ObservableList<Node> getChildren()	Returns the collection of all child nodes that have been added to the VBox. The collection is returned as an ObservableList type, which includes the method addAll, letting you add one or more nodes to the list.
static void setAlignment(Pos alignment)	Sets the alignment for child nodes within the HBox. See Table 4-5 for an explanation of the Pos enumeration. For more information, see the section "Aligning Nodes in a Layout Pane" later in this chapter.

(continued)

TABLE 4-4 *(continued)*

Constructor	Description
`static void setMargin(Node child, Insets value)`	Sets the margins for a given child node. See Table 4-2 for the constructors of the Insets class. For more information, see the section "Adding Space with Margins" earlier in this chapter.
`void setPadding(Insets value)`	Sets the padding around the inside edges of the VBox. See Table 4-2 for the constructors of the Insets class. For more information, see the section "Spacing Things Out" earlier in this chapter.
`static void setVgrow(Node child, Priority priority)`	Sets the growth behavior of the given child node. See Table 4-3 for an explanation of the Priority enumeration. For more information, see the section "Adding Space by Growing Nodes" earlier in this chapter.

The VBox class is defined in the `javafx.scene.layout` package, so you should include the following `import` statement in any program that uses a VBox:

```
import javafx.scene.layout.*;
```

Here's an example that creates three buttons and uses a VBox to arrange them into a column:

```
Button btn1 = new Button("Button One");
Button btn2 = new Button("Button Two");
Button btn3 = new Button("Button Three");
VBox vbox = new VBox(btn1, btn2, btn3);
```

You can accomplish the same thing by using the default constructor and calling the `getChildren` method, as in this example:

```
VBox vbox = new VBox();
vbox.getChildren().addAll(btn1, btn2, btn3);
```

As with the HBox class, you can use spacing, padding, margins, and spacer nodes to control the spacing of nodes within a VBox. Here's an example that sets 10 pixels of vertical space between nodes and 10 pixels of padding on each edge of the pane:

```
Button btn1 = new Button("One");
Button btn2 = new Button("Two");
```

```
Button btn3 = new Button("Three");
VBox vbox = new VBox(10, btn1, btn2, btn3);
vbox.setPadding(new Insets(10));
```

Here's an example that creates a column of three buttons, with one button at the top of the column and two at the bottom, with 10 pixels of spacing and padding:

```
// Create the buttons
Button btn1 = new Button("One");
Button btn2 = new Button("Two");
Button btn3 = new Button("Three");

// Create the spacer
Region spacer = new Region();

// Set the Vgrow for the spacer
VBox.setVgrow(spacer, Priority.ALWAYS);

// Create the VBox layout pane
VBox vbox = new VBox(10, btn1, spacer, btn2, btn3);
vbox.setPadding(new Insets(10));
```

Aligning Nodes in a Layout Pane

Both the HBox and the VBox layout panes have a setAlignment method that lets you control how the nodes that are contained within the pane are aligned with one another. The setAlignment method accepts a single argument, which is one of the constants defined by the Pos enumeration, described in Table 4-5.

The Pos enumeration is defined in the javafx.geometry package, so you should include the following import statement in any program that uses Pos:

```
import javafx.geometry.*;
```

The following example shows how you might create a vertical column of three buttons, centered within the pane:

```
Button btn1 = new Button("Number One");
Button btn2 = new Button("Two");
Button btn3 = new Button("The Third Button");
VBox vbox = new VBox(10, btn1, btn2, btn3);
vbox.setPadding(new Insets(10));
vbox.setAlignment(Pos.CENTER);
```

TABLE 4-5   **The Pos Enumeration**

Constant	Vertical Alignment	Horizontal Alignment
Pos.TOP_LEFT	Top	Left
Pos.TOP_CENTER	Top	Center
Pos.TOP_RIGHT	Top	Right
Pos.CENTER_LEFT	Center	Left
Pos.CENTER	Center	Center
Pos.CENTER_RIGHT	Center	Right
Pos.BOTTOM_LEFT	Bottom	Left
Pos.BOTTOM_CENTER	Bottom	Center
Pos.BOTTOM_RIGHT	Bottom	Right
Pos.BASELINE_LEFT	Baseline	Left
Pos.BASELINE_CENTER	Baseline	Center
Pos.BASELINE_RIGHT	Baseline	Right

When this pane is added to a scene and then shown in a stage, the results resemble the window shown in Figure 4-3.

FIGURE 4-3:
Three buttons centered in a VBox layout pane.

Using the Flow Layout

The flow layout comes in two flavors: horizontal and vertical. A horizontal flow layout arranges its child nodes in a row until the width of the pane reaches a certain size that you can specify. When that size is reached, the layout begins a new row of child nodes beneath the first row. This flow continues, starting a new row each time the size limit is reached, until all the child nodes have been placed.

A vertical flow layout works the same way except that child nodes are laid out in columns until the size limit is reached. When the size limit is reached, a new column immediately to the right of the first column is started.

You use the FlowPane class to create a flow layout. Table 4-6 shows the constructors and most commonly used methods for the FlowPane class.

TABLE 4-6 ## FlowPane Constructors and Methods

Constructor	Description
FlowPane()	Creates an empty horizontal flow layout with both the horizontal and vertical gaps set to zero.
FlowPane(double hgap, double vgap)	Creates an empty horizontal flow layout with the specified horizontal and vertical gaps.
FlowPane(double hgap, double vgap, Node...children)	Creates a horizontal flow layout with the specified horizontal and vertical gaps and populated with the specified child nodes.
FlowPane(Node...children)	Creates a horizontal flow layout with both the horizontal and vertical gaps set to zero and populated with the specified child nodes.

Note: In each of the following constructors, Orientation *can be* Orientation.HORIZONTAL *or* Orientation.VERTICAL.

Constructor	Description
FlowPane(Orientation orientation)	Creates an empty flow layout with the specified orientation and both the horizontal and vertical gaps set to zero.
FlowPane(Orientation orientation, double hgap, double vgap)	Creates an empty flow layout with the specified orientation and the specified horizontal and vertical gaps.
FlowPane(Orientation orientation, double hgap, double vgap, Node... children)	Creates a flow layout with the specified orientation and horizontal and vertical gaps, populated with the specified children.
FlowPane(Orientation orientation, Node... children)	Creates a flow layout with the specified orientation and both the horizontal and vertical gaps set to zero, populated with the specified children.

Method	Description
ObservableList<Node> getChildren()	Returns the collection of all child nodes. The collection is returned as an ObservableList type, which includes the method addAll, letting you add one or more nodes to the list.

(continued)

TABLE 4-6 *(continued)*

Constructor	Description
`void setAlignment(Pos alignment)`	Sets the alignment for nodes within the rows and columns. See Table 4-5 for an explanation of the Pos enumeration. For more information, see the section "Aligning Nodes in a Layout Pane" earlier in this chapter.
`void setColumnHAlignment(Pos alignment)`	Sets the alignment for nodes within the columns. See Table 4-5 for an explanation of the Pos enumeration. For more information, see the section "Aligning Nodes in a Layout Pane" earlier in this chapter.
`void setHgap(double value)`	Sets the horizontal gap. For a horizontal flow layout, this is the amount of space between nodes. For a vertical flow layout, this is the amount of space between columns.
`static void setMargin(Node child, Insets value)`	Sets the margins for a given child node. See Table 4-2 for the constructors of the Insets class. For more information, see the section "Adding Space with Margins" later in this chapter.
`void setOrientation(Orientation orientation)`	Sets the orientation of the flow layout, which can be `Orientation.HORIZONTAL` or `Orientation.VERTICAL`.
`void setPadding(Insets value)`	Sets the padding around the inside edges of the `flow layout`. See Table 4-2 for the constructors of the Insets class. For more information, see the section "Spacing Things Out" earlier in this chapter.
`void setPrefWrapLength(double value)`	Sets the preferred wrap length for the pane. For a horizontal flow layout, this represents the preferred width of the pane; for a vertical flow layout, it represents the preferred height.
`void setRowHAlignment(Pos alignment)`	Sets the alignment for nodes within the rows. See Table 4-5 for an explanation of the Pos enumeration. For more information, see the section "Aligning Nodes in a Layout Pane" earlier in this chapter.
`void setVgap(double value)`	Sets the vertical gap. For a vertical flow layout, this is the amount of space between nodes. For a horizontal flow layout, this is the amount of space between rows.

The FlowPane class is defined in the `javafx.scene.layout` package, so you should include the following `import` statement in any program that uses a flow layout:

```
import javafx.scene.layout.*;
```

The constructors for this class let you specify the horizontal and vertical gaps, which provide the spacing between the horizontal and vertical elements of the layout, the orientation (horizontal or vertical), and the child nodes with which to populate the layout.

To set the limit at which the flow layout wraps, you use the `setPrefWrapLength` method. The wrap length is applied to the dimension in which the pane flows its contents. Thus, for a horizontal flow layout, the wrap length specifies the preferred width of the pane; for a vertical flow layout, the wrap length specifies the pane's preferred height.

Note that regardless of the preferred wrap length, if you don't call this method, the wrap length defaults to 400 pixels.

The following example creates a horizontal layout with 10 pixels of horizontal and vertical gaps, populated by five buttons, and a preferred wrap length of 300 pixels:

```
Button btn1 = new Button("Button One");
Button btn2 = new Button("Button Two");
Button btn3 = new Button("Button Three");
Button btn4 = new Button("Button Four");
Button btn5 = new Button("Button Five");
FlowPane pane = new FlowPane(Orientation.HORIZONTAL,
        10, 10, btn1, btn2, btn3, btn4, btn5);
pane.setPrefWrapLength(300);
```

Figure 4-4 shows how these buttons appear when the layout is added to a scene and the scene displayed in a stage. This figure also shows how the buttons in the flow layout are rearranged when the user resizes the window. Notice that initially, the first three buttons appear on the first row and the next two appear on the second row. When the window is dragged a bit wider, the buttons reflow so that four fit on the first row and just one spills to the second row. Then, when the window is dragged smaller, just two buttons appear on the first two rows and a third row is created for the fifth button.

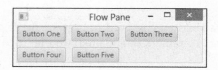

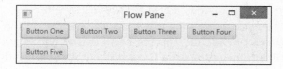

Using the Border Layout

The border layout is a pane that is carved into five regions: Top, Left, Center, Right, and Bottom, as shown in Figure 4-5. When you add a component to the layout, you can specify which of these regions the component goes in.

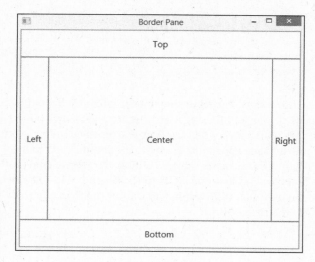

TIP

Border layout is the ideal layout manager for applications that have a traditional window arrangement in which menus and toolbars are displayed at the top of the window, a status bar or OK and Cancel buttons are displayed at the bottom, a navigation pane is displayed on the left, various task panes are displayed on the right, and content is displayed in the middle.

You use the BorderPane class to create a border layout. Table 4-7 lists the constructors and the most commonly used methods for the BorderPane class.

TABLE 4-7   **BorderPane Constructors and Methods**

Constructor	Description
BorderPane ()	Creates an empty border layout.
BorderPane (Node center)	Creates a border layout with the specified center node.
BorderPane (Node center, Node top, Node right, Node bottom, Node left)	Creates a border layout with the specified center, top, right, bottom, and left nodes.
Method	**Description**
void setCenter(Node node)	Sets the center node.
void setTop(Node node)	Sets the top node.
void setRight(Node node)	Sets the right node.
void setBottom(Node node)	Sets the bottom node.
void setLeft(Node node)	Sets the left node.
void setAlignment(Pos alignment)	Sets the alignment for nodes within border pane. See Table 4-5 for an explanation of the Pos enumeration. For more information, see the section "Aligning Nodes in a Layout Pane" earlier in this chapter.
static void setMargin(Node child, Insets value)	Sets the margins for a given child node. See Table 4-2 for the constructors of the Insets class. For more information, see the section "Adding Space with Margins" earlier in this chapter.

The BorderPane class is defined in the javafx.scene.layout package, so you should include the following import statement in any program that uses a border layout:

```
import javafx.scene.layout.*;
```

The default constructor for this class creates an empty border layout, to which you can add nodes later, as in this example:

```
Button btn1 = new Button("Button One");
Button btn2 = new Button("Button Two");
Button btn3 = new Button("Button Three");
VBox vbox = new VBox(btn1, btn2, btn3);
BorderPane pane = new BorderPane();
pane.setCenter(vbox);
```

Here, three buttons are created and added to a VBox. Then, a border layout is created, and the VBox is added to its center region.

Alternatively, you can add a node to the center region via the BorderPane constructor, like this:

```
BorderPane pane = new BorderPane(vbox);
```

The third constructor listed in Table 4-7 lets you add nodes to all five regions at once. The following example assumes that you have already created five panes, named centerPane, topPane, rightPane, bottomPane, and leftPane:

```
BorderPane pane = new BorderPane(centerPane,
    topPane, rightPane, bottomPane, leftPane);
```

TIP

Here are a few additional important points to know about the BorderPane class:

>> **If you don't add a node to a region, that region is not rendered.**

>> **The border layout regions are sized according to their contents.**

Thus, if you add a VBox pane to the right region, the width of the VBox pane will determine the width of the right region.

>> **If the user resizes the window to make it wider, the top, center, and bottom regions will expand in width — the width of the left and right regions remains unchanged.**

Similarly, if the user drags the window to make it taller, the left, center, and right regions expand in height; the height of the top and bottom regions remains the same.

>> **The nodes you add to the regions of a border pane will themselves almost always be other layout panes.**

Using the GridPane Layout

The grid pane layout manager lets you arrange GUI elements in a grid of rows and columns. Unlike a tile pane, the rows and columns of a grid pane do not have to be the same size. Instead, the grid pane layout automatically adjusts the width of each column and the height of each row based on the components you add to the panel.

Here are some important features of the grid pane layout manager:

» You can specify which cell you want each component to go in, and you can control each component's position in the panel.

» You can create components that span multiple rows or columns, such as a button two columns wide or a list box four rows high.

» You can tell GridPane to stretch a component to fill the entire space allotted to it if the component isn't already big enough to fill the entire area. You can specify that this stretching be done horizontally, vertically, or both.

» If a component doesn't fill its allotted area, you can tell the grid pane layout manager how you want the component to be positioned within the area — for example, left- or right-aligned.

The following sections describe the ins and outs of working with grid pane layouts.

Sketching out a plan

Before you create a grid pane layout, draw a sketch showing how you want the components to appear in the panel. Then slice the panel into rows and columns, and number the rows and columns starting with zero in the top-left corner. Figure 4-6 shows such a sketch for an application that lets a user order a pizza.

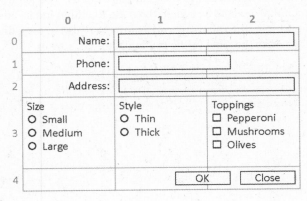

FIGURE 4-6: Sketching out a panel.

After you have the panel sketched out, list the components, their *x* and *y* coordinates on the grid, their alignment, and whether each component spans more than one row or column. Here's an example:

Component	x	y	Alignment	Spans
Label "Name"	0	0	Right	
Label "Phone"	0	1	Right	
Label "Address"	0	2	Right	
Name text field	1	0	Left	2
Phone text field	1	1	Left	2
Address text field	1	2	Left	2
Size radio buttons	0	3	Left	
Style radio buttons	1	3	Left	
Toppings check boxes	2	3	Left	
OK and Close buttons	2	4	Right	

After you lay out the grid, you can write the code to put each component in its proper place.

Creating a grid pane

Table 4-8 shows the most frequently used constructors and methods of the Grid-Pane class, which you use to create a grid pane.

To create a basic grid pane, you first call the GridPane constructor. Then, you use the add method to add nodes to the grid pane's cells. The parameters of the add method specify the node to be added, the node's column index, and the node's row index. For example, the following code snippet creates a label, and then creates a grid pane and adds the label to the cell at column 0, row 0:

```
Label lblName = new Label("Name");
GridPane grid = new GridPane();
grid.add(lblName, 0, 0);
```

TABLE 4-8

GridPane Constructors and Methods

Constructor	Description
GridPane()	Creates an empty grid pane.

Method	Description
void add(Node node, int col, int row)	Adds a node at the specified column and row index.
void add(Node node, int col, int row, int colspan, int rowspan)	Adds a node at the specified column and row index with the specified column and row spans.
void addColumn(int col, Node... nodes)	Adds an entire column of nodes.
void addRow(int row, Node... nodes)	Adds an entire row of nodes.
<ObservableList> getColumnConstraints()	Returns the column constraints. For more information, see to Table 4-6.
<ObservableList> getRowConstraints()	Returns the row constraints. For more information, see Table 4-7.
void setColumnSpan(Node node, int colspan)	Sets the column span for the specified node.
void setRowSpan(Node node, int colspan)	Sets the row span for the specified node.
void setHalignment(Node node, HPos value)	Sets the horizontal alignment for the node. Allowable values are HPos.LEFT, HPos.CENTER, and HPos.RIGHT.
void setValignment(Node node, VPos value)	Sets the vertical alignment for the node. Allowable values are H=VPos.BOTTOM, VPos.CENTER, and VPos.TOP, and VPos.BASELINE.
void setHgap(double value)	Sets the size of the gap that appears between columns.
void setVgap(double value)	Sets the size of the gap that appears between rows.
static void setMargin(Node node, Insets value)	Sets the margin for a particular node. See Table 4-2 earlier in this chapter for an explanation of the Insets class.
void setPadding(Insets value)	Sets the padding around the inside edges of the grid pane. See Table 4-2 earlier in this chapter for an explanation of the Insets class.
void setMinHeight(double value)	Sets the minimum height of the grid pane.
void setMaxHeight(double value)	Sets the maximum height of the grid pane.
void setPrefHeight(double value)	Sets the preferred height of the grid pane.

(continued)

TABLE 4-8 *(continued)*

Constructor	Description
void setMinWidth(double value)	Sets the minimum width of the grid pane.
void setMaxWidth(double value)	Sets the maximum width of the grid pane.
void setPrefWidth(double value)	Sets the preferred width of the grid pane.

The typical way to fill a grid pane with nodes is to call the add method for each node. However, if you prefer, you can add an entire column or row of nodes with a single call to either addColumn or addRow. For example, this example creates a label and a text field, and then creates a grid pane and adds the label and the text field to the first row:

```
Label lblName = new Label("Name");
TextField txtName = new TextField();
GridPane grid = new GridPane();
grid.addRow(0, lblName, txtName);
```

If a node should span more than one column, you can call the setColumnSpan method to specify the number of columns the node should span. For example:

```
GridPane.setColumnSpan(txtName, 2);
```

Here, the txtName node will span two columns. You use the setRowSpan in a similar way if you need to configure a node to span multiple rows.

To control the horizontal alignment of a node, use the setHalignment method as in this example:

```
GridPaNE.setHalignment(lblName, HPos.RIGHT);
```

Here, the lblName node is right-aligned within its column. The setValignment method works in a similar way.

Like other layout panes, the GridPane class has a host of methods for setting spacing and alignment details. You can use the setHgap and setVgap methods to set the spacing between rows and columns so that your layouts won't look so cluttered. You can use the setPadding and setMargins methods to set padding and margins, which work just as they do with other layout panes. And you can set the minimum, maximum, and preferred width and height for the grid pane.

Working with grid pane constraints

You can control most aspects of a grid pane's layouts using methods of the Grid-Pane class, but unfortunately, you can't control the size of individual columns or rows. To do that, you must use the ColumnConstraints or RowConstraints class, as described in Tables 4-9 and 4-10.

TABLE 4-9   **The ColumnConstraints Class**

Constructor	Description
ColumnConstraints()	Creates an empty column constraints object.
ColumnConstraints(double width)	Creates a column constraint with a fixed width.
ColumnConstraints(double min, double pref, double max)	Creates a column constraint with the specified minimum, preferred, and maximum widths.
Method	**Description**
void setMinWidth(double value)	Sets the minimum width of the column.
void setMaxWidth(double value)	Sets the maximum width of the column.
void setPrefWidth(double value)	Sets the preferred width of the column.
void setPercentWidth(double value)	Sets the width as a percentage of the total width of the grid pane.
void setHgrow(Priority value)	Determines whether the width of the column should grow if the grid pane's overall width increases. Allowable values are Priority.ALWAYS, Priority.NEVER, and Priority.SOMETIMES.
void setFillWidth(boolean value)	If true, the grid pane will expand the nodes within this column to fill empty space.
void setHalignment(HPos value)	Sets the horizontal alignment for the entire column. Allowable values are HPos.LEFT, HPos.CENTER, and HPos.RIGHT.

To use column constraints to set a fixed width for each column in a grid pane, first create a constraint for each column. Then, add the constraints to the grid pane's constraints collection. Here's an example:

```
ColumnConstraints col1 = new ColumnConstraints(200);
ColumnConstraints col2 = new ColumnConstraints(200);
ColumnConstraints col3 = new ColumnConstraints(200);
GridPane grid = new GridPane();
grid.getColumnConstraints().addAll(col1, col2, col3);
```

TABLE 4-10 **The RowConstraints Class**

Constructor	Description
RowConstraints()	Creates an empty row constraints object.
RowConstraints(double height)	Creates a column constraint with a fixed height.
RowConstraints(double min, double pref, double max)	Creates a column constraint with the specified minimum, preferred, and maximum heights.

Method	Description
void setMinHeight(double value)	Sets the minimum height of the row.
void setMaxHeight(double value)	Sets the maximum height of the row.
void setPrefHeight(double value)	Sets the preferred height of the row.
void setPercentHeight(double value)	Sets the height as a percentage of the total height of the grid pane.
void setVgrow(Priority value)	Determines whether the height of the row should grow if the grid pane's overall height increases. Allowable values are Priority.ALWAYS, Priority.NEVER, and Priority.SOMETIMES.
void setFillHeight(boolean value)	If true, the grid pane will expand the nodes within this row to fill empty space.
void setValignment(VPos value)	Sets the horizontal alignment for the entire row. Allowable values are VPos.TOP, Pos.CENTER, VPos.BOTTOM., and VPos.BASELINE.

Column constraints are matched to their corresponding columns based on the collection of constraints added to the GridPane. Thus, in the preceding example, the col1 constraint will be applied to the first column, col2 will be applied to the second column, and col3 will be applied to the third column.

One of the most useful features of column constraints is their ability to distribute the width of a grid pane's columns as a percentage of the overall width of the grid pane. For example, suppose the grid pane will consist of three columns and you want them to all be of the same width regardless of the width of the grid pane. The following code accomplishes this:

```
ColumnConstraints col1 = new ColumnConstraints();
col1.setPercentWidth(33);
ColumnConstraints col2 = new ColumnConstraints();
col2.setPercentWidth(33);
ColumnConstraints col3 = new ColumnConstraints();
```

```
col3.setPercentWidth(33);
GridPane grid = new GridPane();
grid.getColumnConstraints().addAll(col1, col2, col3);
```

In this example, each column will fill 33 percent of the grid.

TIP

Several of the attributes that can be set with column or row constraints mirror attributes you can set for individual nodes via the GridPane class. For example, you can set the horizontal alignment of an individual node by calling the setHalignment method on the grid pane. Or, you can set the horizontal alignment of an entire column by creating a column constraint, setting its horizontal alignment, and then applying the column constraint to a column in the grid pane.

Examining a grid pane example

Listing 4-1 shows the code for a program that displays the scene I drew for Figure 4-7, and Figure 4-7 shows how this scene appears when the program is run. Figure 4-7 shows that the final appearance of this scene is pretty close to the way I sketched it.

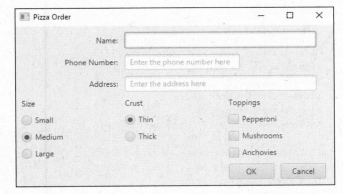

FIGURE 4-7:
The Pizza Order application in action.

LISTING 4-1: **The Pizza Order Application**

```
package com.lowewriter.PizzaOrder;

import javafx.application.*;
import javafx.stage.*;
import javafx.scene.*;
import javafx.scene.layout.*;
import javafx.scene.control.*;
import javafx.geometry.*;
```

(continued)

LISTING 4-1:   *(continued)*

```
public class PizzaOrder extends Application
{
    public static void main(String[] args)
    {
        launch(args);
    }

    Stage stage;
    TextField txtName;
    TextField txtPhone;
    TextField txtAddress;
    RadioButton rdoSmall;
    RadioButton rdoMedium;
    RadioButton rdoLarge;
    RadioButton rdoThin;
    RadioButton rdoThick;
    CheckBox chkPepperoni;
    CheckBox chkMushrooms;
    CheckBox chkAnchovies;

    @Override public void start(Stage primaryStage)
    {
        stage = primaryStage;

        // Create the name label and text field                    →34
        Label lblName = new Label("Name:");
        txtName = new TextField();
        txtName.setMinWidth(100);
        txtName.setPrefWidth(200);
        txtName.setMaxWidth(300);
        txtName.setPromptText("Enter the name here");

        // Create the phone number label and text field            →42
        Label lblPhone = new Label("Phone Number:");
        txtPhone = new TextField();
        txtPhone.setMinWidth(60);
        txtPhone.setPrefWidth(120);
        txtPhone.setMaxWidth(180);
        txtPhone.setPromptText("Enter the phone number here");

        // Create the address label and text field                 →50
        Label lblAddress = new Label("Address:");
        txtAddress = new TextField();
        txtAddress.setMinWidth(100);
        txtAddress.setPrefWidth(200);
        txtAddress.setMaxWidth(300);
        txtAddress.setPromptText("Enter the address here");
```

```
// Create the size pane                                                    →58
Label lblSize = new Label("Size");
rdoSmall = new RadioButton("Small");
rdoMedium = new RadioButton("Medium");
rdoLarge = new RadioButton("Large");
rdoMedium.setSelected(true);
ToggleGroup groupSize = new ToggleGroup();
rdoSmall.setToggleGroup(groupSize);
rdoMedium.setToggleGroup(groupSize);
rdoLarge.setToggleGroup(groupSize);

VBox paneSize = new VBox(lblSize, rdoSmall, rdoMedium, rdoLarge);
paneSize.setSpacing(10);

// Create the crust pane                                                   →72
Label lblCrust = new Label("Crust");
rdoThin = new RadioButton("Thin");
rdoThick = new RadioButton("Thick");
rdoThin.setSelected(true);
ToggleGroup groupCrust = new ToggleGroup();
rdoThin.setToggleGroup(groupCrust);
rdoThick.setToggleGroup(groupCrust);

VBox paneCrust = new VBox(lblCrust, rdoThin, rdoThick);
paneCrust.setSpacing(10);

// Create the toppings pane                                                →84
Label lblToppings = new Label("Toppings");
chkPepperoni = new CheckBox("Pepperoni");
chkMushrooms = new CheckBox("Mushrooms");
chkAnchovies = new CheckBox("Anchovies");

VBox paneToppings = new VBox(lblToppings, chkPepperoni,
    chkMushrooms, chkAnchovies);
paneToppings.setSpacing(10);

// Create the buttons                                                      →94
Button btnOK = new Button("OK");
btnOK.setPrefWidth(80);
btnOK.setOnAction(e -> btnOK_Click() );

Button btnCancel = new Button("Cancel");
btnCancel.setPrefWidth(80);
btnCancel.setOnAction(e -> btnCancel_Click() );

HBox paneButtons = new HBox(10, btnOK, btnCancel);
```

(continued)

LISTING 4-1:   *(continued)*

```
        // Create the GridPane layout                              →105
        GridPane grid = new GridPane();
        grid.setPadding(new Insets(10));
        grid.setHgap(10);
        grid.setVgap(10);
        grid.setMinWidth(500);
        grid.setPrefWidth(500);
        grid.setMaxWidth(800);

        // Add the nodes to the pane                                →114
        grid.addRow(0, lblName, txtName);
        grid.addRow(1, lblPhone, txtPhone);
        grid.addRow(2, lblAddress, txtAddress);
        grid.addRow(3, paneSize, paneCrust, paneToppings);
        grid.add(paneButtons,2,4);

        // Set alignments and spanning                             →121
        GridPane.setHalignment(lblName, HPos.RIGHT);
        GridPane.setHalignment(lblPhone, HPos.RIGHT);
        GridPane.setHalignment(lblAddress, HPos.RIGHT);
        GridPane.setColumnSpan(txtName,2);
        GridPane.setColumnSpan(txtPhone,2);
        GridPane.setColumnSpan(txtAddress,2);

        // Set column widths                                       →129
        ColumnConstraints col1 = new ColumnConstraints();
        col1.setPercentWidth(33);
        ColumnConstraints col2 = new ColumnConstraints();
        col2.setPercentWidth(33);
        ColumnConstraints col3 = new ColumnConstraints();
        col3.setPercentWidth(33);
        grid.getColumnConstraints().addAll(col1, col2, col3);

        // Create the scene and the stage                          →138
        Scene scene = new Scene(grid);
        primaryStage.setScene(scene);
        primaryStage.setTitle("Pizza Order");
        primaryStage.setMinWidth(500);
        primaryStage.setMaxWidth(900);
        primaryStage.show();

    }

    public void btnOK_Click()                                      →148
    {

        // Create a message string with the customer information
        String msg = "Customer:\n\n";
        msg += "\t" + txtName.getText() + "\n";
```

```
        msg += "\t" + txtPhone.getText() + "\n\n";
        msg += "\t" + txtAddress.getText() + "\n";
        msg += "You have ordered a ";

        // Add the pizza size
        if (rdoSmall.isSelected())
            msg += "small ";
        if (rdoMedium.isSelected())
            msg += "medium ";
        if (rdoLarge.isSelected())
            msg += "large ";

        // Add the crust style
        if (rdoThin.isSelected())
            msg += "thin crust pizza with ";
        if (rdoThick.isSelected())
            msg += "thick crust pizza with ";

        // Add the toppings
        String toppings = "";
        toppings = buildToppings(chkPepperoni, toppings);
        toppings = buildToppings(chkMushrooms, toppings);
        toppings = buildToppings(chkAnchovies, toppings);
        if (toppings.equals(""))
            msg += "no toppings.";
        else
            msg += "the following toppings:\n"
                + toppings;

        // Display the message
        Alert a = new Alert(Alert.AlertType.INFORMATION, msg);
        a.setTitle("Order Details");
        a.showAndWait();
    }

    public String buildToppings(CheckBox chk, String msg)
    {
        // Helper method for displaying the list of toppings
        if (chk.isSelected())
        {
            if (!msg.equals(""))
            {
                msg += ", ";
            }
            msg += chk.getText();
        }
        return msg;
    }
```

→189

(continued)

LISTING 4-1: *(continued)*

```
public void btnCancel_Click()                                    →203
{
    stage.close();
}
```
}

The following paragraphs point out the highlights of this program:

» →34 A label and text field are created for the customer's name.

» →42 A label and text field are created for the customer's phone number.

» →50 A label and text field are created for the customer's address.

» →58 A label and three radio buttons are created for the pizza's size. The label and radio buttons are added to a VBox named paneSize.

» →72 A label and two radio buttons are created for the pizza's crust style. The label and radio buttons are added to a VBox named paneCrust.

» →84 A label and three check boxes are created for the pizza's toppings. The label and check boxes are added to a VBox named paneToppings.

» →94 The OK and Cancel buttons are created and added to an HBox named paneButtons.

» →105 The grid pane layout is created. The padding and horizontal and vertical gaps are set to 10, and the width is set to range from 500 to 800.

» →114 The nodes are added to the pane. The name, phone number, and address labels and text fields are added to rows 0, 1, and 2. Then, the size, crust, and toppings VBox panes are added to row 3. Finally, the HBox that contains the buttons are added to column 2 of row 4. (**Remember:** Row and column indexes are numbered from 0, not from 1.)

» →121 Sets the horizontal alignment and column span options.

» →129 Column constraints are created to distribute the column widths evenly.

» →138 The scene is created, and the stage is displayed.

» →148 The btnOK_Click method is called when the user clicks OK. This method creates a summary of the customer's order and displays it using the Alert class.

» →189 buildToppings is simply a helper method that assists in the construction of the message string.

» →203 The stage is closed when the user clicks the Close button.

Chapter **5**

Getting Input from the User

In the first four chapters of this minibook, I discuss how to create JavaFX programs using only two basic JavaFX input controls: labels and buttons. If all you ever want to write are programs that display text when the user clicks a button, you can put the book down now. But if you want to write programs that actually do something worthwhile, you need to use other JavaFX input controls.

In this chapter, you find out how to use some of the most common JavaFX controls. First, you read about the label and controls that get information from the user. You find out more details about the text field control, which gets a line of text, and the text area control, which gets multiple lines. Then I move on to two input controls that get either/or information from the user: radio buttons and check boxes.

Along the way, you discover an important aspect of any JavaFX program that collects input data from the user: data validation. Data validation routines are essential to ensure that the user doesn't enter bogus data. For example, you can use data validation to ensure that the user enters data into required fields or that the data the user enters into a numeric field is indeed a valid number.

Using Text Fields

A *text field* is a box into which the user can type a single line of text. You create text fields by using the TextField class. Table 5-1 shows some of the more interesting and useful constructors and methods of this class.

TABLE 5-1   Handy TextField Constructors and Methods

Constructor	Description
TextField()	Creates a new text field.
TextField(String text, int cols)	Creates a new text field with an initial text value.
Method	**Description**
String getText()	Gets the text value entered in the field.
void requestFocus()	Asks for the focus to be moved to this text field. Note that the field must be in a scene for the focus request to work.
void setEditable(boolean value)	If false, makes the field read-only.
void setMaxWidth(double width)	Sets the maximum width for the field.
void setMinWidth(double width)	Sets the minimum width for the field.
void setPrefColumnCount(int cols)	Sets the preferred size of the text field in columns (that is, the number of average-width text characters).
void setPrefWidth(double width)	Sets the preferred width for the field.
void setPromptText(String prompt)	Sets the field's prompt value. The prompt value will not be displayed if the field has a text value or if the field has focus.
void setText(String text)	Sets the field's text value.

The TextField class is defined in the javafx.scene.control package, so you should include the following import statement in any program that uses a text field:

```
import javafx.scene.control.*;
```

The most common way to create a text field is to call the constructor without arguments, like this:

```
TextField text1 = new TextField();
```

You can set the initial value to be displayed like this:

```
TextField text1 = new TextField("Initial value");
```

Or, if you need to set the value later, you can call the setText method:

```
text1.setText("Text value");
```

To retrieve the value that the user has entered into a text field, call the getText method like this:

```
String value = text1.getText();
```

As with any JavaFX control, managing the width of a text field can be a bit tricky. Ultimately, JavaFX will determine the width of the text field based on a number of factors, including the size of the window that contains the stage and scene and any size constraints placed on the pane or panes that contain the text field. You can set minimum and maximum limits for the text field size by calling the set-MinWidth and setMaxWidth methods, and you can indicate the preferred width via the setPrefWidth method, as in this example:

```
TextField text1 = new TextField();
text1.setMinWidth(150);
text1.setMaxWidth(250);
text1.setPrefWidth(200);
```

Another way to set the preferred width is with the setPrefColumnCount method, which sets the width in terms of average-sized characters. For example, the following line sizes the field large enough to display approximately 50 characters:

```
text1.setPrefColumnCount(50);
```

Note that the setPrefColumnCount method does *not* limit the number of characters the user can enter into the field. Instead, it limits the number of characters the field can display at one time.

Whenever you use a text field, provide a prompt that lets the user know what data he should enter into the field. One common way to do that is to place a label control immediately to the left of the text field. For example:

```
Label lblName = new Label("Name:");
lblName.setMinWidth(75);
TextField txtName = new TextField();
txtName.setMinWidth(200);
HBox pane = new HBox(10, lblName, txtName);
```

Here, a label and a text field are created and added to an HBox pane so they will be displayed side-by-side.

JavaFX also allows you to display a prompt inside of a text field. The prompt is displayed in a lighter text color and disappears when the field receives focus. You use the setPromptText method to create such a prompt:

```
TextField txtName = new TextField();
txtName.setPromptText("Enter the customer's name");
```

Here, the text Enter the customer's name will appear inside the text field.

To retrieve the value entered by the user into a text field, you use the getText method, as in this example:

```
String lastName = textLastName.getText();
```

Here the value entered by the user in the textLastName text field is assigned to the String variable lastName.

Figure 5-1 shows the operation of a simple program that uses a text field to allow the user to enter the name of a character in a play and the name of the actor who will play the role. Assuming the user enters text in both fields, the program then displays an alert box indicating who will play the role of the character. If the user omits either or both fields, an alert box displays to indicate the error.

Figure 5-1 shows what the main stage for this program looks like, as well as the message box windows displayed when the user enters both names or when the user omits a name. The JavaFX code for this program is shown in Listing 5-1.

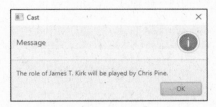

FIGURE 5-1:
The Role Player
application in
action.

LISTING 5-1: **The Role Player Program**

```
package com.lowewriter.RolePlayer;

import javafx.application.*;
import javafx.stage.*;
import javafx.scene.*;
import javafx.scene.layout.*;
import javafx.scene.control.*;
import javafx.geometry.*;

public class RolePlayer extends Application          →10
{
    public static void main(String[] args)
    {
        launch(args);
    }

    TextField txtCharacter;                          →17
    TextField txtActor;
```

(continued)

Getting Input from
the User

LISTING 5-1: *(continued)*

```java
@Override public void start(Stage primaryStage)
{

    // Create the Character                                          →23
    Label lblCharacter = new Label("Character's Name:");
    lblCharacter.setMinWidth(100);
    lblCharacter.setAlignment(Pos.BOTTOM_RIGHT);

    // Create the Character text field                               →28
    txtCharacter = new TextField();
    txtCharacter.setMinWidth(200);
    txtCharacter.setMaxWidth(200);
    txtCharacter.setPromptText(
        "Enter the name of the character here.");

    // Create the Actor label                                        →35
    Label lblActor = new Label("Actor's Name:");
    lblActor.setMinWidth(100);
    lblActor.setAlignment(Pos.BOTTOM_RIGHT);

    // Create the Actor text field                                   →40
    txtActor = new TextField();
    txtActor.setMinWidth(200);
    txtActor.setMaxWidth(200);
    txtActor.setPromptText("Enter the name of the actor here.");

    // Create the OK button                                          →46
    Button btnOK = new Button("OK");
    btnOK.setMinWidth(75);
    btnOK.setOnAction(e -> btnOK_Click() );

    // Create the Character pane                                     →51
    HBox paneCharacter = new HBox(20, lblCharacter, txtCharacter);
    paneCharacter.setPadding(new Insets(10));

    // Create the Actor pane                                         →55
    HBox paneActor = new HBox(20, lblActor, txtActor);
    paneActor.setPadding(new Insets(10));

    // Create the Button pane                                        →59
    HBox paneButton = new HBox(20, btnOK);
    paneButton.setPadding(new Insets(10));
    paneButton.setAlignment(Pos.BOTTOM_RIGHT);

    // Add the Character, Actor, and Button panes to a VBox          →64
    VBox pane = new VBox(10, paneCharacter, paneActor, paneButton);
```

```
    // Set the stage                                             →67
    Scene scene = new Scene(pane);
    primaryStage.setScene(scene);
    primaryStage.setTitle("Role Player");
    primaryStage.show();
}

public void btnOK_Click()                                        →74
{
    String errorMessage = "";                                    →76

    if (txtCharacter.getText().length() == 0)                    →78
    {
        errorMessage += "\nCharacter is a required field.";
    }

    if (txtActor.getText().length() == 0)                        →83
    {
        errorMessage += "\nActor is a required field.";
    }

    if (errorMessage.length() == 0)                              →88
    {
        String message = "The role of "
            + txtCharacter.getText()
            + " will be played by "
            + txtActor.getText()
            + ".";
        Alert a = new Alert(Alert.AlertType.INFORMATION, message);
        a.setTitle("Cast");
        a.showAndWait();
    }
    else
    {
        Alert a = new Alert(Alert.AlertType.WARNING, errorMessage);  →101
        a.setTitle("Missing Data");
        a.showAndWait();
    }
}
}
```

This program isn't very complicated, so the following paragraphs just hit the highlights:

>> →10 The name of the program's main class is RolePlayer.

>> →17 These class variables allow any of the RolePlayer class methods to access the two text fields.

» →23 These lines create a label to identify the Character text box. The field is set to a minimum width of 100 pixels and is right-justified so that the labels for that identify the two text fields will be aligned properly.

» →28 These lines create the Character text field with a minimum and maximum width of 200 pixels. The prompt text is set to Enter the name of the character here. This text will appear within the text field whenever the text field does not have focus, unless the user has entered something else. In Figure 5-1, the Character text field has focus so the prompt text isn't visible.

» →35 These lines create a label to identify the Actor text field. Like the Character label, the Actor label's width is set to 100 pixels and it's right-aligned.

» →40 These lines create the Actor text field, set its width to 200 pixels, and assign prompt text. You can see the prompt text in Figure 5-1 because the Actor text field doesn't have focus.

» →46 These lines create the OK button. The btnOK_Click method is called when the user clicks the button.

» →51 These lines create an HBox pane and add the Character label and text box to it.

» →55 These lines create another HBox pane and add the Actor label and text box to it.

» →59 These lines create a third HBox pane to hold the button.

» →64 Now that all the controls are created and added to HBox panes, the three HBox panes are added to a VBox pane so that the text boxes with their associated labels and the button are stacked vertically.

» →67 These lines create a scene to show the VBox pane and then add the scene to the primary stage and show the stage.

» →74 The btnOK_Click method is called whenever the user clicks OK.

» →76 The errorMessage variable holds any error message that might be necessary to inform the user of missing data.

» →78 This if statement ensures that the user has entered data into the Character text box. If no data is entered, an error message is created.

» →83 This if statement ensures that the user has entered data into the Actor text box. If no data is entered, an error message is appended to the errorMessage field.

» →88 This if statement determines whether any data validation errors have occurred by testing the length of the errorMessage field. If the length is zero, no error has been detected, so the program assembles the message variable to display which actor will be playing which character. Then, an Alert box is used to display the message.

» →101 This line displays the error message if the user forgets to enter data in the Character or Actor text fields.

Validating Numeric Data

You need to take special care if you're using a text field to get numeric data from the user. The getText method returns a string value. You can pass this value to one of the parse methods of the wrapper classes for the primitive numeric types. To convert the value entered in a text box to an int, use the parseInt method:

```
int count = Integer.parseInt(txtCount.getText());
```

Here the result of the getText method is used as the parameter of the parseInt method.

Table 5-2 lists the parse methods for the various wrapper classes. *Note:* Each of these methods throws NumberFormatException if the string can't be converted. As a result, you need to call the parseInt method in a try/catch block to catch this exception.

TABLE 5-2

Methods That Convert Strings to Numbers

Wrapper Class	parse Method
Integer	parseInt(String)
Short	parseShort(String)
Long	parseLong(String)
Byte	parseByte(String)
Float	parseFloat(String)
Double	parseDouble(String)

TIP

If your program uses more than one or two numeric-entry text fields, consider creating separate methods to validate the user's input. The following code snippet shows a method that accepts a text field and a string that provides an error message to be displayed if the data entered in the field can't be converted to an int.

The method returns a Boolean value that indicates whether the field contains a valid integer:

```
private boolean isInt(TextField f, String msg)
{
    try
    {
        Integer.parseInt(f.getText());
        return true;
    }
    catch (NumberFormatException e)
    {
        Alert a = new Alert(Alert.AlertType.WARNING, msg);
        a.setTitle("Invalid Data");
        a.showAndWait();
        return false;
    }
}
```

You can call this method whenever you need to check whether a text field has a valid integer. Here's a method that gets the value entered in a txtCount text field and displays it in message box if the value entered is a valid integer:

```
public void buttonOKClick()
{
    if (isInt(textCount,
        "You must enter an integer."))
    {
        Alert a = new
            Alert(Alert.AlertType.INFORMATION,
                "You entered " +
                Integer.parseInt(textCount.getText()),
                "Your Number");
        a.showAndWait();
    }
    textCount.requestFocus();
}
```

Here the isInt method is called to make sure that the text entered by the user can be converted to an int. If so, the text is converted to an int and displayed in an alert box.

Using Check Boxes

A *check box* is a control that the user can click to check or clear. Check boxes let the user specify a Yes or No setting for an option. Figure 5-2 shows a window with three check boxes.

FIGURE 5-2:
Three check boxes.

TECHNICAL
STUFF

Strictly speaking, a check box can have *three* states: checked, unchecked, and undefined. The undefined state is most often used in conjunction with a TreeView control.

To create a check box, you use the CheckBox class. Its favorite constructors and methods are shown in Table 5-3.

TABLE 5-3   **Notable CheckBox Constructors and Methods**

Constructor	Description
CheckBox()	Creates a new check box that is initially unchecked
CheckBox(String text)	Creates a new check box that displays the specified text

Method	Description
String getText()	Gets the text displayed by the check box
boolean isSelected()	Returns true if the check box is checked or false if the check box is not checked
void setOnAction(EventHandler<ActionEvent> value)	Sets an ActionEvent listener to handle action events
void setSelected(boolean value)	Checks the check box if the parameter is true; unchecks it if the parameter is false
void setText(String text)	Sets the check box text

As with any JavaFX control, if you want to refer to a check box in any method within the program, declare a class variable to reference the control:

```
CheckBox chkPepperoni, chkMushrooms, chkAnchovies;
```

Then you can use statements like these in the start method to create the check boxes and add them to a layout pane (in this case, pane1):

```
chkPepperoni = new CheckBox("Pepperoni");
pane1.add(chkPepperoni);

chkMushrooms = new CheckBox("Mushrooms");
pane1.add(chkMushrooms);

chkAnchovies = new CheckBox("Anchovies");
pane1.add(chkAnchovies);
```

Notice that I didn't specify the initial state of these check boxes in the constructor. As a result, they're initially unchecked. If you want to create a check box that's initially checked, call setSelected method, like this:

```
chkPepperoni.setSelected(true);
```

In an event listener, you can test the state of a check box by using the isSelected method, and you can set the state of a check box by calling its setSelected method. Here's a method that displays a message box and clears all three check boxes when the user clicks OK:

```
public void btnOK_Click()
{
    String msg = "";
    if (chkPepperoni.isSelected())
        msg += "Pepperoni\n";
    if (chkMushrooms.isSelected())
        msg += "Mushrooms\n";
    if (chkAnchovies.isSelected())
        msg += "Anchovies\n";
    if (msg.equals(""))
        msg = "You didn't order any toppings.";
    else
        msg = "You ordered these toppings:\n"
                + msg;
    Alert a = new Alert(Alert.AlertType.INFORMATION,
        msg");
    a.setTitle("Your Order");
    a.showAndWait();
```

```
    chkPepperoni.setSelected(false);
    chkMushrooms.setSelected(false);
    chkAnchovies.setSelected(false);
}
```

Here, the name of each pizza topping selected by the user is added to a text string. If you select pepperoni and anchovies, for example, the following message displays:

```
You ordered these toppings:
Pepperoni
Anchovies
```

TECHNICAL STUFF

If you want, you can add event listeners to check boxes to respond to events generated when the user clicks those check boxes. Suppose that your restaurant has anchovies on the menu, but you refuse to actually make pizzas with anchovies on them. Here's a method you can call in an event listener to display a message if the user tries to check the Anchovies check box; after displaying the message, the method then clears the check box:

```
public void chkAnchovies_Click(){
    Alert a = new Alert(Alert.AlertType.WARNING,
        "We don't do anchovies here.");
    a.setTitle("Yuck!")
    a.showAndWait();
    chkAnchovies.setSelected(false);
}
```

To add this event listener to the Anchovies check box, call its setOnAction method, like this:

```
chkAnchovies.setOnAction(e -> chkAnchovies_Click() );
```

TIP

Add a listener to a check box only if you need to provide immediate feedback to the user when she selects or deselects the box. In most applications, you wait until the user clicks a button to examine the state of any check boxes in the frame.

Using Radio Buttons

Radio buttons are similar to check boxes, but with a crucial difference: They travel in groups, and a user can select only one radio button at a time from each group. When you click a radio button to select it, the radio button within the same group that was previously selected is deselected automatically. Figure 5-3 shows a window with three radio buttons.

FIGURE 5-3:
A frame with
three radio
buttons.

To work with radio buttons, you use two classes. First, you create the radio buttons themselves with the RadioButton class, whose constructors and methods are shown in Table 5-4. Then you create a group for the buttons with the ToggleGroup class and add the radio buttons to the toggle group.

TABLE 5-4 **Various RadioButton Constructors and Methods**

Constructor	Description
RadioButton()	Creates a new radio button with no text
RadioButton(String text)	Creates a new radio button with the specified text
Method	**Description**
String getText()	Gets the text displayed by the radio button
boolean isSelected()	Returns true if the radio button is selected or false if the radio button is not selected
void setOnAction(EventHandler<ActionEvent> value)	Sets an ActionEvent listener to handle action events
void setSelected(boolean value)	Selects the radio button if the parameter is true; de-selects it if the parameter is false
void setText(String text)	Sets the check box text

TECHNICAL STUFF

A ToggleGroup object is simply a way of associating a set of radio buttons so that only one of the buttons can be selected. The toggle group object itself is not a control and is not displayed. To display radio buttons, you add the individual radio buttons, not the toggle group, to a layout pane.

The usual way to create a radio button is to declare a variable to refer to the button as a class variable so that it can be accessed anywhere in the class, as in this example:

```
RadioButton rdoSmall, rdoMedium, rdoLarge;
```

Then, in the start method, you call the RadioButton constructor to create the radio button:

```
rdoSmall = new RadioButton("Small");
```

Thereafter, you can add the radio button to a layout pane in the usual way.

To create a toggle group to group radio buttons that work together, call the ToggleGroup class constructor:

```
ToggleGroup sizeGroup = new ToggleGroup();
```

Then call the setToggleGroup method of each radio button:

```
rdoSmall.setToggleGroup(sizeGroup);
rdoMedium.setToggleGroup(sizeGroup);
rdoLarge.setToggleGroup(sizeGroup);
```

REMEMBER

Toggle groups have nothing to do with how radio buttons display. To display radio buttons, you must still add them to a layout pane. And there's no rule that says that all the radio buttons within a toggle group must be added to the same layout pane. However, it is customary to display all the radio buttons in a single toggle group together on the scene so that the user can easily see that the radio buttons belong together.

TECHNICAL STUFF

If you've worked with radio buttons in Swing, you'll want to note an important distinction between the way JavaFX toggle groups work versus how button groups work in Swing. In JavaFX, radio buttons that are outside a toggle group are independent of one another. In Swing, radio buttons that are outside a button group are all part of a default group. Thus, in JavaFX, always add radio buttons to a toggle group, even if the scene has only a single toggle group.

Chapter **6**

Choosing from a List

A n entire category of JavaFX controls are designed to let the user choose one or more items from a list. This chapter presents three such controls: choice boxes, combo boxes, and lists. Along the way, you discover how to use the `ObservableList` interface, which is used to manage the list of items displayed by a choice box, combo box, or a list view control.

Actually, if you've read along so far, you've already been briefly introduced to the `ObservableList` interface, as it's also used to manage the list of controls that are displayed in a layout pane. In Chapter 5, you read about how to use the `addAll` method of this interface. In this chapter, you read about the additional capabilities of this interface.

You also discover how to add an event listener that can respond when the user changes the current selection.

Using Choice Boxes

A *choice box* is a control that lets the user choose a single item from a drop-down list. Initially, the choice box shows just the item that's currently selected. When the user clicks the choice box, the list of choices reveals. The user can change the selection by clicking any of the items in the list. Figure 6-1 shows a scene with a simple choice box.

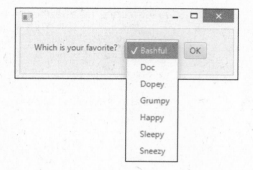

FIGURE 6-1:
A scene with a choice box.

You use the ChoiceBox class to create choice boxes. Table 6-1 lists the most frequently used constructors and methods of this class.

TABLE 6-1   **Common ChoiceBox Constructors and Methods**

Constructor	Description
ChoiceBox<T>()	Creates an empty choice list of the specified type
ChoiceBox<T>(ObservableList<T> items)	Creates a choice list and fills it with the values in the specified list

Method	Description
ObservableList<T> getItems()	Gets the list of items
void setItems(ObservableList<T> items)	Sets the list of items
T getValue()	Returns the currently selected item
void setValue(T value)	Sets the currently selected item
void show()	Shows the list of items
void hide()	Hides the list of items
boolean isShowing()	Indicates whether the list of items is currently visible

Creating a choice box

Creating a choice box is easy. The `ChoiceBox` class is generic, so specify a type for the list that will be associated with the choice box. For example:

```
ChoiceBox<String> choice = new ChoiceBox<String>();
```

Here, a choice box that displays strings is created.

The next step is to add items to the choice box. You can do that by calling the `getItems` method to access the list of items and then calling the `add` method to add an item:

```
choice.getItems().add("Bashful");
choice.getItems().add("Doc");
choice.getItems().add("Dopey");
choice.getItems().add("Grumpy");
choice.getItems().add("Happy");
choice.getItems().add("Sleepy");
choice.getItems().add("Sneezy");
```

Alternatively, you could call the `addAll` method and add all the strings at once, like this:

```
choice.getItems().addAll("Bashful", "Doc", "Dopey",
                         "Grumpy", "Happy", "Sleepy",
                         "Sneezy");
```

TECHNICAL
STUFF

The `getItems` method returns an object of type `ObservableList`, which offers a number of methods that let you work with the list. For more information, see the section "Working with Observable Lists" later in this chapter.

The `ChoiceBox` class also includes a constructor that lets you add an `ObservableList` object when you create the choice box. This lets you create the list before you create the choice box. You see an example of this constructor in action in the section "Working with Observable Lists" in this chapter.

TIP

You can add any kind of object you want to a choice box. The choice box calls the `toString` method of each item to determine the text to display in the choice list. Suppose you have a class named `Astronaut` that represents an astronaut on a space mission:

```
class Astronaut
{
    private String firstName;
    private String lastName;
```

```
    public Astronaut(String FirstName, String LastName)
    {
        firstName = FirstName;
        lastName = LastName;
    }

    public String toString()
    {
        return firstName + " " + lastName;
    }
}
```

Then, you could create a choice box listing the crew of Apollo 13 like this:

```
ChoiceBox<Astronaut> apollo13;
apollo13 = new ChoiceBox<>();
apollo13.getItems().add(new Astronaut("Jim", "Lovell"));
apollo13.getItems().add(new Astronaut(
    "John", "Swigert"));
apollo13.getItems().add(new Astronaut("Fred", "Haise"));
```

If you wish, you can display the contents of a choice box without waiting for the user to click the box. To do that, call the show method, like this:

```
apollo13.show();
```

To hide the list, call the hide method:

```
apollo13.hide();
```

Setting a default value

By default, a choice box has no initial selection when it's first displayed. To set an initial value, call the setValue method, passing it the list object that you want to make the initial selection.

If the choice box contains strings, you can set the initial value by passing the desired string value to the setValue method:

```
choice.setValue("Dopey");
```

If the specified string doesn't exist in the list, the initial value will remain unspecified.

If the choice box contains objects, such as the `Astronaut` objects, illustrated in the preceding section, you must pass a reference to the object you want to be the default choice. For example:

```
Astronaut lovell = new Astronaut("Jim", "Lovell");
Astronaut swigert = new Astronaut("John", "Swigert");
Astronaut haise = new Astronaut("Fred", "Haise");
ChoiceBox apollo13 = new ChoiceBox<Astronaut>();
apollo13.getItems().addAll(lovell, swigert, haise);
apollo13.setValue(lovell);
```

Here, Jim Lovell is set as the default astronaut.

Getting the selected item

You can call the `getValue` method to get the item selected by the user. The type of the value returned depends on the type specified when you created the choice box. For example, if you specified type `String`, the `getValue` method returns strings. If you specified type `Astronauts` for the choice box, the `getValue` method returns astronauts.

The `getValue` method is often used in the action event handler for a button. For example:

```
public void btnOK_Click()
{
    String message = "You chose ";
    message += apollo13.getValue();
    Alert a = new Alert(Alert.AlertType.INFORMATION, message);
    a.setTitle("Your Favorite Astronaut");
    a.showAndWait();
}
```

Working with Observable Lists

As you saw in the previous section, the `ChoiceBox` class does not include methods that let you directly add or remove items from the list displayed by the choice box. Instead, it includes a method named `getItems` that returns an object of type `ObservableList`. The object returned by this method is an *observable list*; it represents the list displayed by the choice box.

To work with the items displayed by a choice box, you must first access the observable list and then use methods of the ObservableList class to access the individual items in the list.

Observable lists are used not only by the ChoiceBox class, but also by other control classes that display list items, such as ComboBox and List, which you can read about later in this chapter. Both of those classes also have a getItems method that returns an ObservableList.

Observable lists are also used by layout panes, such as HBox and VBox, which you can read about in Chapter 4. The getChildren method that's common to all layout classes returns an ObservableList.

So far in this book, I've discussed just two methods of the ObservableList interface: add and addAll, which lets you add items to the observable list. Here's an example of the add method from earlier in this chapter:

```
cbox.getItems().add("Bashful");
```

And here's an example from Chapter 4, which uses the addAll method to add buttons to a layout pane:

```
pane.getChildren().addAll(btnOK, btnCancel);
```

The ObservableList interface has many other methods besides add and addAll. Table 6-2 shows the methods you're most likely to use.

TABLE 6-2   ## Commonly Used ObservableList Methods

Method	Description
void add(E element)	Adds the specified element to the end of the list.
void add(int index, E element)	Adds the specified object to the list at the specified index position.
void addAll(E...elements)	Adds all the specified elements to the end of the list.
void addAll(Collection<E> c)	Adds all the elements of the specified collection to the end of the list.
E set(int index, E elem)	Sets the specified element to the specified object. The element that was previously at that position is returned as the method's return value.
void clear()	Deletes all elements.

Method	Description
void remove(int *fromIndex*, int *toIndex*)	Removes all objects whose index values are between the values specified.
void removeAll(E...elements)	Removes all objects whose index values are between the values specified.
boolean contains(Object *elem*)	Returns a boolean that indicates whether the specified object is in the list.
E get(int *index*)	Returns the object at the specified position in the list.
int indexOf(Object *elem*)	Returns the index position of the first occurrence of the specified object in the list. If the object isn't in the list, it returns −1.
boolean isEmpty()	Returns a boolean value that indicates whether the list is empty.
E remove(int *index*)	Removes the object at the specified index and returns the element that was removed.
boolean remove(Object *elem*)	Removes an object from the list. **Note:** More than one element refers to the object; this method removes only one of them. It returns a boolean that indicates whether the object was in the list.
int size()	Returns the number of elements in the list.
void addListener(ListChangeListener listener)	Adds a ListChangeListener that's called whenever the list changes.

If you're familiar with Java collection classes, such as ArrayList, you may have noticed that many of the methods listed in Table 6-2 are familiar. That's because the ObservableList class extends the List class, which is implemented by classes, such as ArrayList and Vector. As a result, any method that can be used with an ArrayList can also be used with an ObservableList.

For example, you can clear the contents of a choice box in the same way you'd clear the contents of an array list:

```
cbox.getItems().clear();
```

If you need to know how many items are in a choice box, call the size method:

```
int count = cbox.getItems().size();
```

To remove a specific item from the list, use the `remove` method:

```
cbox.getItems().remove("Grumpy");
```

This method returns a `boolean` that indicates whether the string was removed from the list; if `remove` returns `false`, it simply means that `"Grumpy"` was not in the list.

You can easily insert items from an existing Java collection, such as an array list, into a choice box by specifying the collection in the `addAll` method. For example, suppose you already have an array list named `list` that contains the items you want to display in the choice box. You can add the items like this:

```
cbox.getItems().addAll(list);
```

You might be wondering why an observable list is required for the items displayed by list-based JavaFX controls. Why not just use the existing collection classes? The reason is that for list-based controls to work efficiently, the controls themselves need to monitor any changes you might make to the list of items so that the control can automatically update the displayed items. The last method listed in Table 6-2 (`addListener`) provides this capability by allowing you to add a listener that's called whenever the contents of the list changes. You will rarely call this method directly. But the controls that use observable lists *do* call this method create event listeners that automatically update the control whenever the contents of the list changes.

Note: You do *not* use the `addListener` method to respond when the user selects an item in a choice box or other type of list control. Instead, you use an interesting construct called a *selection model* to respond to changes in the selected item, as described in the next section.

Listening for Selection Changes

It's not uncommon to want your program to respond immediately when the user changes the selection of a choice box or other list control, without waiting for the user to click a button to submit the data. For example, you might have a label whose value you want to update immediately whenever the user changes the selection. You might even want to show or hide different controls based on the selection.

Unfortunately, the choice box and other list controls don't generate an action event when the user changes the selection. As a result, the `ChoiceBox` class doesn't have a `setOnAction` method. Instead, you must use a complicated sequence of method calls to set up a different type of event listener, called a *change listener*.

Here's the sequence:

1. Get the selection model by calling the `getSelectionModel` **method on the choice box.**

The `getSelectionModel` method returns the control's *selection model,* which is an object that manages how the user can select items from the list. The selection model is an object that implements one of several classes that extend the abstract `SelectionModel` class. For a choice box, the selection model is always of type `SingleSelectionMode`, which implements a selection model that allows the user to select just one item from the list at a time.

2. Get the `selectedItem` **property by calling the** `selectedItemProperty` **method on the selection model.**

The `SelectionModel` class has a method named `selectedItemProperty` that accesses a property named `selectedItem`, which represents the item currently selected. (A *property* is a special type of JavaFX object whose value can be monitored by a listener that's called whenever the value of the property changes.)

3. Add a change listener by calling the `addListener` **method on the** `selectedItem` **property.**

The listener will be called whenever the value of the `selectedItem` property changes. The change listener implements a functional interface called, naturally, `ChangeListener`. Because `ChangeListener` is a functional interface (that is, it has just one abstract method), you can use a lambda expression to implement the change listener.

You normally do all three of these steps in a single statement, as in this example:

```
choice.getSelectionModel().selectedItemProperty()
    .addListener( (v, oldValue, newValue) ->
        lbl.setText(newValue); );
```

In the preceding example, the change listener sets the value displayed by a label control to the new value selected by the user.

Being a functional interface, `ChangeListener` defines a single function named `changed`, which is called whenever the value of the property changes. The `changed` method receives three arguments:

» `observable`: The property whose value has changed

» `oldValue`: The previous value of the property

» `newValue`: The new value of the property

Choosing from a List

These three parameters are specified in the parentheses at the beginning of the lambda expression. In the body of the lambda expression, the `newValue` parameter is assigned to the text of a label. Thus, the value selected by the user will be displayed by the label, and the label will be updated automatically whenever the user changes the choice box selection.

Using Combo Boxes

A *combo box* is a more advanced sibling to the choice box control. The main improvements you get with a combo box are

>> **A combo box includes the ability to limit the number of items displayed when the list is shown.**

If the number of items in the list exceeds the limit, a scroll bar is added automatically to allow the user to scroll through the entire list.

>> **A combo box includes a text field that lets the user enter a value directly rather than select the value from a list.**

The text field is optional and is not shown by default, but you can add it with a single method call.

Figure 6-2 shows a combo box with the text field shown.

>> **A combo box fires an action event whenever the user changes the selection.**

Thus, setting up an event handler to respond to the user's selection change is easier with a combo box than it is with a choice box.

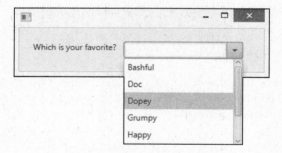

FIGURE 6-2:
A combo box.

You use the `ComboBox` class to create combo boxes. Table 6-3 lists the most frequently used constructors and methods of this class.

TABLE 6-3

Common ComboBox Constructors and Methods

Constructor	Description
`ComboBox<T>()`	Creates an empty combo box of the specified type.
`ComboBox<T>(ObservableList<T> items)`	Creates a combo box and fills it with the values in the specified list.

Method	Description
`void setEditable(boolean value)`	If `true`, a text field is displayed to allow the user to directly edit the selection.
`void setVisibleRowCount(int value)`	Sets the number of items to display.
`void setPromptText(String text)`	Sets the prompt text initially displayed in the text field.
`ObservableList<T> getItems()`	Gets the list of items.
`void setItems(ObservableList<T> items)`	Sets the list of items.
`T getValue()`	Returns the currently selected item.
`void setValue(T value)`	Sets the currently selected item.
`void show()`	Shows the list of items.
`void hide()`	Hides the list of items.
`void setOnAction(EventHandler<ActionEvent> handler)`	Sets an event handler that's called whenever the selection changes.
`boolean isShowing()`	Indicates whether the list of items is currently visible.

Creating combo boxes

Creating a combo box is much like creating a choice box. Because the `ComboBox` is generic, specify a type for the items it will contain, as in this example:

```
ComboBox<String> cbox = new ComboBox<String>();
```

Then you can use the `getItems` method to access the `ObservableList` object that contains the content of the list displayed by the combo box. For example, you can add items to the list like this:

```
cbox.getItems().addAll("Bashful", "Doc", "Dopey",
                       "Grumpy", "Happy", "Sleepy",
                       "Sneezy");
```

For more information about working with the `ObservableList` interface, flip to the section "Working with Observable Lists" earlier in this chapter.

TIP

By default, the user isn't allowed to edit the data in the text field portion of the combo box. If you want to allow the user to edit the text field, use the `setEditable` method, like this:

```
cbox.setEditable(true);
```

Then the user can type a value that's not in the combo box.

If you want, you can limit the number of items displayed by the list by calling the `setVisibleRows` method:

```
cbox.setVisibleRows(10);
```

Here, the list displays a maximum of ten items. If the list contains more than ten items, a scroll is added automatically so the user can scroll through the entire list.

You can also specify a prompt text to display in the text field component of a combo box by calling the `setPromptText` method:

```
cbox.setPromptText("Make a choice");
```

Here, the text `Make a choice` displays in the text field.

Getting the selected item

To get the item selected by the user, use the `getValue` method, just as you do for a choice box. You typically do that in an action event handler that responds to a button click. For example:

```
public void btnOK_Click()
{
    String message = "You chose ";
    message += cbox.getValue();
    Alert a = new Alert(Alert.AlertType.INFORMATION,
        message);
    a.setTitle("Your Choice");
    a.showAndWait();
}
```

TIP

Bear in mind that the value returned by the `getValue` method may not be one of the values in the combo box's list. That's because the user can enter anything he wishes to in the text field of an editable combo box. If you want to know whether the user selected an item from the list or entered a different item via the text field, use the `contains` method of the `ObservableList` class, like this:

```
if (!cbox.getItems().contains(cbox.getValue()))
{
    Alert a = new Alert(Alert.AlertType.INFORMATION,
        "You chose outside the box");
    a.setTitle("Good Thinking!");
    a.showAndWait();
}
```

Here, the alert displays if the user enters an item that's not in the list.

Handling combo box events

When the user selects an item from a combo box, an action event is generated. In most applications, you simply ignore this event because you usually don't need to do anything immediately when the user selects an item. Instead, the selected item is processed when the user clicks a button.

If you want to provide immediate feedback when the user selects an item, you can set up an event handler by calling the combo box's `setOnAction` method. In most cases, the easiest way to do that is to create a method that contains the code you want to execute when the user selects an item and then pass this method to the `setOnAction` method via a lambda expression.

For example, the following method displays a message box that says `My favorite too!` if the user picks Dopey:

```
Public void cbox_Changed()
{
    if (if cbox.getValue().equals("Dopey"))
    {
        Alert a = new
            Alert(Alert.AlertType.INFORMATION,
            "He's my favorite too!",
        a.setTitle("Good Choice");
        a.showAndWait();
    }
}
```

Here's the code to call this method whenever the user changes the combo box selection:

```
cbox.setOnAction (e -> cbo_Changed() );
```

Using List Views

A *list view* is a powerful JavaFX control that displays a list of objects within a box. Depending on how the list is configured, the user can select one item in the list or multiple items. In addition, you have amazing control over how the items in the list display. Figure 6-3 shows a sample scene with a list view.

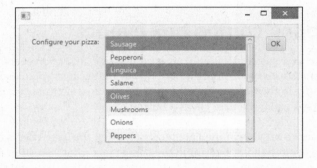

List views and combo boxes have several important differences:

>> **A list view doesn't have a text field that lets the user edit the selected item.** Instead, the user must select items directly from the list view.

>> **The list view doesn't drop down.** Instead, the list items display in a box whose size you can specify.

>> **The items in a list view can be arranged vertically (the default) or horizontally.** Figure 6-4 shows a horizontal list box.

>> **List views allow users to select more than one item.** By default, a list view lets users select just one item, but you can easily configure it to allow for multiple selections.

TIP

To select multiple items in a list, hold down the Ctrl key and click the items you want to select. To select a range of items, click the first item, hold down the Shift key, and click the last item.

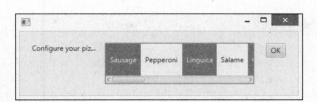

FIGURE 6-4:
A horizontal list view control.

You use the ListView class to create a list view control. Table 6-4 lists the most frequently used constructors and methods of this class.

TABLE 6-4   ## Common ListView Constructors and Methods

Constructor	Description
ListView<T>()	Creates an empty list view of the specified type.
ListView<T>(ObservableList<T> items)	Creates a list view and fills it with the values in the specified list.

Method	Description
ObservableList<T> getItems()	Gets the list of items.
void setItems(ObservableList<T> items)	Sets the list of items.
void setOrientation(Orientation o)	Sets the orientation of the list. The orientation can be Orientation.HORIZONTAL or Orientation.VERTICAL.
MultipleSelectionModel<T> getSelectionModel()	Returns the selection model for the list view control. You can use the selection model to get an observable list of selected items by calling its getSelectedItems method. You can also retrieve the most recently selected item by calling getSelectedItem.

Creating a list view

Creating a list view control is similar to creating a combo box. Here's an example that creates the list view that's shown in Figure 6-3:

```
ListView list = new ListView();
list.getItems().addAll("Sausage", "Pepperoni",
    "Linguica", "Salame", "Olives", "Mushrooms",
    "Onions", "Peppers", "Pineapple", "Spinach",
    "Canadian Bacon", "Tomatoes", "Kiwi",
    "Anchovies", "Gummy Bears");
```

Notice that the list view shown in Figure 6-3 shows only the first eight items in the list. As with a combo box, a scroll bar is automatically added to a list view if the total number of items in the items collection cannot be displayed.

By default, the list view control allows only a single selection to be made. To allow multiple selections, add this line:

```
list.getSelectionModel().setSelectionMode(SelectionMode.MULTIPLE);
```

To arrange the list view horizontally rather than vertically, add this line:

```
list.setOrientation(Orientation.HORIZONTAL);
```

Getting the selected items

Getting the selected items from a list view control is a bit tricky. First, you must get the selection model object by calling the getSelectionModel. Then, you call the selection model's getSelectedItems method. This returns a read-only observable list that contains just the items that have been selected.

Here's an example that builds a string that lists all the items selected by the user for the pizza toppings list view that is shown in Figure 6-3:

```
String tops = "";
ObservableList<String> toppings;
toppings = list.getSelectionModel().getSelectedItems();
for(String topping : toppings)
{
    tops += topping + "\n";
}
```

In the preceding example, the tops string will contain all the toppings selected by the user, separated by new line characters.

Using Tree Views

A *tree view* is a fancy JavaFX control that displays hierarchical data in outline form, which we computer nerds refer to as a tree. Tree structures are very common in the world of computers. The folder structure of your disk drive is a tree, as is a JavaFX scene graph.

Figure 6-5 shows a JavaFX scene that has a tree view control in it. In this example, I use a tree control to represent a few of my favorite TV series, along with series that were spun off from them.

FIGURE 6-5:
A tree view
control.

Before I get into the mechanics of how to create a tree control, you need to know a few terms that describe the elements in the tree itself:

>> **Node:** Each element in the tree is a *node.* Each node in a tree is created from `TreeItem` class. The `TreeItem` class is a generic class, so you can associate a type with it. Thus, you can create a tree using objects of any type you wish, including types you create yourself.

>> **Root node:** A *root node* is the starting node for a tree. Every tree must have one — and only one — root node. When you create a tree component, you pass the root node to the `TreeView` constructor.

>> **Child node:** The nodes that appear immediately below a given node are that node's *child nodes.* A node can have more than one child.

>> **Parent node:** The node immediately above a given node is that node's *parent node.* Every node except the root node must have one — and only one — parent.

>> **Sibling nodes:** *Sibling nodes* are children of the same parent.

>> **Leaf node:** A *leaf node* is one that doesn't have any children.

>> **Path:** A path contains the node and all its ancestors — that is, its parent, its parent's parent, and so on — all the way back to the root.

>> **Expanded node:** An *expanded node* is one whose children are visible.

>> **Collapsed node:** A *collapsed node* is one whose children are hidden.

Building a tree

Before you can actually create a tree view, you must first build the tree it displays. To do that, use the TreeItem class, the details of which I discuss in Table 6-5.

TABLE 6-5   ## The TreeItem Class

Constructor	Description
TreeItem<T> ()	Creates an empty tree node.
TreeItem<T>(T value)	Creates a tree node with the specified value.
Method	**Description**
T getValue()	Returns the tree item's value.
void setValue(T value)	Sets the tree item's value.
ObservableList getChildren()	Returns an ObservableList that represents the children of this tree item.
TreeItem getParent()	Gets this node's parent.
void setExpanded(boolean expanded)	Specify true to expand the node.
boolean isExpanded()	Returns a boolean that indicates whether the tree item is expanded.
boolean isLeaf()	Returns a boolean that indicates whether the tree item is a leaf node (that is, has no children). A leaf node can't be expanded.
TreeItem nextSibling()	Returns the next sibling of this tree item. If there is no next sibling, returns null.
TreeItem prevSibling()	Returns the previous sibling of this tree item. If there is no previous sibling, returns null.

The `TreeItem` class provides three basic characteristics for each node:

>> **The value,** which contains the data represented by the node.

   In my example, I use strings for the user objects, but you can use objects of any type you want for the user object. The tree control calls the user object's `toString` method to determine what text to display for each node. The easiest way to set the user object is to pass it via the `TreeItem` constructor.

>> **The parent of this node,** unless the node happens to be the root.

>> **The children of this node,** represented as an `ObservableList`.

   The list will be empty if the node happens to be a leaf node. You can create or retrieve child nodes using the familiar methods of the `ObservableList` interface. For more information, refer to the section "Working with Observable Lists" earlier in this chapter.

In this section, I build a tree that lists some of the spinoff shows from three popular television shows of the past:

>> *The Andy Griffith Show,* which had two spinoffs: *Gomer Pyle, U.S.M.C.,* and *Mayberry R.F.D.*

>> *All in the Family,* which directly spawned four spinoffs: *The Jeffersons, Maude, Gloria,* and *Archie Bunker's Place.*

   In addition, two of these spinoffs had spinoffs of their own involving the maids: The Jeffersons' maid became the topic of a short-lived show called *Checking In,* and Maude's maid became the main character in *Good Times.*

>> *Happy Days,* which spun off *Mork and Mindy, Laverne and Shirley,* and *Joanie Loves Chachi.*

You can take many approaches to building trees, most of which involve some recursive programming. I'm going to avoid recursive programming in this section to keep things simple, but my avoidance means that you have to hard-code some of the details of the tree into the program. Most real programs that work with trees need some type of recursive programming to build the tree.

The first step in creating a tree is declaring a `TreeItem` variable for each node that isn't a leaf node. For my TV series example, I start with the following code:

```
TreeItem andy, archie, happy,
        george, maude;
```

These variables can be local variables within the `start` method because once you get the tree set up, you won't need these variables anymore. You see why you don't need variables for the leaf nodes in a moment.

Next, I create the root node and set its expanded status to `true` so that it will expand when the tree displays initially:

```
TreeItem root = new TreeItem("Spin Offs ");
root.setExpanded(true);
```

To simplify the task of creating all the other nodes, I use the following helper method, makeShow:

```
public TreeItem<String> makeShow(String title,
    TreeItem<String> parent)
{
    TreeItem<String> show = new TreeItem<String>(title);
    show.setExpanded(true);
    parent.getChildren().add(show);
    return show;
}
```

This method accepts a string and another node as parameters, and returns a node whose user object is set to the `String` parameter. The returned node is also added to the parent node as a child, and the node is expanded. Thus you can call this method to both create a new node and place the node in the tree.

The next step is creating some nodes. Continuing my example, I start with the nodes for *The Andy Griffith Show* and its spinoffs:

```
andy = makeShow("The Andy Griffith Show", root);
makeShow("Gomer Pyle, U.S.M.C.", andy);
makeShow("Mayberry R.F.D.", andy);
```

Here, makeShow is called to create a node for *The Andy Griffith Show,* with the root node specified as its parent. The node returned by this method is saved in the andy variable. Then makeShow is called twice to create the spinoff shows, this time specifying andy as the parent node.

Because neither *Gomer Pyle, U.S.M.C.,* nor *Mayberry R.F.D.* had a spinoff show, I don't have to pass these nodes as the parent parameter to the makeShow method. That's why I don't bother to create a variable to reference these nodes.

Next in my example, I create nodes for *All in the Family* and its spinoffs:

```
archie = makeShow("All in the Family", root);
george = makeShow("The Jeffersons", archie);
makeShow("Checking In", george);
```

```
maude = makeShow("Maude", archie);
makeShow("Good Times", maude);
makeShow("Gloria", archie);
makeShow("Archie Bunker's Place", archie);
```

In this case, *The Jeffersons* and *Maude* have child nodes of their own. As a result, variables are required for these two shows so that they can be passed as the parent parameter to makeShow when I create the nodes for *Checking In* and *Good Times*.

Finally, here's the code that creates the nodes for *Happy Days* and its spinoffs:

```
happy = makeShow("Happy Days", root);
makeShow("Mork and Mindy", happy);
makeShow("Laverne and Shirley", happy);
makeShow("Joanie Loves Chachi", happy);
```

The complete tree is successfully created in memory, so I can get on with the task of creating a TreeView control to show off the tree.

Creating a TreeView control

You use the TreeView class to create a tree component that displays the nodes of a tree. Table 6-6 shows the key constructors and methods of this class.

TABLE 6-6   The TreeView Class

Constructor	Description
TreeView<T>()	Creates an empty tree (not very useful, if you ask me).
TreeView<T>(TreeItem root)	Creates a tree that displays the tree that starts at the specified node.

Method	Description
TreeItem getRoot()	Gets the root node.
void setRoot(TreeItem root)	Sets the root node.
MultipleSelectionModel<T> getSelectionModel()	Returns the selection model for the list view control. You can use the selection model to get an observable list of selected items by calling its getSelectedItems method. You can also retrieve the most recently selected item by calling getSelectedItem.
void setRootVisible(boolean visible)	Determines whether the root node should be visible.

The first step in creating a `TreeView` control is declaring a `TreeView` variable as a class instance variable so that you can access it in any method within your program, as follows:

```
TreeView tree;
```

Then, in the application's `start` method, you call the `TreeView` constructor to create the tree view control, passing the root node of the tree you want it to display as a parameter:

```
tree = new TreeView(root);
```

By default, the user can select just one node from the tree. To allow the user to select multiple nodes, use this strange incantation:

```
tree.getSelectionModel().setSelectionMode(
    SelectionModel.MULTIPLE);
```

Here the `getSelectionModel` method is called to get the selection model that manages the selection of nodes within the tree. This method returns an object of type `MultipleSelectionModel`, which includes a method named `setSelectionMode` that lets you set the selection mode. To allow multiple items to be selected, you must pass this method the `SelectionModel.MULTIPLE`.

That's it! You now have a `TreeView` control that you can add to a layout pane and display in your scene.

TIP

Although the tree displayed by a tree view control must begin with a root node, in many cases the root node is superfluous. For example, in the example you've been looking at, what's the point of showing the root node? The `TreeView` control lets you suppress the display of the root node if you don't want it to be shown. To hide the root node, just call this method:

```
tree.setShowRoot(false);
```

Figure 6-6 shows how the tree appears with the root node hidden.

Getting the selected node

There are several ways to determine which node or nodes are currently selected in a tree view. One way is to access the tree's selection model by calling the `getSelectionModel`. Then, you can call the selection model's `getSelectedItems` method to return a read-only observable list that contains the items that have been selected.

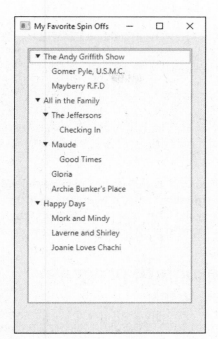

FIGURE 6-6:
A tree view control with the root node hidden.

For example:

```
String msg = "";
ObservableList<TreeItem<String>> shows =
    tree.getSelectionModel().getSelectedItems();
for(TreeItem show : shows)
{
    msg += show.getValue() + "\n";
}
```

In the preceding example, the msg string will contain all the shows that the user has selected from the tree, separated by new line characters.

An alternative is to add an event handler that's called whenever the selection changes. You can do that like this:

```
tree.getSelectionModel().selectedItemProperty()
    .addListener( (v, oldValue, newValue) ->
            tree_SelectionChanged(newValue) );
```

Here, the getSelectionModel method is called to retrieve the selection model. Then, the selectedItemProperty is called to retrieve the selected item property. Finally, an event listener is created for this property by using a lambda expression that calls a method named tree_SelectionChanged. The value of the new selection is passed as a parameter.

Here's what the tree_SelectionChanged method looks like:

```
public void tree_SelectionChanged(TreeItem<String> item)
{
    if (item != null)
    {
        lblShowName.setText(item.getValue());
    }
}
```

Here, a label named lblShowName is updated to display the value of the newly selected item. *Note:* An if statement is used to ensure that the item is not null. That's necessary because if the user deselects an item, the tree_Selection Changed method will be called with a null value as its item parameter.

Looking at a complete program that uses a tree view

Whew! That was a lot of information to digest. In this section, I put it all together.

Listing 6-1 shows the complete program that creates the scene shown in Figure 6-6. This program lets the user select a show from the tree and displays the title of the selected show in a label below the tree.

LISTING 6-1: **The Spinoff Program**

```
package com.lowewriter.SpinOff;

import javafx.application.*;
import javafx.stage.*;
import javafx.scene.*;
import javafx.scene.layout.*;
import javafx.scene.control.*;
import javafx.geometry.*;

public class SpinOffs extends Application
{
    public static void main(String[] args)
    {
        launch(args);
    }

    TreeView<String> tree;                                          →17
    Label lblShowName;
```

```
@Override public void start(Stage primaryStage)
{
    TreeItem<String>  root, andy, archie,                    →22
                      happy, george, maude;

    root = new TreeItem<String>("Spin Offs");                →25
    root.setExpanded(true);

    andy = makeShow(                                         →28
        "The Andy Griffith Show", root);
    makeShow("Gomer Pyle, U.S.M.C.", andy);
    makeShow("Mayberry R.F.D", andy);

    archie = makeShow("All in the Family", root);            →33
    george = makeShow("The Jeffersons", archie);
    makeShow("Checking In", george);
    maude = makeShow("Maude", archie);
    makeShow("Good Times", maude);
    makeShow("Gloria", archie);
    makeShow("Archie Bunker's Place", archie);

    happy = makeShow("Happy Days", root);                    →41
    makeShow("Mork and Mindy", happy);
    makeShow("Laverne and Shirley", happy);
    makeShow("Joanie Loves Chachi", happy);

    tree = new TreeView<String>(root);                       →46
    tree.setShowRoot(false);
    tree.getSelectionModel().selectedItemProperty()          →48
        .addListener( (v, oldValue, newValue) ->
            tree_SelectionChanged(newValue) );

    lblShowName = new Label();

    VBox pane = new VBox(10);
    pane.setPadding(new Insets(20,20,20,20));
    pane.getChildren().addAll(tree, lblShowName);            →56

    Scene scene = new Scene(pane);

    primaryStage.setScene(scene);
    primaryStage.setTitle("My Favorite Spin Offs");
    primaryStage.show();

}

public TreeItem<String> makeShow(String title,              →66
TreeItem<String> parent)
```

(continued)

LISTING 6-1: *(continued)*

```
    {
        TreeItem<String> show = new TreeItem<String>(title);
        show.setExpanded(true);
        parent.getChildren().add(show);
        return show;
    }

    public void tree_SelectionChanged(TreeItem<String> item)          →75
    {
        if (item != null)
        {
            lblShowName.setText(item.getValue());
        }
    }
}
```

All the code in this program has already been shown in this chapter, so I just point out the highlights here:

- » →17 The tree and list models are defined as class instance variables.

- » →22 TreeItem variables are defined for the root node and each show that has spinoff shows.

- » →25 The root node is created with the text Spin-Offs.

- » →28 These lines create the nodes for *The Andy Griffith Show* and its spinoffs.

- » →33 These lines create the nodes for *All in the Family* and its spinoffs.

- » →41 These lines create the nodes for *Happy Days* and its spinoffs.

- » →46 This line creates the TreeView control, specifying root as the root node for the tree. The next line hides the root node.

- » →48 This line creates the event listener for the selected item property. The lambda expression causes the method named tree_SelectionChanged to be called whenever the selection status of the TreeView control changes.

- » →56 The TreeView control and the label are added to a VBox layout pane, which is then added to the scene just before the stage is shown.

- » →66 The makeShow method creates a node from a string and adds the node to the node passed as the parent parameter.

- » →75 The tree_SelectionChanged method is called whenever the selected node changes. It simply displays the title of the selected show in the lblShow-Name label, provided the passed TreeItem is not null.

7

Web Programming

Contents at a Glance

Chapter **1**

Creating Servlets

Servlets are among the most popular ways to develop web applications today. Many of the best-known websites are powered by servlets. In this chapter, I give you just the basics: what a servlet is, how to set up your computer so that you can code and test servlets, and how to create a simple servlet. The next two chapters build on this topic, presenting additional web programming techniques.

Understanding Servlets

Before you can understand what a servlet is and how it works, you need to understand the basics of how web servers work. Web servers use a networking protocol called HTTP to send web pages to users. (*HTTP* stands for *Hypertext Transfer Protocol*, but that won't be on the test.) With HTTP, a client computer uses a uniform resource locator, or URL, to request a document that's located on the server computer. HTTP uses a *request/response model*, which means that client computers (web users) send request messages to HTTP servers, which in turn send response messages back to the clients.

A basic HTTP interaction works something like this:

1. Using a web-browser program running on a client computer, you specify the URL of a file that you want to access.

   In some cases, you actually type the URL of the address, but most of the time you click a link that specifies the URL.

2. Your web browser sends an HTTP request message to the server computer indicated by the URL.

   The request includes the name of the file that you want to retrieve.

3. The server computer receives the file request, retrieves the requested file, and sends the file back to you in the form of an HTTP response message.

4. The web browser receives the file, interprets the contents of the file (usually a mix of HTML, JavaScript, and CSS), and displays the result onscreen.

The most important thing to note about normal web interactions is that they're static. By *static*, I mean that the content of the file sent to the user is always the same. If the user requests the same file 20 times in a row, the same page displays 20 times.

By contrast, a servlet provides a way for the content to be dynamic. A *servlet* is simply a Java program that extends the `javax.servlet.Servlet` class. The `Servlet` class enables the program to run on a web server in response to a user request, and output from the servlet is sent back to the web user as an HTML page.

When you use servlets, Steps 1, 2, and 4 of the preceding procedure are the same; the fateful third step is what sets servlets apart. If the URL specified by the user refers to a servlet rather than a file, Step 3 goes more like this:

3. The server receives the servlet request, locates the Java servlet indicated by the request and passes the request to the servlet. The servlet then generates an appropriate response, sends it back to the client via an HTTP response message, and kicks back until another request is delivered to the servlet.

In other words, instead of sending the contents of a file, the server sends the output generated by the servlet. Typically, the servlet program generates a mixture of HTML, JavaScript, and CSS that's displayed by the browser.

Servlets are designed to get their work done quickly and then end. The first time a servlet is run, it processes one request from a browser, generates one page and sends it back to the browser, and then patiently waits for another request. A single servlet can handle thousands or millions of requests, but each request is treated as an independent process: The request is received and processed, and a result is sent back.

If the server decides that a servlet hasn't seen any requests for a while, it may shut down the servlet. In that case, the servlet will be started up again automatically the next time a request for it is received by the container.

Using Tomcat

Unfortunately, you can't run servlet programs on any old computer. First, you have to install a special program called a *servlet container* to turn your computer into a server that's capable of running servlets. The best-known servlet container is Apache Tomcat, which is available free from the Apache Software Foundation at `http://tomcat.apache.org`. For this chapter, I used Tomcat version 10.

Tomcat can also work as a basic web server. In actual production environments, Tomcat is usually used in combination with a specialized web server, such as Apache's HTTP Server.

Installing Tomcat

Installing Tomcat is a simple as installing any other piece of software: You simply download and run an installer. You can find the installer at `https://tomcat.apache.org/download-10.cgi`. This page lists several downloads. The one you want is found under Binary Distributions/Core and is named 32-bit/ 64-bit Windows Service Installer.

The installer runs like any other Windows installer. You can accept all the defaults if you want, but you may want to change a few defaults along the way — specifically, the following:

>> On the Change Components page, I recommend you switch from Normal to Full installation. This will give you all Tomcat components, including sample programs.

>> On the Configuration page, provide a username and password for the Administrator logon.

>> On the Java Virtual Machine page, you can change the path to the Java JRE if you want. By default, it will pick the Java 8 JRE if that's installed on your system. If you have a more current version, you can select it instead. For example, you can enter **C:\Program Files\Java\jdk-14.0.1** to select JDK 14.

>> On the final page of the installation wizard, you can select Run Apache Tomcat to automatically start the Tomcat service. That way, you won't have to manually start it.

When the installation completes, an icon appears in the System Tray to allow you to control Tomcat. This Tray icon offers commands that let you configure various Tomcat options, start or stop the Tomcat service, and create a Thread Dump, which can be useful for debugging purposes.

Testing Tomcat

To find out whether you installed Tomcat correctly, you can try running the test servlets that are automatically installed when you install Tomcat. Open a web-browser window, and type this address:

```
http://localhost:8080
```

The page shown in Figure 1-1 appears.

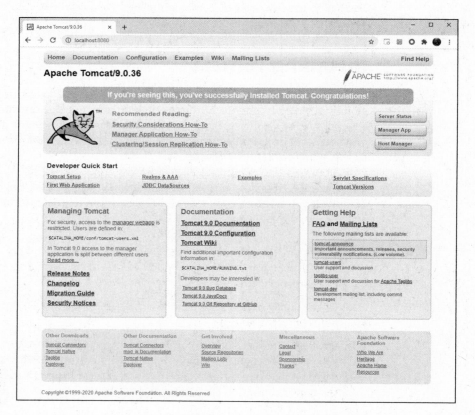

FIGURE 1-1: Tomcat's home page.

You can view interesting information about Tomcat by clicking the Server Status button, located near the upper right of the home page. Figure 1-2 shows the Server Status page. Here, you can see plenty of information about the Tomcat server. One item you may want to note is the JVM version. In Figure 1-2, you can verify that JVM 14.0.1 is active.

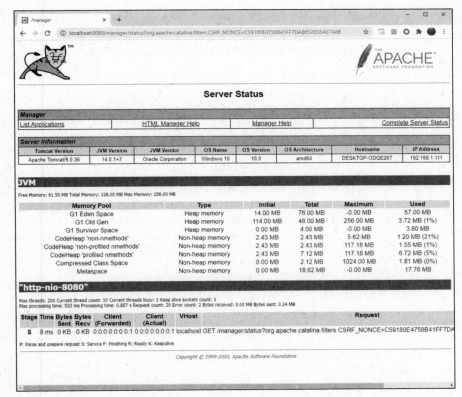

FIGURE 1-2: Tomcat's server status page.

Creating a Simple Servlet

Okay, enough of the configuration stuff; now you can start writing some code. The following sections go over the basics of creating a simple servlet named HelloWorld, which simply displays the text "Hello, World!" in the browser window. This servlet will be housed in a package named com.lowewriter.helloworld.

Creating the folder structure for a servlet

Before you start writing code, take a few minutes to set up a folder structure to hold all the bits and pieces of your application. Start by creating a root folder. You can put this folder anywhere you want, and you can name it anything you want. For this example, I've named the folder HelloWorldServlet.

Next, create the following subfolders within that root folder:

>> src: This folder will hold your .java source files.

>> deploy: This folder will hold the deployment image for the application.

>> web: This folder can hold any static HTML or JSP pages. You should also create a folder named WEB-INF in the web folder as explained next.

>> web\WEB-INF: This folder holds an important configuration file named web. xml. You should also create a folder named classes within the WEB-INF folder, as described next.

>> web\WEB-INF\classes: This folder will hold the .class files for your application. Note that if your application uses a package, you should create a folder hierarchy that corresponds to your package name. For example, if the package is com.lowewriter.helloworld, you'll need to create the corresponding folder com\lowewriter\helloworld within the classes folder.

The resulting folder structure for the HelloWorld servlet looks like this:

```
HelloWorldServlet
    src
    deploy
    web
        WEB-INF
            classes
                com
                    lowewriter
                        helloworld
```

Creating the web.xml file

Every Java servlet application requires a file named web.xml in the web\WEB-INF folder. This file specifies several critical attributes of the servlet application, including the name of the application, the name of the servlet class, and the URL used to invoke the servlet.

For more information about XML, refer to chapter 5 of the bonus content.

The web.xml file for the HelloWorld servlet application is shown in Listing 1-1.

LISTING 1-1: **The web.xml File for the HelloWorld Servlet**

```
<?xml version="1.0" encoding="UTF-8"?>                                        →1
<web-app xmlns="http://xmlns.jcp.org/xml/ns/javaee"                          →2
  xmlns:xsi="http://www.w3.org/2001/XMLSchema-instance"
  xsi:schemaLocation="http://xmlns.jcp.org/xml/ns/javaee
                      http://xmlns.jcp.org/xml/ns/javaee/web-app_4_0.xsd"
  version="4.0"
  metadata-complete="true">

    <display-name>Hello World!</display-name>                               →9

    <servlet>                                                               →11
        <servlet-name>helloworld</servlet-name>
        <servlet-class>com.lowewriter.helloworld.HelloWorld</servlet-class>
    </servlet>

    <servlet-mapping>                                                       →16
        <servlet-name>helloworld</servlet-name>
        <url-pattern>/Hello</url-pattern>
    </servlet-mapping>

</web-app>
```

The following paragraphs describe the key elements of this file:

» →1 This ?xml element is a standard header required at the start of all XML files.

» →2 The web-app element provides basic information about the application. This information is the same for all web.xml files.

» →9 The display-name element provides a descriptive name for the application.

» →11 The servlet tag associates a servlet with a class. The servlet tag has two subelements:

  • servlet-name: This element provides the name of the servlet. In Listing 1-1, the name is helloworld.

  • servlet-class: This element provides the fully qualified name of the class that implements the servlet. In Listing 1-1, the name is com.lowewriter.helloworld.HelloWorld.

» →16 The servlet mapping element associates a URL with a servlet. It has two subelements:

- servlet-name: This element should be the same as the corresponding subelement of the servlet element — in this case, helloworld.

- URL-pattern: This element indicates the URL that will be used to invoke the servlet. Note that the URL here is relative to the application's root. Thus, /Hello means that the user can invoke the HelloWorld servlet by specifying *host*/helloworld/Hello, where *host* identifies the web host. (See "Running a Servlet," later in this chapter, for more information.)

In this simple example, only one servlet is defined for the application, and that servlet is associated with just one URL. However, more complicated applications can include more than one servlet element, and each servlet can have multiple URL-pattern elements that provide aliases for the servlet.

Now that you've set up the folder structure and created the web.xml file, you're ready to get on to writing code. To create the code for a servlet class, create a .java file in the src folder. The next few sections show you how to write the Java code for the HelloWorld servlet.

Importing the servlet packages

Most servlets need access to at least three packages: javax.servlet, javax.servlet.http, and java.io. As a result, you usually start with these import statements:

```
import java.io.*;
import javax.servlet.*;
import javax.servlet.http.*;
```

Depending on what other processing your servlet does, you may need additional import statements.

Extending the HttpServlet class

To create a servlet, you write a class that extends the HttpServlet class. Table 1-1 lists six methods you can override in your servlet class.

Most servlets override at least the doGet method. This method is called by the servlet engine when a user requests the servlet by typing its address in the browser's address bar or by clicking a link that leads to the servlet.

TABLE 1-1 **The HttpServlet Class**

Method	When Called	Signature
doDelete	HTTP DELETE request	`public void doDelete(Http ServletRequest request, HttpServletResponse response)throws IOException, ServletException`
doGet	HTTP GET request	`public void doGet(HttpServletRequest request, HttpServletResponse response) throwsIOException, ServletException`
doPost	HTTP POST request	`public void doPost(HttpServletRequest request, HttpServletResponse response) throws IOException, ServletException`
doPut	HTTP PUT request	`public void doPut(HttpServletRequest request, HttpServletResponse response) throws IOException, ServletException`
init()	First time servlet is run	`public void init() throws ServletException`
destroy()	Servlet is destroyed	`public void destroy()`

Two parameters are passed to the doGet method:

>> An HttpServletRequest object representing the incoming request from the user. You use the request parameter primarily to retrieve data entered by the user in form fields. You find out how to do that later in this chapter, in the section "Getting Input from the User."

>> An HttpServletResponse object representing the response that is sent back to the user. You use the response parameter to compose the output that is sent back to the user. You find out how to do that in the next section.

Printing to a web page

One of the main jobs of most servlets is writing HTML output that's sent back to the user's browser. To do that, you first call the getWriter method of the HttpServletResponse class, which returns a PrintWriter object that's connected to the response object. Thus, you can use the familiar print and println methods to write HTML text.

Here's a doGet method for a simple HelloWorld servlet:

```
public void doGet(HttpServletRequest request,
    HttpServletResponse response)
        throws IOException, ServletException
```

```
{
    PrintWriter out = response.getWriter();
    out.println("Hello, World!");
}
```

Here the `PrintWriter` object returned by `response.getWriter()` is used to send a simple text string back to the browser. If you run this servlet, the browser displays the text `Hello, World!`.

Responding with HTML

In most cases, you don't want to send simple text back to the browser. Instead, you want to send formatted HTML. To do that, you must first tell the response object that the output is in HTML format. You can do that by calling the `setContentType` method, passing the string `"text/html"` as the parameter. Then you can use the `PrintWriter` object to send HTML.

Listing 1-2 shows a basic `HelloWorld` servlet that sends an HTML response.

LISTING 1-2:	The HelloWorld Servlet

```
package com.lowewriter.helloworld;

import java.io.*;
import javax.servlet.*;
import javax.servlet.http.*;
public class HelloWorld extends HttpServlet
{
    public void doGet(HttpServletRequest request,
        HttpServletResponse response)
            throws IOException, ServletException
    {
        response.setContentType("text/html");
        PrintWriter out = response.getWriter();
        out.println("<html>");
        out.println("<head>");
        out.println("<title>HelloWorld</title>");
        out.println("</head>");
        out.println("<body>");
```

```
        out.println("<h1>Hello, World!</h1>");
        out.println("</body>");
        out.println("</html>");
    }
}
```

Here the following HTML is sent to the browser (I added indentation to show the HTML's structure):

```html
<html>
  <head>
    <title>HelloWorld</title>
  </head>

<body>
    <h1>Hello, World!</h1>
  </body>
</html>
```

When run, the HelloWorld servlet produces the page shown in Figure 1-3.

FIGURE 1-3:
The HelloWorld servlet displayed in a browser.

TIP

Obviously, you need a solid understanding of HTML to write servlets. If HTML is like a foreign language, you need to pick up a good HTML book, such as *HTML, XHTML, & CSS For Dummies,* 7th Edition, by Ed Tittel and Jeff Noble (Wiley), before you go much further.

For your reference, Table 1-2 summarizes all the HTML tags that I use in this book.

TABLE 1-2   **Just Enough HTML to Get By**

HTML tag	Description
`<html>`, `</html>`	Marks the start and end of an HTML document.
`<head>`, `</head>`	Marks the start and end of the head section of an HTML document.
`<title>`, `</title>`	Marks a title element. The text between the start and end tags is shown in the title bar of the browser window.
`<body>`, `</body>`	Marks the start and end of the body section of an HTML document. The content of the document is provided between these tags.
`<h1>`, `</h1>`	Formats the text between these tags as a level-1 heading.
`<h2>`, `</h2>`	Formats the text between these tags as a level-2 heading.
`<h3>`, `</h3>`	Formats the text between these tags as a level-3 heading.
`<form action="url", method="method">`	Marks the start of a form. The `action` attribute specifies the name of the page, servlet, or JavaServer Page (JSP) the form is posted to. The `method` attribute can be GET or POST; it indicates the type of HTTP request sent to the server.
`</form>`	Marks the end of a form.
`<input type="type", name="name">`	Creates an input field. Specify `type="text"` to create a text field or `type="submit"` to create a Submit button. The `name` attribute provides the name you use in the program to retrieve data entered by the user.
` `	Represents a nonbreaking space.

Running a Servlet

So exactly how do you run a servlet? Follow these steps:

1. **Compile the** `.java` **file to create a** `.class` **file.**

Use the `javac` command.

2. **Move the** `.class` **file into the correct class directory.**

For example, the directory might be `web\classes\com\lowewriter\helloworld`.

3. **Create a** `.war` **file for the application.**

A `.war` file is a special type of `.jar` file that's used to deploy web applications. The `.war` file will be placed in the application's `deploy` folder.

To create the .war file, open a command prompt and navigate to your application's root folder. Then enter this command:

```
jar cvf deploy\helloworld.war —C web .
```

This command creates the file helloworld.war in the deploy folder. The .war file will contain the entire contents of the web folder. Note that the period at end of the command is required.

4. **Copy the** .war **file into Tomcat's** webapps **folder.**

For example, the folder might be C:\Program Files\Apache Software Foundation\Tomcat 10.0\webapps. When you drop the .war file there, Tomcat automatically installs the application. Within a few moments, you see a folder named helloworld appear in the webapps folder.

5. **Open a web browser and enter the URL for the servlet.**

For the HelloWorld servlet, enter the following URL:

```
http://localhost:8080/helloworld/Hello
```

Improving the HelloWorld Servlet

The HelloWorld servlet, shown in Listing 1-2 earlier in this chapter, isn't very interesting because it always sends the same text. Essentially, it's a static servlet — which pretty much defeats the purpose of using servlets in the first place. You could just as easily have provided a static HTML page.

Listing 1-3 shows the code for a more interesting version called RandomHello. This version uses the random method of the Math class to pick a random number from 1 to 6 and then uses this number to decide which greeting to display.

LISTING 1-3: **The RandomHello Servlet**

```
package com.lowewriter.randomhello;

import java.io.*;
import javax.servlet.*;
import javax.servlet.http.*;
import java.util.*;
public class HelloServlet extends HttpServlet
{
    public void doGet(HttpServletRequest request,
```

(continued)

LISTING 1-3: *(continued)*

```java
        HttpServletResponse response)
            throws IOException, ServletException
    {
        response.setContentType("text/html");
        PrintWriter out = response.getWriter();
        String msg = getGreeting();
        out.println("<html>");
        out.println("<head>");
        out.println("<title>HelloWorld Servlet</title>");
        out.println("</head>");
        out.println("<body>");
        out.println("<h1>");
        out.println(msg);
        out.println("</h1>");
        out.println("</body>");
        out.println("</html>");
    }

    private String getGreeting()
    {
        String msg = "";
        int rand = (int)(Math.random() * (6)) + 1;
        switch (rand)
        {
            case 1:
                return "Hello, World!";
            case 2:
                return "Greetings!";
            case 3:
                return "Felicitations!";
            case 4:
                return "Yo, Dude!";
            case 5:
                return "Whasssuuuup?";
            case 6:
                return "Hark!";
        }
        return null;
    }
}
```

Getting Input from the User

If a servlet is called by an HTTP GET or POST request that came from a form, you can call the getParameter method of the request object to get the values entered by the user in each form field. Here's an example:

```
String name = request.getParameter("name");
```

Here the value entered in the form input field named name is retrieved and assigned to the String variable name.

Working with forms

As you can see, retrieving data entered by the user in a servlet is easy. The hard part is creating a form in which the user can enter the data. To do that, you create the form by using a separate HTML file. Listing 1-4 shows an HTML file named InputServlet.html that displays the form shown in Figure 1-4.

LISTING 1-4: **The InputServlet.html File**

```html
<html>
  <head>
    <title>Input Servlet</title>
  </head>

<body>
    <form action="/servlet/InputServlet"
        method="post">
    Enter your name: 
    <input type="text" name="Name">
    <br><br>
    <input type="submit" value="Submit">
    </form>
  </body>
</html>
```

The action attribute in the form tag of this form specifies that /servlet/InputServlet is called when the form is submitted, and the method attribute indicates that the form is submitted via a POST rather than a GET request.

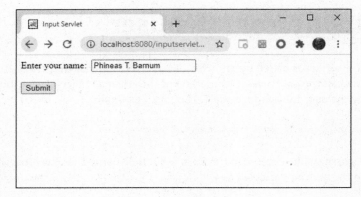

FIGURE 1-4:
A simple input
form.

The form itself consists of an input text field named name and a Submit button.
It's nothing fancy — just enough to get some text from the user and send it to a
servlet.

Using the InputServlet servlet

Listing 1-5 shows a servlet that can retrieve the data from the form shown in
Listing 1-3.

LISTING 1-5: **The InputServlet Servlet**

```
package com.lowewriter.inputservlet;

import java.io.*;
import javax.servlet.*;
import javax.servlet.http.*;

public class InputServlet extends HttpServlet
{
    public void doGet(HttpServletRequest request,
        HttpServletResponse response)
            throws IOException, ServletException
    {
        String name = request.getParameter("Name");
        response.setContentType("text/html");
        PrintWriter out = response.getWriter();
        out.println("<html>");
        out.println("<head>");
        out.println("<title>Input Servlet</title>");
        out.println("</head>");
        out.println("<body>");
```

```
        out.println("<h1>");
        out.println("Hello " + name);
        out.println("</h1>");
        out.println("</body>");
        out.println("</html>");
    }
}
```

As you can see, this servlet really isn't much different from the first `HelloWorld` servlet in Listing 1-2. The biggest difference is that it retrieves the value entered by the user in the `name` field and uses it in the HTML that's sent to the response `PrintWriter` object. If the user enters **Calvin Coolidge** in the `name` input field, for example, the following HTML is generated:

```
<html>
  <head>
    <title>HelloWorld</title>
  </head>

<body>
    <h1>Hello Calvin Coolidge</h1>
  </body>
</html>
```

Thus the message `Hello Calvin Coolidge` is displayed on the page.

Although real-life servlets do a lot more than just parrot back information entered by the user, most of them follow this surprisingly simple structure — with a few variations, of course. Real-world servlets validate input data and display error messages if the user enters incorrect data or omits important data, and most real-world servlets retrieve or update data in files or databases. Even so, the basic structure is pretty much the same.

Using Classes in a Servlet

When you develop servlets, you often want to access other classes that you've created, such as input/output (I/O) classes that retrieve data from files or databases, utility or helper classes that provide common functions such as data validation, and perhaps even classes that represent business objects such as customers or products. If you save all your classes in the same folder, the `javac` compiler will be able to find them when you compile the classes, and the `jar` command will combine them into the `.war` file.

To illustrate a servlet that uses several classes, Figure 1-5 shows the output from a servlet that lists movies read from a text file. This servlet uses three classes:

>> `Movie`: A class that represents an individual movie.

>> `MovieIO`: A class that has a static public method named `getMovies`. This method returns an `ArrayList` object that contains all the movies read from the file.

>> `ListFiles`: The main servlet class. It calls the `MovieIO.getMovies` class to get an `ArrayList` of movies and then displays the movies on the page.

FIGURE 1-5: The ListMovies servlet.

The code for the `Movie` class is shown in Listing 1-6. As you can see, this class doesn't have much: It defines three public fields (`title`, `year`, and `price`) and a constructor that lets you create a new `Movie` object and initialize the three fields. Note that this servlet doesn't use the `price` field.

LISTING 1-6: **The Movie Class**

```
public class Movie
{
    public String title;
    public int year;
    public double price;
    public Movie(String title, int year,
                 double price)
    {
        this.title = title;
```

```
        this.year = year;
        this.price = price;
    }
}
```

Listing 1-7 shows the `MovieIO` class. This class uses the file I/O features that are presented in chapter 2 of the bonus content to read data from a text file. The text file uses tabs to separate the fields and contains these lines:

```
Modern Times→1936→14.95
It's a Wonderful Life→1946→15.95
12 Angry Men→1957→14.95
2001: A Space Odyssey→1968→19.95
Young Frankenstein→1974→16.95
The Princess Bride→1987→16.95
Apollo 13→1995→19.95
Sea Biscuit→2003→12.95
The Imitation Game→2014→17.95
```

Here the arrows represent tab characters in the file. I'm not going to go over the details of this class here, except to point out that `getMovies` is the only public method in the class, and it's static, so you don't have to create an instance of the `MovieIO` class to use it. For details on how this class works, turn to chapter 2 of the bonus content.

LISTING 1-7:   **The MovieIO Class**

```
package com.lowewriter.movie;

import java.io.*;
import java.util.*;

public class MovieIO
{
    public static ArrayList<Movie> getMovies()
    {
        ArrayList<Movie> movies =
            new ArrayList<Movie>();
        BufferedReader in =
            getReader("movies.txt");
        Movie movie = readMovie(in);
        while (movie != null)
        {
            movies.add(movie);
            movie = readMovie(in);
```

(continued)

LISTING 1-7: *(continued)*

```java
        }
        return movies;
    }

    private static BufferedReader getReader(
        String name)
    {
        BufferedReader in = null;
        try
        {
            File file = new File(name);
            in = new BufferedReader(
                new FileReader(file) );
        }
        catch (FileNotFoundException e)
        {
            System.out.println(
                "The file doesn't exist.");
            System.exit(0);
        }
        return in;
    }

    private static Movie readMovie(BufferedReader in)
    {
        String title;
        int year;
        double price;
        String line = "";
        String[] data;
        try
        {
            line = in.readLine();
        }
        catch (IOException e)
        {
            System.out.println("I/O Error");
            System.exit(0);
        }
        if (line == null)
            return null;
        else
        {
            data = line.split("\t");
            title = data[0];
```

```
        year = Integer.parseInt(data[1]);
        price = Double.parseDouble(data[2]);
        return new Movie(title, year, price);
    }
  }
}
```

Listing 1-8 shows the code for the ListMovie servlet class.

LISTING 1-8: **The ListMovie Servlet Class**

```
package com.lowewriter.movie;

import java.io.*;
import javax.servlet.*;
import javax.servlet.http.*;
import java.util.*;

public class ListMovies extends HttpServlet
{
    public void doGet(HttpServletRequest request,          →10
        HttpServletResponse response)
            throws IOException, ServletException
    {
        response.setContentType("text/html");
        PrintWriter out = response.getWriter();
        String msg = getMovieList();
        out.println("<html>");
        out.println("<head>");
        out.println("<title>List Movies Servlet</title>");
        out.println("</head>");
        out.println("<body>");
        out.println("<h1>Some of My Favorites</h1>");
        out.println("<h3>");
        out.println(msg);
        out.println("</h3>");
        out.println("</body>");
        out.println("</html>");
    }
    public void doPost(HttpServletRequest request,         →30
        HttpServletResponse response)
            throws IOException, ServletException
```

(continued)

LISTING 1-8:   *(continued)*

```
    {
        doGet(request, response);
    }

    private String getMovieList()                                      →37
    {
        String msg = "";
        ArrayList<Movie> movies = MovieIO.getMovies();
        for (Movie m : movies)
        {
            msg += m.year + ": ";
            msg += m.title + "<br>";
        }
        return msg;
    }
}
```

The following paragraphs describe what its methods do:

>> →10 The doGet method calls the getMovieList method to get a string that contains a list of all the movies, separated by break (<br>) tags. Then it uses a series of out.println statements to write HTML that displays this list.

>> →30 The doPost method simply calls the doGet method. That way, the servlet works whether it is invoked by a GET or a POST request.

>> →37 The getMovieList method calls the MovieIO.getMovies method to get an ArrayList that contains all the movies read from the file. Then it uses an enhanced for loop to retrieve each Movie object. Each movie's year and title is added to the msg string, separated by <br> tags.

Chapter **2**

Using JavaServer Pages

n Book 7, Chapter 1, you discover how to create servlets that write HTML data directly to a page by using the PrintWriter object accessed through response. out. Although this technique works, it is far from the ideal way to develop web applications. Any sensible web application design separates the business logic (the Java code) from the presentation (the HTML). Generating HTML directly from a Java servlet was fine for introducing how servlets work, but it isn't a viable way to write actual web applications.

That's where JavaServer Pages (a technology usually called JSP for short) comes in. A *JavaServer Page* is an HTML file that has Java servlet code embedded in it in special tags. When you run a JSP, all the HTML is automatically sent as part of the response, along with any HTML that's created by the Java code you embed in the JSP file. As a result, JSP spares you the chore of writing all those out.println statements.

Note: To work with JavaServer Pages, you must first set up a working Tomcat environment. For information on how to do that, please refer to Book 7, Chapter 1.

Understanding JavaServer Pages

A JSP is an HTML document that's saved in a file with the extension `.jsp` instead of `.htm` or `.html`. In many cases, `.jsp` files are stored in the application's home folder. But if your application has a lot of `.jsp` files, you can store them in sub-folders. When you run the `jar` command to create a `.war` file to deploy to Tomcat, the `.jsp` files will be included in the `.war` file so they wind up in the right place on the server. (For more information, refer to Chapter 1 of this minibook.)

A `.jsp` file consists of a mixture of HTML elements and special tags. Here's an incredibly simple example — not particularly useful, but it demonstrates how tags are embedded with HTML:

```
<html>
<body>
    <p>Two plus two equals ${2 + 2}.</p>
   </body>
</html>
```

Here, the embedded tag is `${2 + 2}`. This tag is called an *expression tag.* When the page is rendered, the expression in the tag is evaluated and the result is rendered to the page. So, this JSP page displays the following text on the browser:

```
Two plus two equals 4.
```

Let's take a closer look at the life cycle of a simple JSP page such as the preceding one. Suppose the JSP page is named `Four.jsp` and it lives in a web application named `misc` on your local computer, where Tomcat is installed. You could request the JSP page by entering the following in a browser address bar: **http://localhost:8080/misc/Four.jsp**. Or, you could click a link that references this URL.

Either way, the server receives a `GET` request for the `Four.jsp` file.

Assuming this is the first time `Four.jsp` is requested, its HTML and tags are parsed and converted to a Java servlet, saved in a source file named `Four_jsp.java` and compiled to a class named `Four_jsp.class`. These files are saved within a folder called `work` on the Tomcat server.

The `Four_jsp` servlet is then invoked. It uses an output stream to write HTML code for the page in much the same manner as the programs in Chapter 1 of this minibook wrote HTML using `PrintWriter.out` to write HTML. In fact, if you peek inside the `Four_jsp.java` file, you'll find these lines:

```
out.write("<html>\r\n");
out.write("
<body>\r\n");
out.write("    <p>Two plus two equals ");
out.write((java.lang.String)
org.apache.jasper.runtime.PageContextImpl.proprietaryEvaluate("${2 + 2}",
    java.lang.String.class,
    (javax.servlet.jsp.PageContext)_jspx_page_context, null));
out.write("</p>
\r\n");
out.write("  </body>\r\n");
out.write("</html>");
```

With the exception of the line that calls the proprietaryEvaluate method, these lines simply regurgitate the HTML in the Four.jsp file. The proprietary Evaluate method call handles the ${2 + 2} expression.

Fortunately, you don't have to worry about how the servlet created for a JSP page works — you can simply trust that it does. The point is that every JSP page becomes a servlet, and it's the servlet that actually responds to the request for the JSP page.

Most JSP applications involve not only GET requests but POST or PUT requests as well. A typical JSP presents a form into which the user enters data. When the user submits the form, the data is sent to the server as a POST request. The POST is handled by a servlet often called a *controller*. (The controller servlet is distinct from the servlet automatically generated for the .jsp page; you have to code the controller servlet yourself!)

The controller servlet receives the data entered by the user, validates it, and acts on it, often by retrieving or updating database data. The controller typically adds data to the JSP page by appending it to the request in the form of request attributes, which usually wind up in the request message body as JSON data. Then the controller forwards the request to the JSP page.

When the JSP page receives the request, it renders the response in the form of HTML, which is then sent to the user to be displayed on the browser.

This basic servlet architecture is consistent with the popular Model-View-Controller (MVC) paradigm used for modern web applications. In an MVC application, application components are segregated into three distinct layers:

>> **Model:** The Model handles representation of and access to data objects that drives the application. Typically, the Model layer includes classes that define application objects such as customers, invoices, products, courses, employees,

and so on. In addition, the Model layer may contain classes that manage access to the underlying storage that holds the data objects, such as SQL databases. (In some cases, the access classes are considered to be a separate sublayer, often called the *Repository*.) The Model layer may also house business rules, for example decisions about discounts or shipping methods.

>> **View:** The View handles the presentation to the end user. In a JSP application, the JSP pages comprise the view.

>> **Controller:** The Controller manages the interactions between the Model and View layers. In a JSP application, the servlets, which receive requests, act on them, and then redirect to the JSP pages, are the main elements of the Controller layer.

I explain how to create controller servlets later in this chapter. But first, let's look more closely at the types of tags you can use in a JSP page.

Using UEL Expressions

You can include many different types of tags in JSP pages. In this section, I show you how you can incorporate expressions into your JSP pages using Unified Expression Language (UEL) tags and JSP Standard Tag Library (JSTL) tags.

Unified Expression Language

One of the most common needs in JSP pages is to evaluate expressions. The syntax for evaluating expressions is governed by a standard called UEL. Here's the basic form of a UEL expression:

```
${ expression }
```

Wherever you place a UEL expression in your JSP page, the expression will be evaluated and the expression itself will be replaced by the result of the evaluation.

Here's the example that I present earlier in this chapter:

```
<p>Two plus two equals ${2 + 2}.</p>
```

Here, the UEL tag is ${2 + 2} and the expression within the tag is 2 + 2. When the expression is evaluated, the result (4) replaces the entire tag. So, the text rendered into the page is:

```
Two plus two equals 4.
```

A UEL expression can contain standard mathematical, relational, and logical operators, plus a few others, as shown in Table 2-1.

TABLE 2-1   ## UEL Expression Operators

Operator	Explanation
Arithmetic Operators	
+	Addition
–	Subtraction (or negation when used as a unary operator)
*	Multiplication
/, div	Division
%, mod	Modulus division
Relational Operators	
==, eq	Equal to
!=, ne	Not equal to
>, gt	Greater than
<, lt	Less than
>=, ge	Greater than or equal to
<=, le	Less than or equal to
Logical Operators	
and, &&	And
or, ||	Or
!, not	Not
Conditional Operator	
A ? B : C	Evaluate *A*, then evaluate either *B* or *C* depending on result of *A*
Empty Operator	
empty	Prefix operator that determines if a value is null or empty.

UEL expressions are also used to access objects that are available via request attributes. In your Java servlet, you can create a request attribute by calling the Request object's setAttribute method. Then, the JSP page can retrieve the attribute using a UEL expression. I explain how to do this in the section "Working with Attributes," later in this chapter.

JSP Standard Tag Library

JSTL tags are designed to simplify the task of crafting the user interface in a JSP page. With JST tags, you can easily provide for data formatting, as well as repeating or conditional content.

JSTL is based on XML, so JSTL tags fit naturally in your HTML code. JSTL tags elements are enclosed by ‹ and › symbols, have an element name, and can optionally have attributes and content. A start tag marks the beginning of a JSTL tag, and the end is marked by an end tag that uses the same element name preceded by a slash. If a JSTL tag has no content between the start and end elements, you can combine the start and end tags by ending the tag with a slash. For example:

```
<c:out value = "Two plus two equals ${2+2}."/>
```

This example renders the text Two plus two equals 4.

Here's an example of a JSTL tag that has separate start and end tags with content in between:

```
<c:forEach begin="1" end="10">
    <p>All work and no play makes Jack a dull boy.</p>
</c:forEach>
```

This example renders ten consecutive paragraphs, all saying All work and no play makes Jack a dull boy.

JSTL tags come in five flavors:

>> **Core tags:** Provide the core functionality of JSTL and are likely the ones you'll work with most.

>> **Formatting tags:** Provide special formatting capabilities for numbers, dates, and times.

>> **Function tags:** Provide several common operations, mostly for manipulating strings.

>> **SQL tags:** Let you interact with SQL databases.

>> **XML tags:** Let you interact with XML data.

In the following sections, I show you some of the more common tags in the core and formatting groups.

Deploying the JSTL taglib files

Before you can start using JSTL tags in your JSP pages, you need to download the `.jar` files that implement the tags. You can download the necessary `.jar` files from the Apache Tomcat website at `http://tomcat.apache.org/download-taglibs.cgi`. You should download the following two files:

```
taglibs-standard-impl-1.2.5.jar
taglibs-standard-spec-1.2.5.jar
```

Create a folder named `lib` in your application's `WEB-INF` folder. Then copy the taglib `.jar` files to the new `lib` folder.

If you have the necessary permissions, you can also deploy these files directly to Tomcat so they'll be available to any application. Just copy the files to `web-apps\ROOT\web\WEB-INF\lib`.

Adding the <taglib> directive

Before you can use JSTL tags in a JSP page, you must add a `<taglib>` directive to the JSP page to identify the tag category you want to use (core, formatting, functions, and so on). The `<taglib>` directive sets a prefix that appears before tags for the specific category. For example, you can specify that all core tags begin with `c:`.

Table 2-2 shows the prefix and `<taglib>` directive you should use for each JSTL tag category. Although you can specify any prefix you want, you should stick to the standard prefixes listed in the table.

Here's an example of a simple `.jsp` with `<taglib>` directives for the core and functions categories:

```
<html>
<%@ taglib prefix = "c" uri = "http://java.sun.com/jsp/jstl/core" %>
<%@ taglib prefix = "fmt" uri = "http://java.sun.com/jsp/jstl/fmt" %>
<body>
    <p>This page contains two taglib directives</p>
  </body>
</html>
```

TABLE 2-2　　**JSTL Tag Prefixes and <taglib> Directives**

Category	Prefix	<taglib> Directive
Core	c	`<%@ taglib prefix = "c" uri = "http://java.sun.com/jsp/jstl/core" %>`
Formatting	fmt	`<%@ taglib prefix = "fmt" uri = "http://java.sun.com/jsp/jstl/fmt" %>`
Functions	f	`<%@ taglib prefix = "fn" uri = "http://java.sun.com/jsp/jstl/functions" %>`
SQL	sql	`<%@ taglib prefix = "sql" uri = "http://java.sun.com/jsp/jstl/sql" %>`
XML	x	`<%@ taglib prefix = "x" uri = "http://java.sun.com/jsp/jstl/xml" %>`

As you can see, the taglib directives appear near the top of the file, immediately following the <html> start tag.

Looking at Core Tags

The core tags provide basic functionality, such as displaying the results of expressions, making choices, and processing loops, with a few other goodies thrown in for good measure. Table 2-3 shows all 14 of the available core tags. I look more closely at the most important ones in the sections that follow.

TABLE 2-3　　**Core Tags**

Tag	What It Does
catch	Catches any exceptions that might be thrown within the body of the tag.
choose	A switch-like construct that sets up a selection of choices identified by when tags within the body of the choose tag. You can also include an otherwise tag if none of the when tags is selected.
if	A conditional tag. The body of the tag is evaluated if the condition is true.
import	Gets a resource based on a URL and renders its contents to a page or to a variable or a reader.
forEach	Sets up a for loop. The body of the loop is evaluated once for each iteration of the loop.
forTokens	Iterates over a string of delimited tokens.

Tag	What It Does
out	Evaluates an expression and renders it to the page. It can optionally encode XML tags to guard against script injection attacks.
otherwise	Used in the body of a choice tag. Its body is evaluated if none of the when tags in the choice tag's body is true.
param	Adds a parameter to a URL created by the url tag.
redirect	Redirects to a URL.
remove	Removes a variable created by a set tag.
set	Creates a variable and assigns it a value.
url	Creates a URL to which parameters can be added using the param tag.
when	Used in the body of a choice tag. Its body is evaluated if the expression is true.

Using c:out

One of the most commonly used core tags is out. This tag evaluates an expression and renders it to the page. The expression is supplied via the value attribute, like this:

```
<c:out value="All work and no play makes Jack a dull boy."/>
```

You can include a UEL expression tag in the value. For example:

```
<c:out value = "Two plus two equals ${2+2}."/>
```

And you can include the attribute excapeXML to force the special XML characters <, >, &, ', and " to be rendered by escape sequences such as < or &. This special encoding can help prevent a common type of software attack called *cross-site scripting*, in which an attacker attempts to inject code into the value of a form field.

Working with variables

You can create variables that you can use in your JSP page by using the c:set tag. You specify the name of the variable using the var attribute and the value by using the value attribute.

When you're finished with a variable, you can destroy it by using the c:remove tag.

Here's an example that creates a variable named two, gives it the value 2, uses it in expressions in the value attribute of a c:out tag, and then removes the variable:

```
<c:set var="two" value="2"/>
<c:out value="${two} + ${two} = ${two + two}"/>
<c:remove var="two"/>
```

Getting conditional

You can create the equivalent of an if statement by using the c:if tag. Here's an example:

```
<c:set var="commissionrate" value="0.02"/>
<c:set var="salestotal" value="9886.00"/>

<c:if test="${salestotal >= 10000.00}">
    <c:set var="commissionrate" value="0.05"/>
    <p><c:out value="Your commission rate is ${commissionrate}."/></p>
</c:if>
```

Here, a variable named commissionrate is created and given an initial value of 0.02. Then, a c:if tag is used to see if a variable named salestotal is greater than $10,000. If it is, commissionrate is increased to 0.05 and the lucky salesperson is told about the increase. Alas, in this case, there will be no such happy news, as salestotal has been set to $9,886 — just below the threshold. Better luck next month!

Note that the c:if tag does not have an else clause. You can simulate that by following a c:if with another c:if that tests for the opposite condition, like this:

```
<c:set var="commissionrate"/>
<c:set var="salestotal" value="12955.00"/>
<p><c:out value="Your sales total is ${salestotal}."/></>

<c:if test="${salestotal >= 10000.00}">
    <c:set var="commissionrate" value="0.05"/>
</c:if>

<c:if test="${salestotal < 10000.00}">
    <c:set var="commissionrate" value="0.02"/>
</c:if>`

<p><c:out value="Your commission rate is ${commissionrate}."/></p>
```

In this example, the commissionrate variable is not given an initial value; instead, it is assigned either 0.02 or 0.05 depending on the value of salestotal. This time, the lucky salesperson gets the commission bump because salestotal is set to $12,995.

This same logic could be implemented with c:choice as follows:

```
<c:choose>

    <c:when test="${salestotal < 10000.00}">
        <c:set var="commissionrate" value="0.02"/>
    </c:when>

    <c:when test="${salestotal >= 10000.00}">
        <c:set var="commissionrate" value="0.05"/>
    </c:when>

</c:choose>
```

Here, the first c:when covers the case in which the sales total is less than $10,000 and the second c:when covers the case where the sales total is greater than or equal to $10,000.

Another way to do this would be with c:otherwise, like this:

```
<c:choose>

    <c:when test="${salestotal >= 10000.00}">
        <c:set var="commissionrate" value="0.05"/>
    </c:when>

    <c:otherwise>
        <c:set var="commissionrate" value="0.02"/>
    </c:otherwise>

</c:choose>
```

Creating loops

The last core tag I'll cover in this section is c:forEach, which lets you create simple iterations. You can mimic a standard for statement by setting the begin and end attributes, like this:

```
<c:forEach var="i" begin="1" end="10">
    <p><c:out value="${i}"/></p>
</c:forEach>
```

This loop simply counts from 1 to 10, displaying each number on a separate line.

You can use the step attribute to specify a step other than 1, but unfortunately you can't use a negative value for step. So, there's no way to count backward, at least not without some clever programming. Here's one way to do it:

```
<c:forEach var="i" begin="0" end="9">
    <c:set var="j" value="${10-i}"/>
    <p><c:out value="${j}"/></p>
</c:forEach>
```

Notice here that the forEach actually counts from 0 this time instead of from 1. That's because the subtraction is done before the value is printed, so if you started from 1 you'd end up counting backward from 9.

You can also use c:forEach to iterate over a list. I show you an example of that in the section "The ListMovies Application Meets JSP," later in this chapter.

Formatting Numbers

The formatting tags let you format values as numbers and dates. You can specify detailed formats for numbers, controlling such things as the number of decimal digits to display and whether to use commas or currency symbols or format numbers as percentages. You can also format dates using a variety of common formats.

In this section, I focus on using the fmt:formatNumber command to format numeric values. Table 2-4 lists the attributes of this tag.

Here's an example that formats the variable salestotal as currency and commissionrate as a percentage.

```
<fmt:formatNumber var="totalfmt" value="${salestotal}" type="currency"/>
<fmt:formatNumber var="commfmt" value="${commissionrate}" type="percent"/>
```

In both cases, the formatted values are returned in string variables, which can later be included on the page.

TABLE 2-4

fmt:formatNumber Attributes

Attribute	What It Does
value	The value to be formatted.
var	The variable used to store the result. If omitted, the tag is rendered in place.
type	Specifies one of three types of formatting: number, currency, or percent.
groupingUsed	A true or false value indicating whether digits should be grouped with commas.
maxIntegerDigits	The maximum number of digits shown to the left of the decimal point.
minIntegerDigits	The minimum number of digits shows to the left of the decimal point.
maxFractionDigits	The maximum number of digits shown to the right of the decimal point.
minFractionDigits	The minimum number of digits shows to the right of the decimal point.
pattern	A formatting pattern used to format the value. The pattern string can include symbols such as # for a digit, 0 for digits that should show as zero even if not significant, commas and decimal points, and currency symbols.

Here's an example that uses various attributes to indicate that the numeric value of the variable n should be rendered with at most seven digits and at least one digit to the left of the decimal point, exactly three digits to the right of decimal point, and with digit groups separated by commas:

```
<fmt:formatNumber value="${n}"
                  maxIntegerDigits="7"
                  minIntegerDigits="1"
                  maxFractionDigits="3"
                  minFractionDigits="3"
                  groupingUsed="true"/>
```

With this formatting, the value 1234.56 will be displayed as 1,234.560.

The pattern attribute lets you provide a formatting specification more concisely by using characters such as # and 0 for digits, as well as commas and a decimal point. Here's how the previous formatting example could be done with pattern:

```
<fmt:formatNumber value="${n}" pattern="#,###,##0.000"/>
```

The resulting output is the same.

Considering the Controller Servlet

Up to this point in this chapter, I focus on using expressions and JSTL tags to create JSP pages. The JSP pages are responsible for handling the Presentation layer of a JSP application. In this section, I turn my attention to creating servlets that handle the Controller layer.

A servlet for a JSP application is very different from a stand-alone servlet. A stand-alone servlet is responsible for emitting HTML that will be sent to the client as part of the HTTP response (see Chapter 1 of this minibook). It typically does this by grabbing a PrintWriter instance from the Response object and writing hand-crafted HTML.

That's a serious violation of the MVC design paradigm for modern web applications, in which the presentation logic of the application should be separated from the business and control logic. In a JSP application, the JSP page is solely responsible for crafting the user interface: That's the job of the HTML and the UEL expressions and JSTL tags in the .jsp file.

The servlet in a JSP application is responsible for processing the request received via a GET message, which may involve calling upon classes in the Model layer to retrieve or update database data or to make business decisions. When these tasks have been completed, the servlet redirects the request to the JSP page, which builds the final presentation and sends it off to the client.

There are two important tasks you need to know how to do to create even a simple controller servlet for a JSP application. First, you need to know how to attach data to the Request object so that the data will be passed on to the JSP page; you do that by setting request attributes. Second, you need to know how to redirect the request to the JSP page.

Setting request attributes

To attach data to the Request object, you use the Request.setAttribute method. This method has the following signature:

```
void setAttribute(java.lang.String name, java.lang.Object o)
```

As you can see, just two parameters are required. The first parameter is a String that provides a name for the attribute. The second parameter is the attribute itself, which can be literally any kind of Java object. In other words, you can pass absolutely anything you want to the Presentation layer. It's up to the Presentation layer to render the object in a format suitable for the client.

Here's an example that adds an `ArrayList` named movies using the attribute name movies:

```
request.setAttribute("movies", movies);
```

The JSP can then access the `ArrayList` using the attribute name movies.

In the section "Creating loops," earlier in this chapter, I mention that you can use `c:forEach` to iterate over a collection. Here's an example that retrieves Movie objects from the movies attribute, assigns them to a variable named movie, and then displays the Title property of each Movie:

```
<c:forEach var="movie" items="${movies}">
        <p>${movie.title}</p>
    </c:forEach>
```

You'll see code similar to this in a complete action later in this chapter.

Redirecting to the JSP page

When your controller servlet has attached all the data to request attributes, it's time to redirect the request to the JSP page. You do in the servlet's doGet or doPost method by using a somewhat confusing combination of method calls on the Request object:

```
request.getRequestDispatcher("/ListMovies.jsp").forward(request, response);
```

Here, you first call a Request method named getRequestDispatcher to obtain a RequestDispatcher object. This method requires that you specify a URL for the resource you want to redirect to — in this case, the JSP page named List Movies.jsp.

When you have the RequestDispatcher in hand, you call its forward method, which forwards the request. You must supply both a Request and a Response object, which are typically passed on from the Request and Response method received as arguments to the doGet or doPost method. Keep in mind that the Request object now carries the attributes you assigned. In this way, the JSP page will receive the attributes so it can format the data.

That's about all there is to it, at least for the simplest of JSP pages. In the next section, I put everything together in a complete application.

The ListMovies Application Meets JSP

This section presents a JSP version of the Movies application that I present in Chapter 1 of this minibook. This version uses a servlet called `moviejsp`. Figure 2-1 shows the output displayed by the JSP. Like the version in Chapter 1, this version displays the movies contained in a text file named `movies.txt`. This version also adds a count of the movies in the file, as shown in the header "9 of My Favorite Movies."

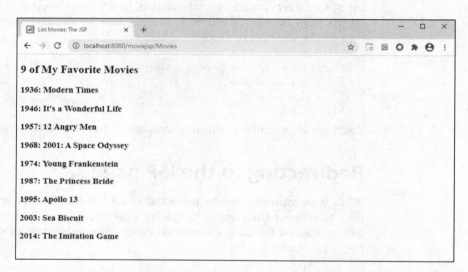

FIGURE 2-1:
The ListMovies
JSP in action.

The application consists of four components:

» `Movie.java`: A class that represents a movie with three properties: `Title`, `Year`, and `Price`. The class uses the accessor pattern to expose these properties. (This class is shown in Listing 2-1.)

» `MovieIO.java`: A class that manages the file I/O necessary to retrieve the movie data from the `Movies.txt` file. This class is identical to the version presented in Chapter 1 of this minibook and is shown in Listing 2-2 for reference.

» `ListMovies.java`: The controller servlet for the application. In the `web.xml` file, this class is mapped directly to a servlet named `movies` with the URL pattern `\Movies`. So, it can be invoked from a browser by navigating to `http://localhost:8080/moviejsp/Movies`. This class is shown in Listing 2-3.

» `ListMovies.jsp`: The JSP page for the application. When a request is made for the `moviejsp` servlet, the `ListMove` class calls upon `MovieIO` to get the list of movies. It then appends the list to the `Request` object and forwards the request to `ListMovies.jsp`. The JSP page then renders the HTML for the response and sends it to the browser. The JSP page is shown in Listing 2-4.

LISTING 2-1: **Movies.java**

```java
package com.lowewriter.moviejsp;                                    →1

public class Movie
{
    private String title;
    private int year;
    private double price;

    public Movie(String title, int year,
                 double price)
    {
        this.title = title;
        this.year = year;
        this.price = price;
    }

    public String getTitle()                                        →17
    {
        return title;
    }

    public int getYear()                                            →22
    {
        return year;
    }

    public double getPrice()                                        →27
    {
        return price;
    }

}
```

Here are the key lines of this class:

>> →1 The Movie class is part of the package com.lowewriter.moviejsp. All the other classes for this application also belong to this package.

>> →17 The getTitle method is the accessor for the title property.

>> →22 getYear is the accessor for the year property.

>> →27 getPrice is the accessor for the price property.

LISTING 2-2: **MovieIO.java**

```java
package com.lowewriter.moviejsp;

import java.io.*;
import java.util.*;

public class MovieIO
{
    public static ArrayList<Movie> getMovies()
    {
        ArrayList<Movie> movies = new ArrayList<Movie>();
        BufferedReader in = getReader("c:\\data\\movies.txt");
        Movie movie = readMovie(in);
        while (movie != null)
        {
            movies.add(movie);
            movie = readMovie(in);
        }
        return movies;
    }

    private static BufferedReader getReader(String name)
    {
        BufferedReader in = null;
        try
        {
            File file = new File(name);
            in = new BufferedReader(new FileReader(file) );
        }
        catch (FileNotFoundException e)
        {
            System.out.println("The file doesn't exist.");
            System.exit(0);
        }
        return in;
    }
```

```
private static Movie readMovie(BufferedReader in)
{
    String title;
    int year;
    double price;
    String line = "";
    String[] data;

    try
    {
        line = in.readLine();
    }
    catch (IOException e)
    {
        System.out.println("I/O Error");
        System.exit(0);
    }

    if (line == null)
        return null;
    else
    {
        data = line.split("\t");
        title = data[0];
        year = Integer.parseInt(data[1]);
        price = Double.parseDouble(data[2]);
        return new Movie(title, year, price);
    }
}
}
```

LISTING 2-3: **ListMovies.java**

```
package com.lowewriter.moviejsp;

import java.io.*;
import javax.servlet.*;
import javax.servlet.http.*;
import java.util.*;

public class ListMovies extends HttpServlet
{
    public void doGet(HttpServletRequest request,                    →10
                    HttpServletResponse response)
        throws IOException, ServletException
```

(continued)

LISTING 2-3:   *(continued)*

```
    {
        ArrayList<Movie> movies = MovieIO.getMovies();                 →14

        request.setAttribute("movies", movies);                        →16
        request.setAttribute("number", movies.size());                 →17
        request.getRequestDispatcher("/ListMovies.jsp")                →18
            .forward(request, response);

    }

}
```

Here are the key points of the ListMovie.java servlet class:

>> →10 The doGet method accepts Request and Response objects.

>> →14 Here, MovieIO.getMovies() is called to obtain a list of movies. The controller servlet is concerned only with the representation of this data as a list; the MovieIO class is the only class that knows that the underlying data is stored in a text file.

>> →16 The list is added to the Request object as an attribute named movies.

>> →17 The number of elements in the list is added as an attribute named number.

>> →18 The request along with its attributes is redirected to the ListMovies. jsp page.

LISTING 2-4:   **ListMovies.jsp**

```
<!doctype html public "-//W3C//DTD HTML 4.0 Transitional//EN">

<%@ taglib prefix="c" uri="http://java.sun.com/jsp/jstl/core" %>      →3

<html>
  <head>
    <title>List Movies: The JSP</title>
  </head>
```

```
<body>
    <h2><c:out value="${number}"/> of My Favorite Movies</h1>        →10
    <h3 >
     <c:forEach var="movie" items="${movies}">                       →12
       <p>${movie.year}: ${movie.title}</p>
     </c:forEach>
    </h3>

  </body>
</html>
```

Following is an explanation of the key lines in this JSP:

» →3 This taglib directive is required to use the JSTL core tags.

» →10 The number attribute is retrieved from the Response object and inserted in the heading.

» →12 This c:forEach tag iterates over all the movie objects in the movies attribute. For each movie, the year and title properties are extracted and displayed on the page.

Chapter **3**

Using JavaBeans

A *JavaBean* is a special type of Java class that you can use in several interesting ways to simplify program development. Some beans are designed to be visual components that you can use in a graphic user interface (GUI) editor to build user interfaces quickly. Other beans, known as *Enterprise JavaBeans* (EJB), are designed to run on special EJB servers and can run the data access and business logic for large web applications.

In this chapter, I look at a more modest type of JavaBean that's designed to simplify the task of building JavaServer Pages (or JSP, which I cover in Book 7, Chapter 3). In a nutshell, you can use the simple JavaBeans to build JavaServer Pages without writing any Java code in the JSP itself. JavaBeans let you access Java classes by using special HTML-like tags in the JSP page.

To use JavaBeans, you must have Tomcat enabled. For more information about installing and configuring Tomcat, please refer to Book 7, Chapter 2.

Getting to Know JavaBeans

Simply put, a JavaBean is any Java class that conforms to the following rules:

» **It must have an empty constructor** — that is, a constructor that accepts no parameters. If the class doesn't have any constructors at all, it qualifies because the default constructor has no parameters. But if the class has at least one constructor that accepts one or more parameters, it must also have a constructor that has no parameters to qualify as a JavaBean.

>> **It must have no** `public` **instance variables.** All the instance variables defined by the class must be either `private` or `protected`.

>> **It must provide methods named** get*Property* **and** set*Property* **to get and set the value of any** *properties* **the class provides, except for** `boolean` **properties that use** is*Property* **to get the property value.** The term *property* isn't an official Java term. In a nutshell (or should that be *in a bean-pod?*), a property is any value of an object that can be retrieved by a `get` method (or an `is` method, if the property is `boolean`) or set with a `set` method. If a class has a property named `lastName`, for example, it should use a method named `getLastName` to get the last name and `setLastName` to set the last name. Or, if the class has a `boolean` property named `taxable`, the method to set it is called `setTaxable`, and the method to retrieve it is `isTaxable`.

Note that a class doesn't have to have any properties to be a JavaBean, but if it does, the properties have to be accessed according to this naming pattern. Also, not all properties must have both a `get` and a `set` accessor. A *read-only property* can have just a `get` accessor, and a *write-only property* can have just a `set` accessor.

REMEMBER

The property name is capitalized in the methods that access it, but the property name itself isn't. Thus `setAddress` sets a property named `address`, not `Address`.

That's all there is to it. More advanced beans can also have other characteristics that give them a visual interface so that they can be used drag-and-drop style in an integrated development environment (IDE). Also, technically beans should implement an interface that allows their state to be written to an output stream so that they can be re-created later. But for the purposes of this chapter, any class that meets the three criteria stated here is a bean and can be used as a bean in JSP pages.

You've already seen plenty of classes that have methods with names like `getCount` and `setStatus`. These names are part of a design pattern called the *Accessor pattern*, which I cover in Book 3, Chapter 2. Thus you've seen many examples of beans throughout this book, and you've probably written many bean classes yourself already.

REMEMBER

Any class that conforms to this pattern is a bean. There's no `JavaBean` class that you have to extend; neither is there a `Bean` interface that you have to implement to create a bean. All a class has to do to be a bean is stick to the pattern.

Looking Over a Sample Bean

Listing 3-1 shows a sample JavaBean class named `Triangle` that uses the Pythagorean theorem to calculate the long side of a right triangle if you know the length of the two short sides. This class defines three properties: `sideA` and `sideB` represent the two short sides of the triangle, and `sideC` represents the long side. The normal way to use this bean is to first use the `setSideA` and `setSideB` methods to set the `sideA` and `sideB` properties to the lengths of the short sides and then use the `getSideC` method to get the length of the long side.

TECHNICAL STUFF

In case you can't remember way back to high school, the long side is equal to the square root of the first short side squared plus the second short side squared.

LISTING 3-1: **The Triangle Bean**

```
package com.lowewriter.calculators;                                →1
public class Triangle
{
    private double sideA;                                          →4
    private double sideB;
    public Triangle()                                              →6
    {
        this.sideA = 0.0;
        this.sideB = 0.0;
    }
    public String getSideA()                                       →11
    {
        return Double.toString(this.sideA);
    }
    public void setSideA(String value)                             →15
    {
        try
        {
            this.sideA = Double.parseDouble(value);
        }
        catch (Exception e)
        {
            this.sideA = 0.0;
        }
    }
    public String getSideB()                                       →26
    {
        return Double.toString(this.sideB);
    }
```

(continued)

Using JavaBeans

LISTING 3-1: *(continued)*

```java
public void setSideB(String value)                              →30
{
    try
    {
        this.sideB = Double.parseDouble(value);
    }
    catch (Exception e)
    {
        this.sideB = 0.0;
    }
}
public String getSideC()                                        →41
{
    if (sideA == 0.0 || sideB == 0.0)
        return "Please enter both sides.";
    else
    {
        Double sideC;
        sideC = Math.sqrt(
            (sideA * sideA) + (sideB * sideB));
        return Double.toString(sideC);
    }
}
}
```

The following paragraphs point out the highlights of this bean class:

» →1 As with most servlet classes, this bean is part of a package. In this case, the package is named com.lowewriter.calculators. (I'm assuming that if you need a bean to calculate the Pythagorean theorem, you probably want other beans to calculate derivatives, prime numbers, Demlo numbers, and the like. You can put those beans in this package, too.)

» →4 This class uses a pair of instance variables to keep track of the two short sides. As per the rules for JavaBeans, these instance variables are declared as private.

» →6 A constructor with no parameters is declared. (Strictly speaking, this constructor doesn't have to be explicitly coded here, because the default constructor does the trick, and the two instance variables are initialized to their default values of 0 automatically.)

» →11 The getSideA method returns the value of the sideA property as a string.

>> →15 The setSideA method lets you set the value of the sideA property with a string. This method uses a try/catch statement to catch the exceptions that are thrown if the string can't be parsed to a double. If the string is invalid, the sideA property is set to 0.

>> →26 The getSideB method returns the value of the sideB property as a string.

>> →30 The setSideB method sets the value of the sideB property from a string. Again, a try/catch statement catches any exceptions and sets the property to 0 if the string can't be parsed to a double.

>> →41 The getSideC method calculates the length of the long side and then returns the result as a string. If either of the values is 0, however, the method assumes that the user hasn't entered any data, so it returns an error message instead. (That's a reasonable assumption, because none of the sides of a triangle can be zero.) Notice that there is no setSideC method. As a result, sideC is a read-only property.

Using Beans with JSP Pages

To work with a bean in a JSP page, you add special tags to the page to create the bean, set its properties, and retrieve its properties. Table 3-1 lists these tags, and the following sections describe the details of using each one.

TABLE 3-1   **JSP Tags for Working with Beans**

Tag	Description
`<jsp:useBean id="name" class="package.class" />`	Establishes a reference to the bean and creates an instance if necessary. The name specified in the id attribute is used by the other tags to refer to the bean.
`<jsp:getProperty name="name" property="property" />`	Retrieves the specified property from the bean identified by the name attribute.
`<jsp:setProperty name="name" property="property" value="value" />`	Sets the specified property to the value specified in the value attribute.
`<jsp:setProperty name="name" property="property" param="parameter" />`	Sets the specified property to the value of the parameter specified in the param attribute. The parameter is usually the name of a form field.
`<jsp:setProperty name="name" property="*" />`	Sets all the properties defined by the bean to corresponding parameter values, provided that a parameter with the correct name exists.

Creating bean instances

To include a bean in a JSP page, you add a special `jsp:useBean` tag to the page. In its simplest form, this tag looks like this:

```
<jsp:useBean id="name" class="package.Class"/>
```

The `id` attribute provides the name that you use elsewhere in the JSP to refer to the bean, and the `class` attribute provides the name of the class, qualified with the package name. Here's a `jsp:useBean` tag to use the `Triangle` bean:

```
<jsp:useBean id="triangle" class="com.lowewriter.calculators.Triangle"/>
```

The `jsp:useBean` tag creates an instance of the bean by calling the empty constructor if an instance doesn't already exist. If the bean does already exist, the existing instance is used instead.

Here are a few additional things you should know about the `jsp:useBean` tag:

>> The `jsp:useBean` tag can appear anywhere in the JSP document, but it must appear before any other tag that refers to the bean.

>> This tag and all bean tags are case-sensitive, so be sure to code them exactly as shown. `<jsp:usebean.../>` won't work.

>> If Tomcat complains that it can't find your bean when you run the JSP, double-check the package and class name — they're case-sensitive, too — and make sure that the bean is stored in a directory under `WEB-INF\classes` that's named the same as the package. You might store the `Triangle` bean's class file in `WEB-INF\classes\calculators`, for example. (For more info on Tomcat, flip to Book 7, Chapter 2.)

>> The `jsp:useBean` element can have a body containing `jsp:setProperty` tags that initialize property values. Then the element is formed more like normal HTML, with proper start and end tags, as in this example:

```
<jsp:useBean id="t1" class="com.lowewriter.calculators.Triangle" >
<jsp:setProperty name="t1" property="sideA" value="3.0" >
<jsp:setProperty name="t1" property="sideB" value="3.0" >

</jsp:useBean>
```

Don't worry about the details of the `jsp:setProperty` tags just yet. Instead, just make a note that they're executed only if a new instance of the bean is actually created by the `jsp:useBean` tag. If an instance of the bean already exists, the `jsp:setProperty` tags are not executed.

>> The `jsp:useBean` tag also has a `scope` attribute, which I explain in the section "Scoping Your Beans," later in this chapter.

Getting property values

To get the value of a bean's property, you use the `jsp:getProperty` tag. The form of this tag is straightforward:

```
<jsp:getProperty name="name" property="property"/>
```

Here's a tag that gets the `sideC` property from the `Triangle` bean created in the preceding section:

```
<jsp:getProperty name="triangle" property="sideC"/>
```

The `name` attribute must agree with the value you specify in the `id` attribute of the `jsp:useBean` tag that created the bean. Also, the `property` attribute is used to determine the name of the getter method — in this case, `getSideC`.

In most cases, you use `jsp:getProperty` to insert the value of a property into a page, but you can also use it to specify the value of an attribute for some other tag in the JSP document, as in this example:

```
<input type="text" name="sideA"
       value="<jsp:getProperty name="triangle"
                 property="sideA" />" >
```

Here the value of the `sideA` property is retrieved and used for the `value` attribute of an input field named `sideA`. As a result, when this input field is sent to the browser, its initial value is the value from the `Triangle` bean.

WARNING

Be extra careful to match up the quotation marks and the open and close brackets for the tags. In this example, the entire `jsp:getProperty` tag is enclosed within the quotation marks that indicate the value of the input field's `value` attribute. The right bracket that appears at the very end closes the input element itself.

Setting property values

To set a property value, you can use one of several variations of the `jsp:setProperty` tag. If you want to set the property to a literal string, you write the tag like this:

```
<jsp:setProperty name="triangle"
                 property="sideA"
                 value="4.0"/>
```

Here the `name` attribute must match the `id` attribute from the `jsp:useBean` tag that created the bean; the `property` attribute is used to determine the name of

Using JavaBeans

the setter method (in this case, setSideA); and the value attribute provides the value to be set.

Although this form of the jsp:setProperty tag is useful, the param form is more useful. It lets you set the property to the value entered by the user in a form field or passed to the JSP by way of a query string. If your JSP contains a form that has an input field named FirstSide, you can assign that field's value to the sideA property like this:

```
<input type="text" name="FirstSide" >
<jsp:setProperty name="triangle"
                property="sideA"
                param="FirstSide"/>
```

Here, if the user enters a value in the FirstSide field, that value is assigned to the bean's sideA property.

In the preceding example, I deliberately use a name other than sideA for the input field so that you won't be confused by the fact that the property and param attributes specify the same value. In actual practice, you usually give the input field the same name as the property it's associated with, like this:

```
<input type="text" name="sideA" >
<jsp:setProperty name="triangle"
                property="sideA"
                param="sideA"/>
```

If your input fields have names that are identical to the property names, you can assign all of them to their corresponding properties with one tag, like this:

```
<jsp:setProperty name="triangle" property="*"/>
```

Here the asterisk (*) in the property attribute indicates that all properties that have names identical to form fields (or query-string parameters) are assigned automatically. For forms that have a lot of fields, this form of the jsp:setProperty tag can save you a lot of coding.

Viewing a JSP page that uses a bean

So that you can see how these tags work together, Listing 3-2 shows a complete JSP page using the bean that was presented in Listing 3-2. This page displays two text input fields and a button. When the user enters the lengths of a triangle's two short sides in the fields and clicks the button, the page displays the sideC property of the bean to show the length of the third side. Figure 3-1 shows how this page appears when the code is run.

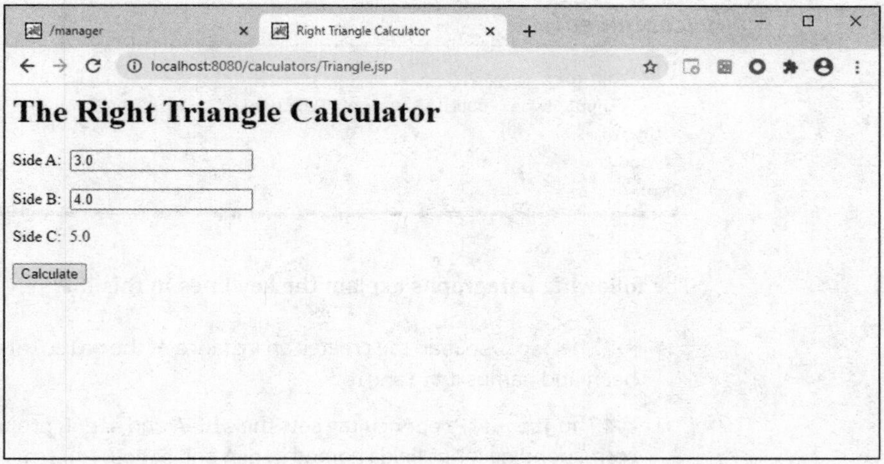

FIGURE 3-1:
The Triangle.jsp
page displayed in
a browser.

LISTING 3-2:	**The Triangle.jsp Page**

```html
<html>
<jsp:useBean id="triangle"                                            →2
      class="com.lowewriter.calculators.Triangle"/>
<jsp:setProperty name="triangle" property="*" />                      →4
  <head>
    <title>Right Triangle Calculator</title>
  </head>
<body>
    <h1>The Right Triangle Calculator</h1>
    <form action="Triangle.jsp" method="post">                        →10
      Side A: 
      <input type="text" name="sideA"                                 →12
          value="<jsp:getProperty
                    name="triangle"
                    property="sideA"/>" >
      <br><br>
      Side B: 
      <input type="text" name="sideB"                                 →18
          value="<jsp:getProperty
                    name="triangle"
                    property="sideB"/>" >
      <br><br>
      Side C: 
      <jsp:getProperty name="triangle"                                →24
          property="sideC"/>
```

(continued)

LISTING 3-2: **(continued)**

```
      <br><br>
      <input type="submit" value="Calculate" >                    →27
    </form>
  </body>
</html>
```

The following paragraphs explain the key lines in this JSP:

>> →2 The jsp:useBean tag creates an instance of the calculators.Triangle bean and names it triangle.

>> →4 The jsp:setProperty tag sets the sideA and sideB properties to the corresponding input fields named sideA and sideB.

>> →10 The form tag creates a form that posts back to the same JSP file using the HTTP POST method.

>> →12 This line creates the first of two input text fields. This one is named sideA, and its initial value is set to the value of the bean's sideA property.

>> →18 The second input text field is named sideB. Its initial value is set to the value of the bean's sideB property.

>> →24 This line is where the sideC property is retrieved, thus calculating the length of side C of the triangle based on the length of sides A and B. The result is simply inserted into the document.

>> →27 The Submit button submits the form so that the Triangle bean can do its thing.

Scoping Your Beans

The *scope* of a JavaBean indicates how long the bean is kept alive. You specify the scope by using the scope attribute of the jsp:useBean tag. The scope attribute can have any of the four values listed in Table 3-2.

The default scope is page, which means that the bean is created and destroyed each time the user requests a new page. The session scope, however, can be very useful for web applications that need to keep track of information about a user from one page to the next. The best-known example is a shopping cart, in which a user can select items that he or she wants to purchase. The contents of the shopping cart can be kept in a session bean.

TABLE 3-2   **Scope Settings**

Scope	Explanation
page	This setting associates the bean with the current page. Thus, every time the user requests the page, a new bean is created. When the page is sent back to the browser, the bean is destroyed. Thus each round trip to the server creates a new instance of the bean.
request	This setting is similar to page, but the bean is available to other pages that are processed by the same request. This scope is useful for applications that use several servlets or JSPs for a single request.
session	This setting associates the bean with a user's session. The first time the user requests a page from the application, a bean is created and associated with the user. Then the same bean is used for subsequent requests by the same user.
application	This setting means that a single copy of the bean is used by all users of the application.

A shopping cart application

Figure 3-2 shows a simple shopping cart application in which the user has the option to purchase three of my recent books by clicking one of the three buttons. When the user clicks a button, an item is added to the shopping cart. If the user has already added the book to the cart, the quantity is increased by 1. In the figure, the user has clicked the button for *Electronics All-in-One For Dummies* once and the button for *Networking All-in-One For Dummies* twice.

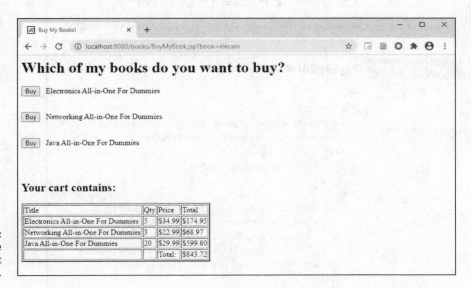

FIGURE 3-2:
A super-simple shopping cart application.

The following paragraphs describe the key techniques that make this shopping cart work:

» The shopping cart itself is a JavaBean that has just two public methods: setBook, which adds a book to the shopping cart, and getList, which returns a string that shows the shopping cart items nicely formatted in an HTML table.

» The shopping cart class contains an inner class that represents a Book object. To keep the application simple, the Book class has the three titles hard-coded into it. In a real shopping cart program, you use a database or some other storage system rather than hard-coding these values.

» The list of products that appears at the top of the page is actually three separate forms, one for each product. Each of these forms specifies a parameter passed via a query string to the JSP on the server. The name of this parameter is book, and its value is the code of the book the user ordered. This parameter is bound to the book property of the shopping cart bean, so when the user clicks one of the buttons, the setBook method is called with the value passed via the book parameter. That's how the shopping cart knows which book the user ordered.

» Below the list of books, the JSP uses a jsp:getProperty tag to get the list property, which displays the shopping cart.

The shopping cart page

Listing 3-3 shows the JSP for the shopping cart page.

LISTING 3-3: **BuyMyBook.jsp**

```
<html>
<jsp:useBean id="cart"                                              →2
 class="com.lowewriter.books.BookCart" scope="session"/>
<jsp:setProperty name="cart" property="*" />                        →4
   <head>
     <title>Buy My Books!</title>
   </head>
<body>
     <h1>Which of my books do you want to buy?</h1>
     <form action="BuyMyBook.jsp?book=elecaio"                      →10
         method="post">
      <input type="submit" value="Buy" >  
      Electronics All-in-One For Dummies<br><br>
     </form>
```

```
        <form action="BuyMyBook.jsp?book=netaio"                           →15
            method="post">
          <input type="submit" value="Buy" >  
          Networking All-in-One Desk Reference
            For Dummies
          <br><br>
        </form>
        <form action="BuyMyBook.jsp?book=wordaio"                          →22
            method="post">
          <input type="submit" value="Buy" >  
          Java All-in-One Desk Reference For Dummies
          <br><br>
        </form>
        <br><h2>Your cart contains:</h2>
        <jsp:getProperty name="cart" property="list" />                    →29
      </body>
</html>
```

The following paragraphs describe the JSP's most important lines:

» →2 The jsp:useBean tag loads the com.lowewriter.books.BookCart JavaBean, specifying that it has session scope. Thus, the bean isn't deleted after each page is requested. Instead, the user works with the same bean instance for his or her entire session.

» →4 The parameter properties are set. The first time the user displays the BuyMyBook.jsp page, there are no parameters, so this method doesn't do anything. But when the user clicks one of the three form buttons, a book parameter is added to the end of the URL that's posted to the server, so the cart's setBook method is called. This causes one copy of the selected book to be added to the cart.

» →10 This line is the form for the first book. Each book has its own form, with a Submit button labeled Buy and a book title. The action attribute specifies that when the Submit button is clicked, the form is posted to BuyMyBook.jsp with the book parameter set to netfd.

» →15 This line is the second book form. This one specifies netaio as the book parameter value.

» →22 This line is the form for the third book. This one specifies wordaio as the value of the book parameter.

» →29 After the forms for the books, a jsp:getProperty tag calls the getList method of the bean. This method returns a string containing an HTML table that displays the current contents of the shopping cart.

The BookCart JavaBean

Now that you've seen the JSP for the shopping cart application, take a look at the Java code for the BookCart bean, shown in Listing 3-4.

REMEMBER

Listing 3-4 contains two classes: a BookCart class and an inner class named Book. So when you compile the code in Listing 3-4, you get two class files: BookCart. class and BookCart$Book.class. To run this application, you have to copy both class files to a Tomcat directory (a WEB-INF\classes\books directory).

LISTING 3-4: **The BookCart JavaBean**

```
package com.lowewriter.books;                                    →1

import java.util.ArrayList;
import java.text.NumberFormat;

public class BookCart
{
    private ArrayList<Book> cart;                                →8
    private NumberFormat cf = NumberFormat.getCurrencyInstance();

    public BookCart()                                            →11
    {
        cart = new ArrayList<Book>();
    }

    public void setBook(String code)                             →16
    {
        boolean found = false;
        for (Book b : cart)
            if (b.getCode().equals(code))
            {
                b.addQuantity(1);
                found = true;
            }
        if (!found)
        {
            Book b = new Book();
            b.setCode(code);
            cart.add(b);
        }
    }

    public String getList()                                      →33
    {
        String list = "<table border=2>";
```

```
            list +="<tr><td>Title</td><td>Qty</td>"
                + "<td>Price</td><td>Total</td></tr>";
            double total = 0.0;
            for (Book b : cart)
            {
                list += "<tr><td>"
                    + b.getTitle() + "</td>"
                    + "<td>" + b.getQuantity()  + "</td>"
                    + "<td>" + cf.format(b.getPrice())
                    + "</td>"
                    + "<td>" + cf.format(b.getTotal())
                    + "</td>"
                    + "</tr>";
                total += b.getTotal();
            }
            list +="<tr><td/><td/><td>Total:</td>"
                + "<td>" + cf.format(total)
                + "</td></tr>";
            list += "</table>>";
            return list;
    }

    private class Book                                        →58
    {
        private String code;                                  →60
        private int quantity;

        public String getCode()                               →63
        {
            return this.code;
        }

        public void setCode(String code)                      →68
        {
            this.code = code;
        }

        public String getTitle()                              →73
        {
            if (code.equals("elecaio"))
                return "Electronics All-in-One For Dummies";
            else if (code.equals("netaio"))
                return "Networking All-in-One For Dummies";
            else if (code.equals("javaaio"))
                return "Java All-in-One For Dummies";
            else
                return "Unknown book";
        }
```

(continued)

LISTING 3-4:   *(continued)*

```
        public double getPrice()                                →85
        {
            if (code.equals("elecaio"))
                return 34.99;
            else if (code.equals("netaio"))
                return 22.99;
            else if (code.equals("javaaio"))
                return 29.99;
            else
                return 0.0;
        }

        public int getQuantity()                                →97
        {
            return this.quantity;
        }

        public void addQuantity(int qty)                        →102
        {
            this.quantity += qty;
        }

        public double getTotal()                                →107
        {
            return this.quantity * this.getPrice();
        }
    }
}
```

The following paragraphs describe the bean's high points:

» →1 The BookCart class lives in the books package.

» →8 The shopping cart itself is kept inside the BookCart bean as a private array list of Book items.

» →11 To be a JavaBean, you need a no-parameter constructor. This one simply initializes the cart array list.

» →16 The setBook method is called to add a book to the shopping cart. The book's code is passed as a parameter. This method first looks at all the books in the array list to see whether the user has already added a book with this code. If so, that book's addQuantity method is called to increase the order quantity for that book by 1. If not, a new book with the specified code is created and added to the cart.

» →33 This method builds a string that contains all the books in the cart presented as an HTML table. If you're not familiar with HTML tables, all you really need to know is that the `<tr>` and `</tr>` tags mark the start and end of each row, and the `<td>` and `</td>` tags mark the start and end of each cell within the row. The table includes one row for each book in the cart. Each row contains cells for the title, quantity, price, and total. If you compare the code in this method with the actual table shown in Figure 3-2, you can get an idea of the HTML that's actually created by this method.

Notice also that the loop that builds each table row keeps a running total for the entire shopping cart, which is displayed in a separate row at the bottom of the table. Also, a row of headings is displayed at the start of the table.

» →58 The Book class is defined as an inner class so that it can represent books in the array list.

» →60 The Book class stores only two items of information for each book: the book code and the quantity. The latter represents the number of books ordered by the user. The other values are calculated by the methods that return them.

» →63 The getCode method simply returns the code variable.

» →68 The setCode method assigns a value to the code variable.

» →73 The getTitle method returns one of three book titles, depending on the code. If the code is not one of the three predefined codes, Unknown book is returned.

» →85 Likewise, the getPrice method returns one of three prices, depending on the code. If the code is not one of the three allowable codes, the book is free!

» →97 The getQuantity method just returns the quantity variable.

» →102 The addQuantity method adds a value to the quantity variable.

» →107 The getTotal method calculates the total by multiplying the price by the quantity.

Chapter **4**

Using HttpClient

The previous chapters in this minibook show you how to create programs specifically designed to run on a web server so they can be consumed by users on the web.

This chapter turns the tables: It shows you how to write Java programs that consume information provided by web servers. This is accomplished via a library that was introduced in Java 11 called the HttpClient.

Understanding HTTP

Before I dive into the details of the HttpClient, I need to explain what HTTP is and how it works. You need at least a basic understanding of HTTP to make the HttpClient work.

HTTP stands for *HyperText Transfer Protocol*. It was created way back in 1997 and quickly became the cornerstone of what we now know as the World Wide Web.

HTTP's initial purpose was to allow users to access HTML files from web servers. It quickly grew beyond that to support a wide variety of data types, such as images, sounds, videos, and so on. Today, HTTP supports the transfer of hundreds of different types of data.

Note that HTTP is *not* involved in rendering HTML files on a screen — that's the job of a web browser. HTTP is responsible for the exchange of data between client and server, not for what the client or the server do with the data being exchanged.

TIP

HTTP is a core protocol used to develop applications that use the modern online design architecture known as REST. For more information, see the nearby sidebar, "Get some REST."

GET SOME REST

In the classic movie *The Princess Bride,* Prince Humperdink complains to his friend Tyron that he has too much to do. "I've got my country's five hundredth anniversary to plan, my wedding to arrange, my wife to murder, and Guilder to frame for it. I'm swamped."

To which Tyron replied, "Get some rest. If you haven't got your health, you haven't got anything."

In the world of HTTP and online applications, REST is a design paradigm that is widely used with applications that have a lot to do. REST stands for *Representational State Transfer.* A REST application is often called *RESTful.*

REST is primarily concerned with how the state of an application is managed between clients and servers. It offers six basic constraints that, if followed carefully, result in applications that perform well, are scalable, are reliable, and are adaptable.

HTTP is closely related to REST. Most RESTful applications are built using HTTP as the transport protocol.

To be legitimately called RESTful, an application's design must follow six key constraints:

- **Client-server architecture:** User interface concerns are separated from data storage concerns. The server is responsible for the data; the client is responsible for the presentation of the data.

- **Statelessness:** Every request from a client contains all the information the server needs to satisfy the request. In other words, the server doesn't keep any context information about the client between requests.

- **Cachability:** When a server responds to a request, it informs the client as to whether the response can be cached. If the response can be cached, the client can save the information for later use. If the response is non-cacheable, the client must request the information each time it needs the information.

- **Uniform interface:** The programming interface of a RESTful system has four key sub-constraints:

  - *Every resource exposed by the application must be identified by a unique identifier called a URI.* You can find more information about URIs in the "URIs and URLs" section, later in this chapter.

  - *Resources are not manipulated directly; they're manipulated via representations.* The most common representations are HTML, XML, and JSON, but there are plenty of other types of representations. The underlying format of the data is managed by the server and is not exposed directly to the client.

  - *Messages are self-descriptive.* The message contains enough information for the receiver to fully understand the message.

  - *Hypermedia is the engine of application state.* This is a fancy way of saying that the application uses hyperlinks rather than fixed paths to resources. So, the client doesn't need to know anything about the server's organization of resources in advance.

- **Layered system:** A complete application may consist of many layers, including things such as proxies, load balancers, and security layers. And the server that actually fulfills the request messages may call upon other servers for help. The client must have no knowledge of these additional layers.

- **Code on demand:** This is the only optional constraint of the REST paradigm. It allows for servers to send executable code back to the client, which the client can then execute. Usually, this code is JavaScript embedded within an HTML response.

Diving into HTTP

The following sections cover some of the key elements of HTTP.

HTTP clients and servers

HTTP clients send requests to *HTTP servers,* which interpret the request, process it, and return a response. In some cases, the client's request is for data, in which case the server returns the data requested by the client (if possible). In other cases, the client sends data to the server and requests that the server store the data. In these cases, the server stores the data (if possible) and returns a response indicating what action was taken.

Resources

A *resource* is an item of content that can be requested by a client. The simplest type of resource is a simple file that lives on a web server. For example, an HTML file that contains information about your company, when placed on a server, becomes a resource. An HTTP client can request this resource from the server. The server, in turn, locates the file and delivers it to the client, which then renders the HTML on the screen so the user can see what she wanted to see (or what you want her to see).

Not all resources are simple files, though. Some resources are actually programs that dynamically generate the content to be returned to the client. The first three chapters in this minibook show you several ways to write such programs.

MIME

MIME, which stands for *Multipurpose Internet Mail Extensions,* is an Internet standard that allows HTTP to send not just text data but data in binary formats as well (such as image files, sound files, video files, and many other data formats). As its name suggests, MIME was originally designed to allow people to send and receive content via email. When HTTP was created, MIME was incorporated into the protocol to allow for delivery of all sorts of content.

URIs and URLs

URI stands for *Uniform Resource Identifier* and *URL* stands for *Uniform Resource Locator.* A URL is just a special type of URI. The difference is that a URI identifies a resource, while a URL indicates where a resource can be found. The terms are often used interchangeably, especially in the context of HTTP.

For information about the syntax of URLs, see the sidebar "What's in a URL?"

Sessions

An HTTP *session* is a complete round trip, starting when a client sends a request to a server and ending when the client receives the response from the server. Sometimes the term *transaction* is used instead of session; the meaning is the same.

Messages

An HTTP *message* is a package of information that is sent from a client to a server or from a server to a client — these two types of messages are called *request* and *response* messages. Messages are sent in the form of text lines, making them conveniently readable by people. (Well, sometimes. If the data is encrypted, the message response will be a bunch of seemingly random characters that look like gibberish.)

TECHNICAL STUFF

WHAT'S IN A URL?

The URL is a vital element of any program that works with HTTP. You work with URLs all the time when you browse the web, but there's more to even the simplest URL than meets the eye.

A URL can consist of five distinct parts, with the following syntax:

```
scheme://host/path?query#fragment
```

Here's what these five parts do:

- *scheme*: The *scheme* identifies the protocol used to access the resource. The scheme is followed by a colon. When working with HTTP, the scheme should be (drumroll, please) `http:`.
- *host*: The *host* identifies the location of the HTTP server. This can be done by providing an IP address, but it's more commonly done using a Domain Name System (DNS) name like `www.dummies.com`, which is then resolved by the Internet's magic DNS system to an IP address.
- *path*: The *path* identifies a specific resource on the host server. It can be the name of a file or, more commonly, a path that lists one or more folders separated by slashes. The path always begins with a slash to separate it from the host name (for example, `files/java/examples.zip`). The resource may also identify a software component such as a servlet.

  A path may end with a filename, but that's not always necessary. If the filename is omitted, the HTTP server may use a default file name such as `index.html`.
- *query*: The query part of a URL is optional but very useful. It provides additional information to the server. The query begins with a question mark and consists of one or more key-value pairs in the form *key=value*. When you need more than one key-value pair in a query, separate the pairs with ampersands. For example, `?month=10&day=31` could be used to represent a date.
- *fragment*: The *fragment* portion of a URL is used less commonly than the other elements but is still useful. It's usually used to refer to a specific part of the resource. For example, if the resource is an HTML file, the fragment may refer to a section within the file. For example, `#references` could be used to specify a references section within an HTML file.

Using HttpClient

Both types of messages have similar formats, consisting of three parts:

» **Start line:** A start line is a single line that is in one of two formats, depending on whether the message is a request or a response.

For a request message, the first line contains the request method (GET, POST, and so on; see the "Methods" section for more information) followed by the HTTP version. In most cases, the version is specified as HTTP/1.1. Here's a typical request first line:

```
GET http://mysite.com/files/hello.txt HTTP/1.1
```

For a response message, the first line contains the HTTP version, a status code, and an explanation of the status code. For example:

```
HTTP/1.0 200 OK
```

The status code 200 indicates that the request has been processed successfully and the response contains the resource you were looking for.

If the resource can't be returned to the client, the server returns a different status code to identify the error. The most familiar error status is the familiar 404 Not Found, but there are dozens of other status codes that represent various types of errors or other conditions.

» **One or more headers:** One or more lines that provide additional information about the message. Each header line is a key/value pair that specifies a name followed by a value, separated by a colon. A blank line is used to mark the end of the headers.

Here are a few common headers:

● Accept: Specifies the MIME type expected by the client.

● Accept-Language: The languages expected by the client.

● Content-type: The MIME type returned by the server.

● Content-length: The length in bytes of the content returned by the server.

» **Body:** One or more lines that contain the resource returned as part of a message.

Methods

The first line of an HTTP request message specifies a *method*, which is a command that indicates what the client is asking the server to do. There are a total of nine different methods, but for most purposes just four will do:

- **>>** GET: Asks the server to return the content of a specified resource.

- **>>** POST: Asks the server to accept data from the client and store it. The POST request provides a URL, but the URL doesn't necessarily identify where the data will be stored. Instead, the server determines the exact location where the data will be stored, subordinate to the provided URL. For example, the server may store the data as a file in a folder specified by the URL. Or the server may add the data to a database specified by the URL. The POST method is often used to provide input data to a program that runs on the server (for example, when a user fills in a form).

- **>>** PUT: Similar to POST, but a PUT request specifies the exact location at which the data must be stored.

- **>>** DELETE: Deletes the resource indicated by the URL.

WHAT THE HECK IS IDEMPOTENCY?

A fun term that gets used a lot when describing HTTP requests is *idempotence*.

A function is idempotent if doing it multiple times has the exact same result as doing it just once. The simplest example of an idempotent function I can think of is multiplying a value by zero. The first time you multiply something by zero, the result is zero. No matter how many times you multiply something by zero after that, the result will still be zero.

PUT is idempotent because no matter how many times you send a PUT request with the same data, only one copy of the data will be stored. In contrast, if you send more than one identical POST request, a separate copy of the data will be stored for each request.

DELETE is idempotent. You can delete something only once. If you try to delete it again, it stays deleted.

GET is also idempotent. This may seem a little counterintuitive, because it's possible that the underlying data may change between two GET requests for the same resource. For example, if a GET requests the current time from the server, you'll get a different value every time. The key point here is that idempotency applies to what the request does to the server's state, not the client's. A GET request doesn't change the resource being requested, so it is inherently idempotent.

I wish my jokes were idempotent — they would be just as funny no matter how many times I told them!

Looking at a simple HTTP exchange

Let's have a look at a typical HTTP conversation, originating with a request message and finishing with the resulting response message.

For this example, we're requesting a file named `hello.txt` that resides in a folder named `files` on a web server named `www.myserver.com`.

The request message would look something like this:

```
GET http://www.myserver.com/files/hello.txt HTTP/1.0
Accept: text/*
Accept-Language: en
```

The message will be sent to `www.myserver.com`, which will look for the specified resource (`/files/hello.txt`). The headers tell the server that the client is expecting a text file in return, but that it expects the file to be in the English language.

If the server finds the file, the server will return a response message that looks something like this:

```
HTTP/1.0 200 OK
Content-Type: text/plain
Content-Length: 13

Hello, world!
```

In this response message, the server has indicated that the file was successfully found (status code 200) and that the content consists of 13 bytes of text data (specifically, the text `Hello, world!`).

If the server can't find the file, it returns an error message like this:

```
HTTP/1.0 404 Not Found
```

Getting Started with Java's HTTP Client Library

Java's HTTP client library, `HttpClient`, was introduced in Java 11 as a way to standardize the task of communicating with HTTP servers. Prior to Java 11, a plethora of independently developed HTTP client libraries resulted in inconsistent ways of working with HTTP. HttpClient provides a standardized way to communicate via HTTP in Java.

The remainder of this chapter shows you how to use the basic features of Http-Client, creating a program you can use to send a GET request to any URI you want and display the text returned in the response message body.

HttpClient defines two classes and two interfaces:

>> HttpClient, a class used to send HTTP requests and receive HTTP responses

>> HttpRequest, a class that represents an HTTP request message

>> HttpResponse, an interface that represents an HTTP response message

>> WebSocket, an interface that provides a more advanced method for communicating with HTTP servers and is not covered in this chapter

You need to provide the appropriate import statements to use HttpClient:

```
import java.net.http.*;
import java.net.http.HttpResponse.*;
import java.net.http.HttpRequest.*;
```

You also need to use the URI class to represent URIs required by HttpResponse, so you need the following import statement:

```
import java.net.URI;
```

In the following sections, I look at these interfaces one by one.

HttpClient

An HttpClient is required to send and receive HTTP messages. Table 4-1 lists the most commonly used methods of this class.

Creating an HttpClient is simple; just call the newHttpClient method, like this:

```
HttpClient client = HttpClient.newHttpClient();
```

That's all there is to it. You've now created an instance of HttpClient.

You'll use the send method to send an HTTP request to a server and receive the response. I have more to say about the send method later in this chapter (see "Using the send method").

TABLE 4-1
The HttpClient Class

Method	Description
`HttpClient newHttpClient()`	Returns a new `HttpClient` instance.
`HttpResponse<T> send(HttpRequest request, HttpResponse.BodyHandler<T> responseBodyHandler)`	Sends a request message and returns the response as a new `HttpResponse` instance. Note that the interface is generic, so you must supply a type. The `responseBodyHandler` argument specifies how the message body will be processed; it specifies the message body type, which corresponds to the type you must specify for the `HttpResponse` result.

HttpRequest

`HttpRequest` is an interface that represents an HTTP request that can be sent to an HTTP server. Table 4-2 shows the commonly used methods of this interface.

TABLE 4-2 ## The HttpRequest and Its Builder Interface

HttpRequest Method	Description
`HttpResponse<T> send(HttpRequest request, BodyHandler<T> responseBodyHandler)`	Sends a request message and returns the response as a new `HttpResponse` instance. The `responseBodyHandler` argument specifies how you want to process the response body.
Static Builder Methods	**Description**
`static HttpRequest.Builder newBuilder()`	Returns an instance of the `HttpRequest`'s `Builder` interface.
`static HttpRequest.Builder GET()`	Sets the method to GET.
`static HttpRequest.Builder POST()`	Sets the method to POST.
`static HttpRequest.Builder uri(URI uri)`	Sets the URI to the specified URI.
`static HttpRequest.Builder header(String name, String value)`	Adds a header.
`static HttpRequest build()`	Builds and returns the `HttpRequest`.

HttpRequest objects are not created using conventional constructors. Instead, they follow a design pattern used commonly in Java called the `Builder` pattern. For more information about this pattern, see the nearby sidebar.

To create an HttpRequest instance, you use a chain of method calls that starts with a call to the static `HttpRequest.newBuilder` method to obtain a `Builder` object, followed by method calls to set the properties of the HttpRequest, and ending with a call to the `build` method, which returns the completed HttpRequest instance. Here's an example:

```
HttpRequest request = HttpRequest.newBuilder()
    .uri(new URI("http://postman-echo.com/time/now"))
    .GET()
.build();
```

Note that the `send` method can throw one of several unchecked exceptions, so you'll need to enclose it in a `try` block to catch the exceptions.

Let's review this piece by piece:

>> The first line declares a variable named request of type HttpRequest, and then calls the newBuilder method to obtain a Builder instance, which will be used to construct a new HttpRequest. The next three lines call methods of Builder, not methods of HttpRequest.

>> The second line calls the uri method to set the URI for the request. This method call passes a URI instance created by the URI class constructor, which accepts the URI as a string. This particular URI refers to a handy HTTP server resource that returns the current time and date as text.

For more information about this URI, see the sidebar "Hey, Mr. Postman!"

>> The third line calls the GET method to set the request method to GET.

>> The final method completes the request and returns it as an HttpRequest instance, which is assigned to the variable named request.

You now have a complete HttpRequest instance, ready to send to an HTTP server.

THE BUILDER PATTERN

Most Java classes are instantiated by calling one of several constructors defined for the class. For example, to create an instance of ArrayList, you call its constructor like this:

```
ArrayList list = new ArrayList();
```

Some constructors accept arguments that set the initial property values for the instance. For example, you can specify the maximum capacity of an ArrayList by using this constructor:

```
ArrayList list = new ArrayList(1000);
```

Standard constructors have served us well for several decades now. But in some cases, constructors can be a nightmare — especially when they have more than two or three arguments. Imagine a constructor that requires a dozen arguments, all of type int. Keeping track of which int is which can be a real problem.

Suppose, for example, that you have an Employee class with a constructor defined like this:

```
public Employee(String name, long age, long wage)
```

Now, if you hire a 45-year old employee named Lucky at an hourly wage of $18 per hour. If you mix up the call to the constructor, you'll end up thinking the employee is 18 years old and you'll pay him $45 per hour — he really is Lucky!

The Builder pattern is an attempt to solve problems like this. It does so by eliminating constructors altogether, and instead uses a special interface called a *builder*, which defines methods that set the various properties of the class. Typically, the builder class is nested within the class being built and is invoked by a static getBuilder method.

The class being built includes a method called newBuilder, which returns an instance of the Builder class. The Builder instance has an internal representation of the property of the class being built. The Builder class defines methods to set the property values; each of these methods updates the internal property value and then returns this.

In addition to the methods that set property values, the Builder class also includes a method named build. The build method returns an instance of the class being built.

When the Builder pattern is used to create a class, a single Java statement can be used to (1) create the builder, (2) set the property values, and (3) build the class instance. For example, the employee age and wage example would construct the `Employee` instance like this:

```
Employee emp = Employee.getBuilder()
                       .name("Lucky")
                       .age(45)
                       .wage(18)
                       .build();
```

The huge advantage of this code is that each of the properties being assigned as the class is constructed is named — you can tell right away that the employee's age is 45 and his wage is 18.

To make this property naming even more obvious, the chain of method calls is usually indented as shown in the preceding example. This makes it even easier to see exactly what properties are being set and what their values are.

HttpResponse

The `HttpResponse` interface represents a response message from an HTTP server. It provides a number of useful messages that break out the various elements of a response, as detailed in Table 4-3.

TABLE 4-3   **The HttpResponse Class**

Method	Description
`T body()`	Returns the message body. T represents the type of the message body data, which is specified by the `responseBodyHandler` argument of the `HttpClient` send method.
`HttpHeaders headers()`	Returns the HTTP headers from the message response.
`HttpRequest request()`	Returns the `HttpRequest` associated with this response. Note that if the request was redirected by the server, the response will represent the redirected request, not the original request.
`int statusCode()`	Returns the status code associated with this response.
`URI uri()`	Returns the URI for this response. Note that if the request was redirected by the server, `uri` will return the redirected URI.
`HttpClient.Version version()`	The HTTP version associated with this response.

Here are two crucial things you need to know about HttpResponse:

>> **HttpResponse has no public constructors.** The only way to create an HttpResponse object is to call the HttpClient's send method, which returns an HttpResponse as a result.

>> **HttpResponse is a generic.** When you declare a variable to reference an HttpResponse, you must specify a type. The type you specify must agree with the type indicated by the responseBodyHandler argument of the send method; this argument indicates the type used to represent the message body. For example, if the send method specifies that the message body will be a String, you must save the HttpResponse instance in a String variable. I explain how this works in the next section, "Using the send method."

TIP

The type returned by the body method matches the type you specified for the HttpResponse instance and specified in the send method's responseBodyHandler argument. More on that in the next section.

Using the send method

Only one piece of the puzzle remains: How to use the send method to send a request message and receive the result message.

The signature for the send method is spelled out in Table 4-1. It's tricky, so I'll repeat it here:

```
HttpResponse<T> send(HttpRequest request, HttpResponse.BodyHandler<T>
    responseBodyHandler)
```

The first argument provides an HttpRequest instance that will be sent to the server. When the server returns a response message, the send method wraps the response message in an HttpResponse instance as the return value for the send method.

Here's where it gets tricky: The primary purpose of the response message is to return the data requested by the client. It does so via the message body. The data can be one of many different MIME types, and the data may be just a few bytes or it could be a large amount of data. We need some way to deal with the many different types of data that may be returned via the message body.

To accomplish this, HttpResponse uses special objects called *body handlers.* A body handler is designed to return data in a form that the program can easily consume. Each body handler returns the message body data as a different type. The simplest body handler returns the entire body as a string. But other body handlers return more complex types, such as input streams that let you consume the message body one line at a time, or byte arrays that treat the returned data as an array. There's even a body handler that simply discards the message body.

Based on the type and size of the data you're expecting, you can choose a body handler that's appropriate to your needs. To select the body handler you want, you use the BodyHandlers class. Table 4-4 shows methods to obtain a few of the more popular body handlers.

TABLE 4-4   **The BodyHandlers Class**

Method	Description
String ofString()	Returns the message body as a string
Stream<String> ofLines()	Returns the message as a stream of strings, each corresponding to one line of the message body
byte[] ofByteArray()	Returns the message body as an array of bytes
InputStream ofInputstream()	Returns the message body as a stream

In this chapter, we return the message body as a string. So, assuming we've created an HttpRequest referenced by a variable named request, we can send the request and receive the response in a variable named response, like this:

```
HttpResponse<String> response = client.send(request, BodyHandlers.ofString());
```

Putting It All Together

Let's summarize the key points of the HttpClient package so far:

>> You use the HttpClient class to send and receive messages over HTTP.

>> You use the HttpRequest message to represent a request message to send to a server.

>> The HttpRequest instance is created using the Builder pattern, not with a conventional constructor.

>> You use the HttpResponse object to represent a response message.

>> The HttpResponse instance is returned by the send method of the HttpClient.

>> The send method specifies the request object, as well as a body handler to process the message body.

>> To return the message body as a string, use BodyHandlers.ofString().

>> The HttpResponse declaration must specify a type that corresponds to the type of the body handler.

Taking all these points into consideration, here's a segment of code that sends a request message and then prints the message body on the console:

```
HttpClient client = HttpClient.newHttpClient();

String url = "http://postman-echo.com/time/now";

HttpRequest request = HttpRequest.newBuilder()
                                 .uri(new URI(url))
                                 .GET()
                                 .build();

HttpResponse<String> response =
    client.send(request, BodyHandlers.ofString());

System.out.println(response.body());
```

The HTTP Tester Program

It's finally time to put all this together in a complete program. Listing 4-1 presents a console program named HTTP Tester, which accepts a URL string from the user, sends a request for the URL, and displays the message body on the console. It then asks the user for another URL, and continues the process until the user enters **exit**.

Here's a sample of what the program looks like when run in a command console, with the parts entered by the user shown in bold:

```
Welcome to the URI tester.

Enter exit to quit.

Enter a URI: http://postman-echo.com/time/now
Sat, 30 May 2020 05:16:26 GMT

Enter a URI: http://postman-echo.com/time.now
Error: status = 404

Enter a URI: postman-echo.com/time/now
That is not a valid URI.

Enter a URI: exit
```

After displaying a welcome message, the program entreats the user to enter a URL. The user enters the following:

    http://postman-echo.com/time/now

This URL returns a simple text string, which shows the date and time. The program displays this information on the console.

The user then enters two incorrect URLs. First, the user incorrectly states the resource name as `time.now` rather than `time/now`. The program responds with the HTTP status code 404 to indicate that the resource could not be found.

Then the user forgets to enter the scheme (`http://`). The program responds by indicating an invalid URI.

Finally, the user enters **exit** to end the program.

LISTING 4-1: **The HTTP Tester Program**

```java
import java.net.http.*;                                              →1
import java.net.http.HttpResponse.*;
import java.net.http.HttpRequest.*;
import java.net.URI;
import java.util.Scanner;

public class HttpTester
{
    public static void main(String[] args) throws Exception        →9
    {

        Scanner sc = new Scanner(System.in);                        →12

        HttpClient client = HttpClient.newHttpClient();             →14

        System.out.println("Welcome to the URI tester.\n");         →16
        System.out.println("Enter exit to quit.\n");

        while (true)                                                →19
        {
            System.out.print("Enter a URI: ");                      →21
            String input = sc.nextLine();

            if (input.toLowerCase().startsWith("exit"))             →24
                System.exit(0);

            try                                                     →27
            {
                HttpRequest request = HttpRequest.newBuilder()      →29
                    .uri(new URI(input))
                    .GET()
                    .build();

                HttpResponse<String> response =                     →34
                        client.send(request, BodyHandlers.ofString());

                if (response.statusCode() == 200)                   →37
                {
                    System.out.println(response.body() + "\n");
                }
                else
                {
                    System.out.println("Error: status = "           →43
                        + response.statusCode()
                        + "\n");
                }
            }
```

```
        catch (IllegalArgumentException ex)                          →49
        {
            System.out.println("That is not a valid URI.\n");
        }
    }
  }
}
```

Here are the key elements of this program:

>> →1 The program begins with the import statements necessary to reference the HttpClient classes. In addition, java.util.Scanner is imported so the program can read input from the console user.

>> →9 The main method throws Exception to allow for any unhandled exceptions in the program.

>> →12 This line creates a new Scanner instance and assigns it to the variable sc.

>> →14 An instance of HttpClient is created and assigned to client.

>> →16 A welcome message is printed.

>> →19 A while(true) loop causes the program to run forever, or at least until the user enters **exit**.

>> →21 The user is prompted to enter a URL. Then the scanner waits for the user's input and stores it in the String variable named input.

>> →24 If the user enters **exit**, or any mixed-case variant (like **EXIT**, **Exit**, or even **ExIt**), the program terminates by calling System.exit(0).

>> →27 The portion of the program that deals with HttpClient is enclosed in a try block to catch any exceptions that may be thrown.

>> →29 An HttpRequest instance is created, using the input string entered by the user as the URI.

>> →34 The HttpClient send method is called to send the request message, returning the result with a String message body in an HttpResponse variable named response.

>> →37 The status code returned by the response is evaluated. If it's 200, indicating that the request was successful, the contents of the message body are printed to the console.

>> →43 If the result is anything other than 200, the status code is printed to the console.

>> →49 If an invalid URL is entered by the user, the call to the URI constructor in line 38 will throw IllegalArgumentException. This line catches the exception and displays an appropriate error.

Index

Symbols

? character, 520
\(character, 521–522
| character, 523–524
!= operator, 130
<= operator, 130
== operator, 130, 131
|| operator, 162
–? option, 29
/* sequence, 54–55
// sequence, 54
\) character, 521–522
> operator, 130
>= operator, 130
+ (addition) operator, 100, 104, 520
+= (addition and assignment) operator, 110–111
-> (arrow operator), 194, 395, 628, 650
-- (decrement) operator, 100, 104, 106–108
/ (division) operator, 100
/= (division and assignment) operator, 110–111
++ (increment) operator, 100, 104, 106–108
* (multiplication) operator, 100, 104, 520
*= (multiplication and assignment) operator, 110–111
% (remainder/modulus) operator, 100, 102–104
%= (remainder and assignment) operator, 110–111
- (subtraction) operator, 100, 104
-= (subtraction and assignment) operator, 110–111

A

aborting countdowns, 506–510
abs( ) function, 114, 115
abstract classes, 307–310
abstract methods, 308

abstract modifier, 308
accessing
  elements in ArrayList class, 437–438
  elements of two-dimensional arrays, 413–414
Accessor pattern, 264, 798
action events, handling, 627–629
ActionEvent class, 638
ActionListener interface, 317
actionPerformed method, 317
active-high/low inputs, 599–600
Adafruit, 580
add( ) method, 433, 437, 447, 450–452, 458, 460, 698
addAll( ) method, 433–434, 447
addFirst( ) method, 447
adding
  elements in ArrayList class, 436–437
  event handlers to pins, 608–609
  event listeners, 721
  fields to interfaces, 314
  items to LinkedList class, 450–452
  jar files to ClassPath, 368–369
  space by growing nodes, 685–687
  space with margins, 684–685
  <taglib> directive, 781–782
addItems method, 466, 469
addition and assignment (+=) operator, 110–111
addition (+) operator, 100, 104, 520
addLast( ) method, 447, 451
AddSubtract1 program, 641–642
AddSubtract2 program, 644–646
AddSubtract3 program, with anonymous inner classes, 647–649
AddSubtract4 program, with lambda expressions, 651–652
AddSubtract5 program, with lambda expressions, 653–654

building *(continued)*

improved versions of else-if example programs, 181–183

instances of objects from classes, 79

interfaces, 311–312

layout panes, 629–630, 679–680

LinkedList class, 450

list views, 739–740

locks, 503–505

loops, 785–786

methods, 198–203

modular JAR files, 379–380

objects from classes, 57–58

packages, 363–365

private fields, 260

programs for experimenting with regular expressions, 512–514

programs with anonymous classes, 357–359

public fields, 260

random numbers, 116–119

Raspberry Pi LED circuits, 581–582

Scanner objects, 94

scenes, 630–631

servlets, 753–774

shallow copies using clone method, 339–341

stages, 631

state change event listeners, 607–608

StringBuilder objects, 396

subclasses, 289–290

threads, 485–488

threads that work together, 493–496

trees, 742–745

TreeView controls, 745–746

two-dimensional arrays, 412–413

web.xml file, 758–760

workspaces, 37

bulk data operations

about, 473–475

using with collections, 471–479

Burd, Barry (author)

Beginning Programming with Java For Dummies, 5th Edition, 7

Java For Dummies, 7th Edition, 7

bus strips, 579

business rules layer, 250–251

Button class, 626–627

Button switcher program, 604–606

buttonClick() method, 627, 628

buttons, creating, 626–627

BuyMyBook.jsp, 808–809

byte nextByte() method, 95

byte ofByteArray() method, 829

byte type, 73, 78, 85

C

C programming language, 11–12, 73, 78, 242

C++ programming language, 73, 78

cachability, RESTful applications and, 816

calculating

classic factorial example, 529–532

dates, 558–560

callbacks, using interfaces for, 316–320

calling

constructors, 268–270

defined, 197

camera serial interface (CSI) (Raspberry Pi), 569

cancel method, 589

capture groups, 523

Car class, 287

caret (^), 520

Cascading Style Sheets (CSS), 617

case keyword, 183, 187–188

case-sensitivity

in identifiers, 53

in Java, 24

in keywords, 48–49

in package names, 363

casting an operand, 101

casting numeric data, 85–86

catch block, 221, 222–223, 226–227, 228

catch clause, 305

catch tag, 782

catch-or-throw rule, 231, 232

categorizing operators, 101

cathode, 576–578

F

factorial, 529

Factory pattern, 309

false keyword, 47

fields

about, 246

adding to interfaces, 314

in class body, 257

members and, 259–260

File class, 533

FileInputStream class, 232

FileNotFoundException, 232–236

files, compiling multiple, 25–26

FileVisitor interface, 533

fill method, 426–427

filling arrays, 426–427

filter method, 474–475, 477

final classes, 296–297

final keyword, 47–48, 70–71, 295–297, 314

final methods, 296

final variables, 70–71

using, 112–113

finalize method, 326

finally block, 221, 228–231

finding string length, 388

flight-simulator programs, 242

float class, 73, 74–76, 78, 85, 105, 119–121, 126

float nextFloat() method, 95

floating port, 600

floating-point types, 74–76, 124–125

floor() function, 120

flowcharts, 185, 251

FlowPane class, 691–694

FlowPane layout pane, 678, 690–694

fmt:formatNumber, 787

folders

creating structure of for servlets, 758

Java Development Kit (JDK), 21–22

setting up for modules, 378–379

for each loop, 471, 474

for loops

about, 164, 392, 429, 453, 530, 540

break statement, 172–173

continue statement, 172–173

counter variable, 166–167

counting backward, 168–169

counting even numbers, 167–168

enhanced, 408–409

expressions, 170–172

format of, 164–166

omitting expressions, 172

using with arrays, 404–405

without bodies, 169–170

for statement, 439, 539

forEach method, 474, 478

forEach tag, 782

FOREVER, 560

ForInit expression, 166

formal type parameters, 459

format() method, 122

formatting

characters for DateTimeFormatter class, 560

dates, 560–562

numbers, 121–123, 786–787

tags, 780

forms, working with, 767–768

forTokens tag, 782

ForUpdate expression, 166

forward slash (/), 369

fragment (URL), 819

function tags, 780

functional interface, 359, 649

functions

abs(), 114, 115

cbrt(), 114

ceil(), 120

exp(), 114

floor(), 120

hypot(), 114, 115

log(), 114

W

About the Author

Doug Lowe has been writing computer programming books since the guys who invented Java were in high school. He's written books on COBOL, FORTRAN, ASP. NET, Visual Basic, IBM mainframe computers, midrange systems, PCs, web programming, and probably a few he's long since forgotten about. He's the author of more than 30 *For Dummies* books, including *Networking For Dummies,* 11th Edition; *Networking All-In-One For Dummies,* 5th Edition; *PowerPoint 2019 For Dummies;* and *Electronics All-In-One For Dummies,* 2nd Edition (all published by Wiley). He lives in that sunny all-American city Fresno, California, where the motto is "Please, Let It Rain!"

Dedication

To my beautiful wife, Kristen Gearhart.

Author's Acknowledgments

Wow, what a challenge to revise a book like this in the middle of a pandemic. I want to throw out special things to everyone at Wiley who was involved in the creation of this book, especially project editor Elizabeth Kuball who did a fantastic job and was incredibly patient with me throughout the entire process. I'd also like to thank Janeice DelVecchio for her amazingly thorough technical review. The book is much better because of the efforts of both Elizabeth and Janeice, as well as all the folks behind the scenes at Wiley who pitched in along the way.

Publisher's Acknowledgments

Executive Editor: Steve Hayes

Project Editor: Elizabeth Kuball

Copy Editor: Elizabeth Kuball

Technical Editor: Janeice DelVecchio

Production Editor: Tamilmani Varadharaj

Cover Image: © Iscatel/Shutterstock

Leverage the power

Dummies is the global leader in the reference category and one of the most trusted and highly regarded brands in the world. No longer just focused on books, customers now have access to the dummies content they need in the format they want. Together we'll craft a solution that engages your customers, stands out from the competition, and helps you meet your goals.

Advertising & Sponsorships

Connect with an engaged audience on a powerful multimedia site, and position your message alongside expert how-to content. Dummies.com is a one-stop shop for free, online information and know-how curated by a team of experts.

- Targeted ads
- Video
- Email Marketing
- Microsites
- Sweepstakes sponsorship

20 MILLION PAGE VIEWS EVERY SINGLE MONTH

15 MILLION UNIQUE VISITORS PER MONTH

43% OF ALL VISITORS ACCESS THE SITE VIA THEIR MOBILE DEVICES

700,000 NEWSLETTER SUBSCRIPTIONS TO THE INBOXES OF

300,000 UNIQUE INDIVIDUALS EVERY WEEK

# of dummies

Custom Publishing

Reach a global audience in any language by creating a solution that will differentiate you from competitors, amplify your message, and encourage customers to make a buying decision.

- Apps
- Books
- eBooks
- Video
- Audio
- Webinars

Brand Licensing & Content

Leverage the strength of the world's most popular reference brand to reach new audiences and channels of distribution.

For more information, visit dummies.com/biz

PERSONAL ENRICHMENT

Staying Sharp
9781119187790
USA $26.00
CAN $31.99
UK £19.99

Facebook
Carolyn Abram
9781119179030
USA $21.99
CAN $25.99
UK £16.99

Guitar
Mark Phillips
Jon Chappell
9781119293354
USA $24.99
CAN $29.99
UK £17.99

Investing
Eric Tyson, MBA
9781119293347
USA $22.99
CAN $27.99
UK £16.99

Beekeeping
Howland Blackiston
9781119310068
USA $22.99
CAN $27.99
UK £16.99

Digital Photography
Julie Adair King
9781119235606
USA $24.99
CAN $29.99
UK £17.99

Meditation
Stephan Bodian
9781119251163
USA $24.99
CAN $29.99
UK £17.99

Pregnancy ALL-IN-ONE
6 Books
9781119235491
USA $26.99
CAN $31.99
UK £19.99

Samsung Galaxy S7
Bill Hughes
9781119279952
USA $24.99
CAN $29.99
UK £17.99

iPhone
Edward C. Baig
Bob "Dr. Mac" LeVitus
9781119283133
USA $24.99
CAN $29.99
UK £17.99

Crocheting
Karen Manthey
Susan Brittain
9781119287117
USA $24.99
CAN $29.99
UK £16.99

Nutrition
Carol Ann Rinzler
9781119130246
USA $22.99
CAN $27.99
UK £16.99

PROFESSIONAL DEVELOPMENT

Windows 10
Andy Rathbone
9781119311041
USA $24.99
CAN $29.99
UK £17.99

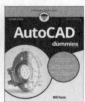
AutoCAD
Bill Fane
9781119255796
USA $39.99
CAN $47.99
UK £27.99

Excel 2016
Greg Harvey, PhD
9781119293439
USA $26.99
CAN $31.99
UK £19.99

QuickBooks 2017
Stephen L. Nelson, MBA, CPA, MS in Taxation
9781119281467
USA $26.99
CAN $31.99
UK £19.99

macOS Sierra
Bob "Dr. Mac" LeVitus
9781119280651
USA $29.99
CAN $35.99
UK £21.99

LinkedIn
Joel Elad, MBAs
9781119251132
USA $24.99
CAN $29.99
UK £17.99

Windows 10 ALL-IN-ONE
10 Books
Woody Leonhard
9781119310563
USA $34.00
CAN $41.99
UK £24.99

SharePoint 2016
Rosemarie Withee
Ken Withee
9781119181705
USA $29.99
CAN $35.99
UK £21.99

Fundamental Analysis
Matt Krantz
9781119263593
USA $26.99
CAN $31.99
UK £19.99

Networking
Doug Lowe
9781119257769
USA $29.99
CAN $35.99
UK £21.99

Office 2016
Wallace Wang
9781119293477
USA $26.99
CAN $31.99
UK £19.99

Office 365
Rosemarie Withee
Ken Withee
Jennifer Reed
9781119265313
USA $24.99
CAN $29.99
UK £17.99

Salesforce.com
Liz Kao
Jon Paz
9781119239314
USA $29.99
CAN $35.99
UK £21.99

Coding
Nikhil Abraham
9781119293323
USA $29.99
CAN $35.99
UK £21.99

dummies.com

dummies
A Wiley Brand

Learning Made Easy

ACADEMIC

9781119293576
USA $19.99
CAN $23.99
UK £15.99

9781119293637
USA $19.99
CAN $23.99
UK £15.99

9781119293491
USA $19.99
CAN $23.99
UK £15.99

9781119293460
USA $19.99
CAN $23.99
UK £15.99

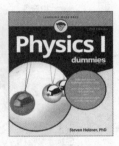

9781119293590
USA $19.99
CAN $23.99
UK £15.99

9781119215844
USA $26.99
CAN $31.99
UK £19.99

9781119293378
USA $22.99
CAN $27.99
UK £16.99

9781119293521
USA $19.99
CAN $23.99
UK £15.99

9781119239178
USA $18.99
CAN $22.99
UK £14.99

9781119263883
USA $26.99
CAN $31.99
UK £19.99

Available Everywhere Books Are Sold

dummies.com

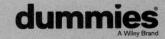

Small books for big imaginations

GETTING STARTED WITH Coding
Get Creative with Code!
Camille McCue, PhD

9781119177173
USA $9.99
CAN $9.99
UK £8.99

MODDING Minecraft™
Build Your Own Minecraft Mods!
Sarah Guthals, PhD
Stephen Foster, PhD
Lindsey Handley, PhD

9781119177272
USA $9.99
CAN $9.99
UK £8.99

MAKING YouTube® VIDEOS
Star in Your Own Video!
Nick Willoughby

9781119177241
USA $9.99
CAN $9.99
UK £8.99

DESIGNING Digital Games
Create Games with Scratch™!
Derek Breen

9781119177210
USA $9.99
CAN $9.99
UK £8.99

GETTING STARTED WITH Raspberry Pi®
Program Your Raspberry Pi™!
Richard Wentk

9781119262657
USA $9.99
CAN $9.99
UK £6.99

EXPERIMENTING WITH Science
Think, Test, and Learn!
Olivia & Mullins, PhD

9781119291336
USA $9.99
CAN $9.99
UK £6.99

CREATING Digital Animations
Animate Stories with Scratch™!
Derek Breen

9781119233527
USA $9.99
CAN $9.99
UK £6.99

GETTING STARTED WITH Engineering
Think Like an Engineer!
Camille McCue, PhD

9781119291220
USA $9.99
CAN $9.99
UK £6.99

WRITING Computer Code
Learn the Language of Computers!
Chris Minnick and Eva Holland

9781119177302
USA $9.99
CAN $9.99
UK £8.99

Unleash Their Creativity

dummies.com

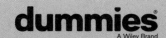